An Integrated Approach to Botany

An Integrated Approach to Botany

Edited by Allen Davies

MURPHY & MOORE
www.murphy-moorepublishing.com

Published by Murphy & Moore Publishing,
1 Rockefeller Plaza,
New York City, NY 10020, USA
www.murphy-moorepublishing.com

An Integrated Approach to Botany
Edited by Allen Davies

International Standard Book Number: 978-1-63987-045-5 (Hardback)

Cataloging-in-Publication Data

An integrated approach to botany / edited by Allen Davies.
p. cm.
Includes bibliographical references and index.
ISBN 978-1-63987-045-5
1. Botany. 2. Plants. 3. Biology. I. Davies, Allen.

QK45.2 .I58 2022
580--dc23

TABLE OF CONTENTS

PREFACE

Botany is a branch of biology that is concerned with the study of plant life. It focuses on the study of plant properties, classification, structure, diseases, biochemical processes and interactions with the environment. The emergence of modern botany has been facilitated by an integration of modern techniques of organic chemical analysis, as well as advances in the understanding of the relationship between the plant genome and its physiology, morphology, biochemistry and behavior. Growth in plant biotechnology has paved the way for the use of plants or plant cell cultures for the synthesis of antibiotics, pesticides and other pharmaceutical products besides the production of genetically modified crops for enhanced yield. The scope of botany today has expanded to investigate the role of plants in the cycling of carbon, oxygen, nitrogen, water and energy, as well as address the global issues of human food security, carbon sequestration, climate change, resource management and sustainability. Different approaches, evaluations, methodologies and advanced studies on botany have been included in this book. A number of latest researches have been included to keep the readers up-to-date with the global concepts in this area of study. This book is an essential guide for both academicians and those who wish to pursue this discipline further.

Significant researches are present in this book. Intensive efforts have been employed by authors to make this book an outstanding discourse. This book contains the enlightening chapters which have been written on the basis of significant researches done by the experts.

Finally, I would also like to thank all the members involved in this book for being a team and meeting all the deadlines for the submission of their respective works. I would also like to thank my friends and family for being supportive in my efforts.

Editor

PREFACE

1

Sorghum stem aerenchyma formation is regulated by *SbNAC_D* during internode development

Anna L. Casto[1,2] | Brian A. McKinley[1] | Ka Man Jasmine Yu[1,3] | William L. Rooney[4] | John E. Mullet[1]

[1]Department of Biochemistry and Biophysics, Texas A&M University, College Station, Texas

[2]Molecular and Environmental Plant Sciences Graduate Program, Texas A&M University, College Station, Texas

[3]Biochemistry and Biophysics Graduate Program, Texas A&M University, College Station, Texas

[4]Department of Soil and Crop Sciences, Texas A&M University, College Station, Texas

Correspondence
John E. Mullet, Department of Biochemistry and Biophysics, Texas A&M University, College Station, TX.
Email: jmullet@tamu.edu

Funding information
This work was funded in part by the DOE Great Lakes Bioenergy Research Center (DOE Office of Science BER DE-FC02-07ER64494, BER DE-SC0018409), US Department of Energy grant no. DE-SC0012629, and the Perry Adkisson Chair.

Abstract

Sorghum bicolor is a drought-resilient C4 grass used for production of grain, forage, sugar, and biomass. Sorghum genotypes capable of accumulating high levels of stem sucrose have solid stems that contain low levels of aerenchyma. The *D*-locus on SBI06 modulates the extent of aerenchyma formation in sorghum stems and leaf midribs. A QTL aligned with this locus was identified and fine-mapped in populations derived from BTx623*IS320c, BTx623*R07007, and BTx623*Standard broomcorn. Analysis of coding polymorphisms in the fine-mapped *D*-locus showed that genotypes that accumulate low levels of aerenchyma encode a truncated NAC transcription factor (Sobic.006G147400, *SbNAC_d1*), whereas parental lines that accumulate higher levels of stem aerenchyma encode full-length NAC TFs (*SbNAC-D*). During vegetative stem development, aerenchyma levels are low in nonelongated stem internodes, internode growing zones, and nodes. Aerenchyma levels increase in recently elongated internodes starting at the top of the internode near the center of the stem. *SbNAC_D* was expressed at low levels in nonelongated internodes and internode growing zones and at higher levels in regions of stem internodes that form aerenchyma. *SbXCP1*, a gene encoding a cysteine protease involved in programmed cell death, was induced in *SbNAC_D* genotypes in parallel with aerenchyma formation in sorghum stems but not in *SbNAC_d1* genotypes. Several sweet sorghum genotypes encode the recessive *SbNAC_d1* allele and have low levels of stem aerenchyma. Based on these results, we propose that *SbNAC_D* is the *D*-gene identified by Hilton (1916) and that allelic variation in *SbNAC_D* modulates the extent of aerenchyma formation in sorghum stems.

KEYWORDS
D-gene, germplasm, programmed cell death, *Sorghum bicolor*

1 | INTRODUCTION

Sorghum bicolor is a highly productive, drought-resilient C4 grass used for production of grain, forage, sugar, and biomass (Mullet et al., 2014; Rooney, 2004; Rooney, Blumenthal, Bean, & Mullet, 2007). Sorghum has a large and diverse germplasm useful for genetic analysis (Mace et al., 2009, 2013; Morris et al., 2013). The species is diploid and mostly inbreeding with a sequenced genome comprised of ten chromosomes that span ~800 Mbp (Mace et al., 2013; McCormick et al., 2018; Paterson et al., 2009). The sorghum research community has developed TILLING (Xin et al., 2008) and transcriptome resources (McCormick et al., 2018; Shakoor et al.,

2014), and sorghum is amenable to transformation (Wu et al., 2014) and genome editing (Jiang et al., 2013). These genetic resources and attributes have facilitated the use of sorghum for hybrid breeding, comparative genomics, and the discovery of numerous important sorghum genes using QTL-assisted gene identification (Hilley, Truong, Olson, Morishige, & Mullet, 2016; Hilley et al., 2017; Klein et al., 2005; Lin et al., 2012; Murphy et al., 2011, 2014).

Plant stems are essential for the transport of water and nutrients to and from roots and leaves. Stems also serve as a temporary storage depot for excess photosynthate and nutrients not immediately needed for basic metabolism, growth, or grain filling (Slewinski, 2012). In grasses, internode numbers and lengths impact canopy architecture, radiation use efficiency, nitrogen, and water use (George-Jaeggli, Jordan, van Oosterom, Broad, & Hammer, 2013; Mullet et al., 2014). High-biomass sorghums develop 4- to 5-m-length stems that can account for up to 84% of harvested biomass (Olson et al., 2012). Sorghum stems are comprised of phytomers produced approximately every 4 days by the shoot apical meristem (Kebrom, McKinley, & Mullet, 2017). Stem internode elongation is repressed during the juvenile phase but increases during the vegetative phase (Kebrom et al., 2017) or following floral initiation (Quinby, 1974). Several genes that regulate sorghum stem growth have been identified including an ABCB1 auxin transporter encoded by *Dw3* (Multani et al., 2003), a plasma membrane protein encoded by *Dw1* (Hilley et al., 2016; Yamaguchi et al., 2016) that is involved in brassinosteroid signaling and cell proliferation (Hirano et al., 2017), and an AGC kinase encoded by *Dw2* (Hilley et al., 2017).

Sorghum stems have the capacity to accumulate high concentrations (~0.5 M) and large amounts of sucrose (up to 40% by weight) similar to sugarcane (Bihmidine, Hunter, Johns, Koch, & Braun, 2013; Godoy & Tesso, 2013; Gutjahr et al., 2013; Lingle, 1987; McKinley, Rooney, Wilkerson, & Mullet, 2016). Stem volume, juiciness, and sugar concentration have been under selection in sweet sorghum in order to maximize sugar yield (Burks, Kaiser, Hawkins, & Brown, 2015; Carvalho & Rooney, 2017; Murray et al., 2008). Traits and QTL that modulate the accumulation of stem sugars have been identified (Burks et al., 2015; Felderhoff et al., 2012; Guan et al., 2011; Han et al., 2015; Hart, Schertz, Peng, & Syed, 2001; Mocoeur et al., 2015; Murray et al., 2008; Shiringani, Frisch, & Friedt, 2010; Srinivas et al., 2009; Xiao-ping, Jin-feng, Cui-ping, & Acharya, 2011), and the biochemical and molecular basis of stem sugar accumulation has been investigated (Calviño, Bruggmann, & Messing, 2008; Gutjahr et al., 2013; Lingle, 1987; McKinley et al., 2016; Milne et al., 2017). The stems of sorghum genotypes that accumulate high levels of sucrose generally lack or accumulate low levels of aerenchyma and maintain functional pith parenchyma with large vacuoles where sucrose is sequestered (Burks et al., 2015; Carvalho & Rooney, 2017; Hilson, 1916). Other sorghum genotypes have stems that convert pith parenchyma into aerenchyma to varying degrees (Carvalho & Rooney, 2017; Hilson, 1916; Stephens & Quinby, 1939).

Genetic variation in the extent of aerenchyma formation in stems and leaf midribs was noted ~100 years ago (Hilson, 1916). Genetic analysis of variation in stem juiciness (vs. pithiness) and the color of leaf midribs (green vs. white) revealed that in some populations, the genetics of aerenchyma formation was simple and that a single dominant gene largely accounted for the appearance of "pithy" stems and white midribs caused by accumulation of aerenchyma (Ayyangar, Ayyare, Rao, & Nambiar, 1936; Hilson, 1916). Swanson and Parker assigned the symbols *D* and *d* to this main effect genetic factor for stem and leaf midrib "dryness" (Swanson & Parker, 1931). Genotypes with pithy stems and white midribs encode an active *D*-allele that causes the accumulation of aerenchyma in stems and leaf midribs (Hilson, 1916; Stephens & Quinby, 1939). A major QTL corresponding to the *D*-gene was mapped to LG-06 by Xu et al. (2000), a location subsequently confirmed using other populations (Hart et al., 2001; Srinivas et al., 2009) and by GWAS (Burks et al., 2015). A stem moisture QTL was also mapped to the same location on LG-06 (Han et al., 2015).

Flooding and hypoxia induce formation of aerenchyma in sorghum roots (Drew, Jackson, & Giffard, 1979). Hypoxia increases the synthesis and accumulation of ethylene in roots, and ethylene induces programmed cell death-mediated root aerenchyma formation (Drew, He, & Morgan, 2000). In species such as rice, some root aerenchyma form in the absence of flooding and additional aerenchyma form in response to flooding or ethylene treatment (Yamauchi, Shimamura, Nakazono, & Mochizuki, 2013). In maize, accumulation of aerenchyma is induced by hypoxia (Drew et al., 2000; Yamauchi et al., 2016). In contrast, the wetland species *Juncus effusus* form root aerenchyma in the absence of ethylene and flooding (Visser & Bogemann, 2006). Programmed cell death is also induced in response to pathogens through the hypersensitive response (Liang et al., 2003). Moreover, programmed cell death occurs in many tissues such as root caps (Fendrych et al., 2014), xylem (Pesquet et al., 2013), the endosperm (Uauy, Distelfeld, Fahima, Blechl, & Dubcovsky, 2006; Young & Gallie, 2000), and during emergence of adventitious roots (Mergemann & Sauter, 2000) and organ senescence (Kim et al., 2014). The molecular events and genes involved in developmental and abiotic stress-induced programmed cell death have been identified in a number of plants (De Clercq et al., 2013; Rajhi et al., 2011; Takahashi, Yamauchi, Rajhi, Nishizawa, & Nakazono, 2015; Yamauchi et al., 2017).

The yield and composition of stem biomass are of critical importance for high-biomass sorghum since stems account for ~80% of the harvested biomass (Olson et al., 2012). Variation in stem aerenchyma formation can impact stem biomass yield and composition by altering the capacity of stems to accumulate sucrose. Therefore, in this study we characterized when and where stem aerenchyma form during stem and internode development and cloned the sorghum *D*-gene to better understand the genetic network that modulates this important stem trait. The analysis showed that the *D*-gene encodes a NAC transcription factor expressed in stems that regulates the expression of a gene involved in developmental programmed cell death and the formation of aerenchyma.

2 | MATERIALS AND METHODS

2.1 | Plant materials and growing conditions

All plants were grown in a glasshouse under 14-hr long days in 14.5-L pots with MetroMix 900 (Sun Gro Horticulture) fertilized as needed with Peters 20-20-20. Plants were thinned to one plant per pot and grown at 10–20 cm spacing. Seed for all genotypes and populations was obtained from the Texas A&M Sorghum Breeding Program (College Station, TX).

2.2 | Plant phenotyping and visualization of aerenchyma by microscopy and CT scanning

The extent of aerenchyma (AER) formation was rated by visual inspection of cross sections of internodes taken approximately at the midpoint between nodes. Ratings were based on a scale from 1 to 5 (1, no AER; 5, high AER) in a manner similar to prior studies (Supporting Information Figure S1) (Carvalho & Rooney, 2017; Nilsen et al., 2017). Results from visual rating were similar to ratings based on a percent of area calculation when data were used for QTL analysis (Carvalho & Rooney, 2017). All AER ratings are the mean of at least three biological replicates unless otherwise indicated.

Multi-slice computed tomography (MSCT) scans of five R07020 stems were collected at the Diagnostic Imaging & Cancer Treatment Center of the Texas A&M Veterinary Medicine & Biomedical Sciences facilities in College Station, Texas. A SOMATOM Definition AS+ (SIEMENS) was used with settings of 120 kVp, 1.024 pixels/mm, at a 0.6 mm slice thickness (Gomez, Carvalho, Shi, Muliana, & Rooney, 2018).

For microscopic imaging, hand sections were made of representative R07020 internodes at different stages of internode development to investigate changes in the distribution and extent of aerenchyma accumulation in stems. The sections were imaged under white light using a Zeiss M^{2}Bio Fluorescence Combination Zoom Stereo/Compound Microscope coupled with a Zeiss AxioCam digital camera (Kramer Scientific). Photographs were captured using the Zeiss AxioVision 3.0.6 software.

2.3 | Quantitative trait locus (QTL) analysis and fine-mapping

Stem aerenchyma QTL were mapped in three populations: BTx623 x IS3620c (RILs, n = 380), BTx623 x R07007 (F5, n = 215), and BTx623 x Standard Broomcorn (F2, n = 133). All plants were grown to anthesis and phenotyped by cutting the internode below the peduncle at a midpoint between the upper and lower node. The cross sections were visually inspected and rated on a scale from 1 (no aerenchyma) to 5 (high aerenchyma) (Supporting Information Figure S1). One plant of each RIL from the BTx623*IS3620c population was grown in the summer of 2014 under long days. Three plants of each line of the BTx623*R07007 population were grown in a glasshouse in the spring of 2015 under long days. Genotyping and genetic map construction were performed as described in Truong, McCormick, Morishige, and Mullet (2014), and DG marker sequences were mapped to version 3.1 of the sorghum reference genome assembly (*Sorghum bicolor* v3.1 DOE-JGI, http://phytozome.jgi.doe.gov/), using BWA (Li & Durbin, 2009). INDEL realignment and joint variant calling were performed with the GATK using the naïve pipeline of the RIG workflow (McCormick et al., 2018). QTL mapping was performed in R/qtl using composite interval mapping (CIM) with 1,000 permutations and α = 0.05 (Broman, Wu, Sen, & Churchill, 2003). BTx623*IS3620c and BTx623*R07007 populations were mapped as RILs, while BTx623*Standard Broomcorn was mapped as an F_2. The BTx623*IS3620c population was used to fine-map *SbNAC_D*. Five plants from each line that had recombination breakpoints in the 2-LOD region spanning the main effect QTL on SBI-06 were grown to anthesis and phenotyped to confirm aerenchyma ratings.

2.4 | DNA sequencing and sequence analysis

Coding region sequence variants between the four parental lines used for QTL analysis were obtained from publicly available whole-genome sequences (Phytozome v12.1.6). The effect of protein sequence variants was analyzed using PROVEAN (Choi, Sims, Murphy, Miller, & Chan, 2012). Genomic DNA was extracted from leaf tissue of diverse sorghum lines according to the Quick-DNA™ Plant/Seed Kit (Zymo Research) instructions. *SbNAC_D* was amplified by polymerase chain reaction (PCR) and was sequenced by Sanger sequencing. All template amplification and sequencing primers are listed in Supporting Information Table S6.

2.5 | Protein sequence analysis

Protein sequences of the closest homologs of *SbNAC_D* in grass species with available genome sequences were identified using BLAST (Phytozome v12). To identify the sugarcane homolog of *SbNAC_D*, the sorghum cDNA sequence was used in a BLAST search of the sugarcane transcriptome (SUCEST-FUN Database, http://sucest-fun.org). Protein sequences were aligned using MUSCLE (Edgar, 2004) and visualized using Jalview (Waterhouse, Procter, Martin, Clamp, & Barton, 2009). Evolutionary trees were inferred using the neighbor-joining method (Jones, Taylor, & Thornton, 1992) in MEGA7 (Kumar, Stecher, & Tamura, 2016). All positions containing gaps and missing data were eliminated.

2.6 | cDNA sequencing and qRT-PCR

Total RNA was extracted from all samples using the Direct-zol™ RNA Miniprep Kit (Zymo Research), and cDNA was synthesized using SuperScript™ III First-Strand Synthesis SuperMix for qRT-PCR (Thermo Fisher Scientific). For analysis of gene expression in *SbNAC_D* and *SbNAC_d1* RILs, the internode subtending the peduncle was collected before aerenchyma formation and 7 – 10 days later when aerenchyma were visible. Internodes were collected from three biological replicates. For the developmental analysis of gene expression in R07020, internodes were collected from three biological replicates at midday. Tissue was collected from the two short nonelongated internodes immediately below the shoot apex that were approximately 1–2 cm in length (int1, int2). Internodes 3 and 4 (int3-X, int4-X) were subsectioned into 1-cm sections. For the fully elongated internodes 5–7 (int5-X, int6-X, int7-X), 2-cm sections were isolated from the top and base of each internodes for analysis. A list of gene targets and primers used for the analyses is provided in Supporting Information Table S6. For all qRT-PCR experiments, relative expression was determined using the comparative cycle threshold (C_t) method. Raw C_t values for each sample were normalized to C_t values for the reference gene *SbUBC* (Sobic.001G526600). Reference gene stability across all samples was determined by NormFinder (Andersen, Jensen, & Ørntoft, 2004). $\Delta\Delta C_t$ values were calculated relative to the sample with the highest expression (lowest Ct value). Relative expression values were calculated with the $2^{-\Delta\Delta C_t}$ method (Pfaffl, 2001). Fold change in gene expression was calculated based on ΔC_t values between the samples with the lowest and highest expression according to the equation $FC = 2^{(\Delta C_t(max) - \Delta C_t(min))}$. Primer specificity was tested by dissociation curve analysis and gel electrophoresis.

For cDNA sequencing, total RNA was extracted from the leaf sheaths of BTx623 and IS3620c and used for cDNA synthesis as described above. The cDNA corresponding to *SbNAC_D* was amplified and sequenced using Sanger sequencing. The primers used for template amplification and sequencing are listed in Supporting Information Table S5.

2.7 | Statistical analyses

All statistical analyses were performed in GraphPad Prism, version 7, for Windows (GraphPad Software, La Jolla California USA, www.graphpad.com).

3 | RESULTS

3.1 | Sorghum genotypes vary in stem aerenchyma accumulation

Sorghum genotypes were grown in a glasshouse and screened for variation in the extent of stem aerenchyma accumulation in order to identify genotypes for further study and parental lines of populations for QTL-assisted identification of genes involved in stem aerenchyma formation. The extent of aerenchyma formation was rated by visual inspection of stem cross sections taken mid-internode on a scale from 1 (no aerenchyma) to 5 (high aerenchyma) (Supporting Information Figure S1). This screen showed that the stems of BTx623 contained low levels of aerenchyma, whereas the stems of IS3620c, R07007, R07020, and Standard broomcorn had higher levels of stem aerenchyma (Figure 1). For most genotypes, aerenchyma levels were similar in fully elongated internodes along the length of the stem. IS3620c was more variable and showed a trend toward higher aerenchyma in the internodes immediately below the peduncle (Int1-3).

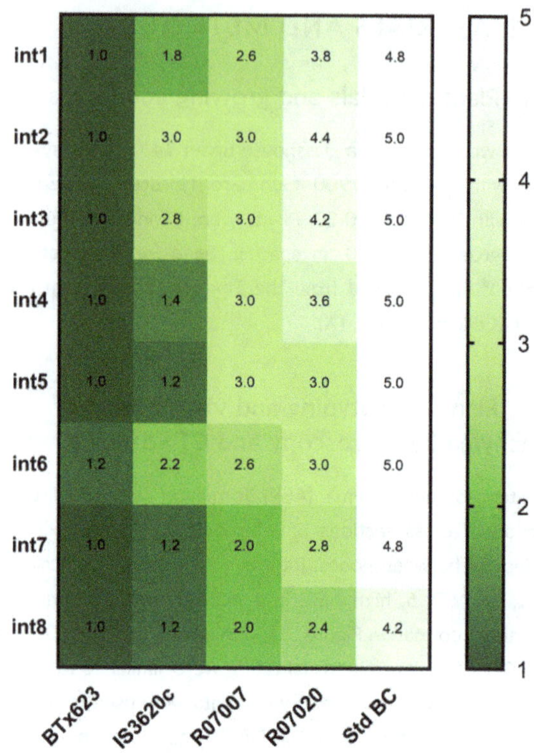

FIGURE 1 Variation in the extent of stem aerenchyma (AER) in diverse sorghum genotypes. Internodes of five sorghum genotypes were screened for the extent of AER formation in eight fully elongated internodes using a scale from 1 (minimal AER, dark green) to 5 (maximum AER, white) (heat map scale to the right). Internode 1 (int1) is the internode at the top of the stem below the peduncle. BTx623, an elite grain line, showed low levels of AER formation in all internodes. Standard Broomcorn (Std BC) had the highest levels of AER across all internodes among all screened genotypes. AER ratings are the average of at least three biological replicates

Standard broomcorn stems contained the highest levels of stem aerenchyma among the genotypes screened.

3.2 | QTL-assisted identification of the *D*-gene

The genetic network that regulates the extent of stem aerenchyma accumulation was investigated using QTL analysis. The results of genotype screening (Figure 1) indicated that populations derived from crosses involving BTx623*IS3620c, BTx623*R07007, and BTx623*Standard broomcorn would be useful for mapping QTL for variation in stem aerenchyma accumulation. Populations derived from each of these crosses were grown in a glasshouse and scored for variation in stem aerenchyma accumulation in internodes of fully elongated stem internodes. The stem aerenchyma phenotypes and pre-existing genetic maps for these populations (Hilley et al., 2016, 2017; Truong et al., 2014) were used to identify QTL that affect the extent of aerenchyma formation in stems. A main effect QTL that modulates variation in stem aerenchyma accumulation was identified on SBI06 at approximately 50 Mbp in all three populations (Figure 2). The LOD scores of the QTL on SBI06 were >20 in the BTx623*IS3620c and BTx623*R07007 RIL populations and ~7.4 in

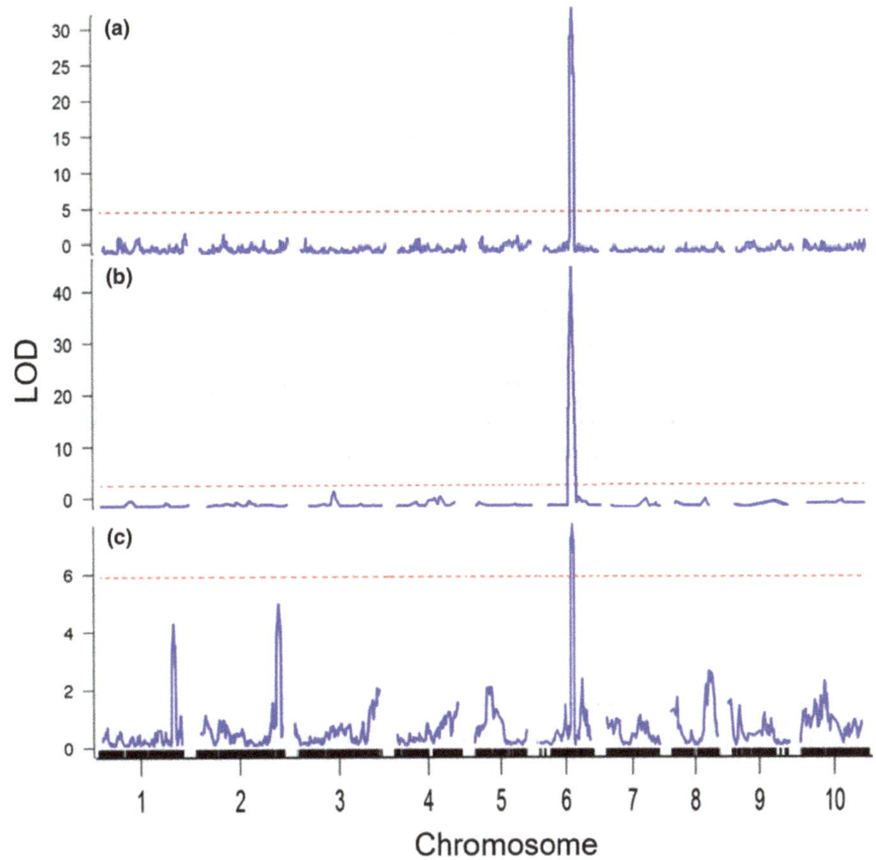

FIGURE 2 QTL associated with variation in aerenchyma (AER) formation were mapped in three biparental populations: BTx623*IS3620c (a, n = 380, RIL), BTx623*R07007 (b, n = 215, F5), and BTx623*Standard Broomcorn (c, n = 133, F2). The extent of AER formation in each line was rated visually in cross sections at a midpoint of fully elongated internodes. Main effect QTL colocalized to chromosome 6 in all three populations where the BTx623 allele reduced stem AER formation. QTL analyses were performed in R/qtl using composite interval mapping. AER ratings in BTx623*IS3620c were from one biological replicate per RIL. AER ratings in BTx623*R07007 are the average of three biological replicates per individual

the BTx623*Standard broomcorn F2 population (Supporting Information Table S1). Two additional QTL with subthreshold LOD scores located on SBI-01 and SBI-02, respectively, were detected using the Standard broomcorn population (Figure 2c). In all three populations, lines with homozygous BTx623 haplotypes spanning the QTL on SBI06 had low stem aerenchyma ratings and lines lacking BTx623 haplotypes accumulated the higher levels of stem aerenchyma (Supporting Information Figure S2).

The QTL located on SBI06 that modulates stem aerenchyma accumulation in the current study is aligned with a QTL on LG-06 previously identified as encoding the *D*-gene (Hart et al., 2001). The prior study was based on the analysis of 132 RILs derived from BTx623*IS3620c and a low-resolution genetic map. To fine-map the *D*-locus, the current study utilized 380 RILs derived from BTx623*IS3620c (Burow et al., 2011) and a high-resolution genetic map constructed by scoring the segregation of ~10,091 DG markers in the expanded RIL population (Truong et al., 2014). A SNP marker located at 50,878,497 on SBI06 was tightly linked to the aerenchyma QTL (Figure 3a). Four RILs with breakpoints immediately flanking this stem aerenchyma QTL marker on SBI06 were identified (Figure 3b). SNP and INDEL markers useful for further fine-mapping breakpoints within the *D*-locus in the four RILs were identified from an alignment of the IS3620c whole-genome sequence and the BTx623 reference sequence (Figure 3b) (McCormick et al., 2018). Sequence analysis of the markers in the four RILs delimited the *D*-locus to a 73-kbp region (Figure 3c, Supporting Information Table S2). Based on gene annotation information in Phytozome (Sbiv3.1), the 73-kbp region encodes a NAC (NAM, ATAF1/2, CUC) transcription factor, three genes encoding threonine aldolase, a MYB-like DNA-binding factor, and three genes encoding predicted (or unknown) proteins that lack functional annotation (Supporting Information Table S2).

The genes in the delimited *D*-locus were analyzed for the presence of sequence variants in coding regions to help identify candidate genes that could correspond to the *D*-gene. Gene intron/exon annotations were checked using RNAseq data (Phytozome v12), and most of the gene annotations were supported by this analysis (McCormick et al., 2018). However, Sobic.006G147400 was annotated with three introns although RNAseq data and information from direct cDNA sequencing (Supporting Information Figure S3) provided evidence for only the two larger introns. Moreover, the incorrectly annotated intron in Sobic.006G147400 encodes amino acid sequences (V32 – E44) that are present in the sorghum cDNA and in homologs of this gene (Figure 4). Therefore, an annotation of Sobic.006G147400 that contains two introns was used for subsequent analysis of coding variants. A proposed intron in Sobic.006G147800 was also not well supported by RNAseq data possibly because expression of this gene was low resulting in incomplete coverage of the gene (Phytozome v12). A gene annotation lacking introns resulted in a continuous coding region through the predicted intron consistent with the annotation of a homolog of Sobic.006G147800 in Setaria

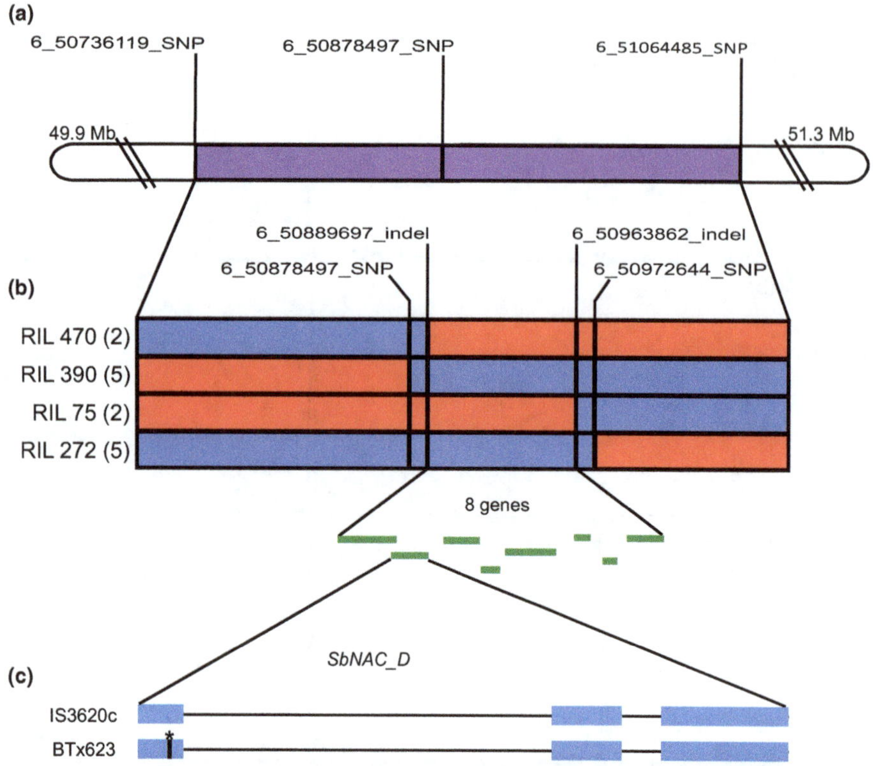

FIGURE 3 Positional cloning of SbNAC_D. (a) The marker 6_50878497_SNP had the highest LOD score in the QTL interval and was used to define a 2-LOD interval spanning the QTL (purple). (b) RILs from BTx623*IS3620c (RIL 470, RIL 390, RIL 75, RIL 272) with breakpoints within the 2-LOD interval were identified and phenotyped for the AER formation at anthesis. AER ratings are indicated in parentheses next to the RIL label to the left of the diagram. Blue haplotypes correspond to IS3620c alleles (+AER), and red haplotypes correspond to BTx623 alleles (−AER). The fine-mapped interval contained eight genes. (c) Sobic.006G147400 (SbNAC_D) was identified as the best candidate for the "D" gene. The BTx623 allele contains a T-to-C polymorphism that causes a Gln to STOP (Q39X) change in the first exon of BTx623 (red line)

that lacks an intron (Phytozome). Based on these results, the annotation of Sobic.006G147800 with and without the proposed intron was analyzed for coding variants.

Genetic analysis in the three populations is consistent with the prediction that there is a sequence variant (or variants) in BTx623 within the *D*-locus that inactivates a gene or regulatory element that normally increases stem aerenchyma formation and that this sequence variant is not present in IS3620c, R07007, and Standard broomcorn. Polymorphisms were identified in the coding regions of the eight genes within the delimited *D*-locus that distinguish the parental lines of the QTL mapping populations (BTx623, IS3620c, R07007, and Standard broomcorn). In addition, coding variants were identified in Rio and SC170, genotypes that have low levels of stem aerenchyma similar to BTx623 (see below and Table 1). Coding variants were further analyzed to determine whether they are predicted to alter protein function (Supporting Information Table S3) (Choi et al., 2012). This analysis showed that three genes in the delimited *D*-locus lacked deleterious coding polymorphisms (Sobic.006g147450, Sobic.006g147500, Sobic.006g148000) that distinguished BTx623, Rio, SC1470, IS3620c, R07007, and Standard broomcorn (Supporting Information Table S3). However, the analysis of coding variants identified a sequence variant in the low aerenchyma lines BTx623, Rio, and SC170 that creates a stop codon (Q39X) in the NAC transcription factor encoded by Sobic.006g147400 (Supporting Information Table S3). In contrast, genotypes that accumulate higher levels of stem aerenchyma (IS3620c, R07007, and Standard broomcorn) encoded full-length NAC proteins. The stop codon (Q39X) in Sobic.006G147400 present in BTx623, Rio, and SC170 removes motifs B-D of the NAC domain (Zhu, Nevo, Sun, & Peng, 2012) and the rest of the C-terminal portion of the protein. Sobic.006G147600, a gene that encodes threonine aldolase, contained a change in amino acid sequence (S218P) in IS3620c and SC170 that was predicted to cause a deleterious change in protein function by PROVEAN analysis (Choi et al., 2012). SC170 has low aerenchyma, while IS3620c has high aerenchyma, so this deleterious change in amino acid sequence is unlikely to be associated with variation aerenchyma formation (Table 1). Two other sequence variants in Sobic.006G147700 present only in Standard broomcorn were predicted to cause deleterious changes in protein sequence. Sequence variants in Sobic.006G147800 were predicted to cause deleterious changes in protein function in R07007 and Standard broomcorn, but these changes were also present in Rio and SC170. Additionally, no coding region mutations in Sobic.006G147800 were identified in IS3620c, so variation coding variants in Sobic.006G147800 were not associated with the stem aerenchyma QTL in the population derived from BTx623*IS3620c. In summary, the analysis identified Sobic.006G147400 (NAC transcription factor) as the only gene with a deleterious coding variant in the low aerenchyma lines (BTx623, Rio, and SC170) that was not present in lines that accumulate higher levels of stem aerenchyma. Taken together, the analysis of coding variants indicated that Sobic.006g147400, a gene encoding a NAC transcription factor, contained a stop codon in BTx623, Rio, and SC170 that was not present in IS3620c, R07007, and Standard broomcorn, consistent with genetic prediction that BTx623 contains a mutation in a gene that is involved in induction of stem aerenchyma formation. As such, we designated the two alleles of this gene as *SbNAC_D* and *SbNAC_d1* (the allele in BTx623) to distinguish this gene from other *NAC*-genes in sorghum.

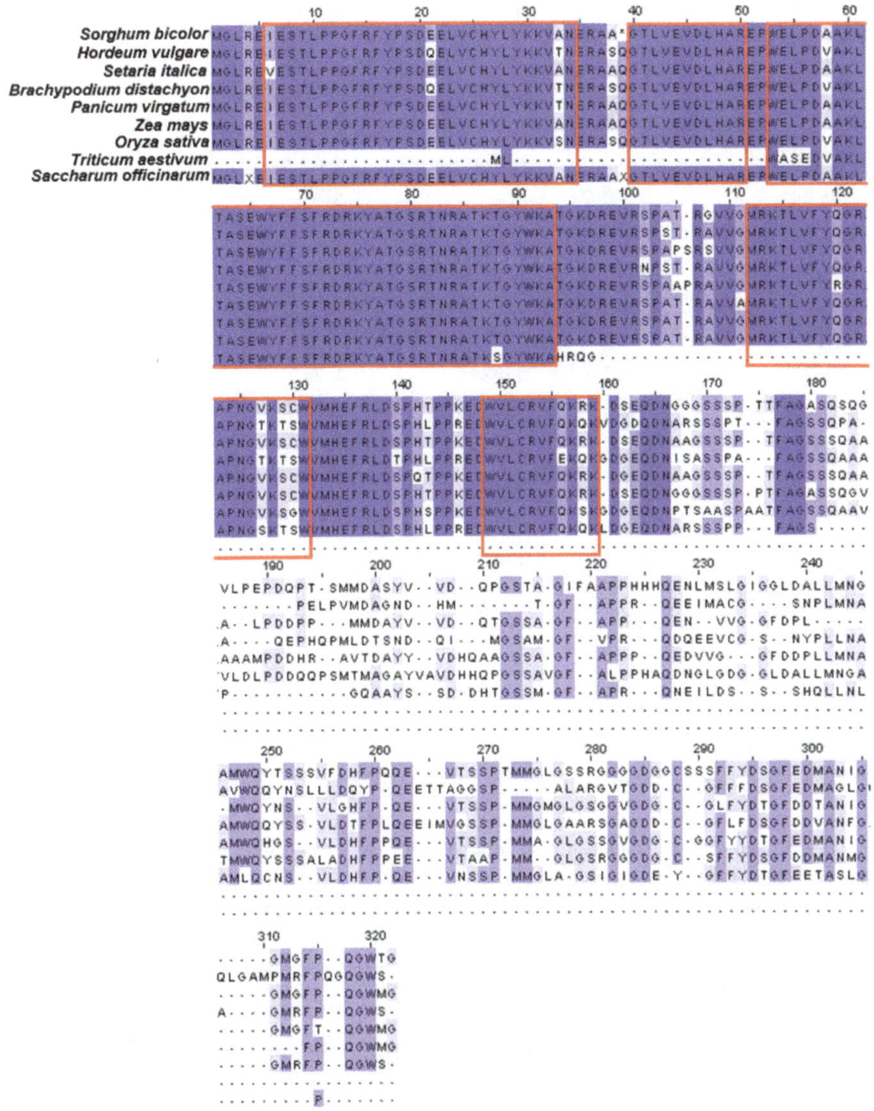

FIGURE 4 Protein sequence alignment of SbNAC_D homologs in select grass species. N-terminal sequences (1–170) of the SbNAC_D homologs are highly conserved, including protein motifs conserved in NAC transcription factors (highlighted by red boxes). Protein sequences were aligned using MUSCLE (Edgar, 2004) and visualized using Jalview (Waterhouse et al., 2009)

3.3 | *D*-alleles in sorghum germplasm

Sequence analysis of Sobic.006g147400 from R07007, R07020, Standard broomcorn, and Feterita showed that these genotypes encode *SbNAC_D*, consistent with the accumulation of relatively high levels of stem aerenchyma in the stems of each genotype (Table 1). The energy sorghum hybrid, TX08001, derived by crossing A/BTx623*R07007 and therefore heterozygous for *NAC_D* alleles (*NAC_D:NAC_d1*) had an intermediate aerenchyma rating (Table 1). The recessive *SbNAC_d1* allele found in BTx623 was also present in other lines used in grain sorghum breeding including Blackhull Kafir, Tx3197, RTx430, SC170, and Hegari, genotypes that accumulate relatively low levels of aerenchyma (Table 1). The sweet sorghums Rio, Della, Wray, Collier, and Grassl, genotypes that accumulate low levels of stem aerenchyma, also encode *SbNAC_d1* (Table 1). In this limited survey of sorghum germplasm, the presence of *SbNAC_D* in genotypes was correlated with high stem aerenchyma and *SbNAC_d1* with lower levels of stem aerenchyma.

3.4 | Phylogenetic analysis of *SbNAC_D*

NAC transcription factors were named for three genes initially discovered to contain NAC domains that contain N-terminal conserved motifs: NAM (no apical meristem), ATAF1 and -2, and CUC (cup-shaped cotyledon) (Souer, van Houwelingen, Kloos, Mol, & Koes, 1996; Zhu et al., 2012). NAC transcription factors are unique to plants, and most plant species including sorghum encode >100 different NAC family members (Kadier et al., 2017). Phylogenetic analysis of grass genes encoding the *SbNAC-D*-like factors generated C4 grass and C3 grass groups (Supporting Information Figure S4). Protein alignment showed that the grass *NAC_D*-like proteins were highly conserved across their N-terminal 170 amino acids that span the NAC domain motifs A-E (Figure 4). Greater divergence was observed across the C-terminal portion of the grass *NAC_D*-like proteins consistent with what has been found for other NAC proteins. The large N-terminal deletion present in a *NAC_D* homolog of wheat may be a consequence of selection for solid stems in this grain crop (Nilsen et al., 2017). A *NAC_D* homolog of sugarcane was also

TABLE 1 *SbNAC_D* alleles identified in various genotypes of *Sorghum bicolor*

Genotype	Allele	Amino acid change	AER rating
R07007	*SbNAC_D*	–	3–5[b]
R07020	*SbNAC_D*	–	4–5
IS3620c[a]	*SbNAC_D*	–	3–5
Standard broomcorn	*SbNAC_D*	–	5
Feterita	*SbNAC_D*	–	3–4
TX08001 (BTx623/R07007)	*SbNAC_D/d1*	–	3
BTx623[a]	*SbNAC_d1*	Q39X	1–2
Blackhull Kafir	*SbNAC_d1*	Q39X	1–2
Tx3197	*SbNAC_d1*	Q39X	1–2[b]
RTx430	*SbNAC_d1*	Q39X	1
SC170	*SbNAC_d1*	Q39X	1
Hegari	*SbNAC_d1*	Q39X	1
Sweet sorghums			
Rio	*SbNAC_d1*	Q39X	1–2[b]
Wray	*SbNAC_d1*	Q39X	1
Della	*SbNAC_d1*	Q39X	1–2[b]
Collier	*SbNAC_d1*	Q39X	1–2
Grassl	*SbNAC_d1*	Q39X	1

Notes. Aerenchyma (AER) ratings were scored on a scale from 1 to 5 (1, no AER; 5, high AER) at a midpoint of an internode. Ratings are an average of at least three biological replicates.
[a]Allele sequence was determined by whole-genome sequences. All others were determined by Sanger sequencing.
[b]AER ratings were adapted from Carvalho & Rooney, 2017.

truncated, which may contribute to reduced aerenchyma formation in stems of sugarcane (Figure 4).

NAC family genes related to *SbNAC_D* were identified in sorghum (Supporting Information Figure S5A) and other plants (Supporting Information Figure S5B) to see whether this would provide information about *SbNAC_D* function. The most closely related *NAC*-gene family member in Arabidopsis, *AtNAC074* (At4G28530), is differentially expressed to some extent in xylem compared to phloem cambium (Zhao, Craig, Petzold, Dickerman, & Beers, 2005). NAC factors such as VND6/7, SND1, and NST1/2 play key roles in secondary cell wall formation and xylem differentiation, developmental processes that involve programmed cell death (PCD) (Nakano, Yamaguchi, Endo, Rejab, & Ohtani, 2015). *NAC*-genes also play a role in PCD-associated senescence (Kim et al., 2014). Therefore, sorghum *NAC*-gene homologs of *NAC*-genes involved in secondary cell wall formation and senescence were identified and compared with *SbNAC_D* homologs from grasses (Supporting Information Figure S6). This showed that grass *NAC_D-like* genes grouped in a cluster distinct from *NAC*-genes involved in secondary cell wall formation and senescence. The analysis also showed that *SbNAC_D* was closely related to ZmNAC22 (GRMZM2G081930), a member of the ZmNAC-h (*NAC1*) subfamily (Peng et al. 2015). This subfamily is distinct from the ZmNAC-j (VND6) and ZmNAC-l (SND/NST1) subfamilies involved in secondary cell wall formation. These results indicate that the function of *SbNAC_D* differs from *NAC* transcription factors involved in xylem secondary cell wall formation/xylem differentiation and senescence.

3.5 | Expression of *D*-locus genes in RILs with BTx623 or IS3620c haplotypes

Sobic.006G147400 was identified as the best candidate gene for the *D*-gene among the genes located in the *D*-locus based on an analysis of coding sequence variants. However, it is possible that there are sequence variants among the parental genotypes used for QTL analysis that modify noncoding regulatory elements in the *D*-locus that affect aerenchyma formation by altering the expression of one or more of the eight genes in the delimited *D*-locus. To check this possibility, the expression of the genes in the delimited *D*-locus was analyzed by qRT-PCR using RNA from stems of three pairs of RILs derived from BTx623*IS3620c with similar flowering times and dwarfing genotypes that either encoded the IS3620c *D*-locus haplotype or the BTx623 *d1*-locus haplotype. RNA was extracted from the top internode of the stem from each RIL shortly after full internode elongation for qRT-PCR analysis. The analysis showed no significant difference in expression of the 8 genes in the three pairs of RILs indicating that polymorphisms distinguishing BTx623 (*d1*) and IS3620c (*D*) in noncoding regions of the *D*-locus were not altering the expression of genes in the *D*-locus in this tissue (data not shown).

3.6 | Developmental timing of stem aerenchyma accumulation and *SbNAC_D* expression

During the analysis above, it was observed that aerenchyma levels were low in internodes that were elongating; however, 7–10 days after internodes reached full elongation, higher levels of aerenchyma were observed in *D*-genotypes. This indicated that stem aerenchyma formation is suppressed in nascent and growing internodes of sorghum stems and that the induction of aerenchyma formation occurs following internode elongation. If *SbNAC_D* is involved in stem aerenchyma formation, then expression of Sobic.006G147400 may increase following internode elongation in parallel with the formation of stem aerenchyma. To test this possibility, internodes from the three pairs of RILs with contrasting *D*-locus/*d1*-locus haplotypes were collected when they first reached full length (time point 1, low aerenchyma) and again 7–10 days later (time point 2) when higher levels of aerenchyma were visible in genotypes encoding *SbNAC_D* (but not in *SbNAC_d1*). The results showed that Sobic.006G147400 expression was significantly higher in internodes that were forming stem aerenchyma (time point 2) in *SbNAC_D* genotypes (Figure 5). In addition, increased expression of Sobic.006G147400 was also observed 7–10 days after internode elongation in *SbNAC_d1* genotypes even though aerenchyma levels did not increase in these genotypes. This indicates that increased expression of Sobic.006G147400 was induced by development and not caused by

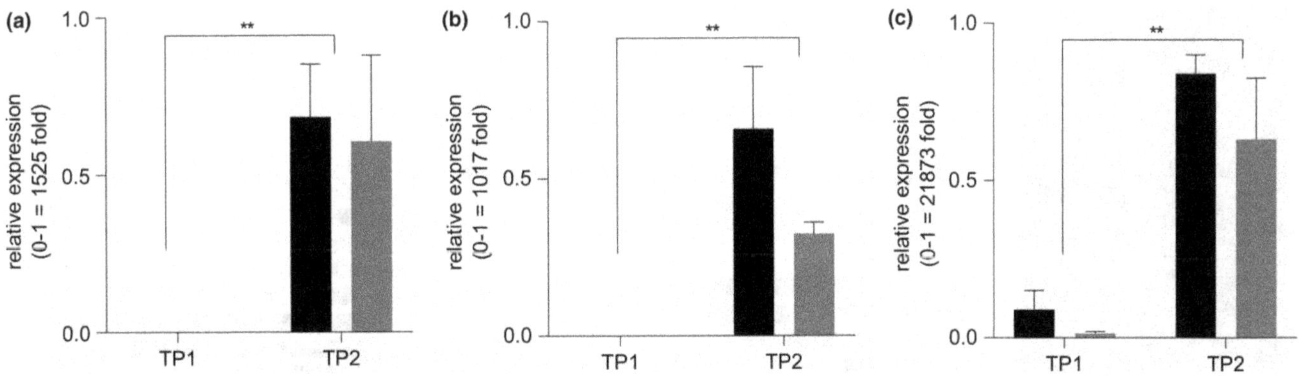

FIGURE 5 *SbNAC_D* expression in three pairs of *SbNAC_D (black bars)/SbNAC_d1 (grey bars)* RILs: (a) RIL226/RIL212, (b) RIL119/RIL474, and (c) RIL392/RIL49. Expression was analyzed in the internode subtending the peduncle via qRT-PCR at time points before aerenchyma (AER) formation (TP1) and 7–10 days later when AER were visible in the internode (TP2). In all three RIL pairs, expression of *SbNAC_D* increased in parallel with AER formation. All gene expression was normalized to the expression of *SbUBC* in each sample. Relative expression was calculated via the $2^{-\Delta\Delta C_t}$ method relative to the internode section with the highest expression of the gene in each RIL pair (Pfaffl et al., 2001). Fold change in expression between the minimum and maximum values on the *y*-axis was calculated based on *SbUBC* normalized values according to $FC = 2^{\Delta C_t(max) - \Delta C_t(min)}$. Expression values are the average of three biological replicates. Differences in expression values between TP1 and TP2 were analyzed by ANOVA in GraphPad Prism v7. **, $p < 0.005$

the formation of aerenchyma. To complete the survey of gene expression, RNA extracted from the internodes from one pair of RILs (RIL49, *SbNACd1*; RIL392, *SbNAC_D*) was used to analyze expression of the seven other genes in the *D*-locus by qRT-PCR (Supporting Information Table S4). Only one of these genes, Sobic.006g148000, showed an increase in expression in stems between time point 1 and time point 2 in RIL392 (*SbNAC_D*) (Supporting Information Table S4). However, no deleterious coding sequence variants were identified in the parental lines used to fine-map *D*-locus QTL (Supporting Information Table S3), suggesting that Sobic.006G148000 plays some other role in internode development.

The developmental timing of *SbNAC_D* expression and aerenchyma accumulation in sorghum stems was examined in more detail during vegetative phase stem growth using R07020, a tall, photoperiod-sensitive, late-flowering, high-biomass genotype. Growth of R07020 in 14 hr day-lengths represses floral initiation allowing analysis of stem aerenchyma formation during an extended period of vegetative growth after completion of the juvenile phase. R07020 internode elongation begins between ~45 and 55 days after germination, and aerenchyma were observed in fully elongated internodes produced during vegetative development. Qualitative visual inspection of cross sections showed that aerenchyma were also present in the leaf midribs and leaf sheaths of R07020 that developed during the vegetative phase.

The timing of aerenchyma formation during growth and development of stem internodes was examined using cross-sectional analysis of R07020 stems (Figure 6a). The stems of rapidly growing R07020 plants contained two short nascent internodes immediately below the shoot apex (Int1, Int2), followed by an elongating internode (Int3) and several older fully elongated internodes (Int4-7) (Figure 6a, left). The two nascent nonelongated internodes (Int1, Int2) and the partially elongated internode 3 contained very low levels of aerenchyma (Figure 6b). Aerenchyma were present in the upper end of fully elongated internode 4 (sections Int4-1 to Int4-6) but not at the basal end of this internode (Int4-9 to Int4-11). Aerenchyma were observed throughout older internodes (Int5-7), although at lower abundance near the base of internode 5 (Figure 6a,b).

Expression of *SbNAC_D* during internode development in R07020 was analyzed using qRT-PCR (Figure 6c). Stem internode sections used for expression analysis were identical to those used for analysis of aerenchyma accumulation (Figure 6b). *SbNAC_D* was expressed at relatively low levels in the two youngest nonelongated internodes located immediately below the shoot apex that lacked aerenchyma (Int1, Int2) (Figure 6c). *SbNAC_D* expression increased to a small extent in internode 3 and reached maximal levels in the upper portion of internode 4 (Figure 6c, sections Int4-1 + 2 to Int4-7 + 8), a region where aerenchyma were visible (Figure 6, Int4). Aerenchyma levels and *SbNAC_D* expression were low at the basal end of internode 4 (i.e., Int4, sections Int4-9–Int4-11). *SbNAC_D* expression was also observed in tissue of older internodes (Int5-7). There was a strong correlation between *SbNAC_D* expression and aerenchyma levels during vegetative stem internode development (ρ = 0.709).

3.7 | SbXCP1 expression is correlated with aerenchyma formation and regulated by *SbNAC_D*

The formation of stem aerenchyma could be mediated by programmed cell death (PCD). Induction of cysteine proteases is associated with PCD in rice (Yin & Xue, 2012) and has been correlated with developmental PCD in Arabidopsis (Olvera-Carillo 2015). Therefore, we identified the sorghum homolog (Sobic.007g172100, *SbXCP1*) of a rice PCD-regulated gene (Os02g48450) that encodes a cysteine protease and characterized its expression to see whether it would be a useful marker for sorghum stem aerenchyma formation. RNA extracted from internodes of the pairs of RILs encoding either

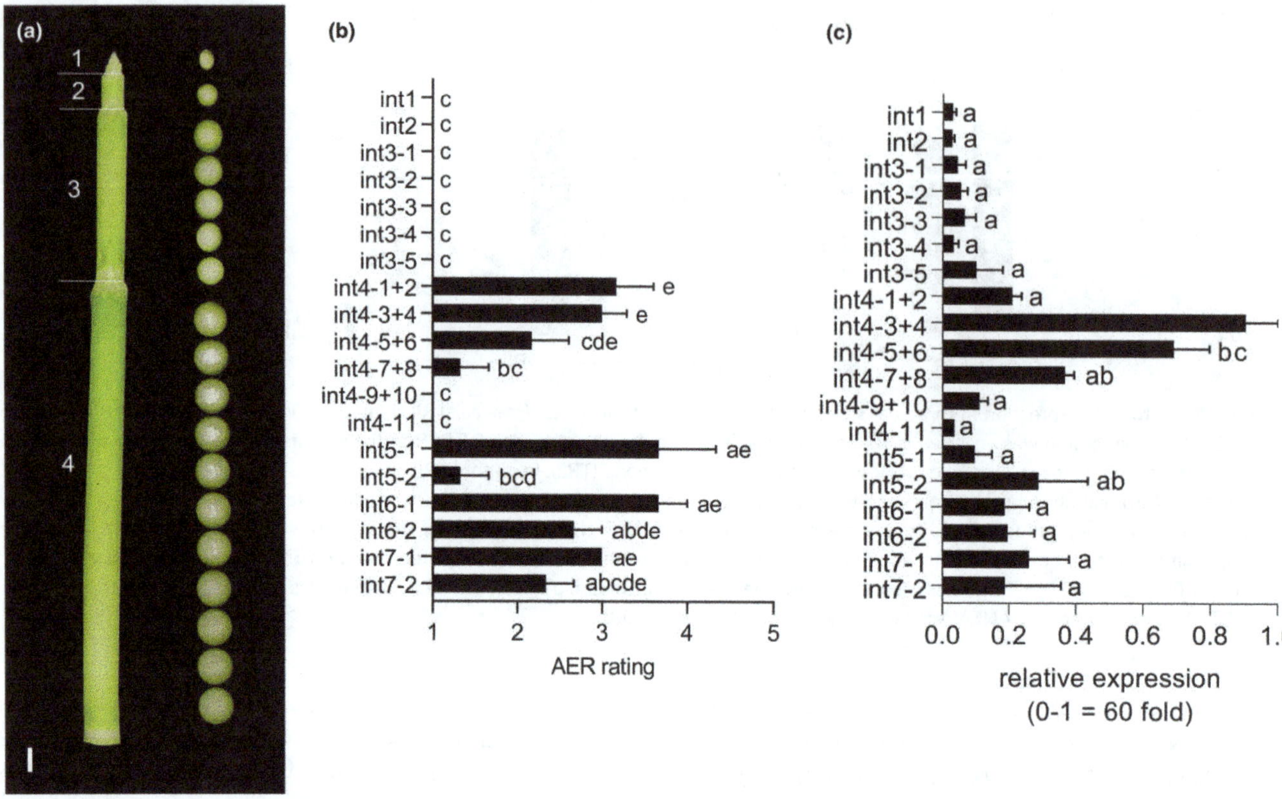

FIGURE 6 Aerenchyma (AER) formation and expression of *SbNAC_D* in stem sections of R07020. (a) Photograph of the top of a R07020 stem. Internode numbers are indicated to the left of the stem. R07020 remained in the vegetative phase throughout these experiments, and internode 1 is the youngest visible internode below the apical dome. Photographs of cross sections of each internode taken every 1 cm are shown to the right of the stem in (a). No AER is visible in internodes 1–3 or in the lower portion of internode 4. AER formation begins in the top of internode 4 in the center of the internode. AER forms in a basipetal manner and spreads radially outward. Scale bar = 1 cm. (b) AER was rated on a scale from 1 (no AER) to 5 (high AER). Ratings of pairs of adjacent sections of internode 4 (int4) were averaged. AER formation of internodes 5–7 was rated 2 cm from the top node of each internode (int5-1, int6-1, int7-1) and 2 cm from the bottom node of each internode (int5-2, int6-2, int7-2). Each rating is the average of five biological replicates. Tukey's multiple comparison test was used to compare ratings of each internode section. Sections with different letters are significantly different from each other, $p < 0.05$. (c) Relative expression of *SbNAC_D* across internode sections of R07020. Expression of *SbNAC_D* is correlated with AER formation (Spearman's $r = 0.709$). Relative expression was calculated via the $2^{-\Delta\Delta C_t}$ method relative to the internode section with the highest expression (Pfaffl, 2001). Expression values are the average of three biological replicates. Fold change in expression between the minimum and maximum values on the *y*-axis was calculated based on SbUBC normalized values according to $FC = 2^{\Delta C_t(max) - \Delta C_t(min)}$

SbNAC_D or *SbNAC_d1* (RIL392/RIL49, RIL474/RIL119, RIL226/RIL212) was used for qRT-PCR analysis of *SbXCP1* gene expression. In the RILs that encode *SbNAC_D*, *SbXCP1* expression increased in parallel with appearance of aerenchyma in internodes (Figure 7a, right). However, *SbXCP1* expression was not significantly induced in RILs that encode *SbNAC_d1* (Figure 7a, left), indicating that *SbNAC_D* is required for induction of *SbXCP1* expression in stems during aerenchyma formation.

The correlation between internode development, aerenchyma formation, and expression of the PCD marker gene *SbXCP1* was also investigated during R07020 stem development (Figure 7b). The relative expression of *SbXCP1* was low in internodes 1 and 2, with a small increase in internode 3, and was high in the mid-to-upper portion of internode 4 (Int4, sections 3–8) (Figure 7b). Elevated but lower expression was observed in internodes 5, 6, and 7. In general, variation in the expression of the PCD marker gene *SbXCP1* was similar to *SbNAC_D* expression and matched the time course and location of aerenchyma formation in stems during internode development. Expression of six of the other genes in the delimited *D*-locus was also analyzed during internode development in R07020 (Supporting Information Figure S7).

3.8 | Spatial analysis of aerenchyma accumulation in sorghum internodes

Microscopic analysis of internode cross sections and CT analysis showed that aerenchyma form first in pith parenchyma located between vascular bundles (VBs) in the center of the internode (Figure 8). As internode development proceeds, aerenchyma accumulation spreads radially, with relative few aerenchyma forming in the rind (Figure 8). Pith parenchyma cells located between VBs were initially converted into aerenchyma leaving islands of VBs and closely

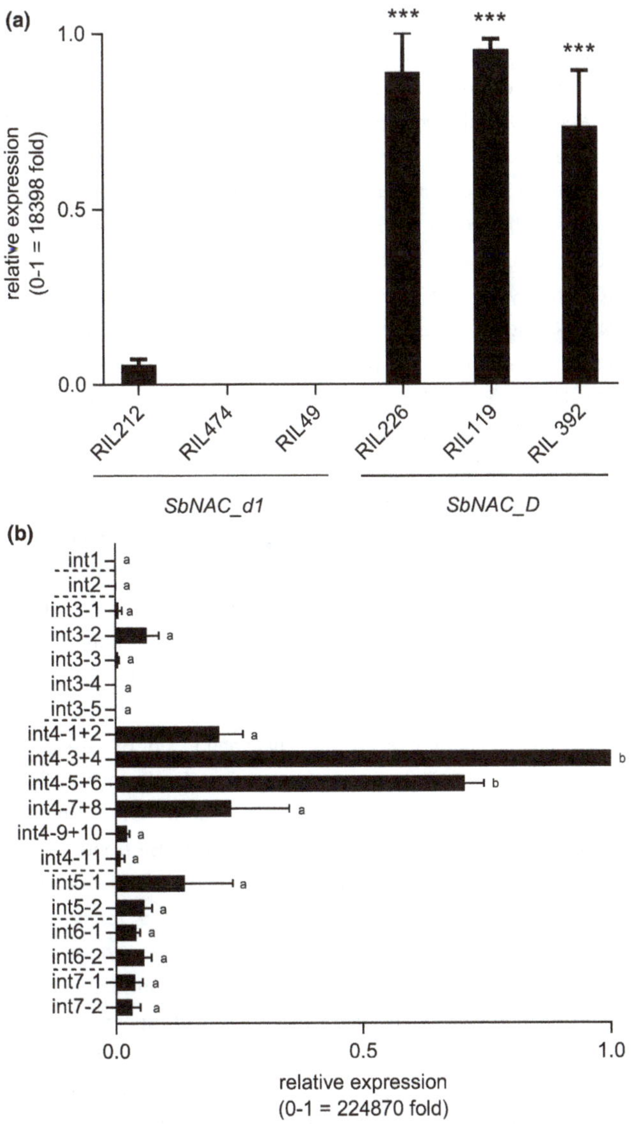

FIGURE 7 Expression of *SbXCP1* in (a) SbNAC_d1/SbNAC_D RILs and (b) R07020 internodes. *SbXCP1* is a developmental programmed cell death (PCD) marker gene. (a) The expression of *SbXCP1* is elevated in *SbNAC_D* RILs (RIL119, RIL392, RIL226) compared to *SbNAC_d1* RILs (RIL212, RIL474, RIL49). The uppermost internode of each RIL was sampled at a time point after aerenchyma (AER) formation had started. Expression differences between RILs were compared by Tukey's multiple comparison test, $p < 0.05$. (b) Expression of *SbXCP1* is low in internode sections where no AER formation is occurring (int1, int2, int3-X). Expression increases in sections of internode 4 (int4-X) where AER are forming. Tukey's multiple comparison test was used to compare ratings of each internode section. Sections with different letters are significantly different from each other, $p < 0.05$. All gene expression was normalized to the expression of *SbUBC* in each sample. Relative expression was calculated via the $2^{-\Delta\Delta C_t}$ method relative to the internode section with the highest expression of each gene (Pfaffl et al., 2001). Fold change in expression between the minimum and maximum values on the *y*-axis was calculated based on *SbUBC* normalized values according to $FC = 2^{\Delta C_t(max) - \Delta C_t(min)}$. Expression values are the average of three biological replicates

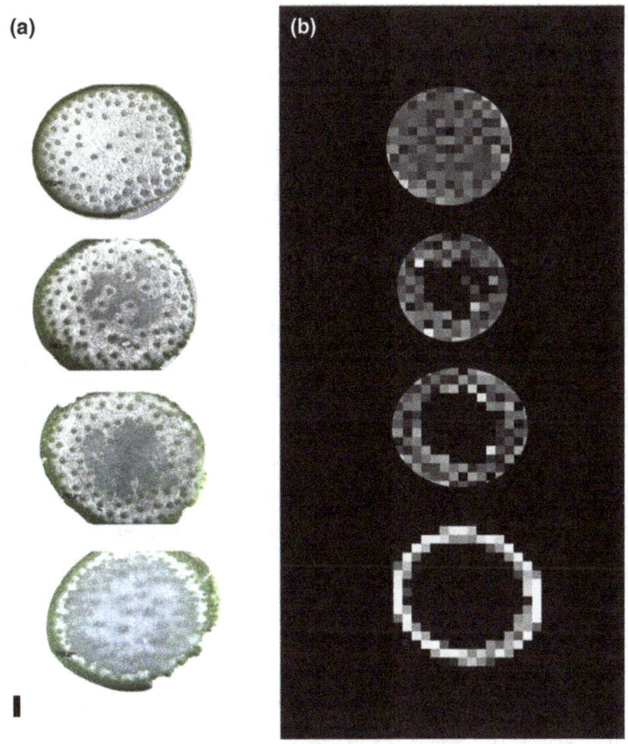

FIGURE 8 Microscopic (a) and CT scan (b) images of R07020 stem cross sections at progressive stages of aerenchyma (AER) formation. From the top down, solid stem with no AER formation to high levels of AER formation across the stem cross section. AER formation begins in the pith parenchyma cells at the center of the stem and spreads radially outward toward the rind. Scale bar = 0.1 cm

associated cells (Figure 8). At later stages of internode development, nearly all of the cells surrounding VBs located in the central region of the stem had been converted into aerenchyma (Figure 8).

4 | DISCUSSION

Genotypic variation in sorghum stem aerenchyma formation was initially documented in 1916 (Hilson, 1916). This genetic determinant of variation in stem/leaf midrib aerenchyma accumulation was named the *D*-locus and further characterized in subsequent studies (Ayyangar et al., 1936; Stephens & Quinby, 1939; Swanson & Parker, 1931). Sorghum genotypes with opaque white leaf midribs were reported to have drier, "pithy" stems, whereas genotypes with green midribs had stems with low levels of aerenchyma that accumulated higher amounts of sugars postflowering. More recent research identified QTL that modulate sorghum leaf midrib color and stem moisture content (Burks et al., 2015; Guan et al., 2011; Han et al., 2015; Hart et al., 2001; Mocoeur et al., 2015; Srinivas et al., 2009; Xiaoping et al., 2011; Xu et al., 2000). In the current study, a main effect QTL on SBI06 was identified in three populations that have a large impact on the accumulation of stem aerenchyma. This QTL aligned with a leaf midrib color QTL previously designated as the *D*-locus ("*D*" for dry stems) (Ayyangar et al., 1936; Hart et al., 2001; Xu et al., 2000). QTL-assisted candidate gene identification and analysis of candidate gene coding variants and gene expression identified Sobic.006g147400, a gene encoding a NAC transcription factor, as the best candidate for the *D*-gene (*SbNAC_D*).

Numerous important sorghum genes involved in fertility restoration, flowering time, seed shattering, and the regulation of stem growth have been cloned using QTL-assisted map-based gene identification (Hilley et al., 2016, 2017; Klein et al., 2005; Lin et al., 2012; Murphy et al., 2011, 2014). This well-established approach was used in the current study to identify a gene and alleles that could explain variation in phenotypes associated with the *D*-locus. QTL that modulate the extent of stem aerenchyma accumulation were mapped in three sorghum populations derived from parents that differed in stem aerenchyma ratings. A single, large-effect QTL on SBI06 was identified in all three populations. Genetic analysis showed that BTx623, the female parent of each population, was the source of sequence variation that reduced aerenchyma accumulation in stems. Fine-mapping delimited the *D*-locus to 73kbp, a region encoding eight genes. The sequences of the eight genes in the delimited *D*-locus derived from the four parental lines used for QTL analysis were examined to identify sequence variants that could affect protein-coding regions. This survey identified a disruptive sequence variant in one of the eight genes in BTx623 that was not present in the other parental lines used for QTL analysis. The key variant introduced a stop codon in the coding region of Sobic.006G147400 resulting in truncation of the NAC transcription factor encoded by this gene. No other disruptive coding variants were identified in BTx623 in genes in the *D*-locus. Disruptive mutations were not present in the coding regions of four of the genes in the delimited *D*-locus, and disruptive coding variants found in the other three genes were present in one or two of the parental lines crossed to BTx623 (IS3620c, R07007, Standard broomcorn) but not all three genotypes. Based on the analysis of coding variants, Sobic.006G147400 was tentatively identified as a candidate for the *D*-gene and the two alleles were named *SbNAC_D* and *SbNAC_d1*. The association between *SbNAC_D* genotype and stem aerenchyma phenotype was extended by screening a panel of diverse sorghum genotypes for *SbNAC_D* alleles and stem aerenchyma accumulation. This analysis showed that genotypes encoding *SbNAC_D* had higher levels of stem aerenchyma relative to genotypes encoding *SbNAC_d1*. Further evidence for the involvement of *SbNAC_D* in the formation of stem aerenchyma was obtained by analyzing the expression of *SbNAC_D* during stages of internode development before and after the onset of aerenchyma accumulation. Analysis of the internode located immediately below the peduncle in plants at flag leaf stage showed that *SbNAC_D* expression was low prior to the accumulation of aerenchyma and much higher 7–10 days later when internodes were accumulating aerenchyma. The correlation between increases in *SbNAC_D* expression and onset of aerenchyma formation was also high in vegetative phase internodes of R07020. In contrast, the expression of the other seven genes located in the delimited *D*-locus showed lower or no correlation with stem aerenchyma formation compared to *SbNAC_D*. In addition, expression of a PCD marker gene encoding a cysteine protease (*SbXCP1*) was highly correlated with expression of *SbNAC_D* and the accumulation of aerenchyma in stems and expression was not induced in stems of genotypes encoding *SbNAC_d1*. The involvement of a NAC transcription factor in stem aerenchyma formation is not surprising since NAC transcription factors have been associated with other types of PCD. For example, the NAC transcription factors VND6/7 play key roles in vascular tissue differentiation, a developmental process that includes PCD (Escamez & Tuominen, 2014). Similarly, NAC transcription factors mediate PCD that occurs during senescence (Kim et al., 2014; Uauy et al., 2006), root cap development (ANAC033) (Fendrych et al., 2014), and PCD associated with oxidative stress (ANAC13) (De Clercq et al., 2013). Phylogenetic analysis showed that the grass homologs of *SbNAC_D* formed a cluster that was differentiated from clusters of NAC transcription factors involved in secondary cell wall formation and senescence. Taken together, these results are consistent with the identification of *SbNAC_D* (Sobic.006g147400) as the *D*-gene, a major modulator of aerenchyma formation in stems.

Sorghum genotypes vary in the timing and extent of aerenchyma formation in leaf midribs and stems during development (Hilson, 1916). In the current study, aerenchyma were not detected in nonelongated nascent internodes of vegetative phase plants and the onset of aerenchyma accumulation was associated with completion of internode elongation. Intercalary meristems and zones of cell elongation are located at the base of elongating internodes in sorghum (Kebrom et al., 2017). In vegetative phase plants, aerenchyma were first observed in the more fully developed upper portion of recently elongated internodes. As internode development proceeds, aerenchyma spread radially from the center of the stem toward the rind/epidermis and in a basipetal manner until most of the internode was filled with aerenchyma. The stem nodes generally lacked or had low levels of aerenchyma throughout development. Rice shows a similar pattern of aerenchyma accumulation during stem and internode development (Steffens, Geske, & Sauter, 2011). The timing of expression of *SbNAC_D* during stem and internode development was consistent with the gene's proposed role in stem aerenchyma formation. *SbNAC_D* was expressed at very low levels in the top two nascent internodes, internode growing zones, and nodes of developing internodes, tissues that accumulate only low levels of aerenchyma. *SbNAC_D* expression increased in the upper portion of fully expanded internodes in the same location where aerenchyma form first during development. Moreover, analysis of internodes derived from *SbNAC_D* and *SbNAC-d1* RILs that differ in the extent of stem aerenchyma formation differentially expressed a gene encoding a sorghum homolog of a cysteine protease similar to vignain (SbXCP1) that is involved in developmental programmed cell death in rice (Yin & Xue, 2012). The expression of this gene increased in parallel with *SbNAC_D* expression and the appearance of stem aerenchyma in stem internodes during development. In contrast, expression of *SbXCP1* remained low in internodes of a sorghum genotype encoding *SbNAC_d1*, indicating that *SbNAC_D* regulates the expression of *SbXCP1* directly or indirectly. Taken together, these results are consistent with the identification of *SbNAC_D* (Sobic.006g147400) as the *D*-gene, a major modulator of aerenchyma formation in stems.

Programmed cell death plays an important role in plant development in response to pathogens and abiotic stress (Kabbage, Kessens, Bartholomay, & Williams, 2017; Van Hautegem, Waters, Goodrich, & Nowack, 2015). The accumulation of aerenchyma in stem pith parenchyma cells provides a way to remobilize nutrients from these cells to support rapid growth of stems and leaves during the vegetative phase. This may be one reason aerenchyma formation is correlated with rapid stem elongation during the vegetative phase. The selective destruction of pith parenchyma cells located in the stem also reduces maintenance respiration. However, the destruction of these stem cells decreases the capacity of stems for sucrose storage that occurs in some genotypes when plants transition to the reproductive phase. By anthesis, the demand for nutrients to support stem and leaf growth has declined and excess sugars accumulate in the vacuoles of pith parenchyma stem cells. A large supply of stem sugars is useful during the grain filling phase, especially as a buffer during periods of water deficit and other stresses that inhibit photosynthesis.

The recessive *SbNAC_d1* allele was present in Blackhull Kafir, Hegari, and Collier, genotypes that were imported from Africa into the United States between 1881 and 1908 (Smith & Frederiksen, 2000). This indicates that the *SbNAC_d1* allele was originated in Africa and was introduced at the onset of sorghum use in the United States consistent with the wide distribution of this allele in U.S. sorghum germplasm (Smith & Frederiksen, 2000). Following introduction into the United States, Blackhull Kafir (*SbNAC_d1*) was used in the development of lines such as BTx3197 (*SbNAC_d1*) and BTx623 (*SbNAC_d1*), which are useful for grain sorghum hybrid production. SC170, a genotype used in breeding R-lines such as RTx430, also encodes *SbNAC_d1*. Therefore, the historically important grain hybrid derived by crossing ATx623*RTx430 was homozygous for *SbNAC_d1*. The identification of *SbNAC_d1* in genotypes used for grain sorghum breeding suggests that reduced levels of stem aerenchyma are beneficial in these crops, possibly by reducing lodging or by increasing stem sugars that could provide a source of carbohydrate for grain filling under conditions of drought or other adverse environmental conditions.

The sweet sorghum genotypes Rio, Della, Wray, Collier, and Grassl all encode *SbNAC_d1* alleles, and the stems of these genotypes had lower levels of aerenchyma compared to genotypes encoding *SbNAC_D*. Reducing stem aerenchyma is important in sweet sorghum because retention of stem pith parenchyma increases the capacity for sucrose accumulation in the large vacuoles of these cells. Two previous sorghum genetic studies that characterized QTL for stem juiciness and sugar yield did not identify QTL aligned with *SbNAC_D* (Felderhoff et al., 2012; Murray et al., 2008). The parental genotypes used in these studies (BTx623*Rio, BTx3197*Rio) all encode *SbNAC_d1* consistent with lack of stem moisture or sugar content QTL aligned with the *D*-locus (Felderhoff et al., 2012; Murray et al., 2008). These previous studies identified other QTL that affect stem sugar content, concentration, and juiciness, indicating that the genetic basis of these traits is complex (Felderhoff et al., 2012; Murray et al., 2008).

Analysis of *NAC_D* homologs in other grasses showed that in the genotypes of wheat and sugarcane, analyzed *NAC_D* homologs contained large deletions suggesting that breeders may have selected for inactive versions of *NAC_D* in these species as occurred in sorghum. In sugarcane, selection for inactive alleles of *NAC_D* is consistent with low levels of stem aerenchyma, high stem juiciness, and high stem sucrose content in that crop (Waclawovsky, Sato, Lembke, Moore, & Souza, 2010). Selection for more solid stems has also occurred in wheat in part to increase resistance to insects that damage stems although the genetic basis of insect resistance is complex (Nilsen et al., 2017).

The identification of alleles of *SbNAC_D* will make marker-assisted selection or genome editing of sorghum genotypes for various end uses easier, although screening for white/green leaf midribs as an indicator of *D*-gene activity is reliable and easy to implement. More importantly, the identification of *SbNAC_D* provides an entry point for the identification of factors that regulate the tissue-specific expression of *SbNAC_D* during development and genes regulated by *SbNAC_D* that modulate aerenchyma formation. Variation in the extent of stem aerenchyma formation was observed in genotypes that encode *SbNAC_D*, and low levels of variation in stem "pithiness" were observed in populations derived from parental lines containing *SbNAC-d1* (Felderhoff et al., 2012). Follow-up studies are needed to more fully elucidate the factors that regulate *SbNAC_D* expression, the developmental PCD pathway that leads to stem aerenchyma formation, and other genes that modulate stem and leaf midrib aerenchyma formation.

ACKNOWLEDGMENTS

The authors thank Dr. Francisco Gomez for his help with CT visualization of stem aerenchyma. This work was funded in part by the DOE Great Lakes Bioenergy Research Center (DOE Office of Science BER DE-FC02-07ER64494, BER DE-SC0018409), US Department of Energy grant no. DE-SC0012629, and the Perry Adkisson Chair.

CONFLICT OF INTEREST

The authors have no conflict of interests to declare.

AUTHOR CONTRIBUTIONS

A.L.C. and J.M. designed experiments and wrote the article; A.L.C. performed most of the experiments; B.M. performed RNAseq data analysis; K.M.J.Y. and A.L.C. performed DNA sequencing; W.L.R. provided seeds and populations; and J.M. supervised experimental design and article preparation.

REFERENCES

Andersen, C. L., Jensen, J. L., & Ørntoft, T. F. (2004). Normalization of real-time quantitative reverse transcription-PCR data: A model-based variance estimation approach to identify genes suited for normalization, applied to bladder and colon cancer data sets. *Cancer Research*, *64*(15), 5245–5250. https://doi.org/10.1158/0008-5472.can-04-0496

Ayyangar, G. N. R., Ayyare, M. A. S., Rao, V. P., & Nambiar, A. K. (1936). Mendelian segregations for juiciness and sweetness in sorghum stalks. *Madras Agricultural Journal*, *24*, 247–248.

Bihmidine, S., Hunter, C., Johns, C., Koch, K., & Braun, D. (2013). Regulation of assimilate import into sink organs: Update on molecular drivers of sink strength. *Frontiers in Plant Science*, *4*(177), https://doi.org/10.3389/fpls.2013.00177

Broman, K. W., Wu, H., Sen, Ś., & Churchill, G. A. (2003). R/qtl: QTL mapping in experimental crosses. *Bioinformatics*, *19*, 889–890.

Burks, P. S., Kaiser, C. M., Hawkins, E. M., & Brown, P. J. (2015). Genomewide association for sugar yield in sweet sorghum. *Crop Science*, *55*(5), 2138–2148. https://doi.org/10.2135/cropsci2015.01.0057

Burow, G. B., Klein, R. R., Franks, C. D., Klein, P. E., Schertz, K. F., Pederson, G. A., … Burke, J. J. (2011). Registration of the BTx623/IS3620C recombinant inbred mapping population of sorghum. *Journal of Plant Registrations*, *5*(1), 141–145. https://doi.org/10.3198/jpr2010.04.0219crmp

Calviño, M., Bruggmann, R., & Messing, J. (2008). Screen of genes linked to high-sugar content in stems by comparative genomics. *Rice*, *1*(2), 166–176. https://doi.org/10.1007/s12284-008-9012-9

Carvalho, G., & Rooney, W. L. (2017). Assessment of stalk properties to predict juice yield in sorghum. *BioEnergy Research*, *10*(3), 657–670. https://doi.org/10.1007/s12155-017-9829-4

Choi, Y., Sims, G. E., Murphy, S., Miller, J. R., & Chan, A. P. (2012). Predicting the functional effect of amino acid substitutions and indels. *PLoS ONE*, *7*(10), e46688. https://doi.org/10.1371/journal.pone.0046688

De Clercq, I., Vermeirssen, V., Van Aken, O., Vandepoele, K., Murcha, M. W., Law, S. R., … Van Breusegem, F. (2013). The membrane-bound NAC transcription factor ANAC013 functions in mitochondrial retrograde regulation of the oxidative stress response in Arabidopsis. *Plant Cell*, *25*(9), 3472–3490. https://doi.org/10.1105/tpc.113.117168

Drew, M. C., He, C. J., & Morgan, P. W. (2000). Programmed cell death and aerenchyma formation in roots. *Trends in Plant Science*, *5*(3), 123–127.

Drew, M. C., Jackson, M. B., & Giffard, S. (1979). Ethylene-promoted adventitious rooting and development of cortical air spaces (aerenchyma) in roots may be adaptive responses to flooding in Zea mays L. *Planta*, *147*(1), 83–88. https://doi.org/10.1007/BF00384595

Edgar, R. C. (2004). MUSCLE: Multiple sequence alignment with high accuracy and high throughput. *Nucleic Acids Research*, *32*(5), 1792–1797. https://doi.org/10.1093/nar/gkh340

Escamez, S., & Tuominen, H. (2014). Programmes of cell death and autolysis in tracheary elements: When a suicidal cell arranges its own corpse removal. *Journal of Experimental Botany*, *65*(5), 1313–1321. https://doi.org/10.1093/jxb/eru057

Felderhoff, T. J., Murray, S. C., Klein, P. E., Sharma, A., Hamblin, M. T., Kresovich, S., … Rooney, W. L. (2012). QTLs for energy-related traits in a sweet x grain sorghum [Sorghum bicolor (L.) Moench] mapping population. *Crop Science*, *52*(5), 2040–2049. https://doi.org/10.2135/cropsci2011.11.0618

Fendrych, M., Van Hautegem, T., Van Durme, M., Olvera-Carrillo, Y., Huysmans, M., Karimi, M., … Nowack, M. K. (2014). Programmed cell death controlled by ANAC033/SOMBRERO determines root cap organ size in Arabidopsis. *Current Biology*, *24*(9), 931–940. https://doi.org/10.1016/j.cub.2014.03.025

George-Jaeggli, B., Jordan, D. R., van Oosterom, E. J., Broad, I. J., & Hammer, G. L. (2013). Sorghum dwarfing genes can affect radiation capture and radiation use efficiency. *Field Crops Research*, *149*, 283–290. https://doi.org/10.1016/j.fcr.2013.05.005

Godoy, J. G. V., & Tesso, T. T. (2013). Analysis of juice yield, sugar content, and biomass accumulation in sorghum. *Crop Science*, *53*(4), 1288–1297. https://doi.org/10.2135/cropsci2012.04.0217

Gomez, F. E., Carvalho, G., Shi, F., Muliana, A. H., & Rooney, W. L. (2018). High throughput phenotyping of morpho-anatomical stem properties using X-ray computed tomography in sorghum. *Plant Methods*, *14*(1), 59. https://doi.org/10.1186/s13007-018-0326-3

Guan, Y.-A., Wang, H.-L., Qin, L., Zhang, H.-W., Yang, Y.-B., Gao, F.-J., … Wang, H.-G. (2011). QTL mapping of bio-energy related traits in Sorghum. *Euphytica*, *182*(3), 431. https://doi.org/10.1007/s10681-011-0528-5

Gutjahr, S., Vaksmann, M., Dingkuhn, M., Thera, K., Trouche, G., Braconnier, S., & Luquet, D. (2013). Grain, sugar and biomass accumulation in tropical sorghums. I. Trade-offs and effects of phenological plasticity. *Functional Plant Biology*, *40*(4), 342–354. https://doi.org/10.1071/FP12269

Han, Y., Lv, P., Hou, S., Li, S., Ji, G., Ma, X., … Liu, G. (2015). Combining next generation sequencing with bulked segregant analysis to fine map a stem moisture locus in Sorghum (Sorghum bicolor L. Moench). *PLoS ONE*, *10*(5), e0127065. https://doi.org/10.1371/journal.pone.0127065

Hart, G. E., Schertz, K. F., Peng, Y., & Syed, N. H. (2001). Genetic mapping of Sorghum bicolor (L.) Moench QTLs that control variation in tillering and other morphological characters. *Theoretical and Applied Genetics*, *103*(8), 1232–1242. https://doi.org/doi 10.1007/s001220100582

Hilley, J., Truong, S., Olson, S., Morishige, D., & Mullet, J. (2016). Identification of Dw1, a regulator of sorghum stem internode length. *PLoS ONE*, *11*(3), e0151271. https://doi.org/10.1371/journal.pone.0151271

Hilley, J. L., Weers, B. D., Truong, S. K., McCormick, R. F., Mattison, A. J., McKinley, B. A., … Mullet, J. E. (2017). Sorghum Dw2 encodes a protein kinase regulator of stem internode length. *Scientific Reports*, *7*(1), 4616. https://doi.org/10.1038/s41598-017-04609-5

Hilson, G. R. (1916). A note on the inheritance of certain stem characters in Sorghum. *Agr Journal India*, *11*, 150–155.

Hirano, K., Kawamura, M., Araki-Nakamura, S., Fujimoto, H., Ohmae-Shinohara, K., Yamaguchi, M., … Sazuka, T. (2017). Sorghum DW1 positively regulates brassinosteroid signaling by inhibiting the nuclear localization of BRASSINOSTEROID INSENSITIVE 2. *Scientific Reports*, *7*(1), 126. https://doi.org/10.1038/s41598-017-00096-w

Jiang, W., Zhou, H., Bi, H., Fromm, M., Yang, B., & Weeks, D. P. (2013). Demonstration of CRISPR/Cas9/sgRNA-mediated targeted gene modification in Arabidopsis, tobacco, sorghum and rice. *Nucleic Acids Research*, *41*(20), e188. https://doi.org/10.1093/nar/gkt780

Jones, D. T., Taylor, W. R., & Thornton, J. M. (1992). The rapid generation of mutation data matrices from protein sequences. *Bioinformatics*, *8*(3), 275–282. https://doi.org/10.1093/bioinformatics/8.3.275

Kabbage, M., Kessens, R., Bartholomay, L. C., & Williams, B. (2017). The life and death of a plant cell. *Annual Review of Plant Biology*, *68*(68), 375–404. https://doi.org/10.1146/annurev-arplant-043015-111655

Kadier, Y., Zu, Y. Y., Dai, Q. M., Song, G., Lin, S. W., Sun, Q. P., … Lu, M. (2017). Genome-wide identification, classification and expression analysis of NAC family of genes in sorghum [Sorghum bicolor (L.) Moench]. *Plant Growth Regulation*, *83*(2), 301–312. https://doi.org/10.1007/s10725-017-0295-y

Kebrom, T. H., McKinley, B., & Mullet, J. E. (2017). Dynamics of gene expression during development and expansion of vegetative stem internodes of bioenergy sorghum. *Biotechnology for Biofuels*, *10*(1), 159. https://doi.org/10.1186/s13068-017-0848-3

Kim, H. J., Hong, S. H., Kim, Y. W., Lee, I. H., Jun, J. H., Phee, B. K., … Lim, P. O. (2014). Gene regulatory cascade of senescence-associated NAC transcription factors activated by ETHYLENE-INSENSITIVE2-mediated leaf senescence signalling in Arabidopsis. *Journal of Experimental Botany*, *65*(14), 4023–4036. https://doi.org/10.1093/jxb/eru112

Klein, R. R., Klein, P. E., Mullet, J. E., Minx, P., Rooney, W. L., & Schertz, K. F. (2005). Fertility restorer locus Rf1 of sorghum (Sorghum bicolor L.) encodes a pentatricopeptide repeat protein not present in the colinear region of rice chromosome 12. *TAG. Theoretical and Applied Genetics.*, *111*(6), 994–1012. https://doi.org/10.1007/s00122-005-2011-y

Kumar, S., Stecher, G., & Tamura, K. (2016). MEGA7: Molecular evolutionary genetics analysis version 7.0 for bigger datasets. *Molecular Biology and Evolution*, *33*(7), 1870–1874. https://doi.org/10.1093/molbev/msw054

Li, H., & Durbin, R. (2009). Fast and accurate short read alignment with Burrows–Wheeler transform. *Bioinformatics*, *25*(14), 1754–1760. https://doi.org/10.1093/bioinformatics/btp324

Liang, H., Yao, N., Song, J. T., Luo, S., Lu, H., & Greenberg, J. T. (2003). Ceramides modulate programmed cell death in plants. *Genes & Development*, *17*(21), 2636–2641. https://doi.org/10.1101/gad.1140503

Lin, Z., Li, X., Shannon, L. M., Yeh, C. T., Wang, M. L., Bai, G., … Yu, J. (2012). Parallel domestication of the Shattering1 genes in cereals. *Nature Genetics*, *44*(6), 720–724. https://doi.org/10.1038/ng.2281

Lingle, S. E. (1987). Sucrose metabolism in the primary culm of sweet sorghum during development. *Crop Science*, *27*(6), 1214–1219. https://doi.org/DOI 10.2135/cropsci1987.0011183X002700060025x

Mace, E. S., Rami, J.-F., Bouchet, S., Klein, P. E., Klein, R. R., Kilian, A., … Jordan, D. R. (2009). A consensus genetic map of sorghum that integrates multiple component maps and high-throughput Diversity Array Technology (DArT) markers. *BMC Plant Biology*, *9*(1), 13. https://doi.org/10.1186/1471-2229-9-13

Mace, E. S., Tai, S., Gilding, E. K., Li, Y., Prentis, P. J., Bian, L., … Wang, J. (2013). Whole-genome sequencing reveals untapped genetic potential in Africa's indigenous cereal crop sorghum. *Nature Communications*, *4*, 2320. https://doi.org/10.1038/ncomms3320

McCormick, R., Truong, S., Sreedasyam, A., Jenkins, J., Shu, S., Sims, D., … Mullet, J. (2018). The Sorghum bicolor reference genome: Improved assembly, gene annotations, a transcriptome atlas, and signatures of genome organization. *The Plant Journal*, *93*(2), 338–354. https://doi.org/10.1111/tpj.13781

McKinley, B., Rooney, W., Wilkerson, C., & Mullet, J. (2016). Dynamics of biomass partitioning, stem gene expression, cell wall biosynthesis, and sucrose accumulation during development of Sorghum bicolor. *The Plant Journal*, *88*(4), 662–680. https://doi.org/10.1111/tpj.13269

Mergemann, H., & Sauter, M. (2000). Ethylene induces epidermal cell death at the site of adventitious root emergence in rice. *Plant Physiology*, *124*(2), 609–614.

Milne, R. J., Reinders, A., Ward, J. M., Offler, C. E., Patrick, J. W., & Grof, C. P. L. (2017). Contribution of sucrose transporters to phloem unloading within Sorghum bicolor stem internodes. *Plant Signaling and Behavior*, *12*(5), e1319030. https://doi.org/10.1080/15592324.2017.1319030

Mocoeur, A., Zhang, Y. M., Liu, Z. Q., Shen, X., Zhang, L. M., Rasmussen, S. K., & Jing, H. C. (2015). Stability and genetic control of morphological, biomass and biofuel traits under temperate maritime and continental conditions in sweet sorghum (Sorghum bicolour). *TAG. Theoretical and Applied Genetics.*, *128*(9), 1685–1701. https://doi.org/10.1007/s00122-015-2538-5

Morris, G. P., Ramu, P., Deshpande, S. P., Hash, C. T., Shah, T., Upadhyaya, H. D., … Kresovich, S. (2013). Population genomic and genome-wide association studies of agroclimatic traits in sorghum. *Proceedings of the National Academy of Sciences*, *110*(2), 453–458. https://doi.org/10.1073/pnas.1215985110

Mullet, J., Morishige, D., McCormick, R., Truong, S., Hilley, J., McKinley, B., … Rooney, W. (2014). Energy sorghum–a genetic model for the design of C4 grass bioenergy crops. *Journal of Experimental Botany*, *65*(13), 3479–3489. https://doi.org/10.1093/jxb/eru229

Multani, D. S., Briggs, S. P., Chamberlin, M. A., Blakeslee, J. J., Murphy, A. S., & Johal, G. S. (2003). Loss of an MDR transporter in compact stalks of maize br2 and sorghum dw3 mutants. *Science*, *302*(5642), 81–84. https://doi.org/10.1126/science.1086072

Murphy, R. L., Klein, R. R., Morishige, D. T., Brady, J. A., Rooney, W. L., Miller, F. R., … Mullet, J. E. (2011). Coincident light and clock regulation of pseudoresponse regulator protein 37 (PRR37) controls photoperiodic flowering in sorghum. *Proceedings of the National Academy of Sciences*, *108*(39), 16469–16474. https://doi.org/10.1073/pnas.1106212108

Murphy, R. L., Morishige, D. T., Brady, J., Rooney, W., Yang, S., Klein, P. E., & Mullet, J. (2014). Ghd7 (Ma(6)) Represses sorghum flowering in long days: Ghd7 alleles enhance biomass accumulation and grain. *Production*, *7*, https://doi.org/10.3835/plantgenome2013.11.0040

Murray, S. C., Sharma, A., Rooney, W., Klein, P., Mullet, J., Mitchell, S. E., & Kresovich, S. (2008). Genetic improvement of sorghum as a biofuel feedstock: I. QTL for stem sugar and grain nonstructural carbohydrates. *Crop Science*, *48*(6), 2165–2179. https://doi.org/10.2135/cropsci2008.01.0016

Nakano, Y., Yamaguchi, M., Endo, H., Rejab, N. A., & Ohtani, M. (2015). NAC-MYB-based transcriptional regulation of secondary cell wall biosynthesis in land plants. *Frontiers in Plant Science*, *6*, 288. https://doi.org/10.3389/fpls.2015.00288

Nilsen, K. T., N'Diaye, A., MacLachlan, P. R., Clarke, J. M., Ruan, Y., Cuthbert, R. D., … Pozniak, C. J. (2017). High density mapping and haplotype analysis of the major stem-solidness locus SSt1 in durum and common wheat. *PLoS ONE*, *12*(4), e0175285. https://doi.org/10.1371/journal.pone.0175285

Olson, S. N., Ritter, K., Rooney, W., Kemanian, A., McCarl, B. A., Zhang, Y., … Mullet, J. (2012). High biomass yield energy sorghum: Developing a genetic model for C4 grass bioenergy crops. *Biofuels, Bioproducts and Biorefining*, *6*(6), 640–655. https://doi.org/doi:10.1002/bbb.1357

Olvera-Carrillo, Y., Van Bel, M., Van Hautegem, T., Fendrych, M., Van Durme, M., Huysmans, M., … Nowack, M. K. (2015). A conserved core of PCD indicator genes discriminates developmentally and environmentally induced programmed cell death in plants. *Plant Physiology*, *169*, 2684–2699.

Paterson, A. H., Bowers, J. E., Bruggmann, R., Dubchak, I., Grimwood, J., Gundlach, H., … Rokhsar, D. S. (2009). The Sorghum bicolor genome and the diversification of grasses. *Nature*, *457*, 551. https://doi.org/10.1038/nature07723

Peng, X., Zhao, Y., Li, X., Wu, M., Chai, W., Sheng, L., … Cheng, B. (2015). Genomewide identification, classification and analysis of NAC type gene family in maize. *J Genet*, *94*, 377–390.

Pesquet, E., Zhang, B., Gorzsas, A., Puhakainen, T., Serk, H., Escamez, S., … Tuominen, H. (2013). Non-cell-autonomous postmortem lignification of tracheary elements in Zinnia elegans. *Plant Cell*, *25*(4), 1314–1328. https://doi.org/10.1105/tpc.113.110593

Pfaffl, M. W. (2001). A new mathematical model for relative quantification in real-time RT–PCR. *Nucleic Acids Research*, *29*(9), e45.

Quinby, J. R. (1974). *Sorghum Improvement and the Genetics of Growth*. Texas Agricultural Experiment Station, College Station, Texas: Texas A&M University Press.

Rajhi, I., Yamauchi, T., Takahashi, H., Nishiuchi, S., Shiono, K., Watanabe, R., … Nakazono, M. (2011). Identification of genes expressed in maize root cortical cells during lysigenous aerenchyma formation using laser microdissection and microarray analyses. *New Phytologist*, *190*(2), 351–368. https://doi.org/doi:10.1111/j.1469-8137.2010.03535.x

Rooney, W. (2004). Sorghum improvement—Integrating traditional and new technology to produce improved genotypes. *Advances in Agronomy*, *83*, 37–109. https://doi.org/10.1016/S0065-2113(04)83002-5

Rooney, W., Blumenthal, J., Bean, B., & Mullet, J. (2007). Designing sorghum as a dedicated bioenergy feedstock. *Biofuels, Bioproducts and Biorefining*, *1*(2), 147–157. https://doi.org/10.1002/bbb.15

Shakoor, N., Nair, R., Crasta, O., Morris, G., Feltus, A., & Kresovich, S. (2014). A Sorghum bicolor expression atlas reveals dynamic genotype-specific expression profiles for vegetative tissues of grain, sweet and bioenergy sorghums. *BMC Plant Biology*, *14*, 35. https://doi.org/10.1186/1471-2229-14-35

Shiringani, A. L., Frisch, M., & Friedt, W. (2010). Genetic mapping of QTLs for sugar-related traits in a RIL population of Sorghum bicolor L. Moench. *Theoretical and Applied Genetics*, *121*(2), 323–336. https://doi.org/10.1007/s00122-010-1312-y

Slewinski, T. L. (2012). Non-structural carbohydrate partitioning in grass stems: A target to increase yield stability, stress tolerance, and biofuel production. *Journal of Experimental Botany*, *63*(13), 4647–4670. https://doi.org/10.1093/jxb/ers124

Smith, C. W., & Frederiksen, R. A. (2000). History of cultivar development in the United States: From "Memoirs of A. B. Maunder - Sorghum Breeder". In C. W. Smith (Ed.), *Sorghum: Origin, History Technology, and Production* (pp. 191–223). Oxford, UK: John Wiley & Sons, Inc.

Souer, E., van Houwelingen, A., Kloos, D., Mol, J., & Koes, R. (1996). The no apical meristem gene of petunia is required for pattern formation in embryos and flowers and is expressed at meristem and primordia boundaries. *Cell*, *85*(2), 159–170. https://doi.org/10.1016/S0092-8674(00)81093-4

Srinivas, G., Satish, K., Madhusudhana, R., Reddy, R. N., Mohan, S. M., & Seetharama, N. (2009). Identification of quantitative trait loci for agronomically important traits and their association with genic-microsatellite markers in sorghum. *TAG. Theoretical and Applied Genetics.*, *118*(8), 1439–1454. https://doi.org/10.1007/s00122-009-0993-6

Steffens, B., Geske, T., & Sauter, M. (2011). Aerenchyma formation in the rice stem and its promotion by H2O2. *New Phytologist*, *190*(2), 369–378. https://doi.org/10.1111/j.1469-8137.2010.03496.x

Stephens, J. C., & Quinby, J. R. (1939). The D Rs P linkage group in sorghum. *Journal of Agricultural Research*, *59*(10), 725–730.

Swanson, A. F., & Parker, J. H. (1931). Inheritance of smut resistance and juiciness of stalk: In the sorghum cross, Red Amber X Feterita. *Journal of Heredity*, *22*(2), 51–56. https://doi.org/10.1093/oxfordjournals.jhered.a103429

Takahashi, H., Yamauchi, T., Rajhi, I., Nishizawa, N. K., & Nakazono, M. (2015). Transcript profiles in cortical cells of maize primary root during ethylene-induced lysigenous aerenchyma formation under aerobic conditions. *Annals of Botany*, *115*(6), 879–894. https://doi.org/10.1093/aob/mcv018

Truong, S. K., McCormick, R. F., Morishige, D. T., & Mullet, J. E. (2014). Resolution of genetic map expansion caused by excess heterozygosity in plant recombinant inbred populations. *G3: Genes|Genomes|Genetics*, *4*(10), 1963–1969. https://doi.org/10.1534/g3.114.012468

Uauy, C., Distelfeld, A., Fahima, T., Blechl, A., & Dubcovsky, J. (2006). A NAC Gene regulating senescence improves grain protein, zinc, and iron content in wheat. *Science*, *314*(5803), 1298–1301. https://doi.org/10.1126/science.1133649

Van Hautegem, T., Waters, A. J., Goodrich, J., & Nowack, M. K. (2015). Only in dying, life: Programmed cell death during plant development. *Trends in Plant Science*, *20*(2), 102–113. https://doi.org/10.1016/j.tplants.2014.10.003

Visser, E. J., & Bogemann, G. M. (2006). Aerenchyma formation in the wetland plant Juncus effusus is independent of ethylene. *New Phytologist*, *171*(2), 305–314. https://doi.org/10.1111/j.1469-8137.2006.01764.x

Waclawovsky, A. J., Sato, P. M., Lembke, C. G., Moore, P. H., & Souza, G. M. (2010). Sugarcane for bioenergy production: An assessment of yield and regulation of sucrose content. *Plant Biotechnology Journal*, *8*(3), 263–276. https://doi.org/10.1111/j.1467-7652.2009.00491.x

Waterhouse, A. M., Procter, J. B., Martin, D. M. A., Clamp, M., & Barton, G. J. (2009). Jalview Version 2—a multiple sequence alignment editor and analysis workbench. *Bioinformatics*, *25*(9), 1189–1191. https://doi.org/10.1093/bioinformatics/btp033

Wu, E., Lenderts, B., Glassman, K., Berezowska-Kaniewska, M., Christensen, H., Asmus, T., … Zhao, Z. Y. (2014). Optimized Agrobacterium-mediated sorghum transformation protocol and molecular data of transgenic sorghum plants. *Vitro Cell Dev Biol Plant*, *50*(1), 9–18. https://doi.org/10.1007/s11627-013-9583-z

Xiao-ping, L., Jin-feng, Y., Cui-ping, G., & Acharya, S. (2011). Quantitative trait loci analysis of economically important traits in Sorghum bicolor × S. sudanense hybrid. *Canadian Journal of Plant Science*, *91*(1), 81–90. https://doi.org/10.1139/CJPS09112

Xin, Z., Wang, M. L., Barkley, N. A., Burow, G., Franks, C., Pederson, G., & Burke, J. (2008). Applying genotyping (TILLING) and phenotyping analyses to elucidate gene function in a chemically induced sorghum mutant population. *BMC Plant Biology*, *8*, 103. https://doi.org/10.1186/1471-2229-8-103

Xu, W., Subudhi, P. K., Crasta, O. R., Rosenow, D. T., Mullet, J. E., & Nguyen, H. T. (2000). Molecular mapping of QTLs conferring stay-green in grain sorghum (Sorghum bicolor L. Moench). *Genome*, *43*(3), 461–469.

Yamaguchi, M., Fujimoto, H., Hirano, K., Araki-Nakamura, S., Ohmae-Shinohara, K., Fujii, A., … Sazuka, T. (2016). Sorghum Dw1, an agronomically important gene for lodging resistance, encodes a novel protein involved in cell proliferation. *Scientific Reports*, *6*, 28366. https://doi.org/10.1038/srep28366

Yamauchi, T., Shimamura, S., Nakazono, M., & Mochizuki, T. (2013). Aerenchyma formation in crop species: A review. *Field Crops Research*, *152*, 8–16. https://doi.org/10.1016/j.fcr.2012.12.008

Yamauchi, T., Tanaka, A., Mori, H., Takamure, I., Kato, K., & Nakazono, M. (2016). Ethylene-dependent aerenchyma formation in adventitious roots is regulated differently in rice and maize. *Plant, Cell and Environment*, *39*(10), 2145–2157. https://doi.org/10.1111/pce.12766

Yamauchi, T., Yoshioka, M., Fukazawa, A., Mori, H., Nishizawa, N. K., Tsutsumi, N., … Nakazono, M. (2017). An NADPH oxidase RBOH functions in rice roots during lysigenous aerenchyma formation under oxygen-deficient conditions. *Plant Cell*, *29*(4), 775–790. https://doi.org/10.1105/tpc.16.00976

Yin, L.-L., & Xue, H.-W. (2012). The MADS29 transcription factor regulates the degradation of the nucellus and the nucellar projection during rice seed development. *The Plant Cell*, *24*(3), 1049–1065. https://doi.org/10.1105/tpc.111.094854

Young, T. E., & Gallie, D. R. (2000). Regulation of programmed cell death in maize endosperm by abscisic acid. *Plant Molecular Biology*, *42*(2), 397–414.

Zhao, C., Craig, J. C., Petzold, H. E., Dickerman, A. W., & Beers, E. P. (2005). The xylem and phloem transcriptomes from secondary tissues of the Arabidopsis root-hypocotyl. *Plant Physiology*, *138*(2), 803–818. https://doi.org/10.1104/pp.105.060202

Zhu, T., Nevo, E., Sun, D., & Peng, J. (2012). Phylogenetic analyses unravel the evolutionary history of NAC proteins in plants. *Evolution*, *66*(6), 1833–1848. https://doi.org/10.1111/j.1558-5646.2011.01553.x

2

Changing frequency of fluctuating light reveals the molecular mechanism for P700 oxidation in plant leaves

Ginga Shimakawa[1] | Chikahiro Miyake[1,2]

[1]Department of Biological and Environmental Science, Faculty of Agriculture, Graduate School of Agricultural Science, Kobe University, Kobe, Japan

[2]Core Research for Environmental Science and Technology, Japan Science and Technology Agency, Tokyo, Japan

Correspondence
Chikahiro Miyake, Department of Biological and Environmental Science, Faculty of Agriculture, Graduate School of Agricultural Science, Kobe University, 1-1 Rokkodai, Nada, Kobe 657-8501, Japan.
Email: cmiyake@hawk.kobe-u.ac.jp

Funding information
This work was supported by the Japan Society for the Promotion of Science (JSPS; grant no. 26450079 to C.M.) and by the Core Research for Evolutional Science and Technology (CREST) division of the Japan Science and Technology Agency (grant no. AL65D21010 to C.M.). G.S. was supported as a JSPS research fellow (grant no. 16J03443).

Abstract

Natural sunlight exceeds the demand of photosynthesis such that it can cause plants to produce reactive oxygen species (ROS), which subsequently cause photo-oxidative damage. Because photosystem I (PSI) is a major source of ROS, plants actively maintain the reaction center chlorophyll of PSI(P700) oxidized under excessive light conditions to alleviate the ROS production. P700 oxidation is universally recognized in photosynthetic organisms as a physiological response to excessive light. However, it is still poorly understood how P700 oxidation is induced in response to fluctuating light with a variety of frequencies. Here, we investigated the relationships of photosynthetic parameters with P700 oxidation in *Arabidopsis thaliana* under a sine fluctuating light with different frequencies. As the photon flux density of the light increased, P700 was oxidized concurrently with the chlorophyll fluorescence parameter qL unless the electron acceptor side of PSI was limited. Conversely, we did not observe a proportional relationship of non-photochemical quenching with P700 oxidation. The mutant *crr-2*, which lacks chloroplast NADPH dehydrogenase, was impaired in P700 oxidation during light fluctuation at high, but not low frequency, unlike the *pgrl1* mutant deficient in PGR5 and PGRL1 proteins, which could not oxidize P700 during light fluctuation at both high and low frequencies. Taken together, our findings suggested that the changing frequency of fluctuating light reveals the tracking performance of molecular mechanisms underlying P700 oxidation.

KEYWORDS
fluctuating light, P700 oxidation, photosynthesis, reactive oxygen species

1 | INTRODUCTION

Plants are vulnerable to photo-oxidative damage by reactive oxygen species (ROS) that originate from the photosynthetic electron transport system. During the light reaction of photosynthesis, photon energy is absorbed by the chlorophyllin photosystem (PS) I and II on the thylakoid membrane, thereby driving photosynthetic linear electron transport from PSII to PSI through the plastoquinone (PQ) pool, the cytochrome b_6/f complex (Cyt b_6/f), and plastocyanin. On the electron acceptor side of PSI, $NADP^+$ is reduced to NADPH with electrons from PSI via ferredoxin (Fd) and Fd-$NADP^+$ reductase. Conversely, in PSII, H_2O is oxidized on the luminal side of the thylakoid membrane; in Cyt b_6/f, the Q-cycle pumps stromal H^+ to the luminal side, both of which establish a proton gradient (ΔpH) across the thylakoid membrane to produce ATP via chloroplast ATP synthase. The resulting NADPH and ATP are utilized for CO_2 assimilation in the Calvin–Benson cycle. Limitation during photosynthetic CO_2 assimilation under low CO_2 and high light intensity can cause accumulation of excess electrons in the photosynthetic electron transport system. The recent study simulated such conditions by applying repetitive short-pulse illumination and found that PSI was rapidly inactivated (i.e. PSI photoinhibition) via ROS, including superoxide anion radical (O_2^-), hydroxyl radical (·OH), and singlet oxygen ($^1O_2$; Sejima et al., 2014; Zivcak, Brestic, Kunderlikova, Sytar, & Allakhverdiev, 2015; Zivcak, Brestic, Kunderlikova, Olsovska, & Allakhverdiev, 2015; Takagi, Takumi, Hashiguchi, Sejima, & Miyake, 2016).

Usually, PSI photoinhibition is rarely observed, and plants can tolerate sunlight (2,000–3,000 μmol photons m^{-2} s^{-1}) that far exceeds the demands of photosynthesis. Recently, Sejima et al. (2014) confirmed that ROS production is suppressed when the reaction center chlorophyll of PSI (P700) is maintained in its oxidized form ($P700^+$), because of the decrease of P700 in the ground state, which is the source of both excess electrons and energy to produce ROS. The oxidation of P700 is recognized as a physiological response to excessive light conditions, such as highlight and low CO_2 (Golding & Johnson, 2003; Klughammer & Schreiber, 1994; Miyake, Miyata, Shinzaki, & Tomizawa, 2005; Shimakawa, Ishizaki et al., 2017). The strategies to oxidize P700 have been utilized and diversified since the origin of oxygenic photosynthesis (Shaku, Shimakawa, Hashiguchi, & Miyake, 2016; Shimakawa, Shaku, & Miyake, 2016). These facts indicate that P700 oxidation is the strategy that prevents photo-oxidative damage caused by ROS in PSI. In nature, plants should be suddenly exposed to an excess light because sunlight is fluctuating with different frequencies (e.g., the range from 0.2 to 50 s, and from 0.004 to 1 Hz; Pfitsch & Pearcy, 1989). Overall, plants develop various molecular mechanisms for P700 oxidation (i.e. the P700 oxidation system) to allow them to tolerate the strongly fluctuating sunlight.

The oxidation of P700 is controlled by diverse molecular mechanisms on both the electron acceptor and donor sides of PSI. On the acceptor side, photosynthetic CO_2 assimilation and O_2^- dependent alternative electron transport function as electron sinks, thereby contributing to P700 oxidation. In C_3 plants, the flux of electrons to O_2 is mainly driven by photorespiration (Hanawa et al., 2017; Sejima et al., 2016; Wiese, Shi, & Heber, 1998). Even although photorespiration is the largest alternative electron sink in land plants (except for C_4 plants; Hanawa et al., 2017), it is not used as the main alternative electron sink by many prokaryotic and eukaryotic algae (Hayashi et al., 2014; Shimakawa et al., 2015; Shimakawa, Akimoto et al., 2016; Shimakawa, Matsuda et al., 2017). Cyanobacteria, which are the ancestors of chloroplasts in land plants, use flavodiiron proteins (FLV) to mediate electron flow to O_2, instead of photorespiration (Helman, Barkan, Eisenstadt, Luz, & Kaplan, 2005; Helman et al., 2003; Shimakawa et al., 2015), and to alleviate photo-oxidative damage (Allahverdiyeva et al., 2013; Shimakawa, Shaku et al., 2016; Zhang, Allahverdiyeva, Eisenhut, & Aro, 2009). The physiological functions of FLV are also conserved in basal land plants such as *Marchantia polymorpha* and *Physcomitrella patens* (Gerotto et al., 2016; Shimakawa, Ishizaki et al., 2017), probably owing to evolutionary and ecological circumstances (Shimakawa, Ishizaki et al., 2017). On the donor side of PSI, the suppression of electron transport into PSI causes P700 oxidation. Basically, photosynthetic linear electron flow is limited in the oxidation of reduced PQ (i.e. plastoquinol, PQH_2) in Cyt b_6/f (Anderson, 1992; Kirchhoff, Horstmann, & Weis, 2000; Schöttler, Tóth, Boulouis, & Kahlau, 2015; Stiehl & Witt, 1969; Yamori et al., 2011). However, in response to short-term environmental fluctuations, the electron transport in Cyt b_6/f is modulated by a regulatory mechanism, which is believed to be strongly associated with ΔpH in plant leaves. The acidification on the luminal side of the thylakoid membrane limits the electron transport in Cytb_6/f (Nishio & Whitmarsh, 1993) and induces the dissipation of excess photon energy as heat at PSII (i.e. qE quenching) reflected as non-photochemical quenching (NPQ) of chlorophyll fluorescence (Ruban, 2016). The formation of ΔpH is coupled with total electron transport activity (Kanazawa & Kramer, 2002) and is adjusted by the proton conductance (gH^+) of the chloroplast ATP synthase (Rott et al., 2011; Takizawa, Kanazawa, & Kramer, 2008) or other ion transporters (Armbruster et al., 2014). Furthermore, ΔpH formation is possibly promoted by cyclic electron transport (CET) around PSI (Allen, 2003; Nandha, Finazzi, Joliot, Hald, & Johnson, 2007). Conversely, the suppression of electron transport into PSI for P700 oxidation(i.e. reduction-induced suppression of electron transport, RISE) has been observed in the cyanobacterium *Synechococcus elongatus* PCC 7942 (Shaku et al., 2016; Shimakawa, Shaku, & Miyake, 2018). In RISE, the suppression of electron transport in Cyt b_6/f is triggered by the reduction of the PQ pool, but not the accumulation of ΔpH,which implies that RISE is caused by the suppression of the Q-cycle in Cytb_6/f because of a shortage of oxidized PQ (Shaku et al., 2016). Furthermore, O_2^- dependent alternative electron transport mediated by terminal oxidases on the thylakoid membrane can contribute to P700 oxidation on the electron donor side of PSI (Feilke et al., 2016; Shimakawa & Miyake, 2018). For the last two decades, P700 oxidation has been observed as a universal physiological response to excess light conditions and reported to be related to a variety of molecular mechanisms. Nevertheless, the molecular mechanisms underlying P700 oxidation in plant leaves are still poorly understood under fluctuating light with different frequencies.

This study aimed to assess the impact of changing frequency of fluctuating light on P700 oxidation and other photosynthetic parameters in the model C_3 plant *Arabidopsis thaliana*. We used a physiological measuring system, in which the photon flux density (PFD) of actinic light (AL) oscillates as a sine-like curve with different frequencies. We measured various photosynthetic parameters in this system and found that P700 oxidation basically exhibited a linear proportional relationship with the chlorophyll fluorescence parameter qL regardless of the frequency of the sine fluctuating light in wild-type *A. thaliana* (Col-0). To investigate the effect of the electron acceptor side limitation in PSI on the relationship, we analyzed the *A. thaliana* mutants, *pgrl1* and *crr-2*. The mutant *pgrl1* is deficient in proton gradient regulation 5 (PGR5)-like 1 protein (PGRL1), the membrane-associated protein, and is impaired in the expression of the soluble protein PGR5 (DalCorso et al., 2008). Previous and recent studies have shown that the lack of PGR5 and PGRL1 causes electron acceptor side limitation in PSI and the inability to oxidize P700, resulting in PSI photoinhibition under excessive light conditions in C_3 plants (DalCorso et al., 2008; Kono & Terashima, 2016;

Munekage et al., 2002; Suorsa et al., 2012; Yamori, Makino, & Shikanai, 2016). Conversely, the mutant *crr-2* is impaired in the expression of the *ndhB* gene and lacks chloroplast NADPH dehydrogenase (NDH; Hashimoto, Endo, Peltier, Tasaka, & Shikanai, 2003). The lack of NDH is known to have little effect on photosynthetic electron transport at steady-state photosynthesis in C_3 plants (Sazanov, Burrows, & Nixon, 1998). However, recent studies suggested that the induction of P700 oxidation is impaired in *crr* mutants of *A. thaliana* and *Oryza sativa*, owing to the severe limitation of electron transport on the acceptor side of PSI, under a rectangular fluctuating light (Kono & Terashima, 2016; Yamori et al., 2016). Based on these findings of these studies, we used *pgrl1* and *crr-2* and subjected them to acceptor side limitation in PSI in *A. thaliana*. In this study, *crr-2* could oxidize P700 following light fluctuation only when the frequency of the fluctuating light was low, indicating that this mutant lacks the ability for the rapid initiation of P700 oxidation under fluctuating light in *A. thaliana*.

2 | METHODS

2.1 | *Arabidopsis thaliana* growth conditions

Plants of *A. thaliana* wild-type (Col-0) and the mutants *pgrl1* and *crr-2* were grown under long-day conditions (16 hr-light, 23°C, 100 μmol photons m^{-2} s^{-1}, white fluorescent lamp/8 hr-dark, 21°C) with a relative humidity of 60 ± 10%. Seeds were planted in pots that contained a 1:1 mix of vermiculite and Metro-Mix 350 (Sun Gro Horticulture, Agawam, MA, USA), and 1000-fold diluted Hyponex solution (Hyponex, Osaka, Japan) was used as a watering solution. Photosynthetic parameters were measured in the rosette leaves of the plants 4 weeks after germination.

2.2 | Measurements of gas exchange, P700, and chlorophyll fluorescence

Gas exchange, $P700^+$ absorbance, and chlorophyll fluorescence were simultaneously measured using a Dual/KLAS-NIR spectrophotometer, a GFS-3000, and a 3010 DUAL leaf cuvette (Walz, Effeltrich, Germany), following previously reported methods (Klughammer & Schreiber, 2016), in which ambient air was saturated with water vapor at 18.0 ± 0.1°C, and the leaf temperature was maintained at 25 ± 2°C. Red AL (630 nm) was supplied using a chip-on-board LED array. Long-term (3–5 hr) dark-adapted plants were used for the measurements.

Exchange of CO_2 and H_2O was measured using GFS-3000 equipped with a 3010-DUAL gas exchange chamber (Walz, Effeltrich, Germany). Atmospheric gas (40 Pa CO_2/21 kPa O_2) was used in the present study.

The redox state of P700 was estimated from four original pulse-modulated dual-wavelength difference signals in the range of near-infrared (785–840, 810–870, 870–970, and 795–970 nm; Klughammer & Schreiber, 2016). The photosynthetic parameters of PSI were calculated from the redox state of P700 as follows (Inoue, Ogawa, & Shibata, 1973; Klughammer & Schreiber, 1994; Schreiber & Klughammer, 2008): quantum yield of photochemical energy conversion in PSI, $Y(I) = (P'_m - P)/P_m$; quantum yield of non-photochemical energy dissipation due to donor side limitation, $Y(ND) = P/P_m$; quantum yield of non-photochemical energy dissipation due to acceptor side limitation, $Y(NA) = (P_m - P'_m)/P_m$; Pm, total amount of photo-oxidizable P700; P'_m, maximum amount of photo-oxidized P700 by a saturation pulse; and *P*, amount of photo-oxidized P700 at steady-state. The sum of the three factors $[Y(I) + Y(NA) + Y(ND)] = 1$. For the determination of P'_m, a short-saturation pulse (8,000 μmol photons m^{-2} s^{-1}, 300 ms) was applied to the plant leaves and, for the determination of P_m, the short-saturation pulse was applied after 10 s illumination with a far red light (740 nm). In this study, we determined P_m before the measurements in Umibozu. Only for the data shown in Figure 4, P_m was determined again at 30 min (in the dark) after 1 hr Umibozu for comparison with the P_m value before the treatment in Umibozu. During Umibozu, AL was temporarily turned off for 1 s just after the application of the short-saturation pulse for the determination of P_o.

The photosynthetic parameters of PSII were calculated from chlorophyll fluorescence (>700 nm; Krause & Weis, 1984) as follows (Baker, 2008): PSII operating efficiency (quantum yield of photochemical energy conversion in PSII), Y(II) or $\phi_{PSII} = (F'_m - F')/F'_m$; non-photochemical quenching, $NPQ = (F_m - F'_m)/F'_m$; fraction of "open" PSII centers (with Q_A oxidized) on the basis of the lake model for the PSII photosynthetic apparatus, $qL = (F'_m - F')/ (F'_m - F_o)$; F_o, minimum fluorescence from a dark-adapted leaf; F'_o, minimum fluorescence from a light-adapted leaf; F_m, maximum fluorescence from a dark-adapted leaf; F'_m, maximum fluorescence from a light-adapted leaf; and F', fluorescence emission from a light-adapted leaf. Pulse-amplitude modulated green measuring light (540 nm, >0.1 μmol photons m^{-2} s^{-1}) was used to determine F_o. To obtained F_m and F'_m, we applied a short-saturation pulse (8,000 μmol photons m^{-2} s^{-1}, 300 ms). During Umibozu, AL was temporarily turned off for 1 s just after the application of the short-saturation pulse for the determination of F'_o. In this study, we determined both F_o and F_m in the dark-adapted leaves before the measurements in Umibozu. Only for the data shown in Supporting Information Figure S2, F_o and F_m were determined again at 30 min (in the dark) after 1 hr Umibozu for the comparison of F_v/F_m with the plant leaves before the treatment in Umibozu.

2.3 | Statistical analysis

We used Student's *t*-tests to detect differences. All statistical analyses were performed using Microsoft Excel 2010 (Microsoft, Washington, USA), Origin 2017 (Lightstone, Tokyo, Japan), and JMP8 (SAS Institute Inc., Tokyo, Japan).

3 | RESULTS

3.1 | Responses of photosynthetic parameters to light fluctuation in *slow* Umibozu

We simultaneously measured CO_2 exchange, P700 absorbance, and chlorophyll fluorescence to calculate various photosynthetic parameters in the dark-adapted leaves of wild-type *A. thaliana* (Col-0) and the mutants, *pgrl1* and *crr-2*, under a sine-like fluctuating light. This physiological measuring system was named Umibozu because the smoothly changing Y(ND) given by the sine curve-like fluctuating light (described below) was like the shape of a traditional Japanese sea monster ('Umibozu', or sea goblin). During *slow* Umibozu, PFD of the light oscillated at a low frequency (6 per hour) like a sine curve in the range between a maximum (1,970 μmol photons m^{-2} s^{-1}) and a minimum PFD (30 μmol photons m^{-2} s^{-1}; Figure 1a). The frequency of fluctuating light in Umibozu is controlled by a software developed by Walz (Effeltrich, Germany) and can be flexibly modified from per second to per day; this enabled the evaluation of the tracking performance of P700 oxidation relative to the speed of light fluctuation. During *slow* Umibozu, photosynthetic parameters were determined every 1 min by the analysis with a saturated short-pulse.

During *slow* Umibozu in Col-0, *pgrl1*, and *crr-2*, the uptake rate of CO_2 was gradually enhanced with an increase in PFD (Figure 1b), indicating that rising PFD activates photosynthesis in Col-0, *pgrl1*, and *crr-2*. Moreover, photosynthesis was induced in response to light in a time-dependent manner because we used dark-adapted plant leaves in this study. That is, the increase in net CO_2 assimilation rate involved the activation of photosynthesis dependent on both light intensity and illumination time, which is the reason why it was not proportional to fluctuating PFD particularly in the first 10 min (Figure 1b). To evaluate photosynthetic parameters before and after the photosynthesis induction, in this study we applied *slow* Umibozu to the plant leaves continuously three times for 30 min, i.e., 3 periods. Whereas in the first 10 min we could not observe any significant differences in the values among these plants, *pgrl1* finally showed the slightly lower CO_2 assimilation rate than the other plants in the end of the measurement (Figure 1b).

Changing PFD in *slow* Umibozu caused the alterations of photosynthetic parameters in PSI in the *A. thaliana* wild-type and mutants. Quantum yield of non-photochemical energy dissipation due to donor side limitation of PSI, Y(ND), indicated the ratio of $P700^+$ to total photo-oxidizable P700 (i.e. P700 oxidation), as determined by spectrophotometric analysis (Harbinson & Hedley, 1993; Inoue et al., 1973; Schreiber & Klughammer, 2008). During *slow* Umibozu, the

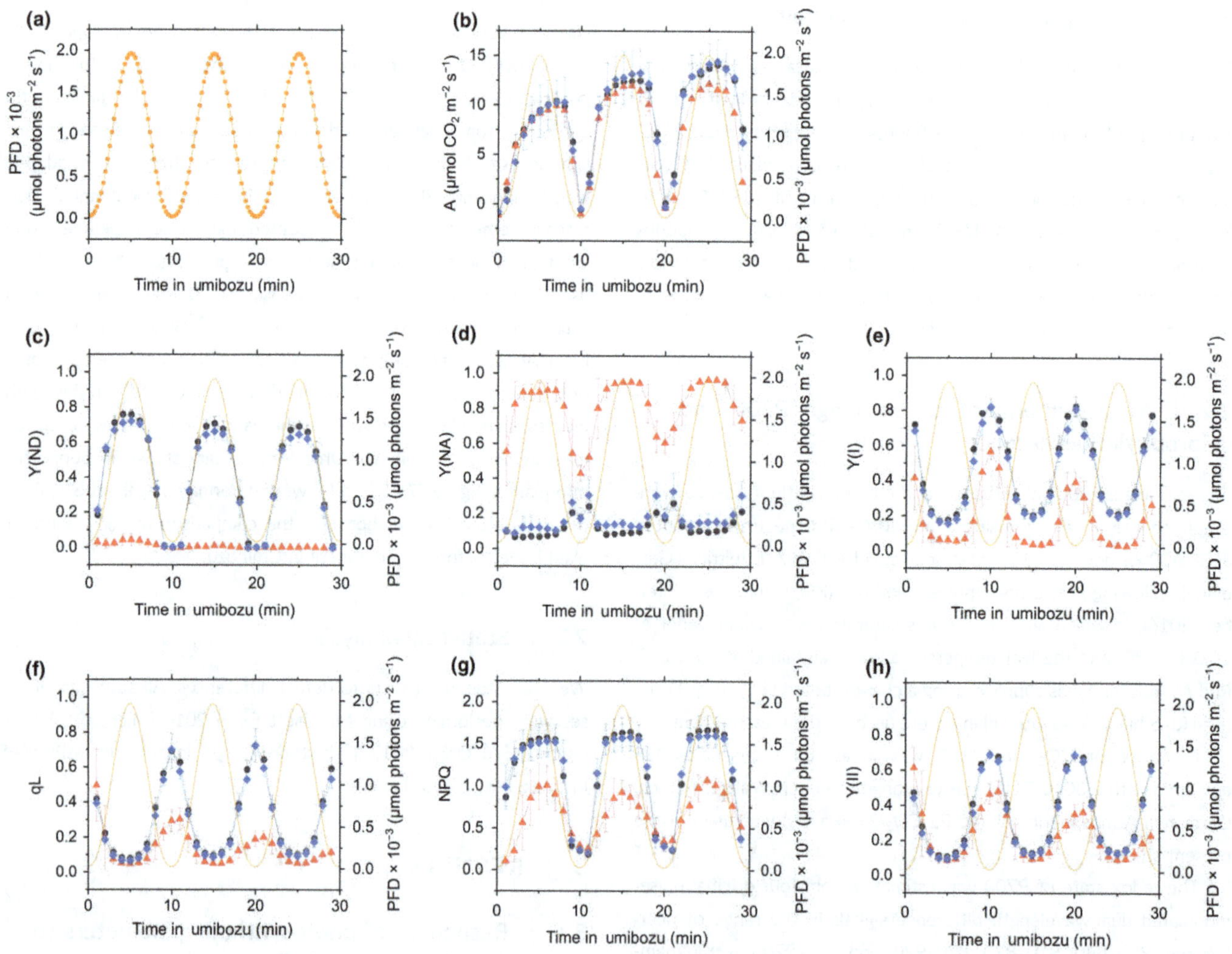

FIGURE 1 Time courses of photon flux density (PFD; a), net CO_2 assimilation rate, A (b), Y(ND) (c), Y(NA) (d), Y(I) (e), qL (f), NPQ (g), Y(II) (h) during *slow* Umibozu in *Arabidopsis thaliana* wild-type (Col-0, black circles) and the mutants, *pgrl1* (red triangles) and *crr-2* (blue diamonds). Orange lines indicate PFD. Data are represented as the means ± SD of three independent measurements

increase in PFD caused smooth increase in Y(ND) in Col-0 and *crr-2* (Figure 1c).We used the *A. thaliana* mutants, *pgrl1* and *crr-2*, as the plants impaired in the electron acceptor side of PSI. As expected, we could not observe P700 oxidation in *pgrl1*; however, unexpectedly, *crr-2* showed P700 oxidation such as Col-0 in *slow* Umibozu (Figure 1c). The quantum yield of non-photochemical energy dissipation due to acceptor side limitation, calculated as Y(NA), was remarkably higher in *pgrl1* than in Col-0 and *crr-2* (Figure 1d). The operating effi ciency of PSI was calculated as the quantum yield of photochemical energy conversion in PSI, Y(I), which decreased with an increase in PFD in Col-0, *pgrl1*, and *crr-2* (Figure 1e). We found lower Y(I) in *pgrl1* than in Col-0 and *crr-2* in *slow* Umibozu (Figure 1e).

From chlorophyll fluorescence analysis, we calculated the photosynthetic parameters around PSII in *slow* Umibozu in Col-0, *pgrl1*, and *crr-2*. The oxidation of the acceptor side of PSII, inferred from photochemical quenching of chlorophyll fluorescence based on the lake model of the antenna system in PSII, qL, basically decreased and then increased inversely against PFD in all the three plants(Figure 1f). We note the possibility of free redox equilibration between PSII, PQ pool, and Cyt b_6/f, which is the reason why qL is questioned as a proper indicator of PQ oxidation (Joliot, Lavergne, & Béal, 1992; Kirchhoff et al., 2000), might remain to be solved in future studies. The qL values were not different among these plants in the first 5 min; however, subsequently, it was lower in *pgrl1* than in Col-0 and *crr-2* (Figure 1f), which was presumably derived from the photo-oxidative damage in PSI in *pgrl1* (Munekage et al., 2002; DalCorso et al., 2008; Suorsa et al., 2012; Kono & Terashima, 2016; Yamori et al., 2016; and Figure 4 in this study). In addition, we calculated NPQ during *slow* Umibozu. In Col-0 and *crr-2*, NPQ was rapidly induced by illumination with AL and then slowly increased until 7 min (Figure 1g). In *pgrl1*, NPQ required considerably longer time to be induced and was observed at lower levels in *slow* Umibozu than in Col-0 and *crr-2* (Figure 1g). The effective quantum yield of PSII, Y(II), was also calculated in *slow* Umibozu in these plants. We observed similar behaviors of Y(II) to those of qL in all the three plants (Figure 1h).

3.2 | Responses of photosynthetic parameters to light fluctuation in *fast* Umibozu

We applied *fast* Umibozu, where PFD of the light oscillated at a high frequency (60 per hour), to the dark-adapted leaves of Col-0, *pgrl1*, and *crr-2*. In this experiment, PFD of AL changed in the range between the maximum (1,970 μmol photons m^{-2} s^{-1}) and minimum values (30 μmol photons m^{-2} s^{-1}) for 30 s (Figure 2a), resulting in 30 periods of a 1-min sine-like light fluctuation during the 30-min measurement in *fast* Umibozu. To follow the rapidly fluctuating PFD during *fast* Umibozu, photosynthetic parameters were determined every 15 s in the first, 5th, 10th, 15th, 20th, 25th, and 30th periods of the 1-min light fluctuation as shown in Figure 2.

As in *slow* Umibozu, we simultaneously measured gas exchange with P700 absorbance and chlorophyll fluorescence during *fast* Umibozu in Col-0, *pgrl1*, and *crr-2* and found little difference in the net CO_2 assimilation rate among these plants (Figure 2b). We observed the lower CO_2 assimilation rate for the first few minutes during *fast* Umibozu in these plants (Figure 2b), likely attributed to photosynthesis induction in the dark-adapted plant leaves.

Unlike in *slow* Umibozu, in *fast* Umibozu, the induction of Y(ND) was hardly observed around the maximum PFD in the first period of the light fluctuation during *fast* Umibozu in Col-0 and *crr-2* (Figure 2c), indicating that the activation of electron sink, i.e., photosynthesis and photorespiration, in the transition from dark to light could not follow the rapid light fluctuation during *fast* Umibozu in this phase. Subsequently, Col-0 could oxidize P700 in response to the rapidly fluctuating light (Figure 2c). However, *crr-2* showed retarded induction of Y(ND) along with the increase in PFD, compared with that in Col-0, throughout the 30-min measurement in *fast* Umibozu (Figure 2c). In *pgrl1*, we could not detect Y(ND) (Figure 2c). The time courses of Y(ND) in these plants were alternatively shown as a raw trace of the relative amount of $P700^+$ (Supporting Information Figure S1), clearly supporting the results in Figure 2c. In *pgrl1*, Y(NA) was remarkably higher throughout *fast* Umibozu, and *crr-2* showed higher Y(NA) in the phases where PFD was rapidly rising, compared with that in Col-0 (Figure 2d). Further, we found that Y(I) was lower in *pgrl1* than in Col-0 and *crr-2* (Figure 2e), as in *slow* Umibozu.

During *fast* Umibozu, qL inversely responded to the changes in PFD, and the values were largely and slightly lower respectively in *pgrl1* and *crr-2* than in Col-0 in the end of the 30-min measurement (Figure 2f). We also measured NPQ during *fast* Umibozu and found that *pgrl1* could not induce NPQ in response to the rapidly fluctuating light, different from Col-0 and *crr-2* (Figure 2g). Furthermore, we found that Y(II) responded to *fast* Umibozu as well as qL in all the three plants (Figure 2h).

3.3 | Effects of the frequency of fluctuating light on the relationships of qL and NPQ with P700 oxidation

We evaluated the relationships of qL and NPQ with Y(ND) during *slow* and *fast* Umibozu in Col-0, *pgrl1*, and *crr-2*. During *slow* Umibozu, we found a linear relationship between qL and Y(ND) in Col-0 (indicated by dashed black line in Figure 3a-c), implying that P700 oxidation was paralleled by the reduction of the electron acceptor side of PSII. The linear relationship was recognized in Col-0 also during *fast* Umibozu except for the data in the first period of the light fluctuation (surrounded by dashed green frame in Figure 3a). Additionally, the relationship between qL and Y(ND) in *crr-2* showed the similar linearity to that in Col-0 during *slow* Umibozu, which was, nevertheless, broken during *fast* Umibozu even after the photosynthesis induction is nearly completed (Figure 3c). That is, *crr-2* showed limited electron transport on the acceptor side of PSI in *fast* Umibozu, and the induction of P700 oxidation could not follow the rapid light fluctuation in *fast* Umibozu. In both *slow* and *fast* Umibozu, we failed to identify a clear relationship between NPQ and Y(ND) in Col-0 and *crr-2* although Y(ND) partially increased along with the induction of NPQ in some phases (Figure 3d,f). In *pgrl1*, we could not evaluate the relationships of these chlorophyll fluorescence

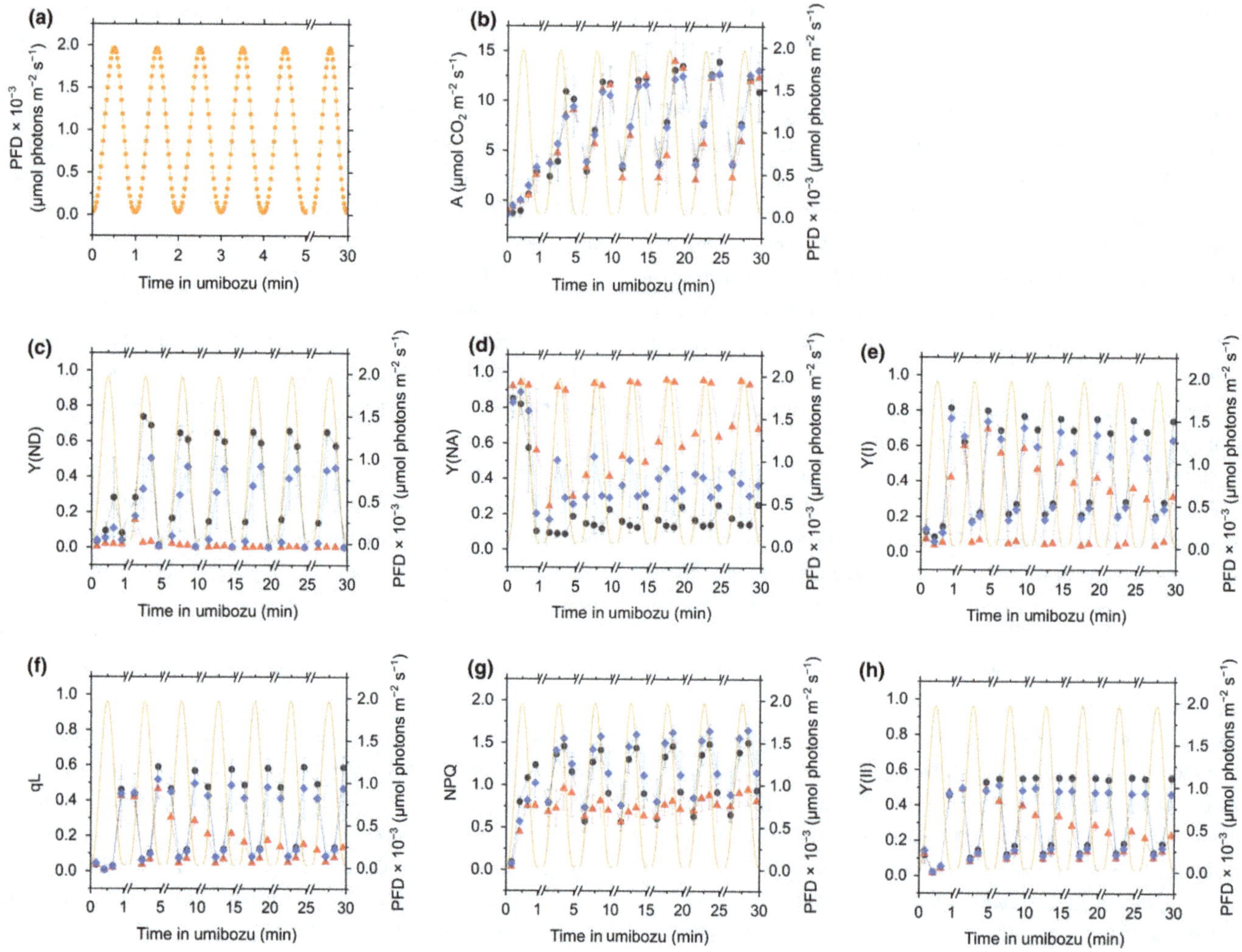

FIGURE 2 Time courses of photon flux density (PFD; a), net CO_2 assimilation rate, A (b), Y(ND) (c), Y(NA) (d), Y(I) (e), qL (f), NPQ (g), Y(II) (h) during *fast* Umibozu in *Arabidopsis thaliana* wild-type (Col-0, black circles) and the mutants, *pgrl1* (red triangles) and *crr-2* (blue diamonds). Orange lines indicate PFD. Data are represented as the means ± SD of three independent measurements

parameters with P700 oxidation (Figure 3b,e), because we did not observe Y(ND) during both *slow* and *fast* Umibozu.

3.4 | Photo-oxidative damage in PSI during Umibozu

Photon energy in the excess of the demand for photosynthesis causes the production of ROS, resulting in the inactivation of PSI in photosynthetic organisms unless P700 oxidation system operates. We exposed the dark-adapted *A. thaliana* plant leaves of Col-0, *pgrl1*, and *crr-2* to Umibozu at three different frequencies (*very slow*, 1; *slow*, 6; and *fast*, 60 per hour) for 1 hr. In this experiment, the PFD of AL fluctuated from the minimum 30 to the maximum 1,970 μmol photons m^{-2} s^{-1} during the 1-hr measurement. In addition, we did not apply any short saturation pulses, because repetitive short-pulse illumination can cause PSI photoinhibition to mask the effect of the sine fluctuating light in Umibozu on PSI in plant leaves (Sejima et al., 2014). We evaluated the residual PSI activity of plant leaves as the ratio of total oxidizable P700 (P_m) measured before and after Umibozu. In Col-0, P_m did not decrease after Umibozu at all three frequencies (Figure 4a), which was in agreement with the experimental fact that Col-0 can maintain P700 oxidized under both *slow* and *fast* Umibozu. Conversely, *pgrl1* showed a severe inactivation of PSI at all frequencies of Umibozu (Figure 4a). The mutant *crr-2* showed inactivated PSI during *fast* Umibozu, but not during *slow* and *very slow* ones (Figure 4a). The decrease in P_m in *pgrl1* and *crr-2* was alleviated by eliminating O_2 to 1 kPa (Figure 4b). These data were consistent with the results of P700 oxidation in Umibozu (Figures 1 and 2) and support that P700 oxidation is required for the protection of PSI against the photo-oxidative damage induced by ROS.

We also evaluated the residual PSII activity of plant leaves as the ratio of maximum quantum efficiency of PSII photochemistry (F_v/F_m) measured before and after Umibozu. To exclude the effects of both qE quenching and state transition on F_v/F_m, we incubated the plant leaves in the dark for 30 min after AL was turned off, and then F_v/F_m values were evaluated. Compared with that in PSI, the decreases in PSII activity were small in these plants (Supporting Information Figure S2).

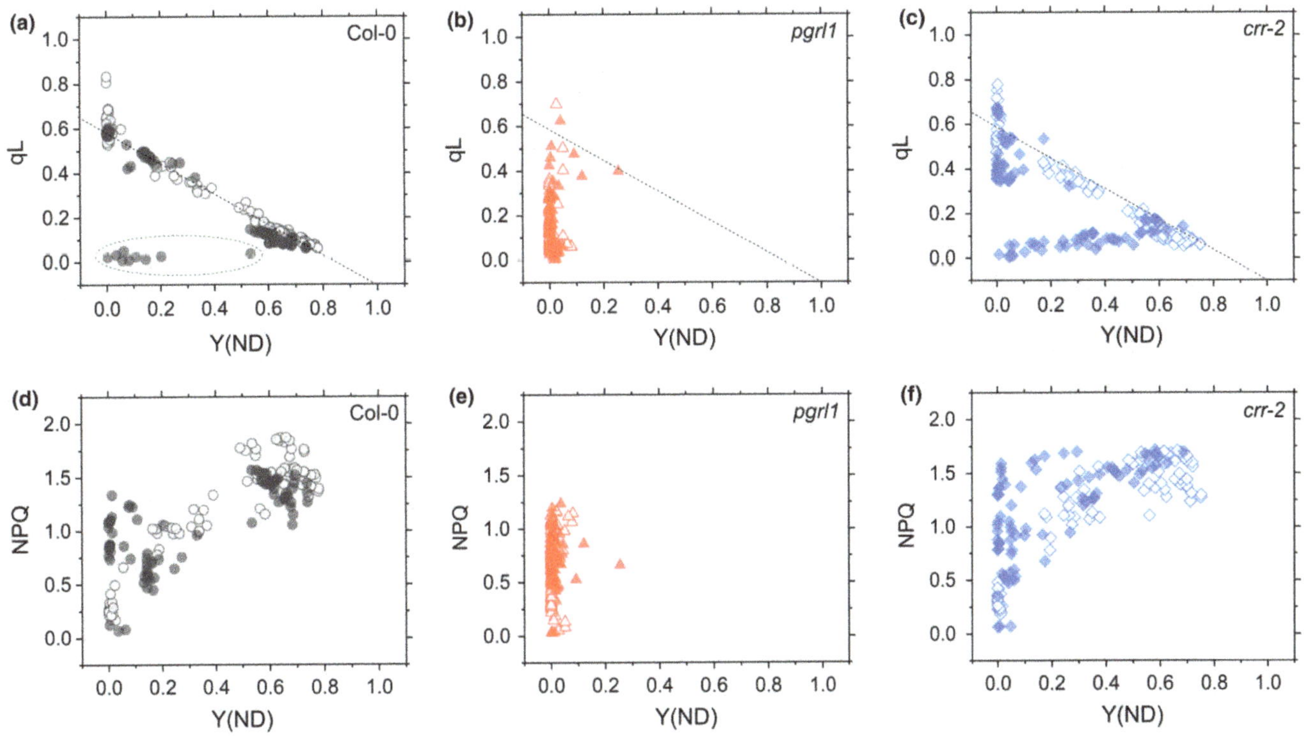

FIGURE 3 Relationships of qL (a-c) and NPQ (d-f) with Y(ND) during *slow* (open symbols) and *fast* (closed symbols) Umibozu in *Arabidopsis thaliana* wild-type (Col-0, a and d) and the mutants, *pgrl1* (b and e) and *crr-2* (c and f). Dashed black line indicates the linear relationship between qL and Y(ND) during *slow* Umibozu in Col-0. Dashed green frame surrounds the data in the first period during *fast* Umibozu in Col-0. Data are derived from three independent measurements

4 | DISCUSSION

In the present study, we sought to investigate the impact of changing frequency of fluctuating light on a variety of photosynthetic parameters in Col-0, *pgrl1*, and *crr-2* during *slow* and *fast* Umibozu and verify the tracking performance of the molecular mechanisms for P700 oxidation in the C_3 plant *A. thaliana*. Since the pioneering work of Eva-Mari Aro's group (Suorsa et al., 2012), rectangular artificial fluctuating lights have been used in photosynthesis research on PSI photoinhibition (Allahverdiyeva et al., 2013; Gerotto et al., 2016; Kono & Terashima, 2016; Yamori et al., 2016). Compared with the rectangular fluctuating light, Umibozu provides a smoothly fluctuating artificial light. The smoothly fluctuating light has been utilized in preceding reports (Lucker, Hall, Zegarac, & Kramer, 2014). In this study, the most significant advantage of the sinusoidal change of PFD is to evaluate the tracking performance of photosynthetic parameters with a variety of light fluctuation by changing the frequency. In the present study, we adjusted the frequency of Umibozu to 6 (*slow*) and 60 per hour (*fast*). In Col-0, P700 oxidation is rapidly induced with the support of donor side mechanisms of PSI during both *slow* and *fast* Umibozu as shown in the strong correlation between qL and Y(ND) except for the moments just after AL was turned on in *fast* Umibozu, i.e., an early phase of photosynthesis induction (Figure 3a). During *slow* Umibozu, we could observe small differences in the photosynthetic parameters between Col-0 and *crr-2* (Figure 1). Nevertheless, the electron acceptor side of PSI in *crr-2* was remarkably limited on exposure to the rapidly rising PFD in *fast* Umibozu as reflected in Y(NA) (Figure 2d), resulting in the retardation of the induction of P700 oxidation (Figure 2c). Finally, the lack of NDH caused PSI photoinhibition only in *fast* Umibozu, but not in *slow* and *very slow* ones (Figure 4). That is, the tracking performance of *A. thaliana* to induce P700 oxidation against rapidly fluctuating light is supported by a NDH-dependent mechanism. Recently, C_3 plant mutants deficient in NDH have been reported to be impaired in P700 oxidation and to show PSI photoinhibition under a rectangular artificial fluctuating light (Kono & Terashima, 2016; Yamori et al., 2016). The rectangular light fluctuation exposes plant leaves to instantaneous changes in PFD of AL, which can be defined as a super-*fast* fluctuating light. Therefore, the inability to oxidize P700 in *crr* mutants under the rectangular fluctuating light also supports the conclusion of the present study that NDH is required for the rapid start of P700 oxidation in C_3 plant leaves. On the other hand, in *pgrl1* P700 oxidation was not observed during both *slow* and *fast* Umibozu (Figures 1 and 2), and the photo-oxidative damage in PSI was not affected by the frequency of the fluctuating light in Umibozu (Figure 4). These data suggest that a PGR5/PGRL1-dependent mechanism plays a critical role to keep P700 oxidized regardless of the frequency of fluctuating light. Overall, changing frequency of fluctuating light in Umibozu provides a novel insight into the mechanism of P700 oxidation in plants under natural environmental variations with different durations and frequencies of light (Pfitsch & Pearcy, 1989).

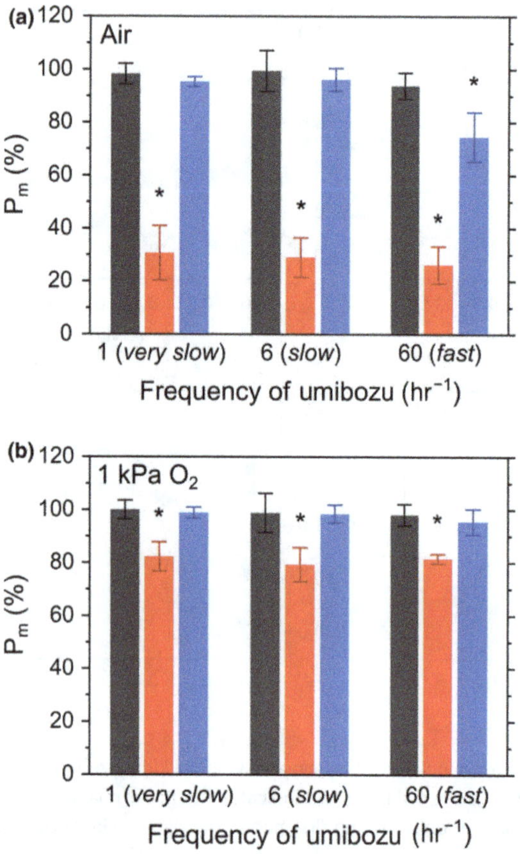

FIGURE 4 Decrease in P_m after Umibozu with different frequencies under ambient air (a) and 1 kPa O_2 (b). P_m was obtained 30 min (in the dark) after 1 hr Umibozu in *Arabidopsis thaliana* wild-type (Col-0, gray) and the mutants, *pgrl1* (red) and *crr-2* (blue). Data are represented as the means ± SD of three independent measurements. Asterisks indicate statistically significant differences ($p < 0.05$) in the residual P_m of the mutants with Col-0 as per Student's *t*-test

Here we rethink the reason why PGR5/PGRL1 and NDH are required for P700 oxidation in *A. thaliana*. The most popular hypothesis regarding the inability to oxidize P700 in *pgrl1* and *crr-2* is likely to be based on the CET model, in which electrons are donated from the acceptor side of PSI back to PQ through the PGR5/PGRL1 and NDH pathways, respectively, to promote lumen acidification (Kono & Terashima, 2016; Nandha et al., 2007; Yamori et al., 2016). Recently, Mosebach et al. (2017) showed that, in the green alga *Chlamydomonas reinhardtii*, the lack of PGR5 and PGRL1 impairs the association of Fd-$NADP^+$ reductase with the thylakoid membrane to lower the reduction rate of $NADP^+$ to NADPH in vitro, which also supports the results of *pgrl1* in this study. Further, in the case of NDH we note another possibility that NDH oxidize Fd in the dark or weak light for the rapid start of P700 oxidation in response to rapidly rising PFD in the process of chlororespiration (Feilke et al., 2016). Under light conditions, Fd is oxidized by Fd-$NADP^+$ reductase to produce NADPH for driving the Calvin–Benson cycle. However, the change of the turnover of the Calvin-Benson cycle requires a longer time, compared with the alteration of photosynthetic electron transport reaction, which might require the oxidation of Fd by chlororespiration for the alleviation of the electron acceptor side limitation in PSI in *fast* Umibozu (Figure 2c). Unlike in *fast* Umibozu, the alteration of the electron sink capacity in the Calvin-Benson cycle might be in time for the induction of P700 oxidation in *slow* Umibozu (Figure 1c).

The linear relationship between qL and Y(ND), which were calculated using completely different experimental approaches (Figure 3a) strongly supports that photosynthetic linear electron flow is basically limited at the step between PSII and PSI, i.e., Cyt b_6/f. Cyt b_6/f is known to be the limiting step of photosynthetic linear electron flow without any specific regulatory mechanism, because the oxidation of PQH_2 is the slowest step in the photosynthetic electron transport system, and the amount of Cyt b_6/f is normally smaller than those of PSII and PSI in plant leaves (Anderson, 1992; Kirchhoff et al., 2000; Schöttler & Tóth, 2014; Schöttler et al., 2015; Stiehl & Witt, 1969; Yamori, Evans, & von Caemmerer, 2010; Yamori et al., 2011). However, under fluctuating environments, plants require regulatory mechanism (i.e. P700 oxidation system) to oxidize P700 for the protection of PSI against ROS damage (Shimakawa, Shaku et al., 2016; Shimakawa, Akimoto et al., 2016; Shimakawa, Ishizaki et al., 2017; Takagi et al., 2017), which is clearly evidenced by the flexible response of Y(ND) to environmental variations. At present, ΔpH across the thylakoid membrane is believed to regulate the oxidation activity of PQH_2 in Cyt b_6/f for the induction of P700 oxidation in response to fluctuating environments. According to existing theories (Kanazawa & Kramer, 2002), photon energy drives photosynthetic linear electron transport to form ΔpH, which then down-regulates photosynthetic electron transport from PSII to PSI by stimulating the suppression of electron transport in Cyt b_6/f and inducing qE quenching (termed as photosynthetic control; Schöttler et al., 2015 and references therein). In addition, the accumulation of H^+ on the luminal side of the thylakoid membrane is also supported by the regulation of CET (Nandha et al., 2007) and g_H^+ (Armbruster et al., 2014; Rott et al., 2011; Takizawa et al., 2008). In C_3 plant leaves, NPQ originates qE quenching, state transition, and PSII photoinhibition, which have different relaxation times (Ruban, 2016). We observed a lag of few minutes between the decreases in PFD and NPQ during *slow* Umibozu in Col-0 (Figure 2e), which is consistent with the common explanation that NPQ in C_3 plants is mainly caused by qE quenching dependent on ΔpH across the thylakoid membrane (Kanazawa & Kramer, 2002; Ruban, 2016). However, we could not identify the proportional linear relationship of NPQ with Y (ND) during both *slow* and *fast* Umibozu in Col-0 (Figure 3d), indicating that the relationships among NPQ, Y(ND), and ΔpH are complicated in the intact plant leaves (Hald, Nandha, Gallois, & Johnson, 2008). Alternatively, a novel regulatory theory for P700 oxidation, RISE (Shaku et al., 2016; Shimakawa et al., 2018), may provide some insights into the understandings of these facts.

Notably, Y(I) in *pgrl1* was lower than those in Col-0 and *crr-2* where Y(II) in *pgrl1* was like those during both *slow* and *fast* Umibozu (Figures 1 and 2). First, we can consider that the Y(I) uncoupled with Y(II) is derived from CET dependent on PGR5/PGRL1. As a second possibility, Y(I) does not properly reflect the efficiency of the electron transport at PSI. Sacksteder and Kramer (2000) suggested that the total electron transport activity in PSI is determined by the summation of the estimations of electron fluxes through cytochromef, plastocyanin, and P700. Finally, the electron transport reaction from the electron acceptor side of PSI back to $P700^+$ possesses μs- or ms-order half times (i.e. the charge recombination of $P700^+$ with F_X exhibits 0.5–1.5 ms of the half time; Semenov et al., 2000), which can cause an overestimation of Y(I), particularly when $P700^+$ is accumulated. The charge recombination of $P700^+$ results in the dissipation of excess electrons on the electron acceptor side of PSI, which might be one of how P700 oxidation protects PSI against ROS damage.

ACKNOWLEDGMENTS

The authors thank Prof. Toshiharu Shikanai (Kyoto University) for supplying the *A. thaliana* seeds of both *pgrl1* and *crr-2* mutants. We would like to thank Editage (www.editage.jp) for English language editing.

CONFLICT OF INTEREST

The authors have no conflict of interest to declare.

AUTHOR CONTRIBUTIONS

C.M. conceived the original screening and research plans; C.M. supervised the experiments; G.S. performed all the experiments; C.M. and G.S. designed the experiments and analyzed the data; C.M. and G.S. conceived the project and drafted the manuscript.

REFERENCES

Allahverdiyeva, Y., Mustila, H., Ermakova, M., Bersanini, L., Richaud, P., Ajlani, G., … Aro, E. M. (2013). Flavodiiron proteins Flv1 and Flv3 enable cyanobacterial growth and photosynthesis under fluctuating light. *Proceedings of the National Academy of Sciences of the United States of America*, *110*, 4111–4116. https://doi.org/10.1073/pnas.1221194110

Allen, J. F. (2003). Cyclic, pseudocyclic and noncyclic photophosphorylation: New links in the chain. *Trends in Plant Science*, *8*, 15–19. https://doi.org/10.1016/S1360-1385(02)00006-7

Anderson, J. M. (1992). Cytochrome b_6f complex: Dynamic molecular organization, function and acclimation. *Photosynthesis Research*, *34*, 341–357. https://doi.org/10.1007/BF00029810

Armbruster, U., Carrillo, L. R., Venema, K., Pavlovic, L., Schmidtmann, E., Kornfeld, A., … Jonikas, M. C. (2014). Ion antiport accelerates photosynthetic acclimation in fluctuating light environments. *Nature Communications*, *5*, 5439. https://doi.org/10.1038/ncomms6439

Baker, N. R. (2008). Chlorophyll fluorescence: A probe of photosynthesis in vivo. *Annual Review of Plant Biology*, *59*, 89–113. https://doi.org/10.1146/annurev.arplant.59.032607.092759

DalCorso, G., Pesaresi, P., Masiero, S., Aseeva, E., Schünemann, D., Finazzi, G., … Leister, D. (2008). A complex containing PGRL1 and PGR5 is involved in the switch between linear and cyclic electron flow in *Arabidopsis*. *Cell*, *132*, 273–285. https://doi.org/10.1016/j.cell.2007.12.028

Feilke, K., Streb, P., Cornic, G., Perreau, F., Kruk, J., & Krieger-Liszkay, A. (2016). Effect of *Chlamydomonas* plastid terminal oxidase 1 expressed in tobacco on photosynthetic electron transfer. *The Plant Journal*, *85*, 219–228. https://doi.org/10.1111/tpj.13101

Gerotto, C., Alboresi, A., Meneghesso, A., Jokel, M., Suorsa, M., Aro, E.-M., & Morosinotto, T. (2016). Flavodiiron proteins act as safety valve for electrons in *Physcomitrella patens*. *Proceedings of the National Academy of Sciences of the United States of America*, *113*, 12322–12327. https://doi.org/10.1073/pnas.1606685113

Golding, A. J., & Johnson, G. N. (2003). Down-regulation of linear and activation of cyclic electron transport during drought. *Planta*, *218*, 107–114. https://doi.org/10.1007/s00425-003-1077-5

Hald, S., Nandha, B., Gallois, P., & Johnson, G. N. (2008). Feedback regulation of photosynthetic electron transport by NADP(H) redox poise. *Biochimica et Biophysica Acta (BBA)-Bioenergetics*, *1777*, 433–440. https://doi.org/10.1016/j.bbabio.2008.02.007

Hanawa, H., Ishizaki, K., Nohira, K., Takagi, D., Shimakawa, G., Sejima, T., … Miyake, C. (2017). Land plants drive photorespiration as higher electron-sink: Comparative study of post-illumination transient O_2-uptake rates from liverworts to angiosperms through ferns and gymnosperms. *Physiologia Plantarum*, *161*, 138–149. https://doi.org/10.1111/ppl.12580

Harbinson, J., & Hedley, C. L. (1993). Changes in P-700 oxidation during the early stages of the induction of photosynthesis. *Plant Physiology*, *103*, 649–660. https://doi.org/10.1104/pp.103.2.649

Hashimoto, M., Endo, T., Peltier, G., Tasaka, M., & Shikanai, T. (2003). A nucleus-encoded factor, CRR2, is essential for the expression of chloroplast *ndhB* in *Arabidopsis*. *Plant Journal*, *36*, 541–549. https://doi.org/10.1046/j.1365-313X.2003.01900.x

Hayashi, R., Shimakawa, G., Shaku, K., Shimizu, S., Akimoto, S., Yamamoto, H., … Miyake, C. (2014). O_2-dependent large electron flow functioned as an electron sink, replacing the steady-state electron flux in photosynthesis in the cyanobacterium *Synechocystis* sp. PCC 6803, but not in the cyanobacterium *Synechococcus* sp. PCC 7942. *Bioscience, Biotechnology, and Biochemistry*, *78*, 384–393. https://doi.org/10.1080/09168451.2014.882745

Helman, Y., Barkan, E., Eisenstadt, D., Luz, B., & Kaplan, A. (2005). Fractionation of the three stable oxygen isotopes by oxygen-producing and oxygen-consuming reactions in photosynthetic organisms. *Plant Physiology*, *138*, 2292–2298. https://doi.org/10.1104/pp.105.063768

Helman, Y., Tchernov, D., Reinhold, L., Shibata, M., Ogawa, T., Schwarz, R., … Kaplan, A. (2003). Genes encoding A-type flavoproteins are essential for photoreduction of O_2 in cyanobacteria. *Current Biology*, *13*, 230–235. https://doi.org/10.1016/S0960-9822(03)00046-0

Inoue, Y., Ogawa, T., & Shibata, K. (1973). Light-induced spectral changes of P700 in the 800-nm region in *Anacystis* and spinach lamellae. *Biochimica et Biophysica Acta (BBA)-Bioenergetics*, *305*, 483–487. https://doi.org/10.1016/0005-2728(73)90194-1

Joliot, P., Lavergne, J., & Béal, D. (1992). Plastoquinone compartmentation in chloroplasts. I. Evidence for domains with different rates of photo-reduction. *Biochimica et Biophysica Acta*, *1101*, 1–12. https://doi.org/10.1016/0167-4838(92)90460-u

Kanazawa, A., & Kramer, D. M. (2002). In vivo modulation of nonphotochemical exciton quenching (NPQ) by regulation of the chloroplast ATP synthase. *Proceedings of the National Academy of Sciences of the United States of America*, *99*, 12789–12794. https://doi.org/10.1073/pnas.182427499

Kirchhoff, H., Horstmann, S., & Weis, E. (2000). Control of the photosynthetic electron transport by PQ diffusion microdomains in thylakoids of higher plants. *Biochimica et Biophysica Acta*, *1459*, 148–168. https://doi.org/10.1016/S0005-2728(00)00143-2

Klughammer, C., & Schreiber, U. (1994). An improved method, using saturating light pulses, for the determination of photosystem I quantum yield via P700$^+$-absorbance changes at 830 nm. *Planta, 192*, 261–268. https://doi.org/10.1007/BF01089043

Klughammer, C., & Schreiber, U. (2016). Deconvolution of ferredoxin, plastocyanin, and P700 transmittance changes in intact leaves with a new type of kinetic LED array spectrophotometer. *Photosynthesis Research, 128*, 195–214. https://doi.org/10.1007/s11120-016-0219-0

Kono, M., & Terashima, I. (2016). Elucidation of photoprotective mechanisms of PSI against fluctuating light photoinhibition. *Plant and Cell Physiology, 57*, 1405–1414. https://doi.org/10.1093/pcp/pcw103

Krause, G. H., & Weis, E. (1984). Chlorophyll fluorescence as a tool in plant physiology. II. Interpretation of fluorescence signals. *Photosynthesis Research, 5*, 139–157. https://doi.org/10.1007/bf00028527

Lucker, B. F., Hall, C. C., Zegarac, R., & Kramer, D. M. (2014). The environmental photobioreactor (ePBR): An algal culturing platform for simulating dynamic natural environments. *Algal Research, 6*, 242–249. https://doi.org/10.1016/j.algal.2013.12.007

Miyake, C., Miyata, M., Shinzaki, Y., & Tomizawa, K. (2005). CO_2 response of cyclic electron flow around PSI (CEF-PSI) in tobacco leaves-relative electron fluxes through PSI and PSII determine the magnitude of non-photochemical quenching (NPQ) of Chl fluorescence. *Plant and Cell Physiology, 46*, 629–637. https://doi.org/10.1093/pcp/pci067

Mosebach, L., Heilmann, C., Mutoh, R., Gäbelein, P., Steinbeck, J., Happe, T., … Hippler, M. (2017). Association of Ferredoxin:NADP$^+$ oxidoreductase with the photosynthetic apparatus modulates electron transfer in *Chlamydomonas reinhardtii*. *Photosynthesis Research, 134*, 291–306. https://doi.org/10.1007/s11120-017-0408-5

Munekage, Y., Hojo, M., Meurer, J., Endo, T., Tasaka, M., & Shikanai, T. (2002). PGR5 is involved in cyclic electron flow around photosystem I and is essential for photoprotection in *Arabidopsis*. *Cell, 110*, 361–371. https://doi.org/10.1016/S0092-8674(02)00867-X

Nandha, B., Finazzi, G., Joliot, P., Hald, S., & Johnson, G. N. (2007). The role of PGR5 in the redox poising of photosynthetic electron transport. *Biochimica et Biophysica Acta, 1767*, 1252–1259. https://doi.org/10.1016/j.bbabio.2007.07.007

Nishio, J. N., & Whitmarsh, J. (1993). Dissipation of the proton electrochemical potential in intact chloroplasts (II. The pH gradient monitored by cytochrome f reduction kinetics). *Plant Physiology, 101*, 89–96. https://doi.org/10.1104/pp.101.1.89

Pfitsch, W. A., & Pearcy, R. W. (1989). Daily carbon gain by *Adenocaulon bicolor* (Asteraceae), a redwood forest understory herb, in relation to its light environment. *Oecologia, 80*, 465–470. https://doi.org/10.1007/BF00380067

Rott, M., Martins, N. F., Thiele, W., Lein, W., Bock, R., Kramer, D. M., & Schöttler, M. A. (2011). ATP synthase repression in tobacco restricts photosynthetic electron transport, CO_2 assimilation, and plant growth by overacidification of the thylakoid lumen. *The Plant Cell, 23*, 304–321. https://doi.org/10.1105/tpc.110.079111

Ruban, A. V. (2016). Nonphotochemical chlorophyll fluorescence quenching: Mechanism and effectiveness in protecting plants from photodamage. *Plant Physiology, 170*, 1903–1916. https://doi.org/10.1104/pp.15.01935

Sacksteder, C. A., & Kramer, D. M. (2000). Dark-interval relaxation kinetics (DIRK) of absorbance changes as a quantitative probe of steady-state electron transfer. *Photosynthesis Research, 66*, 145–158. https://doi.org/10.1023/A:1010785912271

Sazanov, L. A., Burrows, P. A., & Nixon, P. J. (1998). The chloroplast Ndh complex mediates the dark reduction of the plastoquinone pool in response to heat stress in tobacco leaves. *FEBS Letters, 429*, 115–118. https://doi.org/10.1016/S0014-5793(98)00573-0

Schöttler, M. A., & Tóth, S. Z. (2014). Photosynthetic complex stoichiometry dynamics in higher plants: Environmental acclimation and photosynthetic flux control. *Frontiers in Plant Science, 5*, 188. https://doi.org/10.3389/fpls.2014.00188

Schöttler, M. A., Tóth, S. Z., Boulouis, A., & Kahlau, S. (2015). Photosynthetic complex stoichiometry dynamics in higher plants: Biogenesis, function, and turnover of ATP synthase and the cytochrome b_6f complex. *Journal of Experimental Botany, 66*, 2373–2400. https://doi.org/10.1093/jxb/eru495

Schreiber, U., & Klughammer, C. (2008). Saturation pulse method for assessment of energy conversion in PSI. *PAM Application Notes, 1*, 11–14.

Sejima, T., Hanawa, H., Shimakawa, G., Takagi, D., Suzuki, Y., Fukayama, H., … Miyake, C. (2016). Post-illumination transient O_2-uptake is driven by photorespiration in tobacco leaves. *Physiologia Plantarum, 156*, 227–238. https://doi.org/10.1111/ppl.12388

Sejima, T., Takagi, D., Fukayama, H., Makino, A., & Miyake, C. (2014). Repetitive short-pulse light mainly inactivates photosystem I in sunflower leaves. *Plant and Cell Physiology, 55*, 1184–1193. https://doi.org/10.1093/pcp/pcu061

Semenov, A. Y., Vassiliev, I. R., van der Est, A., Mamedov, M. D., Zybailov, B., Shen, G., … Golbeck, J. H. (2000). Recruitment of a foreign quinone into the A1 site of photosystem I: Altered kinetics of electron transfer in phylloquinone biosynthetic pathway mutants studied by time-resolved optical, EPR, and electrometric techniques. *Journal of Biological Chemistry, 275*, 23429–23438. https://doi.org/10.1074/jbc.M000508200

Shaku, K., Shimakawa, G., Hashiguchi, M., & Miyake, C. (2016). Reduction-induced suppression of electron flow (RISE) in the photosynthetic electron transport system of *Synechococcus elongatus* PCC 7942. *Plant and Cell Physiology, 57*, 1443–1453.

Shimakawa, G., Akimoto, S., Ueno, Y., Wada, A., Shaku, K., Takahashi, Y., & Miyake, C. (2016). Diversity in photosynthetic electron transport under [CO_2]-limitation: The cyanobacterium *Synechococcus* sp. PCC 7002 and green alga *Chlamydomonas reinhardtii* drive an O_2-dependent alternative electron flow and non-photochemical quenching of chlorophyll fluorescence during CO_2-limited photosynthesis. *Photosynthesis Research, 130*, 293–305. https://doi.org/10.1007/s11120-016-0253-y

Shimakawa, G., Ishizaki, K., Tsukamoto, S., Tanaka, M., Sejima, T., & Miyake, C. (2017). The liverwort, *Marchantia*, drives alternative electron flow using a flavodiiron protein to protect PSI. *Plant Physiology, 173*, 1636–1647. https://doi.org/10.1104/pp.16.01038

Shimakawa, G., Matsuda, Y., Nakajima, K., Tamoi, M., Shigeoka, S., & Miyake, C. (2017). Diverse strategies of O_2 usage for preventing photo-oxidative damage under CO_2 limitation during algal photosynthesis. *Scientific Reports, 7*, 41022. https://doi.org/10.1038/srep41022

Shimakawa, G., & Miyake, C. (2018). Respiratory terminal oxidases alleviate photo-oxidative damage in photosystem I during repetitive short-pulse illumination in *Synechocystis* sp. PCC 6803. *Photosynthesis Research, 137*, 241–250. https://doi.org/10.1007/s11120-018-0495-y

Shimakawa, G., Shaku, K., & Miyake, C. (2016). Oxidation of P700 in photosystem I is essential for the growth of cyanobacteria. *Plant Physiology, 172*, 1443–1450. https://doi.org/10.1104/pp.16.01227

Shimakawa, G., Shaku, K., Nishi, A., Hayashi, R., Yamamoto, H., Sakamoto, K., … Miyake, C. (2015). FLAVODIIRON2 and FLAVODIIRON4 proteins mediate an oxygen-dependent alternative electron flow in *Synechocystis* sp. PCC 6803 under CO_2-limited conditions. *Plant Physiology, 167*, 472–480. https://doi.org/10.1104/pp.114.249987

Shimakawa, G., Shaku, K., & Miyake, C. (2018). Reduction-induced suppression of electron flow (RISE) is relieved by non-ATP-consuming electron flow in *Synechococcus elongatus* PCC 7942. *Frontiers in Microbiology, 9*, 886. https://doi.org/10.3389/fmicb.2018.00886

Stiehl, H. H., & Witt, H. T. (1969). Quantitative treatment of the function of plastoquinone in phostosynthesis. *Zeitschrift für Naturforschung B, 24*, 1588–1598. https://doi.org/10.1515/znb-1969-1219

Suorsa, M., Jarvi, S., Grieco, M., Nurmi, M., Pietrzykowska, M., Rantala, M., … Aro, E. M. (2012). PROTON GRADIENT REGULATION5 is essential for proper acclimation of *Arabidopsis* photosystem I to naturally and artificially fluctuating light conditions. *The Plant Cell, 24,* 2934–2948. https://doi.org/10.1105/tpc.112.097162

Takagi, D., Ishizaki, K., Hanawa, H., Mabuchi, T., Shimakawa, G., Yamamoto, H., & Miyake, C. (2017). Diversity of strategies for escaping reactive oxygen species production within photosystem I among land plants: P700 oxidation system is prerequisite for alleviating photoinhibition in photosystem I. *Physiologia Plantarum, 161,* 56–74. https://doi.org/10.1111/ppl.12562

Takagi, D., Takumi, S., Hashiguchi, M., Sejima, T., & Miyake, C. (2016). Superoxide and singlet oxygen produced within the thylakoid membranes both cause photosystem I photoinhibition. *Plant Physiology, 171,* 1626–1634. https://doi.org/10.1104/pp.16.00246

Takizawa, K., Kanazawa, A., & Kramer, D. M. (2008). Depletion of stromal Pi induces high 'energy-dependent' antenna exciton quenching (qE) by decreasing proton conductivity at CF_o-CF_1 ATP synthase. *Plant, Cell & Environment, 31,* 235–243.

Wiese, C., Shi, L., & Heber, U. (1998). Oxygen reduction in the Mehler reaction is insufficient to protect photosystems I and II of leaves against photoinactivation. *Physiologia Plantarum, 102,* 437–446. https://doi.org/10.1034/j.1399-3054.1998.1020312.x

Yamori, W., Evans, J. R., & von Caemmerer, S. (2010). Effects of growth and measurement light intensities on temperature dependence of CO_2 assimilation rate in tobacco leaves. *Plant, Cell and Environment, 33,* 332–343. https://doi.org/10.1111/j.1365-3040.2009.02067.x

Yamori, W., Makino, A., & Shikanai, T. (2016). A physiological role of cyclic electron transport around photosystem I in sustaining photosynthesis under fluctuating light in rice. *Scientific Reports, 6,* 20147. https://doi.org/10.1038/srep20147

Yamori, W., Takahashi, S., Makino, A., Price, D. G., Badger, M. R., & von Caemmerer, S. (2011). The roles of ATP synthase and the cytochrome b_6/f complexes in limiting chloroplast electron transport and determining photosynthetic capacity. *Plant Physiology, 155,* 956–962. https://doi.org/10.1104/pp.110.168435

Zhang, P., Allahverdiyeva, Y., Eisenhut, M., & Aro, E. M. (2009). Flavodiiron proteins in oxygenic photosynthetic organisms: Photoprotection of photosystem II by Flv2 and Flv4 in *Synechocystis* sp. PCC 6803. *PLoS ONE, 4,* e5331. https://doi.org/10.1371/journal.pone.0005331

Zivcak, M., Brestic, M., Kunderlikova, K., Olsovska, K., & Allakhverdiev, S. I. (2015). Effect of photosystem I inactivation on chlorophyll *a* fluorescence induction in wheat leaves: Does activity of photosystem I play any role in OJIP rise? *Journal of Photochemistry and Photobiology B: Biology, 152,* 318–324. https://doi.org/10.1016/j.jphotobiol.2015.08.024

Zivcak, M., Brestic, M., Kunderlikova, K., Sytar, O., & Allakhverdiev, S. I. (2015). Repetitive light pulse-induced photoinhibition of photosystem I severely affects CO_2 assimilation and photoprotection in wheat leaves. *Photosynthesis Research, 126,* 449–463. https://doi.org/10.1007/s11120-015-0121-1

3

Thylakoid-integrated recombinant Hcf106 participates in the chloroplast twin arginine transport system

Qianqian Ma[1] | Kristen Fite[2] | Christopher Paul New[1] | Carole Dabney-Smith[1,2]

[1]Graduate Program in Cell, Molecular, and Structural Biology, Miami University, Oxford, Ohio

[2]Department of Chemistry and Biochemistry, Miami University, Oxford, Ohio

Correspondence
Carole Dabney-Smith, Miami University, Oxford, OH.
Email: adabney@miamioh.edu

Present addresses
Qianqian Ma, Johns Hopkins University School of Medicine, Baltimore, Maryland
and
Kristen Fite, Boonshoft School of Medicine, Wright State University, Dayton, Ohio.

Funding information
Department of Energy: Office of Science, Basic Energy Science, Grant/Award Number: DE-SC0014441

Abstract

The chloroplast twin arginine transport (cpTat) system distinguishes itself as a protein transport pathway by translocating fully folded proteins, using the proton-motive force (PMF) as the sole source of energy. The cpTat pathway is evolutionarily conserved with the Tat pathway found in the plasma membrane of many prokaryotes. The cpTat (*Escherichia coli*) system uses three proteins, Tha4 (TatA), Hcf106 (TatB), and cpTatC (TatC), to form a transient translocase allowing the passage of precursor proteins. Briefly, cpTatC and Hcf106, with Tha4, form the initial receptor complex responsible for precursor protein recognition and binding in an energy-independent manner, while a separate pool of Tha4 assembles with the precursor-bound receptor complex in the presence the PMF. Analysis by blue-native polyacrylamide gel electrophoresis (BN-PAGE) shows that the receptor complex, in the absence of precursor, migrates near 700 kDa and contains cpTatC and Hcf106 with little Tha4 remaining after detergent solubilization. To investigate the role that Hcf106 may play in receptor complex oligomerization and/or stability, systematic cysteine substitutions were made in positions from the N-terminal transmembrane domain to the end of the predicted amphipathic helix of the protein. BN-PAGE analysis allowed us to identify the locations of amino acids in Hcf106 that were critical for interacting with cpTatC. Oxidative cross-linking allowed us to map interactions of the transmembrane domain and amphipathic helix region of Hcf106. In addition, we showed that in vitro expressed, integrated Hcf106 can interact with the precursor signal peptide domain and imported cpTatC, strongly suggesting that a subpopulation of the integrated Hcf106 is participating in competent cpTat complexes.

KEYWORDS
cross-linking, oligomerization, receptor complex, truncation, twin arginine transport

1 | INTRODUCTION

The twin arginine transport system (Tat) is one of two protein transport pathways that deliver proteins from the stroma of the chloroplast to the lumen of the plant thylakoid (cpTat); a homologous Tat pathway is also found in a wide range of bacteria (Berks, 2015; Celedon & Cline, 2013; Cline, 2015). The Tat system is distinguished from other protein transport pathways, such as the well-characterized secretory (Sec) system, by transporting fully folded precursor proteins using only the proton-motive force (PMF) for energy (Braun, Davis, & Theg, 2007; Cline & Mori, 2001; Gérard &

Cline, 2007). Proteins transported by the Tat pathway are usually cytosolically synthesized as higher molecular weight precursors containing cleavable N-terminal signal peptides with an obligate twin arginine motif (RR) as is reflected in the name of the system. Tat systems can transport precursor proteins of different sizes ranging from 2 kDa to over 100 kDa, as well as substrates that form oligomers (Celedon & Cline, 2012; Ma & Cline, 2010). A recent stoichiometry study showed that the minimal, functional Tat translocase consists of cpTatC, Hcf106, and Tha4, and is transport active when sufficient Tha4 is present (Celedon & Cline, 2012). The same study showed that in the thylakoid, functional complexes may exist as higher order groups such that up to eight precursors could bind; however, mechanistic detail of how the Tat receptor complex organizes to accommodate the transport of multiple precursor proteins in concert is an important question that needs to be addressed.

In thylakoid (as well as in *Escherichia coli*), three membrane-bound protein components, Tha4 (TatA), Hcf106 (TatB), and cpTatC (TatC), are responsible for the twin arginine-dependent translocation of precursor cargo (Celedon & Cline, 2013; Cline & Mori, 2001). Current models of the Tat system suggest a cyclical mechanism in which the receptor complex is comprised of cpTatC and Hcf106 tightly bound to each other with a loosely associated Tha4 that serves as the initial site of precursor recognition and binding, followed by assembly of additional Tha4 homo-oligomers, which are proposed to form the translocation pore (Aldridge, Storm, Cline, & Dabney-Smith, 2012; Bolhuis, Mathers, Thomas, Barrett, & Robinson, 2001; Cline & Mori, 2001; Dabney-Smith & Cline, 2009; Dabney-Smith, Mori, & Cline, 2006; Pal, Fite, & Dabney-Smith, 2013). Hcf106 is structurally similar to Tha4 in that both contain an amino terminal transmembrane domain (TMD), followed by a hinge region, an amphipathic α-helix (APH), and a loosely structured carboxyl terminus (C-tail). Recently, the structures of Tha4 and Hcf106 bacterial homologs, TatA and TatB, have been solved which agree with previous predictions (Hu, Zhao, Li, Xia, & Jin, 2010; Zhang, Wang, Hu, & Jin, 2014) and even more recently the structure to Tha4 from *Arabidopsis thaliana* (Pettersson et al., 2018). Despite the sequence similarity with Tha4, the two proteins are not interchangeable and thus appear to have distinct functions (Dabney-Smith, Mori, & Cline, 2003).

Although the cpTat receptor complex has been well studied for its essential role in recognizing precursor proteins, few studies in plants have addressed the role of Hcf106 in this process, particularly, how each individual Hcf106 is organized in the multimeric receptor complex as compared to the bacterial homolog TatB (Alami et al., 2003; Behrendt & Bruser, 2014; Bolhuis et al., 2001; Holzapfel et al., 2007; Lee et al., 2006; Rollauer et al., 2012). The aim of the present work was to establish a method allowing the exploration of Hcf106 organization using systematic cysteine substitution. The use of cysteine substitution is a valuable technique that allows the rapid and efficient capture of protein–protein interactions through the oxidation of proximal sulfhydryl groups (–SH) on cysteines to form a disulfide bond. This method has been used previously to determine protein–protein interactions in both the bacterial and plant Tat systems (Alcock et al., 2016; Aldridge, Ma, Gerard, & Cline, 2014; Aldridge et al., 2012; Dabney-Smith et al., 2006; Lee et al., 2006; Pal et al., 2013). As proof of principle of correct integration of in vitro expressed Hcf106, we have used this method to map Hcf106-Hcf106 interactions in thylakoid membranes.

Previous studies showed that in vitro translated Hcf106 can integrate into thylakoid in a manner presumably similar to endogenous Hcf106 and exists in a receptor complex with cpTatC (Gérard & Cline, 2006), which was capable of binding precursor proteins (Gérard & Cline, 2006; Mori & Cline, 2002). What remains unclear from the results of these studies is whether the integrated Hcf106 were simply members of the complex or directly participated in binding either precursor and/or cpTatC. Here, we have used cysteine scanning and disulfide bond formation to systematically map Hcf106 interactions through the TMD to the APH regions, which are known to be of great importance to the organization of the receptor complex. We observed that single cysteine-substituted Hcf106 promoters integrate into isolated thylakoid and that most variants are resistant to alkaline extraction. In addition, they localize in a 700-kDa complex by blue-native PAGE, suggesting that they are fully integrated into the membrane. Interaction sites of Hcf106-Hcf106 were obtained using copper (II)-1, 10-phenanthroline (CuP)-induced cross-linking which provided vital clues for the organization of Hcf106. Using double cysteine substitution in Hcf106, we could detect an Hcf106 oligomer as large as an octamer but could not distinguish if these oligomers were in the receptor complex or part of a separate pool of Hcf106. However, integrated Hcf106 was capable of interacting with transport competent precursor in a specific manner and with exogenous, imported cpTatC. From these data, we conclude that integrated Hcf106 associates with the cpTat translocase.

2 | MATERIALS AND METHODS

2.1 | Preparation of chloroplasts and thylakoid membranes

Intact chloroplasts were prepared from 10- to 12-day-old pea seedlings (*Pisum sativum* L. cv. Laxton's Progress 9 or Little Marvel) as described (Cline, Henry, Li, & Yuan, 1993). Intact, isolated chloroplasts were suspended to 1 mg/ml chlorophyll in import buffer (IB, 50 mM HEPES-KOH, pH 8.0, 330 mM sorbitol) and kept on ice until used. Isolated thylakoid were obtained by osmotic lysis of intact chloroplasts. Briefly, intact chloroplast suspensions were pelleted for 5 min at 1000 *g*, supernatant removed, and suspended at 1 mg/ml chlorophyll in lysis buffer (50 mM HEPES-KOH, pH 8.0, 10 mM $MgCl_2$) with incubation on ice for 5 min. Following lysis, an equal volume of IB, 10 mM $MgCl_2$ was added to the lysate, the thylakoid were then pelleted at 3200 *g* for 8 min, and suspended at 1 mg/ml chlorophyll in IB, 10 mM $MgCl_2$ (Aldridge et al., 2012). For single Cys interaction studies, thylakoid was suspended in 50 mM *N*-ethylmaleimide (NEM) in IB, 10 mM $MgCl_2$ and incubated on ice for 10 min to prevent nonspecific cross-linking from endogenous free sulfhydral groups. NEM-treated thylakoid was subsequently pelleted and washed with three volumes of IB, 10 mM $MgCl_2$ before use.

2.2 | Generation of cysteine-substituted mature Hcf106 and C-tail truncation $Hcf106_{1–107}$

Hcf106 with cysteine substitutions ($Hcf106X_nC$; where amino acid X at position *n* was replaced by cysteine) were generated by Quik-Change mutagenesis (Agilent Technologies) according to manufacturer's instructions. The template used for mutagenesis was the coding sequence for mature Hcf106 (lacking the targeting peptide) in the plasmid pGEM-4Z. The coding sequence for Hcf106 begins with MASLFGVGAPEA.... Cloned constructs were verified by DNA sequencing on both strands at the Center for Bioinformatics and Functional Genomics at Miami University. For the C-tail truncation of $Hcf106_{1–107}$, internal stop codons were inserted via primer-based mutagenesis to generate the C-deletion of 69 amino acids. The size of the truncated Hcf106 is about 17 kD by SDS-PAGE analysis.

2.3 | Preparation of radiolabeled recombinant Hcf106 proteins

Radiolabeled Hcf106 variants were prepared by in vitro translation in a wheat germ extract from capped mRNA in the presence of [^{3}H] leucine (Cline, 1986). The translation products were diluted with an equal volume of 60 mM leucine in 2× import buffer before use. $Hcf106_{1–107}$ translation product was diluted further with 1× import buffer, 30 mM leucine equal to 1:3, 1:6, 1:12, or 1:60 of the initial translation before use.

2.4 | In vitro integration assay

In vitro translated [^{3}H]$Hcf106X_nC$ was integrated into NEM-pretreated thylakoid (100 μg chlorophyll equivalent) for 25 min at 25°C. For $Hcf106_{1–107}$, in vitro translated protein was directly integrated into isolated thylakoid. Reactions were terminated by transfer to 0°C and thylakoid were recovered by centrifugation at 3200 *g* for 8 min. Recovered thylakoid were washed once with two volumes of IB, 10 mM $MgCl_2$.

2.5 | Alkaline extraction assay

Thylakoid with Hcf106 integrated were suspended to 1 ml with 0.2 M Na_2CO_3 or 0.1 M NaOH and incubated for 60 min on ice. Thylakoid were then recovered by centrifugation at 100 000 *g* for 15 min. Pellets were suspended in 30 μl of 20 mM EDTA 1 × IB and mixed with the same volume of 2× reducing sample solubilizing buffer (2× SSB (red); 100 mM Tris-HCl (pH 6.8), 0.2 M DTT, 5% SDS, and 30% glycerol). Samples were analyzed by SDS-PAGE and fluorography.

2.6 | Blue-native gel electrophoresis

Hcf106-integrated thylakoid (~1 mg/ml chlorophyll) were dissolved in 2% digitonin with end-over-end mixing for 1 hr, 4°C and centrifuged at 100 000 *g* for 30 min. Supernatant was mixed with 0.1 vol 10× BN-PAGE sample buffer (5% Serva G, 30% sucrose in 100 mM Bis-Tris-HCl, 500 mM 6-amino-caproic acid, pH 7.0) as described by Cline & Mori (2001). Prior to mixing with BN-PAGE sample buffer, samples were normalized to total protein content by the Pierce BCA Protein Assay kit according to manufacturer's instructions (Thermo-Fisher Scientific). Gels were analyzed by fluorography or subjected to immunoblotting as described (Cline & Mori, 2001). Molecular markers used for blue native gels were dimeric and monomeric ferritin (880 and 440 kDa, respectively) and bovine serum albumin (BSA) dimer (132 kDa).

2.7 | Oxidative cross-linking by disulfide bond formation

Hcf106-integrated were used for cross-linking reactions. A quantity of 1 mM copper (II)-1,10-phenanthroline (CuP from a 150 mM stock) was added as an oxidant to catalyze disulfide formation between proximal cysteine residues. The CuP stock solution (150 mM) contained 150 mM $CuSO_4$ and 500 mM 1, 10-phenanthroline (Dabney-Smith et al., 2006). Cross-linking reactions were carried out for 5 min before stopping with 50 mM ethylmaleimide (NEM, from a 1 M stock in ethanol) and diluted twofold with 1× IB, 10 mM $MgCl_2$. Thylakoids were recovered by centrifugation at 3200 *g* for 8 min, the supernatants removed, and the pellet was suspended in 1× IB, 5 mM EDTA, 10 mM NEM and subjected to centrifugation and supernatant removal. Thylakoid pellets were suspended to ~1 mg/ml chlorophyll and divided into two, centrifuged at 3200 *g* for 8 min, and the supernatants removed. The first half was suspended in 2× nonreducing sample solubilizing buffer (2× SSB (nr); 100 mM Tris-HCl (pH 6.8), 8 M urea, 5% SDS, and 30% glycerol) and the second half was suspended in 2× SSB (red). Samples then were analyzed by SDS-PAGE and fluorography. The intensities of the bands were quantified with ImageJ software (Schneider, Rasband, & Eliceiri, 2012).

2.8 | Disulfide cross-linking of imported cpTatC and integrated Hcf106

The precursor to cpTatC, pre-cpTatCaaaV270C (Kenneth Cline, University of Florida), was used as a source of exogenously integrated cpTatC according to published methods (Aldridge et al., 2014). Three native cysteines were substituted with alanine; hence, the "aaa" designation, and a single cysteine substitution was added at position V270. Radiolabeled in vitro translated pre-cpTatCaaaV270C was incubated with chloroplasts (0.33 mg/ml chlorophyll) and 5 mM Mg-ATP in IB with 100 $\mu E/m^2/s$ of white light in a 25°C water bath for 40 min. After import, intact chloroplasts were treated with thermolysin for 40 min at 4°C and isolated by centrifugation through a 35% Percoll (GE Healthcare) cushion in IB, 5 mM EDTA and washed with IB (Cline et al., 1993). Thylakoid membranes were prepared from isolated chloroplasts by osmotic lysis and centrifugation as described above and suspended in IB, 5 mM $MgCl_2$ to ~1 mg/ml chlorophyll. In vitro translated, unlabeled Hcf106 was integrated into thylakoid as described above. Thylakoids were centrifuged at 3200 *g*, 8 min to remove unincorporated Hcf106 and washed with IB, 10 mM $MgCl_2$. Samples were subjected to cross-linking as describe above.

2.9 | Disulfide cross-linking of integrated Hcf106 and precursor

Radiolabeled, in vitro translated precursor tOE17-25C V-20F, containing an inserted cysteine 25 residues upstream from the signal peptide cleavage site, was incubated with thylakoids which had been preintegrated with in vitro translated, unlabeled Hcf106 in IB, 10 mM $MgCl_2$ at 100 μE/m²/s of white light in a 25°C water bath for 5 min. Samples were subjected to cross-linking as described above.

3 | RESULTS

3.1 | Single cysteine variants of Hcf106 integrate into thylakoid membranes and are resistant to alkaline extraction

Earlier work demonstrated that in vitro translated wild-type Hcf106, which lacks native cysteines, was able to spontaneously integrate into isolated thylakoid and was resistant to alkaline extraction by either 0.2 M carbonate buffer (pH 9.5) or 0.1 M NaOH (pH 11.5) (Fincher, Dabney-Smith, & Cline, 2003). The NaOH treatment extracts proteins that are peripherally associated with membrane or are partially embedded into the membrane via a single transmembrane domain, whereas carbonate extraction is capable of stripping peripherally associated proteins and but not embedded proteins (Rolland et al., 2006).

We generated a series of single-cysteine substitutions from the predicted transmembrane domain to the end of the predicted amphipathic helix in Hcf106 (Figure 1) and investigated whether these cysteine substitutions, especially in the TMH, affected the integration of Hcf106 into the thylakoid membrane as compared to wild type.

Wild-type Hcf106 is largely resistant to alkaline extraction (Fincher et al., 2003). Likewise, each of the Cys-substituted variants of Hcf106 tested was able to integrate into isolated thylakoid and was resistant to 0.2 M carbonate treatment. In addition, most of the variants were also resistant to 0.1 M NaOH extraction (Supporting information Figure S1); the exceptions are I15C, V17C, V18C, and L21C residues which are predicted to be in the hydrophobic core of the membrane. Thus, we concluded that the resistance of most single cysteine substituted Hcf106 to alkaline extraction indicates that the integration of recombinant Hcf106 into isolated thylakoids was not negatively affected.

3.2 | Hcf106 Cys variants can be detected in the 700-kDa receptor complex

Blue-native polyacrylamide gel electrophoresis (BN-PAGE) has shown that the cpTatC/Hcf106 receptor complex migrates as a

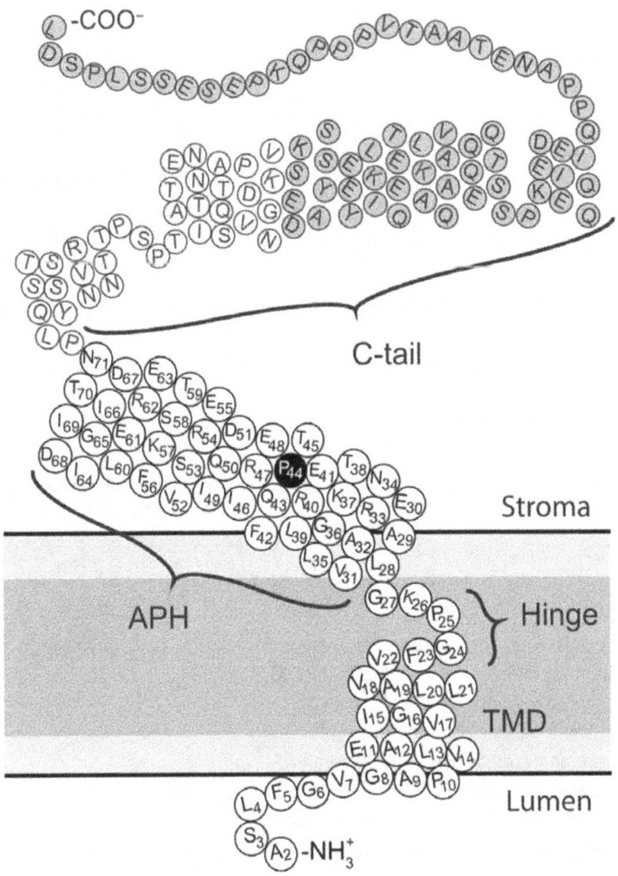

FIGURE 1 Topologic model of Hcf106 based on experimental and predictive structural features. Primary residue sequence of Hcf106 from *Pisum sativum*. Amino acids in regions under investigation in this study are numbered (2–71). Dark gray indicates the hydrophobic core (i.e., acyl chains) of the membrane, while the lighter gray indicates the lipid head groups. Residue P44 (black fill) indicates the arbitrary break point for the two helical wheel projections for the APH of Hcf106 in Figure 5

band at ~700 kDa after solubilization by the detergent digitonin (Cline & Mori, 2001; Fincher et al., 2003). In addition, previous studies showed that recombinant, wild-type, thylakoid-integrated Hcf106 also migrated at 700 kDa, as well as a separate pool of smaller complexes of ~400 to ~200 kDa depending upon the detergent to membrane ratio (Fincher et al., 2003; Mori, Summer, & Cline, 2001). Most of the single cysteine variants of Hcf106 were integrated into isolated thylakoid; therefore, we examined whether the integrated Cys variants would form oligomers when solubilized in detergent as shown by BN-PAGE (Figure 2). We reasoned that if the Cys variants of Hcf106 could interact with cpTatC, then we would expect to find the Cys variants incorporated into a ~700-kDa complex when digitonin-solubilized thylakoid was analyzed by BN-PAGE.

We integrated the Hcf106 variants into isolated thylakoid and solubilized the membranes with digitonin followed with analysis by BN-PAGE. The ratio of detergent to thylakoid chlorophyll content was critical because previous studies demonstrated that an increase in the ratio of detergent to thylakoid resulted in the

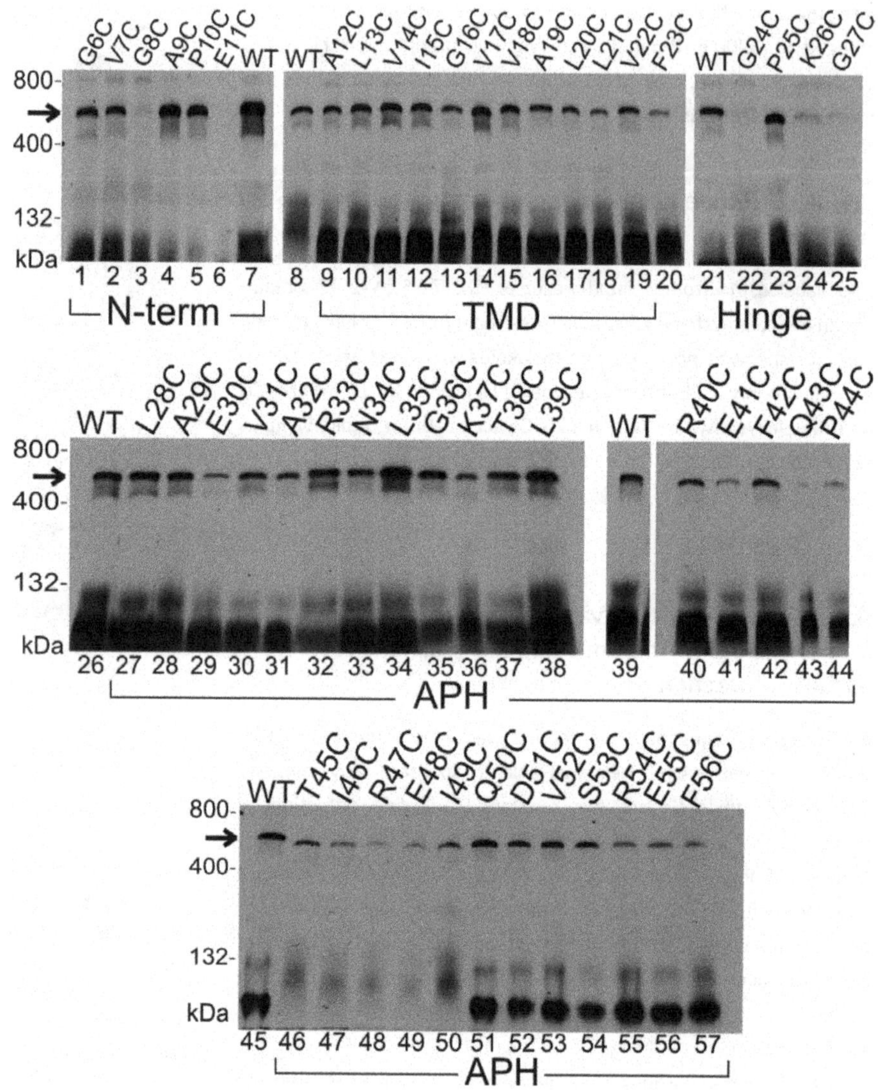

FIGURE 2 Blue-native gel analysis of the integration of recombinant Hcf106 cysteine variants. [^{3}H]Hcf106 single cysteine variant containing complexes were analyzed by BN-PAGE and fluorography. Sample were normalized to total protein (5 μg total protein loaded) (see Section 2). Single Cys substitutions at multiple positions in Hcf106 are indicated across the top of the panels. The 700-kDa complex is indicated by an arrow. Wild-type Hcf106, lacking any cysteine, was used as a control. Molecular mass markers are ferritin (880 and 440 kDa) and BSA (132 kDa). Gels are representative of at least three experiments

persistence of the 700-kDa receptor complex of cpTatC/Hcf106 complex, while the Hcf106 homo-oligomeric complexes between 400 kDa and 200 kDa were disrupted (Cline & Mori, 2001; Fincher et al., 2003). We were most interested in the presence or incorporation of Hcf106 in the 700-kDa receptor complex, so membrane solubilization was done at a ratio of 2% digitonin to 1 mg/ml chlorophyll to minimize the formation of the smaller homo-oligomeric complexes (Fincher et al., 2003). Bands seen at the bottom of the gel indicate the presence of smaller oligomers of Hcf106 (e.g., dimer to tetramer of ~60–120 kDa), but nothing in the 200–400 kDa range. Of the cysteines placed close to the N-terminus, such as G6C, V7C, G8C, A9C, P10C, and E11C, only G8C (Figure 2, lane 3) and E11C (Figure 2, lane 6) were unable to migrate in a 700-kDa complex but did integrate successfully into thylakoid (Supporting information Figure S1), suggesting that key contacts between Hcf106 and cpTatC were disrupted in those variants. Hcf106 variants with a Cys substitution in the transmembrane region such as A12C, L13C, V14C, I15C, G16C, V17C, V18C, A19C, L20C, L21C, and V22C (Figure 2, lanes 9–19), did migrate as a 700-kDa complex just like wild-type Hcf106 (Figure 2, lanes 7–8, 21, 26, 39, 45) suggesting incorporation into the receptor complex. Most of the variants in the hinge region, that is, F23C, G24C, K26C, or G27C (Figure 2, lanes 20, 22, 24, and 25) failed to incorporate into the receptor complex but were able to integrate into thylakoid (Supporting information Figure S1). P25C was able to incorporate into the receptor complex (Figure 2, lane 23). Cys variants in the APH region of Hcf106 showed a similar pattern to wild type when analyzed by BN-PAGE in that all single cysteine substitutions in this region did not abolish incorporation into the 700-kDa complex (Figure 2, lanes 26–57). However, certain cysteine substitutions, for example, E30C, A32C, K37C, E41C, Q43C, and P44C (Figure 2, lanes 29, 31, 36, 41, 43, and 44), consistently showed a lower intensity at the 700-kDa band, suggesting that cysteines in this region of the APH may negatively affect Hcf106 interaction with cpTatC (i.e., incorporation into the 700-kDa complex), but did not affect the integration and membrane stability of the variant (Supporting information Figure S1).

3.3 | C-terminal of Hcf106 is dispensable for cpTatC-Hcf106 receptor complex formation

Hcf106 contains a loosely structured C-tail that was shown to not be required for receptor complex formation in the *E. coli* homolog, TatB (Maldonado et al., 2011). If the truncated Hcf106 could interact with cpTatC, we reasoned that truncated protein could be used to demonstrate that the Integrated Hcf106 is, in fact, incorporating into receptor complexes with endogenous cpTatC (Figure 3). We incubated increasing amounts of $[^3H]Hcf106_{1-107}$, lacking 69 amino acids from the C terminus, with isolated thylakoid and analyzed the membranes by BN-PAGE. As the concentration of in vitro translated $[^3H]Hcf106_{1-107}$ increased, two lower bands of ~600 kDa (Figure 3, lanes 2–4) and 500 kDa (Figure 3, lanes 3–4) appeared, suggesting that in vitro integrated $[^3H]$Hcf106 was competing with the endogenous Hcf106 for binding to endogenous cpTatC. The smaller complexes also contain cpTatC (Figure 3, lanes 9–12), indicating the C-tail does not play a critical role in receptor complex formation. Immunodetection of Hcf106 using the same thylakoid samples demonstrated that full length Hcf106 also migrated into smaller complexes due to the presence of the truncated variant. With these data, the insensitivity to alkaline extraction, and the incorporation into a 700-kDa complex, we conclude that exogenously added Hcf106 is properly inserted into the thylakoid membrane, allowing us to use these Cys-substituted variants to probe the organization of Hcf106 in the receptor complex.

3.4 | The transmembrane domain and amphipathic helix regions of Hcf106 form self-contacts

To characterize the organization of Hcf106-containing complexes, we looked at the organization of Hcf106 by studying interactions between neighboring Hcf106 proteins. We reasoned that interactions between Hcf106 proteins would indicate the organization of Hcf106 in the receptor complex by identifying sites specific for self-interactions as well as provide insight into the organization of the separate pool of Hcf106. We took a cysteine scanning approach, which allowed us to map interactions between neighboring single cysteine substituted Hcf106 proteins or other cpTat components by formation of disulfide bonds between cysteines within ~5 Å of each. Hcf106 proteins containing single cysteine substitutions in the TMD or APH were integrated into isolated thylakoid. In the presence of an oxidant such as copper (II)-1,10-phenanthroline (CuP), free sulfhydryl groups on cysteines in close proximity will form stable disulfide bonds causing a mobility shift from ~28 to ~56 kDa of the cross-linked proteins when analyzed by SDS-PAGE.

Residues close to the N-terminus of Hcf106 were in close proximity to the same residue of a neighboring Hcf106 (Figure 4, lanes 1–5) showing a significant amount of dimer formation, demonstrating that these amino acids are sufficiently close to form a disulfide bond or that this region is very flexible. On the other hand, when the Cys was placed in the TMD, that is, E11C-L21C (Figure 4, lanes 6–16), dimer formation demonstrated a position-specific interaction, indicating a regular face of interaction in this portion of the transmembrane domain. As the cysteine substitution moved out of the TMD through the hinge and into the APH, that is, V22C, F23C, and K26C (Figure 4, lanes 17, 18, 21) formed a dimer, whereas G24C, P25C, and G27C (Figure 4, lanes 19, 20, 22) did not.

The amphipathic helix of Hcf106 showed two different types of interaction. The N-terminal proximal portion of the APH, that is, L28C-P44 (Figure 4, lanes 23–39), showed no interaction overall, with the exceptions of L28C, L35C, E41C, and F42C (Figure 4, lanes 23, 30, 36, 37), compared to the C-terminal proximal portion of the APH, T45C-G65C (Figure 4, lanes 40–60), which demonstrated a stronger and position-specific cross-linking. For example, dimer was detectable for all Cys substitutions in this segment, but dimers formed with cysteines at I49C or Q50C were more intense (Figure 4, compare lanes 44–45 with lanes 36–43) based on equal chlorophyll

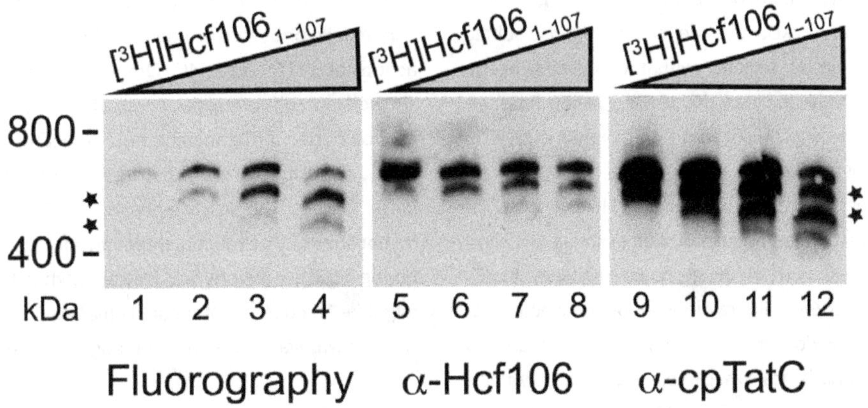

FIGURE 3 Truncated $Hcf106_{1-107}$ assembles into a complex with endogenous cpTatC. An increasing concentration of $Hcf106_{1-107}$ was integrated into thylakoid and subjected to digitonin solubilization, BN-PAGE, and analysis by fluorography (left panel) as in Figure 2. To detect whether this truncated form can generate a different size of the receptor complex, samples were also subjected to immunoblotting with anti-Hcf106 (middle panel), anti-cpTatC (right panel) antibodies. The amount of $Hcf106_{1-107}$ added to thylakoid corresponds to dilutions of the in vitro translation reaction (i.e., 1:60, 1:12, 1:6, and 1:3). Gels are representative of at least three separate experiments. As $Hcf106_{1-107}$ increases, there is a concomitant increase in a smaller complex also containing cpTatC (stars).

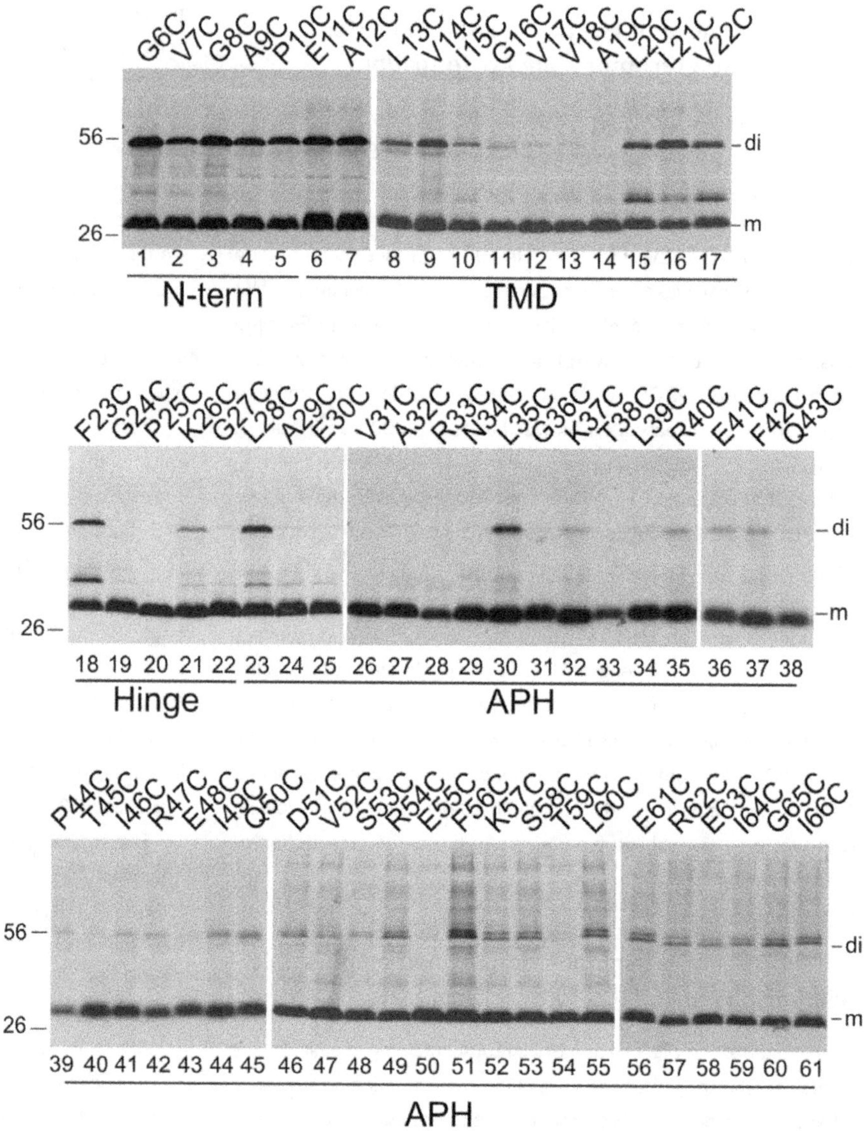

FIGURE 4 Hcf106 forms a dimer when cysteine substitutions are in the N-terminus, central TMD, and APH regions. In vitro translated Hcf106 variants were integrated into NEM pretreated thylakoid membranes and subjected to oxidizing conditions (1 mM CuP) as described (Section 2). Samples were resolved by SDS-PAGE and protein bands were visualized by fluorography. The amino acid position of each cysteine substitution is shown above each panel. Hcf106 monomer (m) and dimer (di) forms are indicated at the right of the panels. Gels are representative of at least three separate experiments

loading. The relative proportion of dimer to total protein was quantified by densitometric analysis (Supporting information Figure S3). The presence of a band at 56 kDa was dependent upon disulfide formation because treatment of samples with the reducing agent, dithiothreitol (DTT), effectively depleted the dimer species. Interestingly, L21C and V22C were not reduced to the monomeric form (Supporting information Figure S3).

The results of the specific interactions on the Hcf106 TMD or APH are plotted on helical wheels (Figure 5). For example, a helical wheel projection of the TMD with the hinge region (residues A12C-G27C) emphasizes that Hcf106 self-interactions occur along no particular face (Figure 5a). The APH domain (~40 amino acids) is too large to be clearly evaluated with one helical wheel, so we generated wheels for the N-terminal half (i.e., L28-P44; (Figure 5b) and the C-terminal half (T45-G65; Figure 5c). The proline at position 44 was arbitrarily determined as the halfway point because it would serve to break the helix. The helical wheel of the N-terminal portion of the APH shows a preferred face for Hcf106 self-interaction because L28, L35, and F42 fall along the hydrophobic face of the helix (Figure 5b), although we do see interactions at E41C and P44C, which do not fall on the same face of the predicted helix. However, we envision the APH could be very mobile or dynamic in the membrane, which may explain the interactions at E41C and P44C. Other residues, for example, A29C-T38C, showed no self-interactions indicating that these faces of the helix may be buried in the membrane, interacting with other cpTat components, or positioned such that they are not near each other on neighboring helices. The helical wheel of the C-terminal portion of the APH demonstrated that roughly three of four faces can interact with neighboring Hcf106 (Figure 5c), indicating that this portion of the helix might be flexible or that it could emanate from the membrane similarly to what was seen with Tha4 (Aldridge et al., 2012).

3.5 | Double-cysteine substitutions suggest a high-ordered Hcf106 complex

Single cysteine substitutions can only detect dimers. To investigate the formation of higher ordered oligomers, we constructed double cysteine substituted Hcf106 to detect the formation of higher

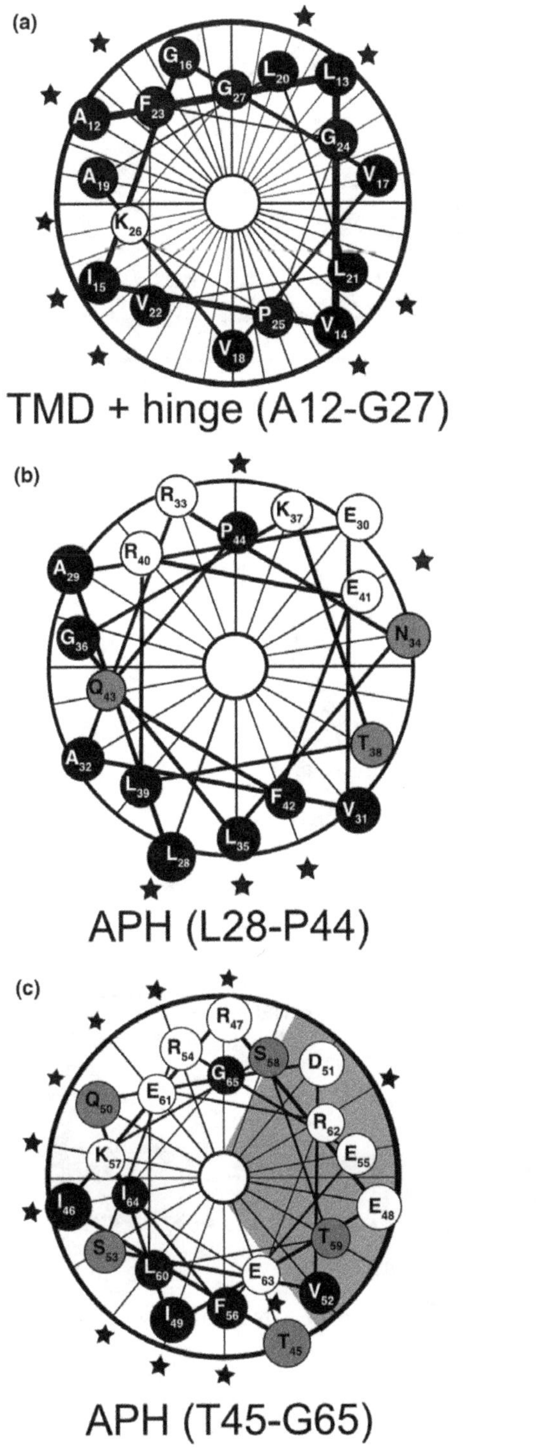

FIGURE 5 Helical wheel projection of the TMD and APH regions of Hcf106 reveals periodic interactions. Helical wheel projections were generated using the Protean module of DNAStar (Lasergene, Madison WI). (a) The TMD and Hinge of Hcf106, residues A12C-G27C; (b) the N-terminal proximal portion of the APH of Hcf106, residues L28C-P44C; and (c) the C-terminal proximal portion of the APH of Hcf106, residues T45C-G65C. Hydrophobic (black), polar uncharged (gray), and charged (white) amino acid residues are indicated. Stars indicate the presence of a dimer that is >10% of total (see Supporting information Figure S2)

ordered protomers. We substituted two cysteines in the transmembrane helix of Hcf106, for example, L13CL21C and V14CL20C, and the resulting proteins were integrated into thylakoid followed by oxidative cross-linking. Initially, when compared to the single cysteine variants L21C, the double Cys-substituted proteins L13CL21C and V14CL20C showed roughly half the integration into thylakoid as L21C (Figure 6a, compare lanes 7, 10 to 9, 12), while double cysteine variants in the APH, e.g., I49CE61C and R54CL60C (Figure 6a, compare lanes 8, 11 to lanes 9, 12), were comparable to the amount of L21C integrated. Decreased integration of the TMD double Cys variants was likely due to the introduction of two cysteines into the helix, decreasing the hydrophobicity of the TMD. Unexpectedly, the cross-linking of these four cysteine variants did not show obvious Hcf106 multimers (Figure 6a, lanes 1–2, 4–5). Furthermore, when the V14CL20C variant was compared to its single cysteine parent, V14C, the double Cys variant also showed about 30% less integration into thylakoids (Figure 6b, lanes 1–2). However, when one cysteine was placed in the N terminus and the other in the TMD, for example, G6CV14C, we observed Hcf106 multimers as high as octamers when analyzed by 10–20% acrylamide gradient gel PAGE (Figure 6, lanes 3–4).

To further investigate if the cross-linked Hcf106 multimers are part of the 700-kDa complex, we subjected the cross-linked G6CV14C double Cys variant to BN-PAGE (Figure 6c). Different cross-linking conditions were used including pretreatment of the translation product with NEM to block the free cysteine (Figure 6, lane 1), the presence or absence of CuP (Figure 6, lanes 2 and 3), cross-linking without quenching (Figure 6, lane 4), and prolonged cross-linking (30 min) (Figure 6, lane 5). Samples were also subjected to nonreducing SDS-PAGE to monitor the status of cross-linking (Figure 6, lane 7–11). On BN-PAGE, we observed a significant amount of 700-kDa complex when the double cysteine variant was cross-linked (Figure 6, lane 2–5). Additional bands were also observed (above 800 kDa and <400 kDa, indicated with arrows), especially when cross-linking in the absence of NEM quenching (Figure 6, lane 4). The bands above the 800 kDa could be nonspecific interactions between the 700-kDa complex and other species. It could also be due to the different mobility of the complex caused by cross-linking induced conformational change. We interpret the <400 kDa band to be indicative of the G6CV14C variant in the separate Hcf106 pool, which may also form higher oligomers. Overall, these results indicate Hcf106 has a strong tendency to form oligomers and that Hcf106 self-oligomerization might be present in both the receptor complex and the free pool of Hcf106.

3.6 | Integrated Hcf106 forms contacts with both imported cpTatC and precursor proteins

To validate whether integrated Hcf106 can participate in the transport process directly, we looked for contacts between integrated

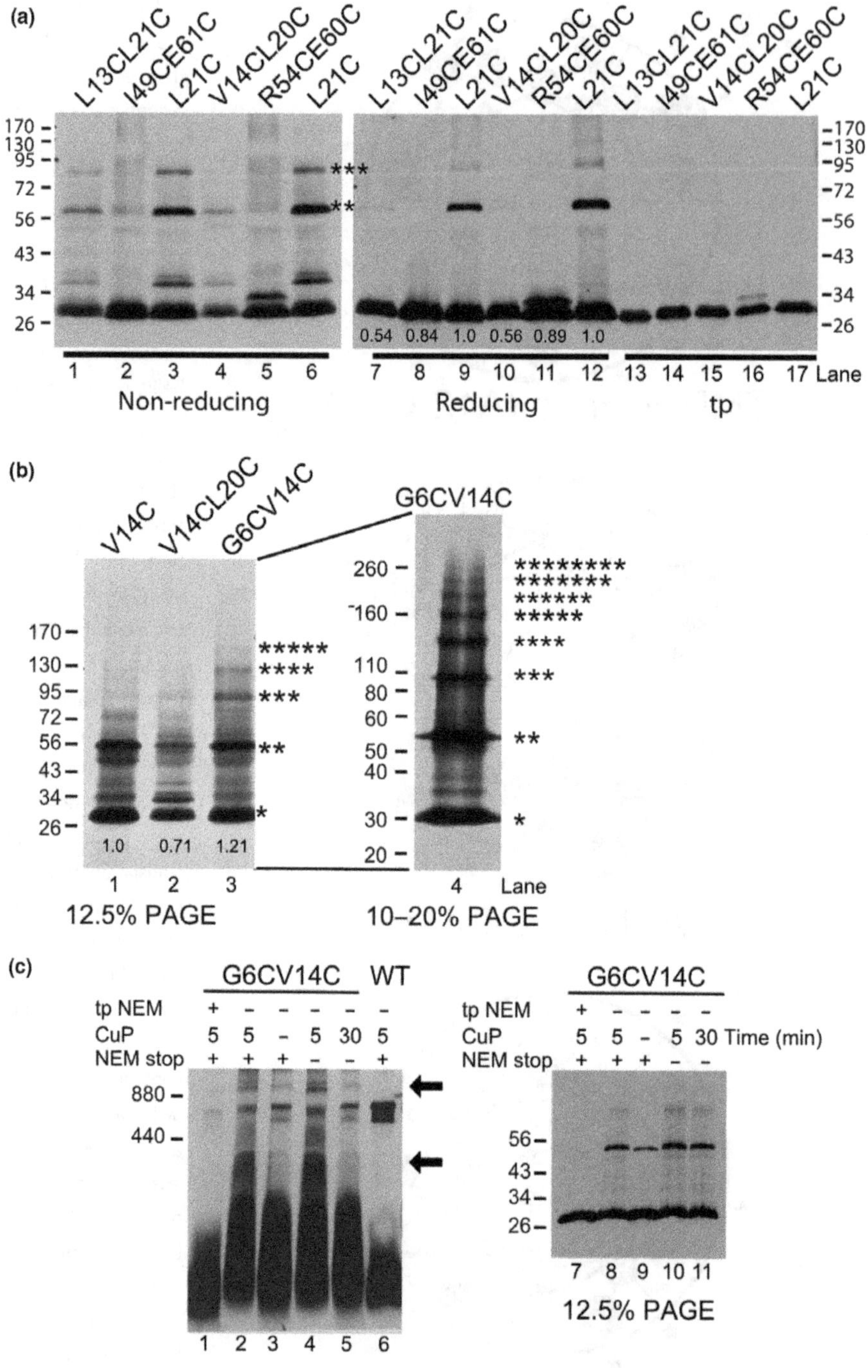

FIGURE 6 Double cysteine substitutions demonstrate a higher order Hcf106 complex. (a) Lanes 1–6 show the cross-linking of double cysteine variants L13CL21C, I49CE61C, V14CL20C, and R54CL60C, and the single Cys variant L21C under nonreducing conditions. Lanes 7–12 show the same samples under reducing conditions. The amount of protein integrated relative to the parent single Cys variant, L21C, is shown below lanes 7–12. Translation products diluted 1:40 are shown in lanes 13–17. Molecular weight markers are indicated on the right and left. Asterisks indicate the number of Hcf106 monomers in the oligomer as determined by size. (b) Cross-linking with the G6CV14C double cysteine variant was compared to the single Cys mutant V14C and double mutant V14CL20C (lanes 1–3). G6CV14C showed higher ordered Hcf106 complexes up to octamers when analyzed by 10–20% SDS-PAGE (lane 4). The amount of protein integrated relative to the parent single Cys variants, V14C, is shown below the lanes. (c) Cross-linked G6CV14C double Cys variant has reduced capability to incorporate into the receptor complex. Lane 2 is the standard cross-linking condition with 5 min incubation of CuP and the reaction was quenched with 50 mM NEM. Lane 1: translation product was pretreated with NEM and subjected to standard cross-linking. Lanes 3–6: cross-linking under variable time or with/without NEM quenching. Lane 6: WT Hcf106 under standard cross-linking conditions. The same samples from lanes 1–5 were also subjected to non-denaturing SDS-PAGE (lane 7–11). Arrows indicate additional bands of cross-linked G6CV14C above and below the 800 kD complex. Gels in all panels represent at least three separate experiments

Hcf106 and imported cpTatC or cpTat pathway precursor proteins. Previous studies found that endogenous Hcf106 in a functional Tat receptor complex was cross-linked to the signal peptide of precursor proteins, containing the photoactivatable TMD(Phe) cross-linker (Gérard & Cline, 2006). Additionally, integrated, recombinant Hcf106 was shown to localize to a ~700-kDa complex that was capable of binding Tat precursor proteins (Cline & Mori, 2001).

Earlier work demonstrated that replacing the original cpTatC transit peptide with the transit peptide of the precursor to the small subunit of RuBisCO resulted in higher efficiency of pre-cpTatC import, mature cpTatC localization to thylakoid that could be detected in direct contact with both the RR proximal region on a Tat pathway precursor signal peptide and Tha4 (Aldridge et al., 2012, 2014; Ma & Cline, 2013). To confirm Cys-substituted Hcf106 participation in the receptor complex, we selected five Hcf106 Cys variants from the transmembrane domain and looked to see if they interact with imported cpTatCV270C. We initially chose transmembrane locations for cysteine substitution because recently the cpTatC bacterial homolog, TatC, was found to form cross-links with the TatB (Hcf106) transmembrane domain via TatC TM5 (Kneuper et al., 2012; Rollauer et al., 2012). Imported cpTatC270C was previously used to map Tha4 binding (Ma & Cline, 2013). We observed [^{3}H]cpTatC at an apparent molecular weight of ~28 kDa, which matches the apparent molecular weight for endogenous cpTatC. Samples with integrated Hcf106, containing Cys substitutions in the hydrophobic core of the Hcf106 transmembrane domain, for example, A12C, L13C, and V14C, formed interactions with cpTatCV270C, showing a ~56-kDa adduct that is not found in the control lane (Figure 7a). For the control lane, no Hcf106 was integrated into the thylakoid. In contrast, L20C, which is closer to the stromal side of the membrane, showed a similar result as the control lane. The data here demonstrate that integrated Hcf106 is in close contact with cpTatC.

Alternatively, to confirm that integrated, recombinant, Cys-substituted Hcf106 was able to bind precursor, we integrated wild-type Hcf106 or various Cys-substituted Hcf106 into thylakoid and used these membranes in precursor binding assays. The precursor, tOE17-25C/V-20F, was shown to bind tightly to the first cpTatC stromal loop (Gérard & Cline, 2007) and was used previously cpTatC cross-linking assays (Ma & Cline, 2010). It is a modified precursor of the 17-kDa subunit of the oxygen-evolving complex, containing a truncated signal peptide (tOE17), a cysteine inserted on the N-terminal side of the twin arginine motif, 25 amino acids from the signal peptide cleavage site (−25C), and a phenylalanine substituted for the valine at position −20 from the signal peptide cleavage site (V-20F) [Figure 7, see reference (Ma & Cline, 2010). We subjected tOE17-25C/V-20F to cross-linking with the Hcf106 C-terminal APH region as it interacts with cpTatC stromal loop 1 (Figure 7c). This loop of cpTatC was also identified as interacting with tOE17-25C/V-20F. tOE17-25C/V-20F showed an interaction with Hcf106T38C, E48C, R54C, E63C, and I69C by the presence of a higher molecular weight adduct when analyzed by non-reducing SDS-PAGE. The strongest interactions involved Hcf106R54C, E63C, and I69C (Figure 7c, lanes 3–7), while wild-type Hcf106, lacking cysteine, was not able to generate an adduct (Figure 7c, lane 2). Taken together these data suggest that Hcf106 variants are incorporating into receptor complexes and can play a functional role in precursor binding.

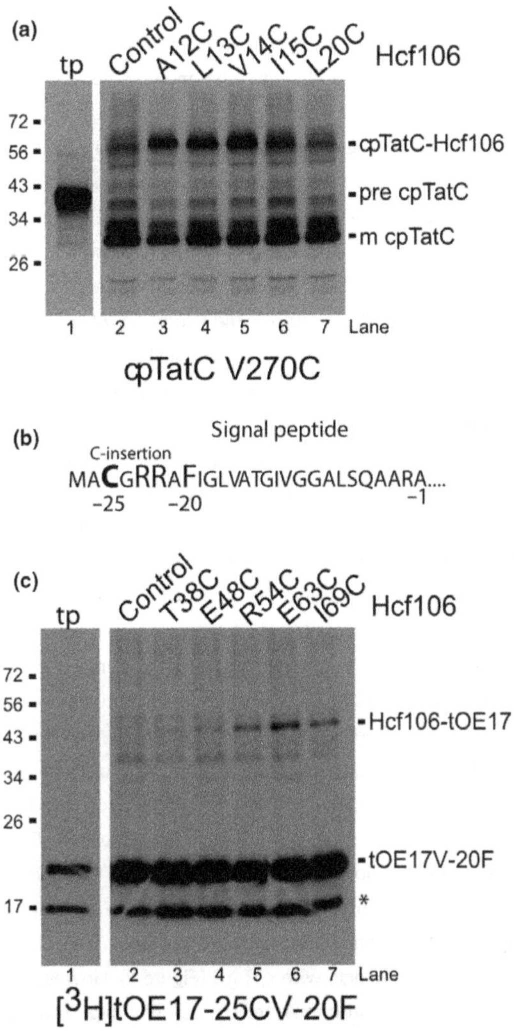

FIGURE 7 Integrated Hcf106 interacts directly with imported cpTatC and the signal peptide of the precursor. (a) In vitro translated [^{3}H]cpTatCaaaV270C was imported into intact chloroplasts. Intact chloroplasts were isolated over a 35% Percoll cushion (Section 2) and recovered thylakoid were used for integration of unlabeled Hcf106 Cys variants, as indicated across the top of the panels, or the same volume of IB, 10 mM $MgCl_2$ as control. Hcf106 TMD shows cross-linking with cpTatC TM5 in a ~56 kDa band that migrates higher than the nonspecific bands in the control lane. Translation products (tp) diluted 1:40 are shown. (b) Residue sequence of the signal peptide of tOE17-25CV-20F showing the −25C substitution, the RR motif, and the V-20F substitution. (c) In vitro translated unlabeled Hcf106 Cys variants, as indicated across the top of the panel, were integrated into isolated thylakoids. Thylakoid were then washed to remove unintegrated protein (Section 2) and incubated with either precursor, [^{3}H]tOE17-25CV-20F, or the same volume of IB, 10 mM $MgCl_2$ as control. The asterisk indicates a mistranslation of OE17 resulting in a smaller form of the precursor. Gels in both panels are representative of at least three separate experiments

4 | DISCUSSION

Recently, structural insights of the individual prokaryotic Tat components were revealed which lay a solid foundation for deciphering the mechanism of the Tat system (Hu et al., 2010; Ramasamy, Abrol, Suloway, & Clemons, 2013; Rollauer et al., 2012; Zhang et al., 2014). Additionally, Blummel, Haag, Eimer, Muller, & Frobel (2015) used photo-cross-linking to elucidate the architecture of TatBC oligomers during the initial assembly step, but due to the nature of the technique could only provide specific residue location for one binding partner. Here, cysteine scanning by disulfide bond formation was used to provide a precise map of interactions between the two homologous proteins in thylakoid. Previous work using isolated thylakoids demonstrated a substrate-gated docking of Tha4 into the cpTatC cavity initiating translocase assembly (Aldridge et al., 2014) and challenging the receptor complex model from *E. coli* in which TatB forms a ring-like structure in the center with TatC occupying the peripheral positions (Cline, 2015; Maurer, Panahandeh, Jungkamp, Moser, & Müller, 2010). However, how Hcf106-cpTatC are arranged to accommodate Tha4 docking and whether there are organizational differences between TatBC in *E. coli* and Hcf106-cpTatC in thylakoid have not been determined yet due to a lack of methodology to study the role of Hcf106 in isolate membranes.

Despite sharing significant sequence similarity, Tha4 and Hcf106 have distinct roles in protein transport (Cline & Mori, 2001; Dabney-Smith et al., 2003). Unlike Tha4, Hcf106 is typically found tightly complexed with cpTatC. Therefore, the use of an α-Hcf106 antibody is not viable as a sequestrant for native Hcf106 as it was for Tha4 (Dabney-Smith et al., 2003) due to this tight interaction between Hcf106 and cpTatC (Cline & Mori, 2001). Furthermore, *hcf106* mutant *Arabidopsis thaliana* plants are seedling lethal (personal communication), making complementation by biochemical methods (as was done for Tha4) not feasible. A previous experiments showed that the in vitro translated [^{3}H]Hcf106 assembled into a 700-kDa receptor complex as analyzed by BN-PAGE (Fincher et al., 2003), setting the stage for a possible involvement of exogenously Hcf106 in the receptor complex. Here, we further characterize integration of in vitro expressed Hcf106 using cysteine substitutions and disulfide bond formation to demonstrate self-interactions and interactions with imported cpTatC and precursor. Introduction of cysteine substitutions into Hcf106 largely did not affect integration into the membrane or presence in the 700-kDa complex (Figure 2). Exceptions include Cys substitutions in areas such as the TMD hydrophobic core, which lowered integration overall, likely due to a decrease in helix hydrophobicity, and in the hinge region, which may also be buried or involved in contacts with cpTatC (Figure 2). However, most of the Hcf106 Cys substitutions tested do integrate into thylakoid allowing the study of the organization of the protein in a native membrane (Figure 2). In *E. coli*, in the absence of TatC, TatB formed a ladder of bands of about 100 kDa to over 880 kDa, suggesting that TatB has oligomeric properties on its own when it is not associated with TatC (Behrendt, Lindenstrauss, & Bruser, 2007; Cleon et al., 2015). In the present study, we see ladders of full length and truncated Hcf106, which may be indicative of the separate pool of Hcf106; however, we also see formation of ladders of cpTatC that correspond in a linear manner to the integration of the truncated version of Hcf106 (Figures 2, 3). We conclude, therefore, that a substantial fraction of the incorporated, truncated Hcf106 is a part of the receptor complex with endogenous cpTatC.

We also identified key residues in Hcf106 that, when replaced by cysteine, severely inhibit its ability to assemble with the receptor complex. For example, in the hinge region, for example, G24C G27C, Cys-substituted Hcf106 lost its ability to assemble into a 700-kDa receptor complex as analyzed by BN-PAGE, suggesting important contacts have been disrupted (Figure 6). This is in agreement with observations in *E. coli* TatB hinge where the Gly/Pro residues have been identified as essential for efficient substrate export (Barrett, Mathers, & Robinson, 2003). There are at least three possible explanations for the absence of the 700-kDa complex when cysteines are substituted into the hinge of Hcf106. The first is that the substitution with Cys at those residues inhibited Hcf106-Hcf106 or Hcf106-cpTatC interactions, resulting in the absence of the 700-kDa complex. Second, Cys substitutions at those residues did not inhibit assembly *per se* but impacted the stability of the 700-kDa complex resulting in a receptor complex unable to withstand digitonin solubilization. Third, these substitutions resulted in decreased integration and therefore decreased participation in the 700-kDa complex. The third explanation is unlikely because Cys substitutions in the hinge region did not appear to inhibit integration and resistance to alkaline extraction of Hcf106 (Figure 2, Supporting information Figure S1). However, the data do suggest that the N terminus around residues G8 and E11 and the hinge region around residues G24 and G27 are of great structural significance to the assembly or stability of the interaction between Hcf106 and cpTatC (Figure 2, Supporting information Figure S1). Glycine often introduces more flexibility to protein structure, and so in replacing glycine with cysteine, the overall flexibility of the hinge region might be changed to interfere with an interaction with cpTatC. In addition to glycine, residue charge has been known to play an important role in membrane protein stability, possibly by forming a salt bridge and oligomeric structure solubilization (Wimley, Gawrisch, Creamer, & White, 1996). For example, Tha4 contains a glutamate (E10) in the transmembrane region that has been shown to be critical for Tha4 function, possibly by stabilization of oligomers through salt bridge formation (Dabney-Smith et al., 2003). However, whether E11 in Hcf106 has a similar role is unclear. Other Hcf106 residues, such as E41, Q43, and P44, may also affect interaction with cpTatC as shown by a decrease in the presence of those Hcf106 Cys variants in the 700-kDa complex (Figure 2). By replacing residues with Cys, Hcf106 contacts with cpTatC would be altered. Further studies are needed to clarify which region of cpTatC interacts with the N-terminus and hinge region of Hcf106. Currently, due to the presence of endogenous Hcf106 in the receptor complex, we can deduce that the presence of the cysteine variants in the receptor complex does not severely inhibit complex functionality.

Truncated $Hcf106_{1-107}$ can assemble with endogenous cpTatC indicating that the C-tail is less important as compared with the TMD and APH for receptor complex formation (Figure 3). This is also in agreement with the observation in *E. coli* that significant transport was observed when 70 residues were removed from C-terminus of TatB (Lee, Buchanan, Stanley, Berks, & Palmer, 2002). We designed these experiments to maximize the observable amount of the 700-kDa receptor complex relative to homo-oligomeric Hcf106 complexes by adjusting the detergent:chlorophyll ratio. Based on our earlier BN-PAGE data, the homo-oligomeric forms should not appear on the gel. This allows us to interpret the lower bands seen when the truncated Hcf106 is incorporated into the receptor complex. As the amount of in vitro translated [^{3}H]$Hcf106_{1-107}$ increases, more of the truncated protein assembled into a complex of lower molecular weight with cpTatC, indicating that $Hcf106_{1-107}$ has the ability to compete with endogenous Hcf106 and as the concentration of $Hcf106_{1-107}$ increases, more cpTatC assembled with the truncated Hcf106 (Figure 3). However, based on the immunoblotting data, the proportion of full length Hcf106 in a 700-kDa complex is less (Figure 3). The apparent ~600 kDa and ~500 kDa bands, together with the double Cys cross-linking data, strongly suggest that integrated recombinant Hcf106 may form oligomers in a separate pool and may compete with the endogenous Hcf106 for association with cpTatC (Figures 3, 6). If the receptor complex contains eight copies of both Hcf106 and cpTatC as predicted (Celedon & Cline, 2012; Mori et al., 2001), the ~600-kDa complexes may indicate approximately four endogenous Hcf106 were replaced with the truncated version while the ~500-kDa complexes indicates most, if not all, of the endogenous Hcf106 were replaced by truncated Hcf106 (Figure 3).

Hcf106 shares structural similarity with Tha4. Both biochemical labeling (Aldridge et al., 2012) of Tha4 as well as solution NMR and computer simulation modeling (Rodriguez et al., 2013) of *E. coli* homolog TatA show that the TMD is tilted and the N-terminal APH is partially embedded in the membrane, rather than the TMD inserted vertically (e.g., TMD parallel to the bilayer normal) and the APH laying on the surface of membrane (e.g., perpendicular to the bilayer normal). The cross-linking data of Hcf106 presented here also suggest a similar topology for Hcf106 (Figure 4). We saw limited interactions between residues L28C to R40C (N-proximal region of the APH), which could be explained by being in the low dielectric environment of the hydrophobic core of the membrane.

In summary, to systematically study Hcf106, a critical component in the chloroplast Tat system, a recombinant library of single Cys substitutions from the N-terminus to the end of APH was generated. Here, we demonstrate that exogenous, recombinant Hcf106 was able to insert into thylakoid and participates directly in the cpTat receptor complex, likely by replacing endogenous, Cys-less versions of Hcf106. This library not only helped clarify Hcf106 self-contacts but also demonstrated that exogenously integrated Hcf106 interacts with the precursor signal peptide via the Hcf106 APH and that the TMD was identified to form close contacts with cpTatC TM5 (Figure 7). These interaction data further confirm the capability of in vitro integrated Hcf106 to function in the thylakoid membrane system and demonstrates a new tool to evaluate the organization of cpTat complexes.

ACKNOWLEDGMENTS

This work was supported by the Department of Energy: Office of Science Basic Energy Science (Award DE-SC0014441 to CDS). The authors thank Ken Cline (University of Florida) for the cpTatC clones and members of the Dabney-Smith lab for critical reading of the manuscript.

AUTHOR CONTRIBUTIONS

QM and CDS designed the research; QM and KF performed research; QM, KF, CPN, and CDS analyzed the data; QM, CPN, and CDS wrote the manuscript.

REFERENCES

Alami, M., Luke, I., Deitermann, S., Eisner, G., Koch, H. G., Brunner, J., & Müller, M. (2003). Differential interactions between a twin-arginine signal peptide and its translocase in Escherichia coli. *Molecular Cell*, *12*, 937–946. https://doi.org/10.1016/S1097-2765(03)00398-8

Alcock, F., Stansfeld, P. J., Basit, H., Habersetzer, J., Baker, M. A., Palmer, T., ... Berks, B. C. (2016). Assembling the Tat protein translocase. *Elife*, *5*, e20718. https://doi.org/10.7554/eLife.20718

Aldridge, C., Ma, X., Gerard, F., & Cline, K. (2014). Substrate-gated docking of pore subunit Tha4 in the TatC cavity initiates Tat translocase assembly. *Journal of Cell Biology*, *205*, 51–65. https://doi.org/10.1083/jcb.201311057

Aldridge, C., Storm, A., Cline, K., & Dabney-Smith, C. (2012). The chloroplast twin arginine transport (tat) component, tha4, undergoes conformational changes leading to tat protein transport. *Journal of Biological Chemistry*, *287*, 34752–34763. https://doi.org/10.1074/jbc.M112.385666

Barrett, C. M., Mathers, J. E., & Robinson, C. (2003). Identification of key regions within the *Escherichia coli* TatAB subunits. *FEBS Letters*, *537*, 42–46. https://doi.org/10.1016/S0014-5793(03)00068-1

Behrendt, J., & Bruser, T. (2014). The TatBC complex of the Tat protein translocase in *Escherichia coli* and its transition to the substrate-bound TatABC complex. *Biochemistry*, *53*, 2344–2354. https://doi.org/10.1021/bi500169s

Behrendt, J., Lindenstrauss, U., & Bruser, T. (2007). The TatBC complex formation suppresses a modular TatB-multimerization in *Escherichia coli*. *FEBS Letters*, *581*, 4085–4090. https://doi.org/10.1016/j.febslet.2007.07.045

Berks, B. C. (2015). The twin-arginine protein translocation pathway. *Annual Review of Biochemistry*, *84*, 843–864. https://doi.org/10.1146/annurev-biochem-060614-034251

Blummel, A. S., Haag, L. A., Eimer, E., Muller, M., & Frobel, J. (2015). Initial assembly steps of a translocase for folded proteins. *Nature Communications*, *6*, 7234. https://doi.org/10.1038/ncomms8234

Bolhuis, A., Mathers, J. E., Thomas, J. D., Barrett, C. M., & Robinson, C. (2001). TatB and TatC form a functional and structural unit of the twin- arginine translocase from *Escherichia coli*. *Journal of Biological Chemistry*, *276*, 20213–20219. https://doi.org/10.1074/jbc.M100682200

Braun, N. A., Davis, A. W., & Theg, S. M. (2007). The chloroplast Tat pathway utilizes the transmembrane electric potential as an energy source. *Biophysical Journal*, *93*, 1993–1998. https://doi.org/10.1529/biophysj.106.098731

Celedon, J. M., & Cline, K. (2012). Stoichiometry for binding and transport by the twin arginine translocation system. *Journal of Cell Biology*, *197*, 523–534. https://doi.org/10.1083/jcb.201201096

Celedon, J. M., & Cline, K. (2013). Intra-plastid protein trafficking: how plant cells adapted prokaryotic mechanisms to the eukaryotic condition. *Biochimica Et Biophysica Acta-Molecular Cell Research*, *1833*, 341–351. https://doi.org/10.1016/j.bbamcr.2012.06.028

Cleon, F., Habersetzer, J., Alcock, F., Kneuper, H., Stansfeld, P. J., Basit, H., ... Palmer, T. (2015). The TatC component of the twin-arginine protein translocase functions as an obligate oligomer. *Molecular Microbiology*, *98*, 111–129. https://doi.org/10.1111/mmi.13106

Cline, K. (1986). Import of proteins into chloroplasts. Membrane integration of a thylakoid precursor protein reconstituted in chloroplast lysates. *Journal of Biological Chemistry*, *261*, 14804–14810.

Cline, K. (2015). Mechanistic aspects of folded protein transport by the twin arginine translocase (Tat). *Journal of Biological Chemistry*, *290*, 16530–16538. https://doi.org/10.1074/jbc.R114.626820

Cline, K., Henry, R., Li, C., & Yuan, J. (1993). Multiple pathways for protein transport into or across the thylakoid membrane. *EMBO Journal*, *12*, 4105–4114. https://doi.org/10.1002/j.1460-2075.1993.tb06094.x

Cline, K., & Mori, H. (2001). Thylakoid DeltapH-dependent precursor proteins bind to a cpTatC-Hcf106 complex before Tha4-dependent transport. *Journal of Cell Biology*, *154*, 719–729. https://doi.org/10.1083/jcb.200105149

Dabney-Smith, C., & Cline, K. (2009). Clustering of C-terminal stromal domains of Tha4 homo-oligomers during translocation by the Tat protein transport system. *Molecular Biology of the Cell*, *20*, 2060–2069. https://doi.org/10.1091/mbc.e08-12-1189

Dabney-Smith, C., Mori, H., & Cline, K. (2003). Requirement of a Tha4-conserved transmembrane glutamate in thylakoid Tat translocase assembly revealed by biochemical complementation. *Journal of Biological Chemistry*, *278*, 43027–43033. https://doi.org/10.1074/jbc.M307923200

Dabney-Smith, C., Mori, H., & Cline, K. (2006). Oligomers of Tha4 organize at the thylakoid Tat translocase during protein transport. *Journal of Biological Chemistry*, *281*, 5476–5483. https://doi.org/10.1074/jbc.M512453200

Fincher, V., Dabney-Smith, C., & Cline, K. (2003). Functional assembly of thylakoid deltapH-dependent/Tat protein transport pathway components in vitro. *European Journal of Biochemistry*, *270*, 4930–4941. https://doi.org/10.1046/j.1432-1033.2003.03894.x

Gérard, F., & Cline, K. (2006). Efficient twin arginine translocation (Tat) pathway transport of a precursor protein covalently anchored to its initial cpTatC binding site. *Journal of Biological Chemistry*, *281*, 6130–6135. https://doi.org/10.1074/jbc.M512733200

Gérard, F., & Cline, K. (2007). The thylakoid proton gradient promotes an advanced stage of signal peptide binding deep within the Tat pathway receptor complex. *Journal of Biological Chemistry*, *282*, 5263–5272. https://doi.org/10.1074/jbc.M610337200

Holzapfel, E., Eisner, G., Alami, M., Barrett, C. M., Buchanan, G., Luke, I., … Muller, M. (2007). The entire N-terminal half of TatC is involved in twin-arginine precursor binding. *Biochemistry*, *46*, 2892–2898. https://doi.org/10.1021/bi062205b

Hu, Y., Zhao, E., Li, H., Xia, B., & Jin, C. (2010). Solution NMR structure of the TatA component of the twin-arginine protein transport system from gram-positive bacterium *Bacillus subtilis*. *Journal of the American Chemical Society*, *132*, 15942–15944. https://doi.org/10.1021/ja1053785

Kneuper, H., Maldonado, B., Jager, F., Krehenbrink, M., Buchanan, G., Keller, R., … Palmer, T. (2012). Molecular dissection of TatC defines critical regions essential for protein transport and a TatB-TatC contact site. *Molecular Microbiology*, *85*, 945–961. https://doi.org/10.1111/j.1365-2958.2012.08151.x

Lee, P. A., Buchanan, G., Stanley, N. R., Berks, B. C., & Palmer, T. (2002). Truncation analysis of TatA and TatB defines the minimal functional units required for protein translocation. *Journal of Bacteriology*, *184*, 5871–5879. https://doi.org/10.1128/JB.184.21.5871-5879.2002

Lee, P. A., Orriss, G. L., Buchanan, G., Greene, N. P., Bond, P. J., Punginelli, C., … Palmer, T. (2006). Cysteine-scanning mutagenesis and disulfide mapping studies of the conserved domain of the twin-arginine translocase TatB component. *Journal of Biological Chemistry*, *281*, 34072–34085. https://doi.org/10.1074/jbc.M607295200

Ma, X., & Cline, K. (2010). Multiple precursor proteins bind individual Tat receptor complexes and are collectively transported. *EMBO Journal*, *29*, 1477–1488. https://doi.org/10.1038/emboj.2010.44

Ma, X., & Cline, K. (2013). Mapping the signal peptide binding and oligomer contact sites of the core subunit of the pea twin arginine protein translocase. *Plant Cell*, *25*, 999–1015. https://doi.org/10.1105/tpc.112.107409

Maldonado, B., Kneuper, H., Buchanan, G., Hatzixanthis, K., Sargent, F., Berks, B. C., & Palmer, T. (2011). Characterisation of the membrane-extrinsic domain of the TatB component of the twin arginine protein translocase. *FEBS Letters*, *585*, 478–484. https://doi.org/10.1016/j.febslet.2011.01.016

Maurer, C., Panahandeh, S., Jungkamp, A. C., Moser, M., & Müller, M. (2010). TatB functions as an oligomeric binding site for folded Tat precursor proteins. *Molecular Biology of the Cell*, *21*, 4151–4161. https://doi.org/10.1091/mbc.e10-07-0585

Mori, H., & Cline, K. (2002). A twin arginine signal peptide and the pH gradient trigger reversible assembly of the thylakoid [Delta]pH/Tat translocase. *Journal of Cell Biology*, *157*, 205–210. https://doi.org/10.1083/jcb.200202048

Mori, H., Summer, E. J., & Cline, K. (2001). Chloroplast TatC plays a direct role in thylakoid (Delta)pH-dependent protein transport. *FEBS Letters*, *501*, 65–68. https://doi.org/10.1016/S0014-5793(01)02626-6

Pal, D., Fite, K., & Dabney-Smith, C. (2013). Direct interaction between a precursor mature domain and transport component Tha4 during twin arginine transport of chloroplasts. *Plant Physiology*, *161*, 990–1001. https://doi.org/10.1104/pp.112.207522

Pettersson, P., Ye, W., Jakob, M., Tannert, F., Klosgen, R. B., & Maler, L. (2018). Structure and dynamics of plant TatA in micelles and lipid bilayers studied by solution NMR. *FEBS Journal*, *285*, 1886–1906. https://doi.org/10.1111/febs.14452

Ramasamy, S., Abrol, R., Suloway, C. J., & Clemons, W. M. Jr (2013). The glove-like structure of the conserved membrane protein TatC provides insight into signal sequence recognition in twin-arginine translocation. *Structure*, *21*, 777–788. https://doi.org/10.1016/j.str.2013.03.004

Rodriguez, F., Rouse, S. L., Tait, C. E., Harmer, J., De, R. A., Timmel, C. R., … Schnell, J. R. (2013). Structural model for the protein-translocating element of the twin-arginine transport system. *Proceedings of the National Academy of Sciences of the United States of America*, *110*, E1092–E1101. https://doi.org/10.1073/pnas.1219486110

Rolland, N., Ferro, M., Ephritikhine, G., Marmagne, A., Ramus, C., Brugiere, S., … Garin, J. (2006). A versatile method for deciphering plant membrane proteomes. *Journal of Experimental Botany*, *57*, 1579–1589. https://doi.org/10.1093/jxb/erj162

Rollauer, S. E., Tarry, M. J., Graham, J. E., Jaaskelainen, M., Jager, F., Johnson, S., … Lea, S. M. (2012). Structure of the TatC core of the twin-arginine protein transport system. *Nature*, *492*, 210–214. https://doi.org/10.1038/nature11683

Schneider, C. A., Rasband, W. S., & Eliceiri, K. W. (2012). NIH Image to ImageJ: 25 years of image analysis. *Nature Methods*, *9*, 671–675. https://doi.org/10.1038/nmeth.2089

Wimley, W. C., Gawrisch, K., Creamer, T. P., & White, S. H. (1996). Direct measurement of salt-bridge solvation energies using a peptide model system: Implications for protein stability. *Proceedings of the National Academy of Sciences of the United States of America*, *93*, 2985–2990. https://doi.org/10.1073/pnas.93.7.2985

Zhang, Y., Wang, L., Hu, Y., & Jin, C. (2014). Solution structure of the TatB component of the twin-arginine translocation system. *Biochimica et Biophysica Acta*, *1838*, 1881–1888. https://doi.org/10.1016/j.bbamem.2014.03.015

4

Several phased siRNA annotation methods can frequently misidentify 24 nucleotide siRNA-dominated *PHAS* loci

Seth Polydore[1,2] | Alice Lunardon[2] | Michael J. Axtell[1,2]

[1]Genetics Ph.D. Program, Huck Institutes of the Life Sciences, The Pennsylvania State University, University Park, Pennsylvania

[2]Department of Biology, The Pennsylvania State University, University Park, Pennsylvania

Correspondence
Michael J. Axtell, Department of Biology and Huck Institutes of the Life Sciences, The Pennsylvania State University, University Park, PA.
Email: mja18@psu.edu

Funding information
NSF, Grant/Award Number: 1339207; NIH Bioinformatics and Statistics CBIOS Predoctoral Training Program, Grant/Award Number: 1T32GM102057-0A1; the Henry W. Popp Graduate Assistantship Fund

Abstract

Small RNAs regulate key physiological functions in land plants. Small RNAs can be divided into two categories: microRNAs (miRNAs) and short interfering RNAs (siRNAs); siRNAs are further subdivided into transposon/repetitive region-localized heterochromatic siRNAs and phased siRNAs (phasiRNAs). PhasiRNAs are produced from the miRNA-mediated cleavage of a Pol II RNA transcript; the miRNA cleavage site provides a defined starting point from which phasiRNAs are produced in a distinctly phased pattern. 21–22 nucleotide (nt)-dominated phasiRNA-producing loci (*PHAS*) are well represented in all land plants to date. In contrast, 24 nt-dominated *PHAS* loci are known to be encoded only in monocots and are generally restricted to male reproductive tissues. Currently, only one miRNA (miR2275) is known to trigger the production of these 24 nt-dominated *PHAS* loci. In this study, we use stringent methodologies in order to examine whether or not 24 nt-dominated *PHAS* loci also exist in *Arabidopsis thaliana*. We find that highly expressed heterochromatic siRNAs were consistently misidentified as 24 nt-dominated *PHAS* loci using multiple *PHAS*-detecting algorithms. We also find that *MIR2275* is not found in *A. thaliana*, and it seems to have been lost in the last common ancestor of Brassicales. Altogether, our research highlights the potential issues with widely used *PHAS*-detecting algorithms which may lead to false positives when trying to annotate new *PHAS*, especially 24 nt-dominated loci.

KEYWORDS
heterochromatic siRNA, *MIR2275*, phased siRNA, small RNA

1 | INTRODUCTION

Small RNAs regulate key physiological functions in land plants, ranging from organogenesis (Boualem et al., 2008; Kutter, Schöb, Stadler, Meins, & Si-Ammour, 2007; Laufs, Peaucelle, Morin, & Traas, 2004; Williams, Grigg, Xie, Christensen, & Fletcher, 2005) to gametogenesis (Grant-Downton, Hafidh, Twell, & Dickinson, 2009). Three major protein families are involved in the biogenesis of small RNAs. The first family is the DICER-LIKE (DCL) protein family. Consisting of four paralogs in the *Arabidopsis thaliana* genome (Baulcombe, 2004; Chapman & Carrington, 2007), DCL proteins hydrolyze RNA precursors into 20–24 nt double-stranded RNA fragments (Millar & Waterhouse, 2005). These double-stranded RNA fragments are then loaded into ARGONAUTE (AGO) proteins, the second protein family, and one strand of the RNA is discarded (Ender & Meister, 2010). Upon Watson-crick binding to other RNA transcripts in the cell, the AGO/single-stranded RNA complex represses other RNA transcripts

(Baumberger & Baulcombe, 2005; Qi, Denli, & Hannon, 2005). Overall, 10 AGOs are encoded in the *A. thaliana* genome (Tolia & Joshua-Tor, 2007). RNA DEPENDENT RNA polymerases (RDRs) are the third family of proteins involved in the biogenesis of many small RNAs. RDRs convert single-stranded RNAs into double-stranded RNAs by synthesizing the complementary strand of the RNA molecule (Willmann, Endres, Cook, & Gregory, 2011). Six *RDRs* are encoded in the *A. thaliana* genome (Willmann et al., 2011).

Small RNAs can be divided into two major categories: microRNAs (miRNAs) which are precisely processed from single-stranded RNA with a hairpin-like secondary structure (Millar & Waterhouse, 2005), and short interfering RNAs (siRNAs), which are derived from double-stranded RNA precursors (Axtell, 2013). siRNAs are further divided into several different groups, including phased siRNAs (phasiRNAs) and heterochromatic siRNAs (hc-siRNAs). Predominantly 24 nts in length, hc-siRNAs function to repress transcription of deleterious genomic elements such as transposable elements or repetitive elements (Ahmed, Sarazin, Bowler, Colot, & Quesneville, 2011) and the promoters of certain genes (Baev et al., 2010) by reinforcing the presence of heterochromatin in targeted areas (Baulcombe, 2004; Sugiyama, Cam, Verdel, Moazed, & Grewal, 2005). Biogenesis of hc-siRNAs begins with transcription by the plant-specific, holoenzyme DNA DEPENDENT RNA POLYMERASE IV (Pol IV) (Onodera et al., 2005). The resulting transcript is then converted into double-stranded RNA by RDR2 and this double-stranded transcript is hydrolyzed by DCL3 (Matzke, Kanno, Daxinger, Huettel, & Matzke, 2009). phasiRNAs are derived from DNA-dependent RNA polymerase II (Pol II) transcripts that have been targeted by miRNAs (Fei, Xia, & Meyers, 2013). Upon miRNA-mediated hydrolysis, the RNA transcript is converted into double-stranded RNA by RDR6 (Cuperus et al., 2010). The resulting double-stranded RNA is then cleaved into 21 nt double-stranded RNA fragments by DCL4 (and less frequently DCL2) (Axtell, Jan, Rajagopalan, & Bartel, 2006).

21–22 nt phasiRNA-producing loci (*PHAS*) are clearly represented in all land plants that have been sequenced thus far (Fei et al., 2013; Zheng, Wang, Wu, Ding, & Fei, 2015). However, 24 nt-dominated *PHAS* loci are only currently described in rice (Song et al., 2012), maize (Zhai, Bischof, et al., 2015; Zhai, Zhang, et al., 2015), and other non-grass monocots (Kakrana et al., 2018). Much like 21 nt-dominated *PHAS*, the biogenesis of these 24 nt-dominated *PHAS* loci begins with the Pol II-dependent transcription of a single-stranded RNA precursor which is then targeted by miR2275 and hydrolyzed. To date, miR2275 is the only miRNA known to trigger the production of 24 nt-dominated phasiRNAs (Fei et al., 2013). The resulting RNA transcript is then converted into a double-stranded RNA molecule by RDR6 (Zhai, Zhang, et al., 2015). However, these phasiRNA precursors are then hydrolyzed by DCL5 (a DCL3 homolog sometimes called DCL3b) to produce 24 nt phasiRNAs (Fei et al., 2013).

Aside from the combination of their size and biogenesis patterns, 24 nt-dominated *PHAS* loci are distinct in various ways. These loci as well as their triggering miRNA, miR2275, are very specifically expressed in the tapetum during early meiosis and quickly recede in expression in other stages of male gametogenesis in rice and maize (Tamim et al., 2018). The AGO protein that loads these phasiRNAs is unknown; however, in maize, *AGO18b* expression levels match those of the 24 nt-dominated *PHAS* loci quite closely and is therefore the most likely candidate to load 24 nt phasiRNAs (Komiya et al., 2014; Zhang, Xia, Meyers, & Walbot, 2015). The targets of these 24 nt phasiRNAs are unknown, but they are apparently necessary for proper male gametogenesis (Ono et al., 2018). 24 nt-dominated *PHAS* loci were also described in the non-grass monocots asparagus, lily, and daylily (Kakrana et al., 2018). These phasiRNAs are produced from processing of inverted repeat (IR) RNAs, instead of the double-stranded RNA precursors observed in rice and maize (Kakrana et al., 2018). Although the 24 nt-dominated *PHAS* loci from non-grass monocots are still expressed most greatly in male reproductive tissue, in asparagus they are also expressed in female reproductive tissue (Kakrana et al., 2018).

We set out to search for evidence of 24 nt-*PHAS* loci in plants besides monocots. We searched for 24 nt *PHAS* loci in the *A. thaliana* genome using small RNA-seq data. Currently, several distinct algorithms are available to calculate the "phasing" of a sRNA-producing locus (Dotto et al., 2014; Guo, Qu, & Jin, 2015; Zheng, Wang, & Sunkar, 2014). In general, these algorithms calculate the number of reads that are "in-phase" against those that are "out-of-phase" in order to determine the likelihood that a particular locus truly produces phasiRNAs (Axtell, 2010). However, 24 nt-dominated siRNA loci are very numerous in angiosperms, and therefore are a potential source of false-positives during searches for *PHAS* loci. We therefore carefully examined *A. thaliana* 24 nt-dominated loci that consistently passed *PHAS*-locus detecting algorithms using multiple methods and find that they are likely just heterochromatic siRNAs (hc-siRNAs). We also use two other methods to examine the presence of 24 nt-dominated *PHAS* loci in the *A. thaliana* genome. We searched for *rdr6*-dependent, 24 nt-dominated loci and found 18 such loci. We also examined homology of the miR2275 which triggers 24 nt phasiRNA biogenesis in rice and maize but found that the Brassicales clade contains no potential homologs for this miRNA. Overall, our results suggest that there are no true 24 nt *PHAS* loci in *A. thaliana*. Furthermore, our analysis shows that existing phasing score algorithms to detect novel *PHAS* loci can lead to false positives.

2 | MATERIALS AND METHODS

2.1 | Finding potential *MIR2275* homologs in angiosperms

The Phytozome (ver 12.1)-curated angiosperm genome sequences were downloaded. The mature *Oryza sativa* miR2275a sequence was downloaded from miRBase (ver 21) (Griffiths-Jones, Saini, van Dongen, & Enright, 2008) and searched against all the other genomes using Bowtie v1.0 (Langmead, Trapnell, Pop, & Salzberg, 2009) allowing for two mismatches.

In order to determine the predicted secondary structures of the Bowtie results of interest, the sequences corresponding to the Bowtie result, plus 200 nucleotides upstream and downstream, were extracted from the genome. The secondary structures of the sequences were predicted using the mFOLD web server (Zuker, 2003) and visually inspected to determine if the sequence formed a hairpin structure consistent with accepted norms for miRNA biogenesis (Axtell & Meyers, 2018). The sequences were aligned using ClustalX ver. 2 with default parameters (Larkin et al., 2007).

For those species for which publicly available small RNA-seq data existed, we downloaded, trimmed, and merged the small RNA libraries. The merged libraries were collapsed to non-redundant reads and investigated using CD-HIT (ver 4.6.8) using the options -n 4, -d 0, and -g 1. The *O. sativa* miR2275a sequence was used as a query.

2.2 | Determination of potential 24 nt-dominated *PHAS* loci in different wild-type, inflorescence libraries

Wild-type biological triplicate small RNA libraries (GSE105262) (Polydore & Axtell, 2018) were merged and aligned against the *A. thaliana* (TAIR 10) genome using ShortStack (ver 3.8) (Johnson, Yeoh, Coruh, & Axtell, 2016) using 27 known *A. thaliana PHAS* loci as a query file (Supporting information Table S1). Three distinct *PHAS*-detecting algorithms (Dotto et al., 2014; Johnson et al., 2016; Zheng et al., 2014) were used to determine phase scores in the merged run. These scores were used as the basis to determine cutoffs for calling significantly phased loci (Supporting information Figures S3–S4). Phase scores of each known *PHAS* locus for each of the three algorithms in the eight wild-type libraries used is listed in Supporting information Dataset S4. Note that we did not use the multiple testing correction for *PHAS* loci *p*-values as done in Dotto et al., 2014 as we wished to test the three algorithms against each other, and the algorithms that yield phase scores could not be adjusted for multiple testing.

Wild-type and *rdr1-1/2-1/6-15* (*rdr1/2/6*) triple mutant small RNA libraries (GSE105262) (Polydore & Axtell, 2018) were aligned against the *A. thaliana* (TAIR 10) genome using ShortStack (ver 3.2) with option –pad 75 and option –min_cov 0.5 rpm (Supporting information Table S2). With these settings, small RNA loci are found as follows: All distinct genomic intervals containing one or more primary sRNA-seq alignments within 75 nts of each other were obtained, and then filtered to remove loci where the total sRNA-seq abundance with a locus was less than 0.5 reads per million. This produced a final set of distinct, non-overlapping small RNA loci. Differential expression to determine downregulated loci was performed as previously described (Polydore & Axtell, 2018). Downregulated, 24 nt-dominated small RNA loci were catalogued into a list. Eight wild-type, inflorescence small RNA libraries (Supporting information Table S2) were run individually against the *A. thaliana* (TAIR 10) genome utilizing ShortStack (ver 3.8.1) using the results of our wild-type and *rdr1/2/6* libraries run as a query file. The phase scores of loci corresponding to the down-regulated, 24 nt-dominated small RNAs loci were evaluated using the binary sequence alignment (BAM)-formatted alignments from each run. ShortStack and an in-house Python script were used to perform the three phase score calculations. For *Brassica rapa*, *Cucumis sativus*, *Phaseolus vulgaris*, and *Solanum tuberosum*, the previous Shortstack-merged small RNA alignments were analyzed in the same way.

For the four other species besides *A. thaliana*, we downloaded publicly available small RNA libraries (Supporting information Dataset S2) for *B. rapa*, *C. sativus*, *P. vulgaris*, and *S. tuberosum* and merged and aligned them against their respective genomes using ShortStack (ver 3.8.1) with default options. The genomes used were ver 1.0 for *B. rapa* (Wang et al., 2011), ver 1.0 for *P. vulgaris* (Schmutz et al., 2014), ver. 2 for *C. sativus* (Huang et al., 2009), and dm_v404 for *S. tuberosum* (Hardigan et al., 2016).

2.3 | AGO immunoprecipitation, genetic dependency, and properties of loci analyses

Calculating the lengths, proportion of multimapping reads, small RNA expression levels (in Reads Per Million (RPM)), determining the genetic dependencies, and the AGO enrichments of various loci were performed as previously described (Polydore & Axtell, 2018). In order to compare the properties of 24 nt-dominated loci that passed the *PHAS*-detection algorithms to those that did not, 10 subsets of 20 loci were randomly selected from the 24 nt-dominated loci that did not pass the *PHAS*-detection algorithms with replacement.

2.4 | Examining *PHAS*-test passing loci for potential miRNA targeting

Sequences corresponding to the *A. thaliana* loci that consistently passed the *PHAS*-detection algorithms plus 200 base pairs up-stream and downs-stream were extracted from the *A. thaliana* genome. Mature miRNA sequences were downloaded from miRBase (ver 21) and aligned against these sequences using the Generic Small RNA-Transcriptome Aligner. A miRNA was considered to be potentially targeting a sequence if it aligned with an Allen et al. score of 3 or less.

3 | RESULTS

3.1 | miR2275 is not found in the *Brassicales* clade

Currently known 24 nt-dominated phasiRNA precursors are known to be targeted only by a single miRNA family, miR2275 (Song et al., 2008; Zhai, Zhang, et al., 2015). We examined all available angiosperm genomes on Phytozome (ver 12.1) for potential homologs of *MIR2275*. In monocots, all but *Brachypodium stacei*, *Spirodela polyrhiza*, and *Zostera marina* had potential miR2275 homologs (Figure 1; Supporting information Figure S1). Twelve eudicots had potential miR2275 homologs based on sequence similarity (Figure 1). We therefore interrogated small RNA libraries for these species to see if a mature miR2275 homolog was expressed. Five eudicots had evidence of mature miR2275 accumulation (Figure 1). All of the monocots for which we obtained small RNA-seq data expressed mature miR2275, except *Musa acuminata* (Figure 1).

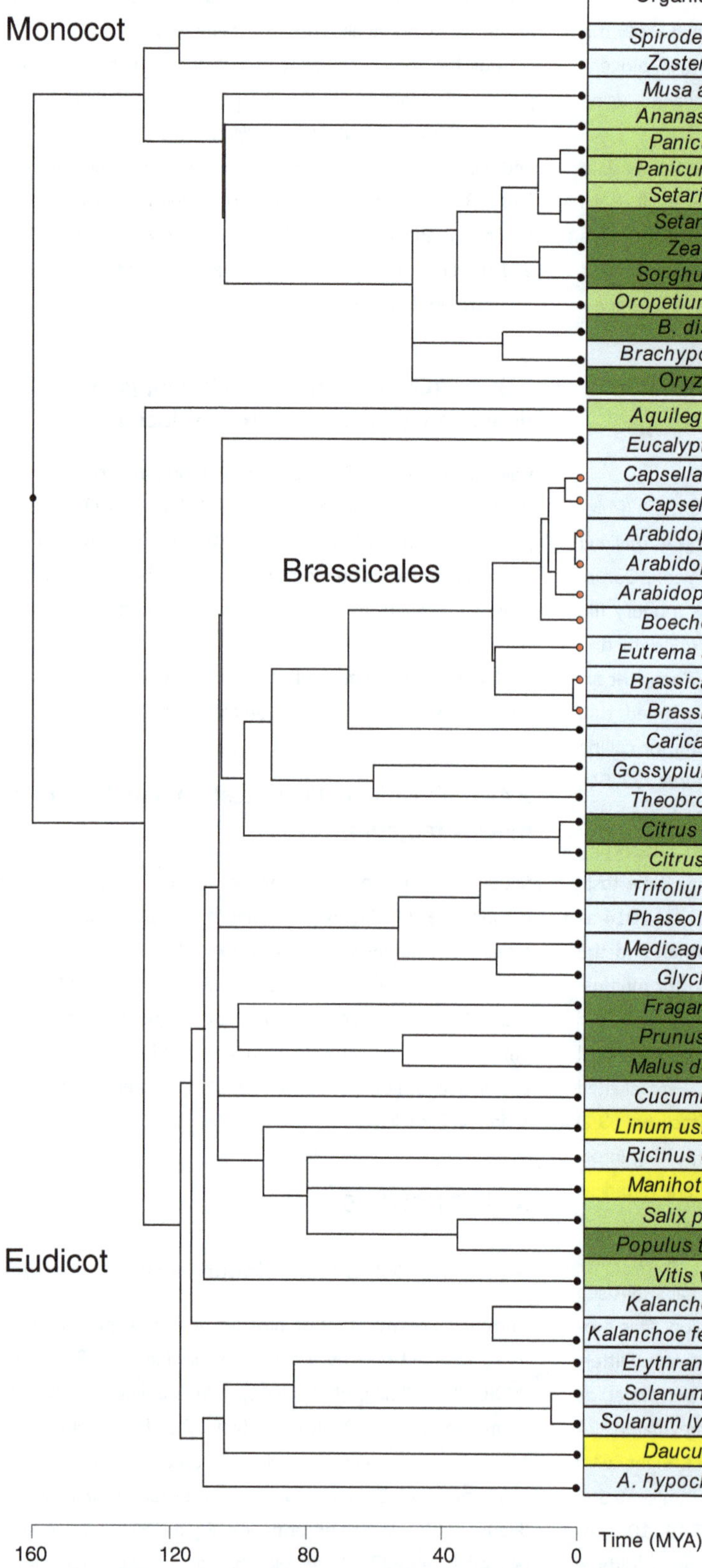

Organism Name	Homo-log(s)?	Mature miRNA Expressed?
Spirodela polyrhiza	No	–
Zostera marina	No	–
Musa acuminata	No	No
Ananas comosus	Yes	–
Panicum hallii	Yes	–
Panicum virgatum	Yes	–
Setaria viridis	Yes	–
Setaria italica	Yes	Yes
Zea mays	Yes	Yes
Sorghum bicolor	Yes	Yes
Oropetium thomaeum	Yes	–
B. distachyon	Yes	Yes
Brachypodium stacei	No	–
Oryza sativa	Yes	Yes
Aquilegia coerulea	Yes	–
Eucalyptus grandis	No	–
Capsella grandiflora	No	–
Capsella rubella	No	–
Arabidopsis halleri	No	–
Arabidopsis lyrata	No	–
Arabidopsis thaliana	No	No
Boechera stricta	No	–
Eutrema salsugineum	No	–
Brassica oleracea	No	No
Brassica rapa	No	No
Carica papaya	No	No
Gossypium raimondii	No	–
Theobroma cacao	No	–
Citrus clementina	Yes	Yes
Citrus sinensis	Yes	–
Trifolium pratense	No	–
Phaseolus vulgaris	No	No
Medicago truncatula	No	No
Glycine max	No	No
Fragaria vesca	Yes	Yes
Prunus persica	Yes	Yes
Malus domestica	Yes	Yes
Cucumis sativus	No	No
Linum usitatissimum	Yes	No
Ricinus communis	No	–
Manihot esculenta	Yes	No
Salix purpurea	Yes	–
Populus trichocarpa	Yes	Yes
Vitis vinifera	Yes	–
Kalanchoe laxiflora	No	–
Kalanchoe fedtschenkoi	No	–
Erythranthe guttata	No	–
Solanum tuberosum	No	No
Solanum lycopersicum	No	No
Daucus carota	Yes	No
A. hypochondriacus	No	–

No homolog found. | Either homolog found, but no mature miRNA verified or vice-versa. | Homolog found but no small RNA libraries are available. | Homolog and mature miRNA are both verified.

FIGURE 1 Conservation of *MIR2275*. Evidence for existence of *MIR2275* homologs in angiosperms. Phylogeny depicts estimated divergence times per TimeTree of Life (Kumar, Stecher, Suleski, & Hedges, 2017)

Alignment of the *MIR2275* loci in species for which potential homologs could be identified shows strong conservation of the mature miRNA and miRNA* sequences (Supporting information Figure S2), suggesting that these loci evolved from a common ancestor. Importantly, because of the high specificity of miR2275 expression in developing anthers (Tamim et al., 2018), it is possible that our analysis includes false negatives, especially in situations where no reproductive tissue small RNA libraries were available. Altogether, our data suggest that miR2275 is not found in *A. thaliana* and that this loss apparently occurred before the last common ancestor for Brassicales.

3.2 | Three *A. thaliana* hc-siRNA loci consistently pass *PHAS*-detecting algorithms

Although *A. thaliana* lacks miR2275, it is possible that 24 nt *PHAS* loci exist in *A. thaliana* and are triggered by a different small RNA. We reasoned that true 24 nt *PHAS* loci would be dependent on one or more of the well-described *A. thaliana RDR* genes: *RDR1*, *RDR2*, or *RDR6*. We thus examined a previously described differential expression analysis that identified *A. thaliana* small RNA loci that were downregulated in an *rdr1/rdr2/rdr6* triple mutant (Polydore & Axtell, 2018). The phase scores of *rdr1/rdr2/rdr6*-dependent, 24 nt-dominated loci were calculated in eight independent small RNA libraries using three different algorithms (Figure 2a). Reasonable cutoffs for *PHAS* loci detection were determined by examining the phase score distributions in the three merged wild-type libraries when well-known 21 nt *PHAS* loci were analyzed (Supporting information Figure S3). These cutoffs were consistent when a larger number of sRNA-seq libraries were examined (Supporting information Figure S4). Of the 31,750 loci examined, only three (Supporting information Figure S5; Dataset S1) passed the *PHAS* loci algorithms consistently in all 8 libraries examined (Figure 2a).

We were interested in determining why these three loci consistently pass the *PHAS*-detection algorithms. As phasiRNA precursors are known to be targeted by miRNAs, we predicted whether or not any known *A. thaliana* miRNAs could target these three loci. We were unable to find any obvious miRNA target sites at these loci. Although miRNA target sites were not apparent at these loci, it is possible that other siRNAs might target them and initiate siRNA phasing. We therefore attempted to determine the most common phase register at each of these loci in the eight different wild-type libraries (Supporting information Figure S6). If these loci are truly phased, then we would expect the same phase register to predominate in each library examined. While this was the case for *TAS2* (a positive control), none of the three 24 nt loci passing the *PHAS*-detection algorithms had consistent phase registers (Supporting information Figure S6). We also note that these three loci did not have the majority of their mapped reads falling into a single phase register in any of the libraries examined, as one would expect for a true *PHAS* locus (Supporting information Figure S6).

We then determined where in the genome these loci were encoded. Similar to hc-siRNA loci, these three loci primarily overlap with repeat- and transposable elements (Figure 2b). In comparison, many known *A. thaliana PHAS* loci are primarily derived from genes (Figure 2b). Rice and maize annotated 24 nt *PHAS* loci are derived from long noncoding RNAs that are encoded in regions of the genome that are devoid of protein-coding genes, transposable elements, or repeats (Song et al., 2008; Zhai, Bischof, et al., 2015; Zhai, Zhang, et al., 2015). We also note that sRNA accumulation from these three loci is downregulated in *nrpd1-3* (NRPD is the largest subunit of Pol IV), *nrpe* (NRPE is the largest subunit of Pol V), and *rdr2* (Figure 2c). hc-siRNAs have the same genetic requirements (Matzke et al., 2009), which strongly suggests that these three loci produce hc-siRNAs. Notably, these loci are not down-regulated in *dcl3* backgrounds. This is in line with past data that show that DCL4 and DCL2 can partially complement production of hc-siRNAs in *dcl3* backgrounds (Gasciolli, Mallory, Bartel, & Vaucheret, 2005).

The Pol IV transcripts from which hc-siRNAs are derived are around 26–50 nts in length, and accumulate to detectable levels in the *dcl2-1/3-1/4-2t* (*dcl234*) triple mutant (Ye et al., 2016; Zhai, Bischof, et al., 2015). We therefore analyzed the lengths of reads mapping to our three putative 24 nt *PHAS* loci in *dcl234* triple mutants. We found most reads were less than 40 nts long, indicating that these putative 24 nt *PHAS* loci are associated with short precursors similar to hc-siRNA loci (Figure 2d). In comparison, reads mapping to the known *TAS* loci had a wider range of sizes, ranging from 24 to 80 nts (Figure 2e). Overall, the size profile of the precursor RNAs further delineate these three loci from known *PHAS* loci.

Finally, we examined the AGO enrichments of small RNAs from these loci. Canonical hc-siRNAs are loaded into AGO4 in order to repress other loci transcriptionally (Mi et al., 2008). While sRNAs from our three putative 24 nt *PHAS* loci are not particularly enriched in AGO4-immunoprecipitation libraries, known 21 nt *PHAS* loci are quite depleted in the same dataset (Supporting information Figure S7). We also note that sRNAs from the three putative 24 nt *PHAS* loci are depleted in AGO1 immunoprecipitation libraries (Supporting information Figure S7), probably owing to the lack of 21 nt sRNAs, which AGO1 primarily loads, produced at these loci (Mi et al., 2008).

3.3 | The three *PHAS*-Test passing loci have distinct characteristics

hc-siRNAs are known to produce small RNAs in a very imprecise manner, unlike the largely precise processing of phasiRNAs (Axtell, 2013). We were interested in how three hc-siRNAs could consistently pass three different *PHAS*-detecting algorithms. One simple explanation is possible erroneous placement of ambiguously mapped reads. These multimapping reads could cause the *PHAS*-detecting algorithms to overestimate the number of "in phase" reads. We therefore determined the proportion of multimapping reads in each locus, but these loci have low proportion of multimapping reads compared to other 24 nt-dominated loci (Figure 3a). We note that these three loci are more highly expressed

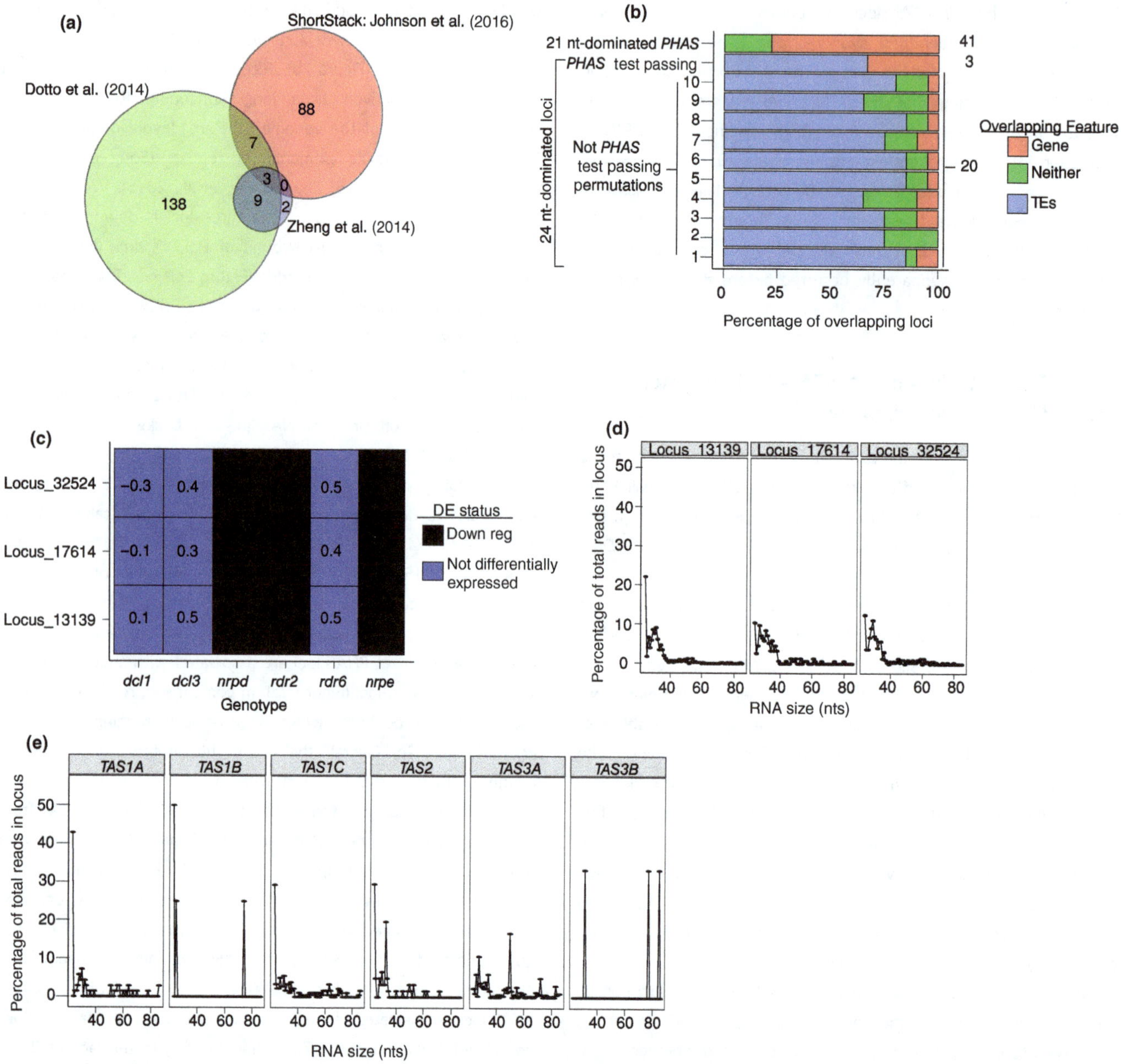

FIGURE 2 Properties of three *Arabidopsis thaliana* 24 nt-dominated small RNA loci that were called 'phased' by three different methods. (a) Venn diagram shows numbers of 24 nt-dominated loci that were called 'phased' by the indicated algorithms. (b) Percentage of the three *PHAS*-test passing loci overlapping either genes or transposable elements. The percentage is calculated as: (number of loci intersecting a feature/total number of loci in category)*100. The total number of loci in each category is given on the right. For 24 nt not-*PHAS* loci, 10 randomly selected cohorts of 20 loci each are shown. (c) Accumulation of the three *PHAS*-test passing loci in different genetic backgrounds. Numbers represent the ratio of small RNA accumulation in the indicated genotypes over that in corresponding wild-type library as computed by DESeq2. The differential expression status was determined via DESeq2 at an FDR of 0.1. (d) Percentage of short RNAs from *dcl2/dcl3/dcl4* triple mutant libraries by read length for the three *PHAS*-test passing loci. (e) Same as in Panel d, except for five known *A. thaliana TAS* loci

than most other 24 nt-dominated loci and even known *PHAS* loci in *A. thaliana* (Figure 3b). These three loci accumulate to nearly 1,000 RPM, while most other 24 nt-dominated loci and known *PHAS* loci only produce around 1 RPM (Figure 3b). Another interesting feature of these loci is their length. All three of these loci are around 10,000 nts in length, far greater than the 200–800 nt length of canonical *PHAS* loci and most other 24 nt-dominated loci (Figure 3c). These results indicate that these loci are simply highly expressed and particularly long hc-siRNA producing loci (Supporting information Figure S3).

3.4 | A small number of *A. thaliana rdr6*-dependent, 24 nt-dominated siRNA loci exist, but are not phased

Canonical monocot 24 nt-dominated *PHAS* loci are produced in part through the biochemical activity of RDR6 (Zhai, Zhang, et al., 2015).

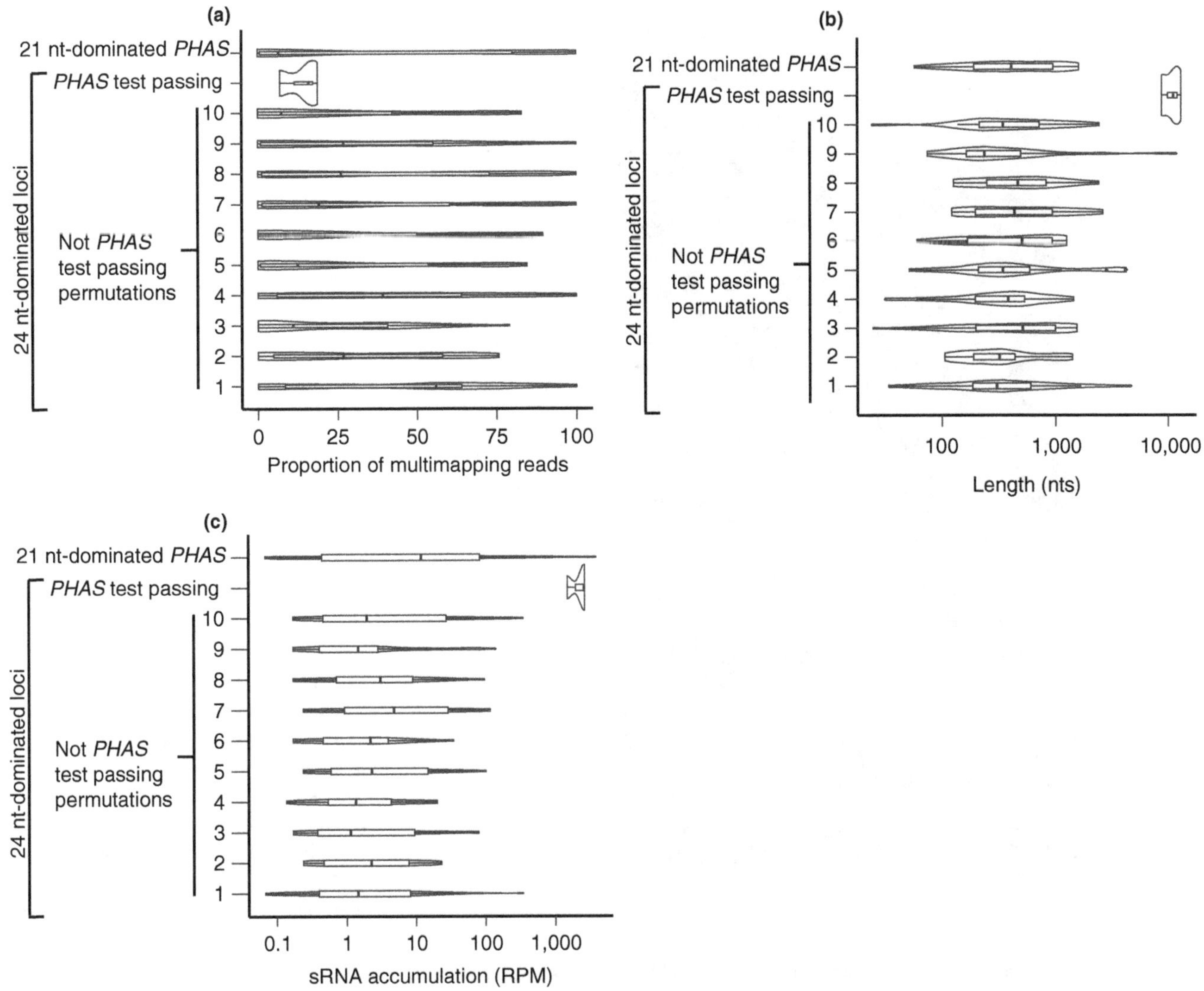

FIGURE 3 The three *PHAS*-test passing small RNA loci have distinct properties compared to other 24 nt-dominated loci. (a) The proportion of multimapped reads in three different types of small RNA loci. For 24 nt-dominated loci that were not called PHAS loci, 10 cohorts comprising 20 randomly selected loci were used as controls. The width of the density plot shows the frequency. The inset boxes show medians (horizontal lines), the 1st–3rd quartile range (boxes), other data out to 1.5 times the interquartile range (whiskers) and outliers (dots). (b) Same as panel a except showing small RNA accumulation. (c) Same as panel a except showing small RNA locus length

We therefore used an alternative strategy to identify possible 24 nt *PHAS* loci by first identifying *rdr6*-dependent, 24 nt-dominated siRNA loci. We found 18 such loci (Supporting information Figure S8; Dataset S1). None of these loci were consistently phased in the eight wild-type, inflorescence *A. thaliana* libraries (Supporting information Table S2) tested according to any of the three algorithms we used (Supporting information Dataset S3). We determined the genetic dependencies of these loci and found that these loci are down-regulated in *nrpd*, *nrpe*, and *rdr2* backgrounds (Figure 4). Furthermore, these loci either overlap transposable elements or are mostly found in otherwise intergenic regions (Supporting information Figure S9a). As these are typical features of hc-siRNAs, it is possible that hc-siRNAs simply erroneously placed at these loci. Again, we examined the proportion of ambiguously mapped reads in these *rdr6*-dependent, 24 nt-dominated loci compared to other 24 nt-dominated loci and found that these loci had low proportions of multimapping reads (Supporting information Figure S9b). Our result argues against this hypothesis.

rdr6-dependent, 24 nt-dominated loci are strongly expressed relative to other 24 nt-dominated loci (Supporting information Figure S9c). These *rdr6*-dependent loci have median expression level of nearly 100 RPM compared to the nearly 1 RPM median expression level of the other 24 nt-dominated loci (Supporting information Figure S9c). This result is similar to the three *A. thaliana* loci that passed our *PHAS*-detection algorithms (Figure 3b). However, the lengths of the *rdr6*-dependent, 24 nt-dominated loci are similar to other 24 nt-dominated loci (Supporting information Figure S9d). Overall, while do find evidence for a small number of *rdr6*-dependent, 24 nt-dominated siRNAs in *A. thaliana*, they are not phased,

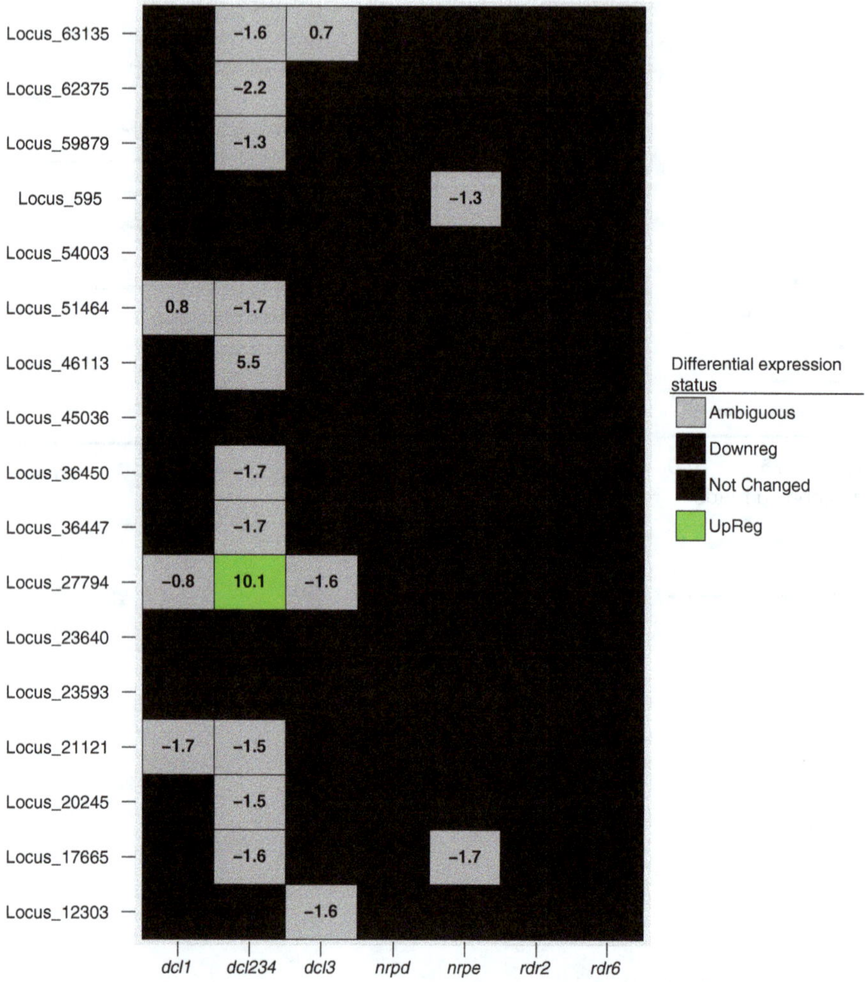

FIGURE 4 Accumulation of 24 nt-dominated, *rdr6*-dependent small RNA loci in different mutant backgrounds. Numbers represent the log2-transformed ratios of small RNA accumulation in the indicated genotypes over that in corresponding wild-type library as computed by DESeq2. The differential expression status was determined via DESeq2 at an FDR of 0.1

nor do they appear readily discernable from more typical hc-siRNA loci. Thus, we find no evidence of 24 nt-dominated *PHAS* loci in *A. thaliana*.

3.5 | Erroneous detection of 24 nt-dominated *PHAS* loci occurs in other land plant species

We were interested in determining if erroneous annotation of *PHAS* loci was unique to *A. thaliana* small RNAs or if these results could be replicated in other species. We therefore interrogated publicly available *B. rapa*, *C. sativus*, *P. vulgaris*, and *S. tuberosum* small RNA libraries (all eudicots) using the three algorithms, searching for putative 24 nt-dominated *PHAS* loci. We specifically chose these four species because they all lacked a potential *MIR2275* homolog (Figure 1). As miR2275 is the only miRNA known to trigger 24 nt-dominated phasiRNAs, any 24 nt-dominated loci called as *PHAS* loci in these species are likely false positives. Each species had 24 nt-dominated small RNAs that were misannotated as *PHAS* loci (Supporting information Figure S10). We first determined where in the genome the 24 nt-dominated loci are encoded. Like in *A. thaliana*, 24 nt-dominated loci that passed the *PHAS*-detection algorithm seem to come predominantly from transposable elements. This was true in all the species examined except for *P. vulgaris* (Figure 5a). Curiously, the *PHAS*-test passing loci had a slightly higher proportion of ambiguously mapped reads compared to other 24 nt-dominated loci in *B. rapa*, *P. vulgaris*, and *S. tuberosum* (Figure 5b). For *C. sativus*, the proportion of multimapping reads in *PHAS*-test passing 24 nt-dominated loci was substantially higher than other 24 nt-dominated loci (Figure 5b). It is still unlikely that multimapping reads contribute significantly to phasing at these loci as the proportion of ambiguously mapped to both types of 24 nt-dominated loci are similar in three of the four species tested (Figure 5b).

In the four species tested, the *PHAS*-test passing 24 nt-dominated loci had both significantly greater lengths (Figure 5c) and expression on average (Figure 5d). We observed this same trend in *A. thaliana* (Figure 3b,c), which suggests that long lengths and high expression levels of 24 nt-dominated loci are conducive to *PHAS* loci misannotations, even in other distantly related species.

4 | DISCUSSION

4.1 | *rdr6*-dependent, 24 nt-dominated loci in the *A. thaliana* genome

We found a handful of *rdr6*-dependent, 24 nt-dominated loci to be encoded in the *A. thaliana* genome. However, these loci have

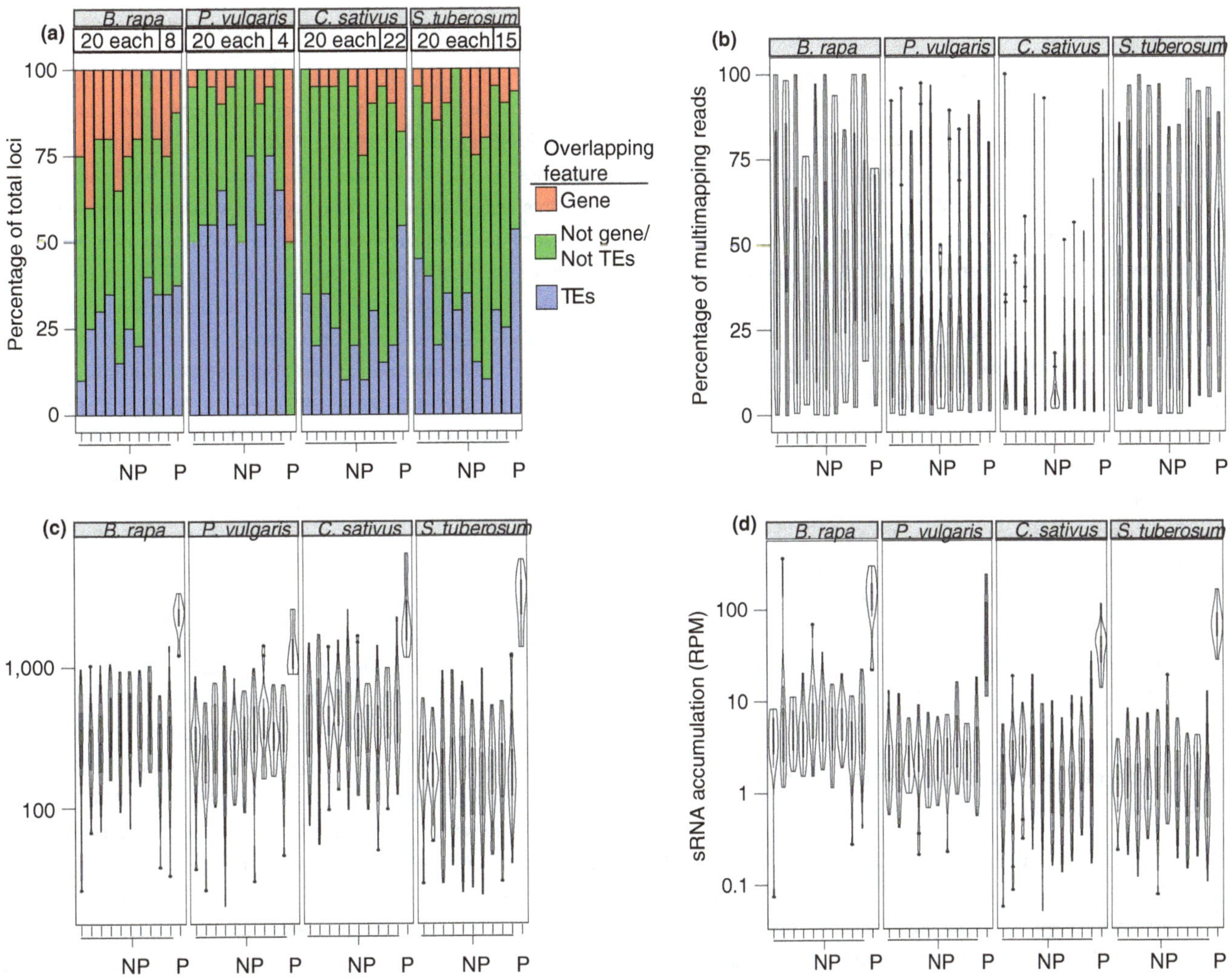

FIGURE 5 Four divergent species also contain 24 nt-dominated loci that passed the three *PHAS*-detecting algorithms. (a) Percentage of 24 nt-dominated Loci overlapping genes, transposable elements, or neither. The species name is shown in the top grey boxes. P: *PHAS*-test passing locus; NP: Not *PHAS*-test passing locus. For NP loci, 10 cohorts of 20 randomly selected loci each were used as negative controls. The number of loci in each category is shown above the bar graphs. (b) Same as panel a except showing the proportion of multimapping reads. (c) Same as panel a except showing length of the small RNA locus. (d) Same as panel a except showing small RNA accumulation

the same genetic dependencies as hc-siRNAs (Figure 4) and are frequently derived from transposable elements (Supporting information Figure S9a). The only other known *rdr6*-dependent, transposon-overlapping small RNA loci encoded in *A. thaliana* are epigenetically activated short interfering RNAs (easiRNAs) (Creasey et al., 2014). easiRNAs are derived from transcriptionally active transposable elements that are hypothesized to be targeted and cleaved by miRNAs (Creasey et al., 2014). Through the biochemical actions of RDR6 and DCL4, the miRNA cleavage product is converted into 21–22 nt double-stranded RNA molecules (Creasey et al., 2014). Furthermore, these easiRNAs are thought to direct initial repression of transposons (He, Yang, Wu, & Zheng, 2015). It is possible that these *rdr6*-dependent, 24 nt-dominated loci could represent a transitory stage of hc-siRNA targeting, in which the genetic machinery of hc-siRNA are used but with a remaining initial dependency on RDR6.

4.2 | Possible losses of the *MIR2275*-generated phasing in eudicots

We examined all Phytozome (ver 12.1)-curated angiosperm genomes (Goodstein et al., 2012) for the presence of *MIR2275* homologs. *MIR2275* shows remarkable conservation in monocots (Figure 1), showing absence only in *Musa accuminata* and *Brachypodium distachyon*. This could be because these are aquatic plants with divergent morphology, and thus there is relaxed selective pressure for *MIR2275*. Another explanation is that the genome assemblies for these species could be incomplete. The lack of

MIR2275 in eudicots was more extensive (Figure 1), with no plants in the Brassicales clade containing a verified *MIR2275* homolog (Figure 1). Interestingly, *Aquilegia coerulea*, a basal eudicot (Sharma & Kramer, 2014), contained a *MIR2275* homolog (Figure 1; Supporting information Figure S1) suggesting that the last common ancestor for eudicots may have contained *MIR2275*, and that the lack of detected putative *MIR2275* homologs in many eudicot plant species could be due to loss of *MIR2275*.

We only examined whether or not a possible *MIR2275* homolog could be detected by homology, if its predicted secondary structure is conducive to *MIRNA* processing (Supporting information Figure S1), and in some species whether or not the putative *MIR2275* homolog was expressed in available small RNA libraries (Figure 1). We did not, however, determine if these species produce true 24 nt-dominated phasiRNAs. It is possible that the potential *MIR2275* homolog in these species is an orphaned *MIRNA* or has taken on a new function. Furthermore, although some putative *MIR2275* homologs had *MIRNA*-like predicted hairpins, we note that they contain mismatches in the stem of the secondary structure which may hinder miRNA biogenesis (Supporting information Figure S1). Further research is necessary to determine if these species truly encode 24 nt-dominated *PHAS* loci.

4.3 | The difficulties of annotating 24 nt-dominated *PHAS* loci

The discovery of 24 nt-dominated *PHAS* loci in maize (Zhai, Bischof, et al., 2015; Zhai, Zhang, et al., 2015), rice (Song et al., 2008), asparagus, daylily, and lily (Kakrana et al., 2018) opened up the possibility that these loci exist in other species, even distantly related eudicots. However, 24 nt-dominated hc-siRNA loci are widespread in land plant genomes (Ghildiyal & Zamore, 2009). This is true even in *A. thaliana*, although 24 nt sRNAs are thought to repress transposable elements (Matzke et al., 2009) and only about 20–30% of the *A. thaliana* genome consists of transposon/repetitive elements (Barakat, Matassi, & Bernardi, 1998). In contrast, other plant genomes consist of as much as 80% transposons/repetitive elements (Springer et al., 2009). The sheer number of 24 nt-dominated sRNA loci could mean that several of them could meet various annotation criteria simply by chance; this is something we have previously noted to occur in natural antisense siRNA annotation (Polydore & Axtell, 2018).

The supposed 24 nt-dominated *PHAS* loci examined in this study consistently showed higher levels of expression than other 24 nt-dominated sRNA loci and derived from very long loci. This is not particularly surprising because the accumulation of reads (in- and out-of-phase) is a factor in most *PHAS*-detection algorithms (Supporting information Figure S3). However, it seems that particularly highly expressed 24 nt-dominated small RNAs are able to consistently pass *PHAS*-detection algorithms because of this. It's possible that the sheer number of reads produced at these loci means that these loci produce enough "in-phase" reads by chance to score highly in *PHAS*-detecting algorithms.

4.4 | Annotating 24 nt-dominated *PHAS* loci in the future

Our study demonstrates that when annotating novel 24 nt-dominated *PHAS*, more than utilizing *PHAS*-detecting algorithms is necessary for robust annotation. First, examining different available mutants, especially of genes involved in small RNA biogenesis, can be critical in determining type of small RNA in question (Figure 2c). All types of phasiRNAs are known to be reliant on the biochemical activity of RDR6 and Pol II (Cuperus et al., 2010; Song et al., 2008; Zhai, Bischof, et al., 2015; Zhai, Zhang, et al., 2015). While it is entirely possible that non-canonical phasiRNAs that are reliant on RDR2 and Pol IV may be described eventually, such a study must verify with robust methodologies that the *PHAS* loci are not false positives due to the sheer number of Pol IV/RDR2 –dependent reads produced in the land plant genome.

We searched for *MIR2275* homologs because miR2275 is the only miRNA known to trigger 24 nt phasiRNAs. If a species lacks a *MIR2275* homolog, then one should be skeptical of any 24 nt-dominated small RNA locus that is annotated as *PHAS* locus in that specie. An organism producing a mature miR2275 small RNA homolog is not sufficient to show that 24 nt-dominated *PHAS* loci are produced in that species. One should also take care to ascertain if any 24 nt-dominated small RNA locus called as *PHAS* loci also contains a miR2275 target site that is "in phase" with the phasiRNAs produced from the transcript. miR2275 is also very specifically expressed in the tapetum of male floral tissue. In asparagus, 24 nt-dominated phasiRNAs have also been shown to be expressed in female floral tissue. However, 24 nt-dominated phasiRNAs have not been demonstrated to be expressed outside of reproductive tissue as of yet. Therefore, any 24 nt-dominated *PHAS* loci annotated in libraries not produced from reproductive tissues are more likely to be mis-annotations.

Despite the fact that several small RNA loci in *A. thaliana* are able to consistently pass the *PHAS*-detecting algorithms in different libraries, reproducibility among distinct small RNA libraries is of utmost importance. It could also be useful to employ several different *PHAS*-test algorithms as well. In *A. thaliana*, three loci were able to consistently pass our *PHAS*-detecting algorithms (Supporting information Figure S5), but the number of loci that each algorithm detected on its own was generally much higher (Figure 2a). Had we not used stringent *PHAS*-detecting methods, it would've been plausible to assume we had found 24 nt-dominated *PHAS* loci in *A. thaliana* based on the number of loci that consistently pass alone. Utilizing multiple small RNA libraries should become easier to accomplish as more and more libraries in different treatments/genotypes become available for socioeconomically relevant species. Certain *PHAS*-detecting programs, such as *PHASIS* (Kakrana et al., 2017), automatically evaluate potential small RNA loci in different libraries individually before merging the results of each library.

It is not always possible to determine the small RNA that targets the phasiRNA precursor transcript. Indeed, several reads may be predicted to target a certain transcript by chance due to the sheer number of unique reads in a small RNA library. However, due to hydrolysis of the precursor transcript following small RNA-mediated targeting, phasiRNAs have well-defined termini from which they are produced (Axtell et al., 2006; Cuperus et al., 2010). Therefore, a true *PHAS* locus should have a large proportion of its reads reproducibly falling into a particular phase register. This is what we observed with well-characterized *PHAS* locus, *TAS2*, but not with the three loci that pass our rigorous *PHAS*-annotation regime in *A. thaliana* (Supporting information Figure S6).

As 21–22 nt-dominated *PHAS* loci are far more common in land plants, especially outside of the monocots, one can conservatively limit their discovery of new *PHAS* loci to 21–22 nt-dominated small RNAs and not employ as rigorous methodology as the ones used in this study. However, one should still employ post *PHAS*-discovery quality controls to ensure these 21–22 nt-dominated *PHAS* loci are genuine (such as determining if the predominant phase registers are reproducibly dominant at these loci (Supporting information Figure S6)). 24 nt-dominated *PHAS* loci on the other hand seem to be have undergone loss in many angiosperms. However, even in the species in which they are conserved, they seem to play very specific, reproductive-associated roles as evidenced by their expression patterns. Great caution should be used for annotating 24 nt-dominated *PHAS* loci in the future.

ACKNOWLEDGMENTS

This research was funded by NSF Award 1339207 to MJA, the NIH Bioinformatics and Statistics (CBIOS) Predoctoral Training Program (1T32GM102057- 0A1) to SP, and the Henry W. Popp Graduate Assistantship Fund to SP.

AUTHOR CONTRIBUTIONS

SP conceived of the project. AL aligned all the small RNA data other than the *A. thaliana* small RNA libraries. SP performed all the other experiments and analyses. MJA and SP wrote the manuscript.

REFERENCES

Ahmed, I., Sarazin, A., Bowler, C., Colot, V., & Quesneville, H. (2011). Genome-wide evidence for local DNA methylation spreading from small RNA-targeted sequences in Arabidopsis. *Nucleic Acids Research*, *39*(16), 6919–6931. https://doi.org/10.1093/nar/gkr324

Axtell, M. J. (2010). A method to discover phased siRNA loci. In C. M., Blake & J. G., Pamela (Eds.), *Plant MicroRNAs* (pp. 59–70). Totowa, NJ: Humana Press. https://doi.org/10.1007/978-1-60327-005-2_5

Axtell, M. J. (2013). Classification and comparison of small RNAs from plants. *Annual Review of Plant Biology*, *64*(1), 137–159. https://doi.org/10.1146/annurev-arplant-050312-120043

Axtell, M. J., Jan, C., Rajagopalan, R., & Bartel, D. P. (2006). A two-hit trigger for siRNA biogenesis in plants. *Cell*, *127*(3), 565–577. https://doi.org/10.1016/j.cell.2006.09.032

Baev, V., Naydenov, M., Apostolova, E., Ivanova, D., Doncheva, S., Minkov, I., & Yahubyan, G. (2010). Identification of RNA-dependent DNA-methylation regulated promoters in Arabidopsis. *Plant Physiology and Biochemistry*, *48*(6), 393–400. https://doi.org/10.1016/j.plaphy.2010.03.013

Barakat, A., Matassi, G., & Bernardi, G. (1998). Distribution of genes in the genome of Arabidopsis thaliana and its implications for the genome organization of plants. *Proceedings of the National Academy of Sciences of the United States of America*, *95*(17), 10044–10049.

Baulcombe, D. (2004). RNA silencing in plants. *Nature*, *431*(7006), 356–363. https://doi.org/10.1038/nature02874

Baumberger, N., & Baulcombe, D. C. (2005). Arabidopsis ARGONAUTE1 is an RNA Slicer that selectively recruits microRNAs and short interfering RNAs. *Proceedings of the National Academy of Sciences, USA*, *102*(33), 11928–11933. https://doi.org/10.1073/pnas.0505461102

Boualem, A., Laporte, P., Jovanovic, M., Laffont, C., Plet, J., Combier, J.-P., & Frugier, F. (2008). MicroRNA166 controls root and nodule development in Medicago truncatula. *Plant Journal: For Cell and Molecular Biology*, *54*(5), 876–887. https://doi.org/10.1111/j.1365-313X.2008.03448.x

Chapman, E. J., & Carrington, J. C. (2007). Specialization and evolution of endogenous small RNA pathways. *Nature Reviews Genetics*, *8*(11), 884–896. https://doi.org/10.1038/nrg2179

Creasey, K. M., Zhai, J., Borges, F., Van Ex, F., Regulski, M., Meyers, B. C., & Martienssen, R. A. (2014). miRNAs trigger widespread epigenetically activated siRNAs from transposons in Arabidopsis. *Nature*, *508* (7496), 411–415. https://doi.org/10.1038/nature13069

Cuperus, J. T., Carbonell, A., Fahlgren, N., Garcia-Ruiz, H., Burke, R. T., Takeda, A., & Carrington, J. C. (2010). Unique functionality of 22-nt miRNAs in triggering RDR6-dependent siRNA biogenesis from target transcripts in *Arabidopsis*. *Nature Structural & Molecular Biology*, *17*(8), 997–1003. https://doi.org/10.1038/nsmb.1866

Dotto, M. C., Petsch, K. A., Aukerman, M. J., Beatty, M., Hammell, M., & Timmermans, M. C. P. (2014). Genome-wide analysis of leafbladeless1-regulated and phased small RNAs underscores the importance of the TAS3 ta-siRNA pathway to maize development. *PLOS Genetics*, *10*(12), e1004826. https://doi.org/10.1371/journal.pgen.1004826

Ender, C., & Meister, G. (2010). Argonaute proteins at a glance. *Journal of Cell Science*, *123*(11), 1819–1823. https://doi.org/10.1242/jcs.055210

Fei, Q., Xia, R., & Meyers, B. C. (2013). Phased, secondary, small interfering RNAs in posttranscriptional regulatory networks. *Plant Cell Online*, *25*(7), 2400–2415. https://doi.org/10.1105/tpc.113.114652

Gasciolli, V., Mallory, A. C., Bartel, D. P., & Vaucheret, H. (2005). Partially redundant functions of Arabidopsis DICER-like enzymes and a role for DCL4 in producing trans-acting siRNAs. *Current Biology*, *15*(16), 1494–1500. https://doi.org/10.1016/j.cub.2005.07.024

Ghildiyal, M., & Zamore, P. D. (2009). Small silencing RNAs: An expanding universe. *Nature Reviews. Genetics*, *10*(2), 94–108. https://doi.org/10.1038/nrg2504

Goodstein, D. M., Shu, S., Howson, R., Neupane, R., Hayes, R. D., Fazo, J., … Rokhsar, D. S. (2012). Phytozome: A comparative platform for green plant genomics. *Nucleic Acids Research*, *40*(Database issue), D1178–D1186. https://doi.org/10.1093/nar/gkr944

Grant-Downton, R., Hafidh, S., Twell, D., & Dickinson, H. G. (2009). Small RNA pathways are present and functional in the angiosperm male gametophyte. *Molecular Plant*, *2*(3), 500–512. https://doi.org/10.1093/mp/ssp003

Griffiths-Jones, S., Saini, H. K., van Dongen, S., & Enright, A. J. (2008). miRBase: Tools for microRNA genomics. *Nucleic Acids Research*, *36* (suppl_1), D154–D158. https://doi.org/10.1093/nar/gkm952

Guo, Q., Qu, X., & Jin, W. (2015). PhaseTank: Genome-wide computational identification of phasiRNAs and their regulatory cascades. *Bioinformatics*, *31*(2), 284–286. https://doi.org/10.1093/bioinformatics/btu628

Hardigan, M. A., Crisovan, E., Hamiltion, J. P., Kim, J., Laimbeer, P., Leisner, C. P., … Buell, C. R. (2016). Genome reduction uncovers a large dispensable genome and adaptive role for copy number variation in asexually propagated *Solanum tuberosum*. *Plant Cell*, *28*, 388–405. https://doi.org/10.1105/tpc.15.00538

He, H., Yang, T., Wu, W., & Zheng, B. (2015). Small RNAs in pollen. *Science China Life Sciences*, *58*(3), 246–252. https://doi.org/10.1007/s11427-015-4800-0

Huang, S., Li, R., Zhang, Z., Li, L., Gu, X., Fan, W., & Li, S. (2009). The genome of the cucumber, *Cucumis sativus* L. *Nature Genetics*, *41*(12), 1275–1281. https://doi.org/10.1038/ng.475

Johnson, N. R., Yeoh, J. M., Coruh, C., & Axtell, M. J. (2016). Improved placement of multi-mapping small RNAs. *G3: Genes, Genomes, Genetics*, *6*(7), 2103–2111. https://doi.org/10.1534/g3.116.030452

Kakrana, A., Li, P., Patel, P., Hammond, R., Anand, D., Mathioni, S., & Meyers, B. (2017). PHASIS: a computational suite for de novo discovery and characterization of phased, siRNA-generating loci and their miRNA triggers. *bioRxiv*, 158832. https://doi.org/10.1101/158832

Kakrana, A., Mathioni, S. M., Huang, K., Hammond, R., Vandivier, L., Patel, P., ... Meyers, B. C. (2018). Plant 24-nt reproductive phasiRNAs from intramolecular duplex mRNAs in diverse monocots. *Genome Research*, *28*, 1333–1344. https://doi.org/10.1101/gr.228163.117

Komiya, R., Ohyanagi, H., Niihama, M., Watanabe, T., Nakano, M., Kurata, N., & Nonomura, K.-I. (2014). Rice germline-specific Argonaute MEL1 protein binds to phasiRNAs generated from more than 700 lincRNAs. *Plant Journal*, *78*(3), 385–397. https://doi.org/10.1111/tpj.12483

Kumar, S., Stecher, G., Suleski, M., & Hedges, S. B. (2017). TimeTree: A resource for timelines, timetrees, and divergence times. *Molecular Biology and Evolution*, *34*(7), 1812–1819. https://doi.org/10.1093/molbev/msx116

Kutter, C., Schöb, H., Stadler, M., Meins, F., & Si-Ammour, A. (2007). MicroRNA-mediated regulation of stomatal development in Arabidopsis. *Plant Cell*, *19*(8), 2417–2429. https://doi.org/10.1105/tpc.107.050377

Langmead, B., Trapnell, C., Pop, M., & Salzberg, S. L. (2009). Ultrafast and memory-efficient alignment of short DNA sequences to the human genome. *Genome Biology*, *10*(3), R25. https://doi.org/10.1186/gb-2009-10-3-r25

Larkin, M. A., Blackshields, G., Brown, N. P., Chenna, R., McGettigan, P. A., McWilliam, H., ... Higgins, D. G. (2007). Clustal W and Clustal X version 2.0. *Bioinformatics*, *23*(21), 2947–2948. https://doi.org/10.1093/bioinformatics/btm404

Laufs, P., Peaucelle, A., Morin, H., & Traas, J. (2004). MicroRNA regulation of the CUC genes is required for boundary size control in Arabidopsis meristems. *Development*, *131*(17), 4311–4322. https://doi.org/10.1242/dev.01320

Matzke, M., Kanno, T., Daxinger, L., Huettel, B., & Matzke, A. J. (2009). RNA-mediated chromatin-based silencing in plants. *Current Opinion in Cell Biology*, *21*(3), 367–376. https://doi.org/10.1016/j.ceb.2009.01.025

Mi, S., Cai, T., Hu, Y., Chen, Y., Hodges, E., Ni, F., & Qi, Y. (2008). Sorting of small RNAs into Arabidopsis argonaute complexes is directed by the 5′ terminal nucleotide. *Cell*, *133*(1), 116–127. https://doi.org/10.1016/j.cell.2008.02.034

Millar, A. A., & Waterhouse, P. M. (2005). Plant and animal microRNAs: Similarities and differences. *Functional & Integrative Genomics*, *5*(3), 129–135. https://doi.org/10.1007/s10142-005-0145-2

Ono, S., Liu, H., Tsuda, K., Fukai, E., Tanaka, K., Sasaki, T., & Nonomura, K.-I. (2018). EAT1 transcription factor, a non-cell-autonomous regulator of pollen production, activates meiotic small RNA biogenesis in rice anther tapetum. *PLOS Genetics*, *14*(2), e1007238. https://doi.org/10.1371/journal.pgen.1007238

Onodera, Y., Haag, J. R., Ream, T., Nunes, P. C., Pontes, O., & Pikaard, C. S. (2005). Plant nuclear RNA polymerase IV mediates siRNA and DNA methylation-dependent heterochromatin formation. *Cell*, *120*(5), 613–622. https://doi.org/10.1016/j.cell.2005.02.007

Polydore, S., & Axtell, M. J. (2018). Analysis of RDR1/RDR2/RDR6-independent small RNAs in Arabidopsis thaliana improves MIRNA annotations and reveals unexplained types of short interfering RNA loci. *Plant Journal: For Cell and Molecular Biology*, *94*(6), 1051–1063. https://doi.org/10.1111/tpj.13919

Qi, Y., Denli, A. M., & Hannon, G. J. (2005). Biochemical specialization within Arabidopsis RNA silencing pathways. *Molecular Cell*, *19*(3), 421–428. https://doi.org/10.1016/j.molcel.2005.06.014

Schmutz, J., McClean, P. E., Mamidi, S., Wu, G. A., Cannon, S. B., Grimwood, J., & Jackson, S. A. (2014). A reference genome for common bean and genome-wide analysis of dual domestications. *Nature Genetics*, *46*(7), 707–713. https://doi.org/10.1038/ng.3008

Sharma, B., & Kramer, E. M. (2014). The MADS-box gene family of the basal eudicot and hybrid aquilegia coerulea "origami" (Ranunculaceae). *Annals of the Missouri Botanical Garden*, *99*(3), 313–322.

Song, X., Li, P., Zhai, J., Zhou, M., Ma, L., Liu, B., & Cao, X. (2012). Roles of DCL4 and DCL3b in rice phased small RNA biogenesis. *Plant Journal*, *69*(3), 462–474. https://doi.org/10.1111/j.1365-313X.2011.04805.x

Springer, N. M., Ying, K., Fu, Y., Ji, T., Yeh, C.-T., Jia, Y., & Schnable, P. S. (2009). Maize inbreds exhibit high levels of copy number variation (CNV) and presence/absence variation (PAV) in genome content. *PLOS Genetics*, *5*(11), e1000734. https://doi.org/10.1371/journal.pgen.1000734

Sugiyama, T., Cam, H., Verdel, A., Moazed, D., & Grewal, S. I. S. (2005). RNA-dependent RNA polymerase is an essential component of a self-enforcing loop coupling heterochromatin assembly to siRNA production. *Proceedings of the National Academy of Sciences of the United States of America*, *102*(1), 152. https://doi.org/10.1073/pnas.0407641102

Tamim, S., Cai, Z., Mathioni, S., Zhai, J., Teng, C., Zhang, Q., & Meyers, B. C. (2018). Cis-directed cleavage and nonstoichiometric abundances of 21-nt reproductive phasiRNAs in grasses. *bioRxiv*, 243907. https://doi.org/10.1101/243907

Tolia, N. H., & Joshua-Tor, L. (2007). Slicer and the Argonautes. *Nature Chemical Biology*, *3*(1), 36–43. https://doi.org/10.1038/nchembio848

Wang, X., Wang, H., Wang, J., Sun, R., Wu, J., Liu, S., & Zhang, Z. (2011). The genome of the mesopolyploid crop species *Brassica rapa*. *Nature Genetics*, *43*(10), 1035–1039. https://doi.org/10.1038/ng.919

Williams, L., Grigg, S. P., Xie, M., Christensen, S., & Fletcher, J. C. (2005). Regulation of Arabidopsis shoot apical meristem and lateral organ formation by microRNA miR166g and its AtHD-ZIP target genes. *Development*, *132*(16), 3657–3668. https://doi.org/10.1242/dev.01942

Willmann, M. R., Endres, M. W., Cook, R. T., & Gregory, B. D. (2011). The functions of RNA-dependent RNA polymerases in Arabidopsis. *Arabidopsis Book*, *9*, e0146. https://doi.org/10.1199/tab.0146

Ye, R., Chen, Z., Lian, B., Rowley, M. J., Xia, N., Chai, J., & Qi, Y. (2016). A dicer-independent route for biogenesis of siRNAs that direct DNA methylation in Arabidopsis. *Molecular Cell*, *61*(2), 222–235. https://doi.org/10.1016/j.molcel.2015.11.015

Zhai, J., Bischof, S., Wang, H., Feng, S., Lee, T., Teng, C., & Jacobsen, S. E. (2015). One precursor one siRNA model for Pol IV- dependent siRNAs biogenesis. *Cell*, *163*(2), 445–455. https://doi.org/10.1016/j.cell.2015.09.032

Zhai, J., Zhang, H., Arikit, S., Huang, K., Nan, G.-L., Walbot, V., & Meyers, B. C. (2015). Spatiotemporally dynamic, cell-type–dependent premeiotic and meiotic phasiRNAs in maize anthers. *Proceedings of the National Academy of Sciences, USA*, *112*(10), 3146–3151. https://doi.org/10.1073/pnas.1418918112

Zhang, H., Xia, R., Meyers, B. C., & Walbot, V. (2015). Evolution, functions, and mysteries of plant ARGONAUTE proteins. *Current Opinion in Plant Biology*, *27*, 84–90. https://doi.org/10.1016/j.pbi.2015.06.011

Zheng, Yun, Wang, S., & Sunkar, R. (2014). Genome-wide discovery and analysis of phased small interfering RNAs in Chinese sacred lotus. *PLoS ONE*, *9*(12), e113790. https://doi.org/10.1371/journal.pone.0113790

Zheng, Yi, Wang, Y., Wu, J., Ding, B., & Fei, Z. (2015). A dynamic evolutionary and functional landscape of plant phased small interfering RNAs. *BMC Biology*, *13*, 32. https://doi.org/10.1186/s12915-015-0142-4

Zuker, M. (2003). Mfold web server for nucleic acid folding and hybridization prediction. *Nucleic Acids Research*, *31*(13), 3406–3415.

Photosynthetic response to increased irradiance correlates to variation in transcriptional response of lipid-remodeling and heat-shock genes

Roxanne van Rooijen[1,2,†] | Jeremy Harbinson[2] | Mark G. M. Aarts[1]

[1]Laboratory of Genetics, Wageningen University and Research, Wageningen, The Netherlands

[2]Horticulture and Product Physiology, Wageningen University and Research, Wageningen, The Netherlands

Correspondence
Mark G. M. Aarts, Laboratory of Genetics, Wageningen University and Research, Droevendaalsesteeg 1, 6708 PB Wageningen, The Netherlands.
Email: mark.aarts@wur.nl

Present address
Roxanne van Rooijen, Cluster of Excellence on Plant Science, Heinrich Heine University, Düsseldorf, Germany.

Funding information
This project was carried out within the research programme of BioSolar Cells, cofinanced by the Dutch Ministry of Economic Affairs.

Abstract

Plants have evolved several mechanisms for sensing increased irradiance, involving signal perception by photoreceptors (cryptochromes), and subsequent biochemical (reactive oxygen species, ROS) and metabolic clues to transmit the signals. This results in the increased expression of heat-shock response genes and of the transcription factor LONG HYPOCOTYL 5 (HY5, mediated by the cryptochrome photoreceptor 1, CRY1). Here, we show the existence of another response pathway in Arabidopsis. This pathway evokes the SPX1-mediated expression activation of the transcription factor PHR1 and leads to the expression of several galactolipid biosynthesis genes. Gene expression analysis of accessions Col-0, Ga-0, and Ts-1, showed activated expression of the SPX1/PHR1-mediated gene expression activation pathway acting on galactolipids biosynthesis genes in both Ga-0 and Col-0, but not in Ts-1. The activation of the SPX1/PHR1-mediated response pathway can be associated with lower photosynthesis efficiency in Ts-1, compared to Col-0 and Ga-0. Besides the accession-associated activation of the SPX1/PHR1-mediated response pathway, comparing gene expression in the accessions showed stronger activation of several heat responsive genes in Ga-0, and the opposite in Ts-1, when compared to Col-0, in line with the differences in their efficiency of photosynthesis. We conclude that natural variation in activation of both heat responsive genes and of galactolipids biosynthesis genes contribute to the variation in photosynthesis efficiency in response to irradiance increase.

KEYWORDS
Arabidopsis thaliana, high light stress, microarray, natural genetic variation, photosynthesis efficiency, photosystem II

1 | INTRODUCTION

The light-use efficiency of photosynthesis is the result of the molecular, structural, and physiological state of the plant (Eberhard, Finazzi, & Wollman, 2008; Foyer, Neukermans, Queval, Noctor, & Harbinson, 2012; Zhu, Long, & Ort, 2008), which depends on many environmental factors. One of these, the level of irradiance, has a direct relation with photosynthesis light-use efficiency as irradiance is the driving force for photosynthesis. At low irradiances, photosynthesis is fully light-limited and photosynthetic light-use efficiency is maximal. Beyond light-limitation, light-use efficiency decreases with increasing irradiances, resulting in the overall phenomenon of light-saturation of photosynthesis at elevated irradiances (Long, Humphries, & Falkowski, 1994; Sinclair & Muchow, 1999). The decrease in light-use efficiency from its maximum under wholly light-limiting conditions implies that irradiance exceeds the capacity for photosynthetic metabolism (Long et al., 1994). A consequence of increased irradiance is an increase in the rate of damaging side reactions of photosynthesis that occur as a result of the reactive nature of many intermediates formed. These reactive intermediates come from redox signals mediated via changes in the degree of thioredoxin and plastoquinone reduction and increased formation of reactive oxygen species (ROS) (Vass, 2012). Reactive oxygen species (singlet oxygen [$^1O_2$], superoxide [O_2^-], the hydroxyl radical [•OH] and hydrogen peroxide [H_2O_2]) are the most conspicuously damaging by-products of photosynthesis (Asada, 2006; Vass, 2012). The regulatory responses of photosynthesis, which are more active during stress (for example q_E-type nonphotochemical quenching or down-regulation of electron transport at the level of the cytochrome b_6/f complex) appear to reduce the formation of ROS, especially under high growth irradiances (Scheibe, Backhausen, Emmerlich, & Holtgrefe, 2005; Suzuki, Koussevitzky, Mittler, & Miller, 2012). The state of the photosynthesis apparatus – its composition, organization, and regulation – is thus under complex control. An increase in irradiance leading to excess will initially provoke rapid, physiological responses, including q_E type quenching, down-regulation of electron transport at the cytochrome b_6/f complex and increased CO_2 fixation activity (Demmig-Adams & Adams, 1992; Li, Wakao, Fischer, & Niyogi, 2009; Niyogi, 1999). If persistent, the increased irradiance will result in longer term acclimation of the photosynthetic apparatus, which can be detected at transcript and protein levels (Li et al., 2009; Walters, 2005). Different species and genotypes display different capacities to acclimate their photosynthetic apparatus to an irradiance increase so it is reasonable to infer that this is at least partly genetically determined (van Rooijen, Aarts, & Harbinson, 2015; van Rooijen et al., 2017).

In the rapid physiological regulatory response to irradiance increase, two major biochemical signals are known to be involved: a pH-change within the chloroplast lumen, occurring within milliseconds after the irradiance increase; and redox changes, acting via changes in reduced thioredoxin levels, oxidized/reduced plastoquinone ratios, and a build-up of ROS. These two classes of processes interact so changes in lumen pH regulation can affect the extent of redox changes and vice versa. Increases in ROS not only result from the irradiance increase itself but also from the concomitant temperature increase (Apel & Hirt, 2004). This combined irradiance and temperature effect causes the activation of heat-shock factors, initiating further acclimation responses (Jung et al., 2013).

The transcriptome changes provoked by an increase in irradiance are not easy to interpret, being partly caused by the temperature increase rather than the irradiance increase (Swindell, Huebner, & Weber, 2007). Increased irradiance and increased temperature together induce a stress response at the site of photosynthesis, the chloroplast, mediated through a biochemical and molecular signaling cascade going from the chloroplast to the nucleus, known as retrograde signaling (Dietz, Wesemann, Wegener, & Seidel, 2018; Fey et al., 2005; Kleine & Leister, 2016). The main retrograde signals in response to increased temperature are the expression of heat-shock proteins through the generation of ROS originating from the chloroplast, which is also thought to be a major retrograde signal required for the photosynthetic acclimation response to excess light (Jung et al., 2013; Rossel, Wilson, & Pogson, 2002; Vanderauwera et al., 2005). The retrograde signaling function of singlet oxygen and H_2O_2 has been largely unraveled (Dietz et al., 2018). Singlet oxygen and H_2O_2 modulate the activity of redox-sensitive components of metabolic pathways such as the HDS-MEcPP (HDS = HYDROXY-2-METHYL-2-(E)-BUTENYL 4-DIPHOSPHATE (HMBPP) SYNTHASE; MEcPP = 2-C-Methyl-D-erythritol-2,4-cyclodiphosphate) and the SAL1-PAP (SAL1 is an enzyme with 3′(2′), 5′-biphosphate nucleotidase and inositol polyphosphate 1-phosphatase activities; PAP= 3′-phosphoadenosine 5′-phosphate) pathways. Both pathways lead to posttranscriptional gene silencing in the nuclear genome (Estavillo et al., 2011; Xiao et al., 2012). Besides modulating transcriptional networks, singlet oxygen and H_2O_2 also modulate translation of transcripts via the inhibition of conserved TARGET OF RAPAMYCIN COMPLEX (TORC) serine/threonine kinases and the simultaneous activation of SUCROSE NONFERMENTING-RELATED KINASE 1 (SnRK1), which regulate the balance between energy status and growth regulation (Robaglia, Thomas, & Meyer, 2012). Another regulatory mechanism is the ROS-mediated control of mRNA stability through the formation of stress granules and processing (p-)bodies (Dietz et al., 2018; Uniacke & Zerges, 2008).

In addition to the indirect detection of increase in irradiance through increased temperature and the formation of ROS, plants directly detect increases in irradiance through their cryptochrome photoreceptors (Kleine, Kindgren, Benedict, Hendrickson, & Strand, 2007). Of the two plant cryptochromes in *Arabidopsis thaliana* (Arabidopsis), CRY1 and CRY2, only CRY1 initiates a transcriptional response to increased irradiance, mediated by the transcription factor LONG HYPOCOTYL 5 (HY5) (Kleine et al., 2007; Lee et al., 2007). The gene expression response to increased irradiance initiated by the heat-shock factors and the response initiated via the HY5 transcription factor are considered to be two distinct gene activation pathways, both in function and in time (Yamamoto et al., 2004). However, it was recently unraveled that HY5 also negatively controls the unfolded protein response in Arabidopsis, in which a set of molecular chaperones is expressed to assist the endoplasmic reticulum in protein folding during stress situations (Nawkar et al., 2017).

As a last indirect sensing mechanism, plants sense increased irradiance through an accumulation of the phytohormone abscisic acid (ABA) in leaves (Galvez-Valdivieso et al., 2009). ABA, high light, and oxidative stress similarly down-regulate *Lhcb* genes; hydrogen peroxide produced in chloroplasts under high light conditions interacts with the ABA signaling network to regulate *Lhcb* expression (Borisova-Mubarakshina et al., 2015; Staneloni, Rodriguez-Batiller, & Casal, 2008).

Until now, the response of Arabidopsis to increased irradiance has only been studied in one accession, Col-0. In our study we include two additional, natural, accessions of Arabidopsis, with contrasting photosynthesis responses to increased irradiance (van Rooijen et al., 2015), to reveal common, as well as genotype-specific, transcriptional responses associated with the acclimation of photosynthesis efficiency. Analysis of different time-points within one study allows the detection of transient expression patterns throughout the acclimation response. Next to the induction of genes involved in the protection against a temperature increase, and genes involved in the generation and perception of ROS - mediated through (retrograde) signaling - we expect the increased irradiance to temporarily arrest plant growth, which should lead to a decrease in the expression of genes involved in growth and development. In the course of acclimation, expression of these genes is expected to return to the status before the increase in irradiance.

2 | MATERIALS AND METHODS

2.1 | Plant material and growth conditions

The Arabidopsis accessions Columbia-0 (Col-0, CS76113), Gabelstein-0 (Ga-0, CS76133) and Tossa de Mar-1 (Ts-1, CS76268) were grown as previously described (van Rooijen et al., 2015). In short, the plants were grown under controlled conditions in the Phenovator (Flood et al., 2016), on rockwool, automatically watered with nutrient solution, in a 10 hr/14 hr day/night cycle, with the temperature set at 20/18°C (day/night), and relative humidity set at 70%. CO_2 levels were ambient. Under controlled conditions, plants in the light were kept at a constant, low, irradiance of 100 μmol m^{-2} s^{-1} (Philips 610 fluorescent tubes, MASTER TL5 HO, 80W). In the increased irradiance treatment, the irradiance was increased to 550 μmol m^{-2} s^{-1} at the onset of the photoperiod on the 25th day after sowing. In the increased temperature treatment, the irradiance was kept at 100 μmol m^{-2} s^{-1}, but the temperature was increased from 20°C to 30°C during the day. Photosynthesis efficiency was measured as previously described (van Rooijen et al., 2015, 2017).

2.2 | RNA sample preparation

For all three genotypes, rosettes of nine plants per genotype were harvested on the day of the irradiance increase, 1 hr after the start of the photoperiod and flash-frozen in liquid nitrogen. For the Col-0 accession, nine rosettes were additionally harvested on the day of the irradiance increase 3.5 hr after the start of the photoperiod, and again 1 day later, at 1 hr after the start of the photoperiod, so 25 hr after increasing the irradiance to 550 μmol m^{-2} s^{-1}. As a control, for all three genotypes rosettes of nine plants per genotype that were kept in low light (100 μmol m^{-2} s^{-1}) were harvested at 1 hr after the start of the photoperiod, and for Col-0 also at the additional time points, at 3.5 and 25 hr. The nine rosettes were subdivided in three groups of three rosettes and treated as three biological replicates. For each biological replicate for each accession-treatment combination, the three rosettes were pooled as one sample for RNA isolation.

Total RNA was extracted using the Direct-zol RNA mini-prep kit from Zymo Research (www.zymoresearch.com). The RNA quality control, labeling, microarray hybridization, and data extraction were performed at ServiceXS B.V. (part of GenomeScan B.V., Leiden, The Netherlands). The RNA concentration was measured using a DropSense96 spectrophotometer (Trinean N.V., Gentbrugge, Belgium). The RNA quality and integrity was determined using Lab-on-Chip analysis on the Agilent 2100 BioAnalyzer (Agilent TEchn9ologies, Inc., Santa Clara, CA, USA). Single-strand-cDNA (ss-cDNA) was prepared using the Ovation® PicoSL WTA System V2 (NuGEN Technologies, Inc., San Carlos, CA, USA) according to the manufacturer's specifications, with an input of 50 ng total RNA. Labeling and fragmentation of the ss-cDNA was performed with the Encore™ Biotin Module (NuGEN Technologies, Inc., San Carlos, CA, USA).

2.3 | Microarray hybridization, scanning, and data analysis

Per sample, 2.5 μg of the labeled ss-cDNA was hybridized onto an AraGene 1.1ST Array plate (Affymetrix, Santa Clara, CA, USA). Hybridization and scanning was performed on a GeneTitan (Affymetrix, Santa Clara, CA, USA). Image analysis and extraction of raw expression data was performed with the Affymetrix Expression Console™ v1.2.1. with "Gene-level Default: RMA-Sketch" settings. Raw data were analysed using the Bioconductor packages in the statistical programming language R (http://www.r-project.org) (Gentleman et al., 2004). The Affymetrix AraGene 1.1ST microarray oligonucleotide probe annotation list TAIRT v19 cdf file was downloaded from Brainarray (http://brainarray.mbni.med.umich.edu/Brainarray/Database/CustomCDF/19.0.0/tairt.asp) and gene expression was normalized using the Robust Multi-array Average (RMA) algorithm (Irizarry et al., 2003), after which a linear model was fitted for every gene. The empirical Bayes method was used to determine significant differences between the samples and the Benjamini and Hochberg method was used for adjustment of the *p* values for multiple testing (Benjamini & Hochberg, 1995; Efron & Tibshirani, 2002). The limma package was used to conduct log2 fold changes between control (low light, LL) conditions and experimental (high light, HL) conditions. Genes were selected with log2 fold change ≤0.7, or with log2 fold change ≥1.4. The limma package was also used to determine differential responsiveness of genes between accessions or between time points with a *p*-value cut-off of 0.05. The microarray dataset has been deposited at NCBI with accession number GSE115051, and can be accessed through http://www.ncbi.nlm.nih.gov/geo/query/acc.cgi?acc=GSE115051.

2.4 | Quantitative reverse transcription PCR (qRT-PCR)

In a separate experiment, but with identical experimental set-up as used for the microarray experiment, nine rosettes were similarly harvested, on the 25th day after sowing, for all accession-treatment combinations at 1 and 3.5 hr after the start of the photoperiod, and immediately frozen in liquid nitrogen. As a control, rosettes of nine plants per accession that were grown in low light (100 μmol m^{-2} s^{-1}) were harvested at the same time points. Three rosettes for each accession-treatment combination were pooled as one sample for RNA isolation, in three biological replications. RNA was extracted according to Onate-Sánchez and Vicente-Carbajosa (2008). After normalization of RNA concentrations, cDNA was synthesized using the Iscript cDNA synthesis kit from Bio-RAD (www.bio-rad.com). Quantitative real-time reverse transcriptase PCR (qRT-PCR) was performed with three technical replicates for each biological replicate using the SYBR-green mastermix from Bio-RAD. Three reference genes were used for normalization of samples: *UBIQUITIN7* (*UBQ7*, At2 g35635), *CYTOCHROME B5 ISOFORM E* (*CB5E*, At5 g53560), and *UBIQUITIN THIOESTERASE* (At1 g28120); expression levels of *UBQ7* and *CB5E* were previously found to be constant under excess light (Jung et al., 2013; Wunder et al., 2013), and expression levels of *UBIQUITIN THIOESTERASE* were previously found to be stable in several genome-wide expression studies involving irradiance changes (Genevestigator database (Hruz et al., 2011), https://genevestigator.com/gv/). The primers used for qRT-PCR are listed in Supporting Information Table S7.

3 | RESULTS

3.1 | Heat-shock response and RNA binding protein genes are consistently responding to an increase in irradiance in Arabidopsis

Arabidopsis accessions Col-0, Ga-0 and Ts-1 were grown for 24 days under 100 μmol m^{-2} s^{-1} growth irradiance (low light, LL), and from the next day onwards exposed to an irradiance of 550 μmol m^{-2} s^{-1} (high light, HL), which is saturating for photosynthesis in these low-light grown plants (van Rooijen et al., 2015). Upon exposure to increased irradiance, the accessions display different photosynthetic light-use efficiencies (determined by measuring Φ_{PSII}, the light-use efficiency of photosystem II (PSII) electron transport, also known as Fq′/Fm′) (Figure 1a). Ts-1 has a generally lower Φ_{PSII} than Col-0 and Ga-0, independent of the growth irradiance level. After the switch from low light to high light, all accessions show a decrease in Φ_{PSII} with Ga-0 showing the smallest drop and highest recovery. Ts-1 recovers most, reaching the same level of Φ_{PSII} as it had before the irradiance increase after 4 days, which was not achieved by the other two accessions. Col-0 Φ_{PSII} recovers most during the day, whereas in Ga-0 it recovers most during the night. These accession-specific differences in Φ_{PSII} response are reflected in measurements of the chlorophyll reflectance index and the photochemical reflectance index (Supporting Information Figure S1) (Datt, 1999; Gamon, Serrano, & Surfus, 1997). Indirectly related to photosynthetic light-use efficiency is the anthocyanin reflectance index, a proxy for anthocyanin content (Gitelson, Merzlyak, & Chivkunova, 2001). Also the anthocyanin reflectance index shows accession-specific differences (Supporting Information Figure S1). However, we did not find an accession-correlated relation between anthocyanin reflectance index and photosynthetic light use efficiency. A rosette transcriptome analysis was performed on the accessions Ts-1, Col-0, and Ga-0 to identify common and accession-specific responsive genes, at 1 hr after the lights were switched on at HL. For Col-0, a time-series transcriptome analysis was performed to investigate the temporal effect on gene expression.

Hierarchical clustering distinguishes all treatments, except for the Col-0 LL control treatments at three consecutive days (1 hr after the lights were switched on), which treatments were all very similar (Figure 1b). The accession differences are most prominent and the effect

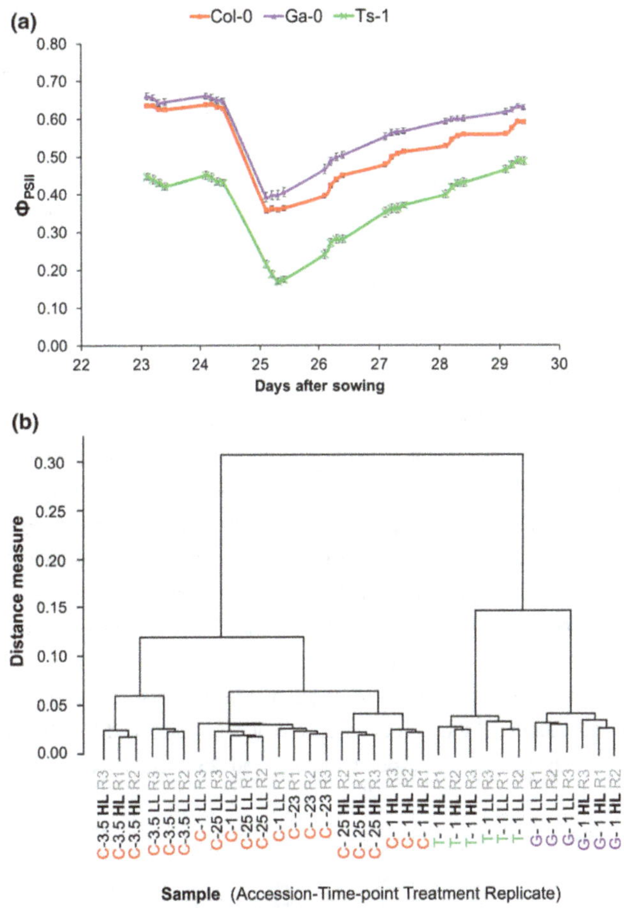

FIGURE 1 The effect of genotype on the photosynthetic response over time to increased irradiance in Arabidopsis. (a) Representative photosynthetic (Φ_{PSII}) phenotypes for Arabidopsis accessions Col-0, Ga-0, and Ts-1, grown for 24 days in 100 μmol m^{-2} s^{-1} growth irradiance and subsequently 6 days in 550 μmol m^{-2} s^{-1} growth irradiance, measured from day 23 until day 30 after sowing, at four time-points per day; (b) Dendrogram of Pearson distance measures, based on microarray-based transcriptome data representing gene expression in rosettes of Arabidopsis accessions Col-0, Ts-1 and Ga-0 exposed to low light (LL; 100 μmol m^{-2} s^{-1} growth irradiance) control conditions or increased irradiance (high light, HL; 550 μmol m^{-2} s^{-1} growth irradiance), sampled at 23 hr before the irradiance increase (−23), and 1, 3.5 and 25 hr after the irradiance increase (1, 3.5, 25), in three experimental replicates (R1, R2 and R3)

of time seems to be larger than the effect of the HL treatment, suggesting that the LL-HL comparison will provide a specific set of HL-responsive genes and a general set of stress-responsive genes. Genes that were statistically significantly more than 1.4-fold up- or down-regulated when comparing the increased irradiance treatment with the control treatment, will be further refer to as "responsive genes" (Figure 2).

Over all, a total of 752 genes were more than 1.4-fold up-regulated, and 440 genes were more than 1.4-fold down-regulated in at least one of the three accessions 1 hr after the irradiance increase from 100 to 550 μmol m^{-2} s^{-1}, when compared to untreated plants (Figure 2a). Of these responsive genes, 161 up-regulated and 59 down-regulated genes were shared among all three accessions (Figure 2a). In Col-0, a total of 1,664 genes were more than 1.4 fold up-regulated, and 1,356 genes were more than 1.4-fold down-regulated at any of the three time-points after the switch, of which 115 were up-regulated at all three time-points, and 120 were down-regulated at all three time-points (Figure 2b). The genes common to both comparisons (i.e. responding in all accessions and at all time-points in Col-0) were considered to represent a core set of 88 Arabidopsis genes (53 up-regulated, and 35 down-regulated) that respond to an irradiance increase (Supporting Information Tables S1 and S2).

Gene ontology analysis regarding the biological processes associated with this core set of 88 genes shows that up-regulated genes are enriched for involvement in "heat-shock response", "photosynthesis" and "RNA binding" processes; while the down-regulated genes show only minor enrichment for "photosynthesis" processes (Figure 3a). When comparing HL and LL conditions the enriched biological processes comprise 39 differentially expressed genes (Figure 3b). The majority of these encode heat-shock proteins (HSPs), which are up-regulated. Only three photosynthesis-associated genes are identified among the 39 differentially expressed genes; these three all encode transcription factors and are all down-regulated.

Comparison of the fold-changes of the core set of responsive genes between the three accessions (Ga-0, Ts-1, and Col-0), or between the three time-points (1, 3.5, and 25 hr after exposure to irradiance increase), revealed differences in responsiveness for some genes (Supporting Information Tables S1 and S2). Of these, only two differed in their expression when comparing accessions. These were the *DEHYDRATION RESPONSIVE ELEMENT BINDING 2A (DREB2A;* At5 g05410*)* gene, which belongs to the drought-related ERF/AP2 transcription factor family, and the *SERINE-ARGININE RICH RNA BINDING PROTEIN 45a (SR45a;* At1 g07350), which is involved in alternative RNA splicing. Twenty-seven genes differed in a time-point comparison, of which all were heat-shock response genes. Nine core genes differed in both the interaccession and time-point comparisons, and these comprised eight heat-shock response genes and one DREB-like drought-related transcription factor other than DREB2A, belonging to the ERF/AP2 transcription factor family.

The expression of four core responsive genes showed a high correlation ($r > 0.5$) with the photosynthesis efficiency of the accessions (measured as Φ_{PSII}); these are the *CELL DIVISION CYCLE 48B* (*CDC48B*) gene, an AAA-type ATPase (At3 g53230), the *HSP60* heat-shock protein gene (At3 g23990), and the *SIAMESE-RELATED 5* (*SMR5*) cell cycle inhibitor gene (At1 g07500; Supporting Information Table S1). The expression of four other genes was negatively correlated with photosynthesis efficiency ($r < -0.5$); *DREB2A* (At5 g05410), the *HSFA2* heat-shock factor gene (At2 g26150), an *HSP20-like* heat-shock protein chaperone gene (At1 g07400), and the *MULTIPROTEIN BRIDGING FACTOR 1C* (*MBF1C*) heat-responsive transcriptional coactivator gene (At3 g24500) (Supporting Information Table S1).

3.2 | Heat-shock response and lipid-remodeling genes differ in response to high light when comparing accessions

Next to the shared, common responses, the three accessions also showed differential responses. 1 hr after the irradiance increase a total of 1,192 genes were more than 1.4-fold up-regulated or down-regulated in at least one of the three accessions, when compared to untreated plants. Of these, 752 genes were up-regulated, and 440

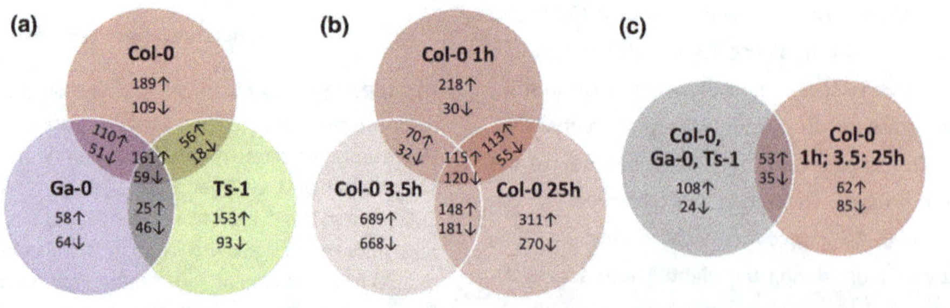

FIGURE 2 Arabidopsis accession- and time-specific gene expression responses to increased irradiance. Venn diagrams displaying the number of significantly (p = 0.05) differentially (more than 1.4-fold up- or down-regulated) expressed genes when comparing (a) rosettes of plants grown at low light (100 μmol m^{-2} s^{-1} growth irradiance) of three accessions (Col-0, Ga-0, or Ts-1; control) to those of plants exposed for the first day to high light (550 μmol m^{-2} s^{-1} growth irradiance), 1 hr after lights were switched on; and (b) rosettes of control Col-0 plants grown at low light (100 μmol m^{-2} s^{-1} growth irradiance) at 1, 3.5 or 25 hr after lights on, to those of Col-0 plants at 1, 3.5, or 25 hr after exposure to increased irradiance (550 μmol m^{-2} s^{-1} growth irradiance); and (c) the differentially expressed genes shared by all accessions indicated in (a) with those shared by all time-points indicated in (b). Arrows indicate up- and down-regulation compared to controls

(a)

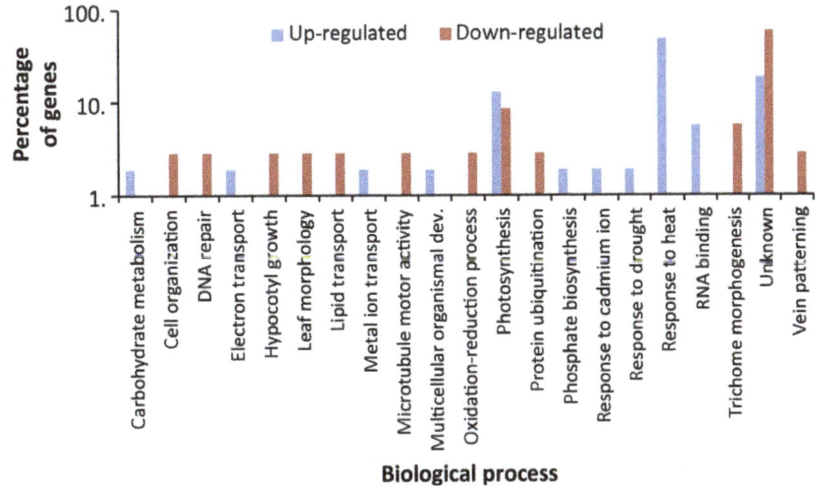

(b)

Molecular function	Gene ID	Gene name	G 1	T 1	C 1	C 3.5	C 25
Chlorophyll binding / flavonoid biosynthesis	At3g22840	EARLY LIGHT-INDUCIBLE PROTEIN 1 (ELIP1)					
	At4g14690	EARLY LIGHT-INDUCIBLE PROTEIN 2 (ELIP2)					
H_2O_2 scavenging	At1g07890	ASCORBATE PEROXIDASE 1 (APX1)					
	At4g31870	GLUTATHIONE PEROXIDASE 7 (GPX7)					
Transcription factor activity	At2g20880	ERF/AP2 FAMILY, ETHYLENE RESPONSE FACTOR 53 (ERF53)					
	At4g28140	ERF/AP2, A6 FAMILY					
	At5g05410	ERF/AP2 FAMILY, DEHYDRATION RESPONSE ELEMENT BINDING 2A (DREB2A)					
	At5g07580	ERF/AP2, B3 FAMILY					
	At5g25190	ERF/AP2, B6 FAMILY					
	At5g44190	GOLDEN2-LIKE2 (GLK2)					
Heat-shock chaperones	At1g07400	HSP20-like chaperones superfamily protein					
	At1g53540	HSP20-like chaperones superfamily protein					
	At1g54050	HSP20-like chaperones superfamily protein					
	At1g59860	HSP20-like chaperones superfamily protein					
	At2g19310	HSP20-like chaperones superfamily protein					
	At2g29500	HSP20-like chaperones superfamily protein					
	At5g51440	HSP20-like chaperones superfamily protein					
	At2g46240	A member of Arabidopsis BAG (Bcl-2-associated athanogene) proteins					
	At5g48570	Tetratricopeptide repeat (TPR) protein, interacts with Hsp90/Hsp70 as co-chaperones					
Heat-shock factor	At2g26150	HEAT-SHOCK TRANSCRIPTION FACTOR A2 (HSFA2)					
	At4g36990	HEAT-SHOCK FACTOR 4 (HSF4)					
Heat-shock protein	At1g74310	HEAT-SHOCK PROTEIN 101 (HSP101)					
	At2g20560	DNAJ heat-shock family protein					
	At2g25140	Casein lytic proteinase/heat-shock protein 100 (Clp/Hsp100) family.					
	At3g08970	J domain protein localized in ER lumen, shows similarity to HSP40 proteins					
	At3g12580	HEAT-SHOCK PROTEIN 70 (HSP70)					
	At3g23990	HSP60; mitochondrial chaperonin HSP.					
	At3g46230	HEAT-SHOCK PROTEIN 17.4 (HSP17.4)					
	At4g12400	Hop3, a tetratricopeptide repeat (TPR) protein, interacts with Hsp90/Hsp70					
	At4g25200	MITOCHONDR.-LOCALIZED SMALL HEAT-SHOCK PROTEIN 23.6 (HSP23.6-MITO)					
	At5g09590	Heat-shock protein 70 (Hsc70-5)					
	At5g12020	17.6 KDA CLASS II HEAT-SHOCK PROTEIN (HSP17.6II)					
	At5g12030	HEAT-SHOCK PROTEIN 17.6A (HSP17.6A)					
	At5g52640	HEAT-SHOCK PROTEIN 90.1 (HSP90.1)					
Unknown	At1g30070	unknown					
	At5g64510	TUNICAMYCIN INDUCED 1 (TIN1), a plant-specific ER stress-inducible protein					
Alternative splicing	At1g07350	Serine-arginine rich RNA binding protein (SR45a) involved in alternative splicing					
	At1g09140	Serine-arginine rich RNA binding protein (SR30) involved in alternative splicing					
Translation elongation	At5g12110	Translation elongation factor EF1B/ribosomal protein S6					

Fold Change HL vs LL: 24, 22, 20, 18, 16, 14, 12, 10, 8, 6, 4, 2, 1, 0.9, 0.8, 0.7, 0.6, 0.5, 0.4, 0.3, 0.2, 0.1, 0

FIGURE 3 Gene ontology and functional annotation of a core set of Arabidopsis genes expressed in rosettes, showing a transcriptional response upon exposure to increased irradiance in all three tested accessions (Ga-0, Ts-1 and Col-0) and at all examined time-points (after 1, 3.5 and 25 hr) when compared to plants grown in control conditions. (a) Gene ontology enrichment for biological process of Arabidopsis genes differentially expressed in rosettes at 1, 3.5, or 25 hr after exposure to increased irradiance, when comparing three accessions, Ga-0, Ts-1, and Col-0 (G, T, C), and three time-points (1, 3.5 and 25; in Col-0 only) to untreated controls; (b) Heat map displaying fold changes in gene expression in increased irradiance (HL) conditions versus low irradiance (LL) control conditions, of core genes with biological functions in photosynthesis (yellow highlighted), response to heat (gray highlighted), and RNA binding (green highlighted), as indicated in (a). All listed genes were statistically significantly more than 1.4-fold up- or down-regulated at all time-points and in all accessions when compared to control conditions. For the complete gene list see Supporting Information Tables S1 and S2

genes were down-regulated (Figure 2a). Of the 1,192 responsive genes, 153 genes were identified as accession-specific responsive genes (Supporting Information Tables S3 and S4), meaning they were differentially responsive ($p = 0.05$) between two or all three accessions (i.e. Col-0, Ga-0 and Ts-1). Of these 153 genes, 22 genes were also differentially expressed between accessions irrespective of the irradiance level, while the others responded specifically differentially to the HL treatment (Supporting Information Tables S3 and S4). Of the 153 accession-specific genes, 60 genes were positively correlated with photosynthesis efficiency ($r > 0.5$) and 55 were negatively correlated ($r < -0.5$; Supporting Information Tables S3 and S4).

Gene ontology analysis of the 153 accession-specific genes showed enrichment for "heat-shock response", "lipid-remodeling", "carbohydrate metabolism", and "photosynthesis" among the up-regulated genes; and for "cell organization", "photosynthesis" and "RNA binding" among the down-regulated genes (Figure 4a). Most prominent among the 51 genes classified in one of the enriched processes are the heat-shock response genes (Figure 4b).

Remarkably, we could not confirm the up-regulation of genes belonging to the well-studied CRY1/HY5 pathway (Kleine et al., 2007). The up-regulation of *HY5* was below the threshold we set in our micro array study; the average fold-changes were: in Ga-0 up-regulated 1.3-fold after 1 hr, in Ts-1 up-regulated 1.7-fold after 1 hr, in Col-0 up-regulated 1.2-fold after 1 hr, 1.5-fold after 3.5 hr, and 0.8-fold after 25 hr in this study. Although the up-regulation in Ts-1 was above our threshold of 1.4-fold, it was not selected for the accession-specific up-regulated genes, because it was not statistically different from the fold-change in Col-0 and Ga-0 ($p = 0.05$).

Additionally, we did not find up-regulated gene clusters involved in ABA responsiveness in this study. Gene clusters involved in ABA responsiveness are known to be induced by high light with reduced infra-red but repressed by high light containing infra-red (Rossel et al., 2002). We probably did not find up-regulation of this cluster in this study, because the ABA signaling network causes down-regulation of *Lhcb* genes only after several days of high light treatment, and not during the first hours of high light treatment (Alsharafa et al., 2014).

3.3 | Differential expression in genes for photosynthetic response and heat-shock response when comparing sequential time-points after increased irradiance in Col-0

To identify transient expression patterns throughout the acclimation response, and to study the effect of diurnal rhythm on gene expression responses during longer term acclimation, time-point-specific expression responses were characterized in Col-0 at three time-points after the irradiance increase. When compared to untreated plants, a total of 3,020 genes were significantly more than 1.4-fold up- or down-regulated in at least one of the three time-points after the irradiance increase (Figure 2b). When focusing on the 384 most responsive genes, 229 genes were more than 2.0-fold up-regulated and 155 genes more than 2.0-fold down-regulated (Supporting Information Tables S5 and S6). Gene ontology analysis of the 384 time-point-specific genes showed that those in the "photosynthesis", "heat-shock response", "lipid-remodeling", "RNA binding", and "carbohydrate metabolism" classes were enriched among the up-regulated genes, while the classes "cell organization", "photosynthesis", and "lipid-remodeling" were enriched among the down-regulated genes (Figure 5a). Figure 5b lists the time-point-specific classification of the responsive genes based on their associated biological process, and their expression difference when compared to control plants. Of the 384 time-specific genes, 58 genes correlated with an $r > 0.5$ to the photosynthesis efficiency measured in the same accession in the same time points and 78 correlated with an $r < -0.5$ (Supporting Information Tables S5 and S6). The majority of the photosynthesis related time-point specifically regulated genes in Col-0 are the genes involved in flavonoid/anthocyanin biosynthesis as well as the transcription factor encoding genes belonging to the ERF/AP2 family.

3.4 | Light response versus heat response

In order to distinguish between the direct effect on gene expression of an increase in irradiance and the indirect effect caused by the increase in temperature due to the increased irradiance, the expression of nine genes was determined using quantitative reverse transcriptase PCR (qRT-PCR). In this set, genes were included that classified as being involved in any of the overrepresented biological processes (Figures 3, 4 and 5). Additionally, the genes were either highly differentially expressed or among the genes for which expression was correlated with Φ_{PSII}. The plants were grown at either an increased irradiance, which also resulted in an increased temperature (HL), or an increased temperature (but with no increase in irradiance) (HT) and compared to plants grown under control conditions (no increased irradiance, no increased temperature; LL, LT). Plants were sampled 1 and 3.5 hr after the start of irradiance.

We investigated in qRT-PCR two genes involved in alternative splicing *SR45a* (At1 g07350) and *SR30* (At1 g09140); two transcription factors *GLK2* (At5 g44190) and *DREB2A* (At5 g05410), three heat-shock genes *HSFA2* (At2 g26150), *HOP3* (At4 g12400), and *CPN60BETA2* (At3 g13470); and two genes involved in lipid-remodeling *SPX1* (At5 g20150) and *GDPD1* (At3 g02040). Of these genes, all but *CPN60Beta2* were indeed confirmed to be responsive to increased irradiance (Figure 6), with *GLK2*, *SPX1*, *GDPD1*, and *DREB2A* to be specifically responsive to the irradiance increase and not to the temperature increase. Expression of *GLK2*, *SPX1*, *GDPD1*, and *DREB2A* was accession-specific, with little or no response in Ts-1. Also the four genes responsive to both the irradiance and temperature increases (*SR30*, *SR45a*, *HSFA2*, and *HOP3*) were accession-specific, again with little or no response in Ts-1. The accession-specific effect was more pronounced after the irradiance increase than after the temperature increase (Figure 6).

(a)

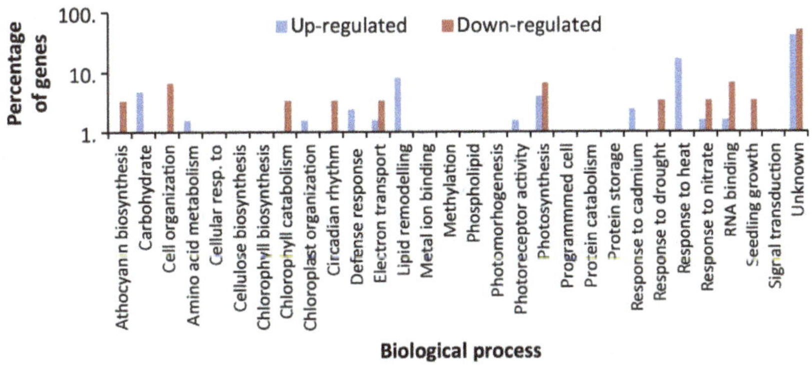

(b)

Molecular function	Gene ID	Gene name	G1	T 1	C1
Anthocyanin biosynthesis	At1g56650	PRODUCTION OF ANTHOCYANIN PIGMENT 1 (PAP1)			
Photoprotection	At4g04020	Fibrillin precursor protein, involved in abscisic acid-mediated photoprotection			
NAD(H) kinase	At3g21070	NAD(H) kinase			
Transcription factor activity	At2g20880	ERF/AP2 FAMILY, ETHYLENE RESPONSE FACTOR 53 (ERF53)			
	At4g25470	ERF/AP2 FAMILY, DREB subfamily A-1 (CBF2)			
	At5g05410	ERF/AP2 FAMILY, DEHYDRATION RESPONSE ELEMENT BINDING 2A (DREB2A)			
	At5g61590	ERF/AP2, B3 FAMILY			
Chaperones	At1g71000	Chaperone protein dnaJ-related			
	At2g46240	A member of Arabidopsis BAG (Bcl-2-associated athanogene) proteins			
	At3g13470	Subunit of chloroplasts chaperonins CHAPERONIN-60BETA2 (CPN60BETA2)			
	At5g43260	Chaperone protein dnaJ-related			
Heat-shock factor	At3g51910	HEAT-SHOCK TRANSCRIPTION FACTOR A7 A (HSFA7A)			
Heat-shock protein	At1g53540	HSP20-like chaperones superfamily protein			
	At2g32120	Heat-shock protein 70T-2 (HSP70T-2)			
	At3g08970	J domain protein localized in ER lumen, shows similarity to HSP40 proteins			
	At3g46230	HEAT-SHOCK PROTEIN 17.4 (HSP17.4)			
	At4g12400	Hop3, a tetratricopeptide repeat (TPR) protein, interacts with Hsp90/Hsp70			
	At4g25200	MITOCHONDRION-LOCALIZED SMALL HEAT-SHOCK PROTEIN 23.6 (HSP23.6-MITO)			
	At5g09590	Heat-shock protein 70 (Hsc70-5)			
	At5g37670	HSP20-like chaperones superfamily protein			
	At5g51440	HSP20-like chaperones superfamily protein			
	At5g56030	HEAT-SHOCK PROTEIN 81-2 (HSP81-2)			
Unknown	At1g03070	Apoptosis-promoting Bax inhibitor-1 family protein			
	At1g17870	ETHYLENE-DEPENDENT GRAVITROPISM-DEFICIENT AND YELLOW-GREEN-LIKE 3 (EGY3)			
	At1g66510	AAR2 protein family			
	At4g21320	heat-stress-associated 32-kD protein			
	At5g13200	GRAM domain family protein			
	At5g64510	TUNICAMYCIN INDUCED 1 (TIN1), a plant-specific ER stress-inducible protein			
Alternative splicing	At1g07350	Serine-arginine rich RNA binding protein (SR45a) involved in alternative splicing			
Unknown	At2g18510	Embryo defective 2444 (emb2444)			
	At3g09160	RNA-binding (RRM/RBD/RNP motifs) family protein			
	At4g03110	unknown RNA binding protein			
Ascorbic acid synthesis	At1g67070	PHOSPHOMANNOSE ISOMERASE 2 (PMI2), also known as DARK INDUCED 9 (DIN9)			
Mannose binding lectin protein	At1g78820	D-mannose binding lectin protein			
Cell wall synthesis	At5g49360	a bifunctional (beta)-D-xylosidase/(alpha)-L-arabinofuranosidase			
Sugar transport	At5g57100	Nucleotide/sugar transporter family protein			
HY5 target, chloroplast organization	At5g23730	REPRESSOR OF UV-B PHOTOMORPHOGENESIS 2 (RUP2)			
	At5g52250	REPRESSOR OF UV-B PHOTOMORPHOGENESIS 1 (RUP1)			
Lipid transfer protein	At2g38530	LIPID TRANSFER PROTEIN 2 (LTP2)			
MGDG synthesis	At2g11810	MONOGALACTOSYLDIACYLGLYCEROL SYNTHASE 3 (MGD3)			
	At5g20410	MONOGALACTOSYLDIACYLGLYCEROL SYNTHASE 2 (MGD2)			
Phosphatase	At3g17790	PURPLE ACID PHOSPHATASE 17 (PAP17)			
Phospholipid catabolism	At3g02040	GLYCEROPHOSPHODIESTER PHOSPHODIESTERASE 1 (GDPD1)			
SQDG synthesis	At4g33030	SULFOQUINOVOSYLDIACYLGLYCEROL 1			
	At5g01220	SULFOQUINOVOSYLDIACYLGLYCEROL 2			
Transcription factor activity	At3g03790	Ankyrin repeat family protein / regulator of chromosome condensation			
Unknown	At2g26660	SPX DOMAIN GENE 2 (SPX2)			
	At5g20150	SPX DOMAIN GENE 1 (SPX1)			
Transcription factor activity	At4g00150	Belongs to one of the LOM (LOST MERISTEMS) genes			
	At4g37180	Myb family transcription factor, contains Pfam domain			
Unknown	At2g40610	Alpha-Expansin Gene Family. Inv. in the form. of nematode-induced syncytia in roots			

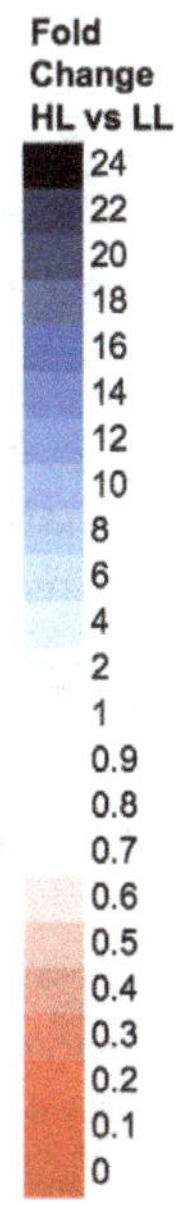

4 | DISCUSSION

Different natural accessions of Arabidopsis vary in their acclimation of photosynthetic efficiency (measured as Φ_{PSII}) to an increase in growth irradiance (van Rooijen et al., 2015, 2017). This study has been designed to identify transcriptional response pathways that are associated with natural variation in Φ_{PSII} acclimation to an increase in growth irradiance between three Arabidopsis accessions. Average levels of growth irradiance is sensed by the plant's nuclear-encoded photoreceptors (phytochromes) to induce appropriate

FIGURE 4 Gene ontology and functional annotation of Arabidopsis genes expressed in rosette leaves, showing an accession-specific transcriptional response upon exposure to increased irradiance. (a) Gene ontology enrichment for biological process of Arabidopsis rosette-expressed genes >1.4 fold differentially expressed at 1 hr after exposure to increased irradiance (HL) when compared to low irradiance (LL) plants (control), and differentially expressed between at least two of the three accessions Ga-0, Ts-1, and Col-0; (b) Heat map presenting gene expression fold changes of genes differentially expressed at 1 hr after exposure to increased irradiance when compared to low irradiance, control, plants, and differentially expressed between at least two of the three tested accessions (G 1, T 1, C 1), annotated to be involved in one of the enriched biological processes as indicated in (a): photosynthesis (yellow highlight), response to heat (gray highlight), RNA binding (olive highlight), carbohydrate metabolism (blue highlight), photoreceptor activity (pink highlight), lipid-remodeling (light-green highlight), and cell organization (orange highlight). For the complete list of differentially expressed genes, see Supporting Information Tables S3 and S4

photomorphogenic development, but at high levels the irradiance is sensed through plastid-localized retrograde signaling pathways (Martin et al., 2016). This study confirms some of the known transcriptional response pathways to increased irradiance (Figure 7a), and it identifies a transcriptional response pathway involved in lipid-remodeling to be up-regulated after irradiance increase.

4.1 | Expression variation in lipid-remodeling genes associates with natural variation in photosynthesis response to increased irradiance

Gene ontology enrichment analysis of the accession-specific responsive genes reveals lipid-remodeling as a physiological response prevalent after irradiance increase (Figure 7a). Closer analysis of these "lipid-remodeling" irradiance responsive genes reveals that especially genes involved in phosphate-dependent lipid-remodeling are accession-specifically up-regulated upon the irradiance increase (Figures 4 and 5). Phosphate is an essential cofactor for photosynthesis and after an irradiance increase there is an increase in the pool size of phosphorylated intermediates accompanying the increased assimilation rate, requiring that more inorganic phosphate (Pi) be made available (Badger, Sharkey, & von Caemmerer, 1984; Caemmerer & Edmondson, 1986; Sivak & Walker, 1988; Walker & Sivak, 1986). A decrease in Pi such as occurs in end-product (or sink) limitation of photosynthesis, is associated with an overall decrease in assimilation rate (Paul & Foyer, 2001; Williams, 1998). Overall, therefore, photosynthesis, cytosolic sugar metabolism, and phosphate are interconnected (Stitt, 1991). In addition, phosphate deficiency responsive genes are shown to be sugar inducible (Müller, Morant, Jarmer, Nilsson, & Hamborg Nielsen, 2007). This transcriptional correlation is thought to be important for either (a) maintenance of the ratio of available Pi and carbon by reducing the cellular sucrose content via galactolipid synthesis; or (b) for the supply of galactolipids as components of the plasma membrane to support enhance growth under sucrose supplementation (Murakawa et al., 2014). Neither Müller et al. (2007) nor Murakawa et al. (2014) have found a link between increased irradiance, photosynthetic acclimation, and phosphate deficiency. This study implicates that increased expression of phosphate-deficiency-initiated lipid-remodeling genes (for the synthesis of galactolipids) in the response to an irradiance increase is a way to provide the extra Pi needed for the up-regulation of photosynthetic and downstream carbohydrate (sucrose) metabolism (Figures 4 and 5). When comparing the three accessions (Figure 4b) the lipid-remodeling genes are most up-regulated in Col-0, suggesting genetic variation in this response, associated with the variation in photosynthetic efficiency response to increased irradiance. An earlier genome-wide association study (GWAS) among 344 Arabidopsis accession for the photosynthetic response to the same increase in irradiance confirms a role for lipid-remodeling in photosynthetic acclimation, as the GWAS associated genomic variation in *DGD1 SUPRESSOR 1* (*DGS1*; At5 g12290) with phenotypic variation in the photosynthetic efficiency response (van Rooijen et al., 2017). *DGS1* encodes a mitochondrial outer membrane protein involved in galactolipid biosynthesis. The expression of *DGS1* is not responsive to increased irradiance, suggesting that the genomic association found in the GWAS acts on the post-transcriptional level rather than on the transcriptional level. However, the fact that GWAS identifies a gene that is involved in phosphate-dependent lipid remodeling strengthens the observation that variation in lipid-remodeling genes (either genomic or expression-based) causes natural variation in the photosynthesis efficiency response to increased irradiance.

One differentially up-regulated lipid-remodeling gene in the present study is *SPX1*, characterized before to be involved in phosphate deficiency response and lipid-remodeling (Puga et al., 2014). SPX1 shares the SPX domain with yeast inorganic phosphate (Pi) sensors and plant Pi starvation signaling proteins, and is a phosphate-dependent inhibitor of the PHOSPHATE STARVATION RESPONSE 1 (PHR1) transcription factor (Puga et al., 2014). *SPX1* is functionally redundant with *SPX2*, also found to be accession-specifically up-regulated in this study (Figure 4). The physical interaction between SPX1/SPX2 and PHR1 is reduced in the absence of inorganic phosphate (Pi), leading to transcription of many Pi-starvation induced genes by PHR1. As *SPX1/2* expression itself depends on PHR1, the SPX1/2-PHR1 interaction results in a regulatory negative feedback loop (Bustos et al., 2010; Puga et al., 2014). This means the increased expression of *SPX1/2* we observe is likely to be the consequence of a release of activated PHR1 protein, which can then induce the expression of Pi-starvation responsive genes, including the build-up of the negative regulator SPX1/2 itself to modulate the expression of the Pi-starvation genes again upon increased Pi availability. Initially transcription of *PHR1* itself would not need to change as long as sufficient PHR1 is activated to initiate the Pi-starvation response. *SPX1* is up-regulated in Col-0 and, to a lesser extent, in Ga-0, but not in Ts-1, 1 hr after the irradiance increase, and not in response to the temperature increase (Figure 6g). Rossel et al. (2002), who exposed Col-0 plants to increased irradiance filtered to

(a)

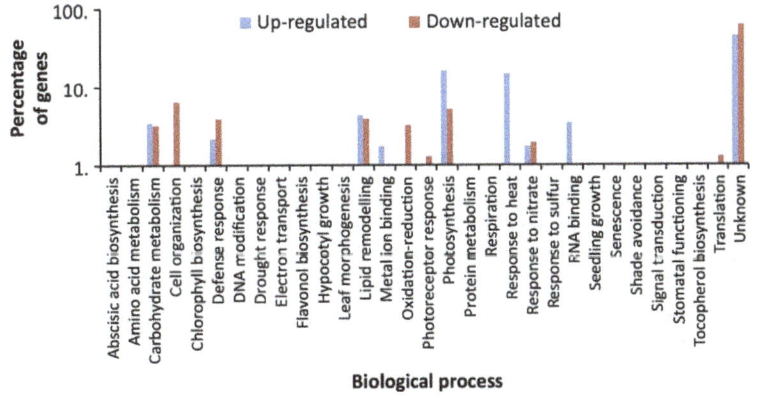

(b)

Molecular function	Gene ID	Gene name	C 1	C 3.5	C 25
Chlorophyll binding	At3g22840	EARLY LIGHT-INDUCABLE PROTEIN (ELIP1)			
	At4g14690	EARLY LIGHT-INDUCIBLE PROTEIN 2 (ELIP2)			
Chlorophyll biogenesis	At4g27440	light-dependent NADPH:protochlorophyllide oxidoreductase B			
Chloroplast biogenesis	At1g69200	fructokinase-like protein			
	At3g54090	fructokinase-like protein			
Chloroplast protein import	At2g28900	involved in plastid import of protochlorophyllide oxidoreductase A			
	At3g13470	subunit of chloroplasts chaperonins			
Chloroplast transcription/ transl.	At2g04530	protein with RNAse Z activity			
Flavonoid biosynthesis	At2g23910	NAD(P)-binding Rossmann-fold superfamily protein			
	At2g24550	unknown protein			
	At3g29590	malonyl-CoA:anthocyanidin 5-O-glucoside-6"-O-malonyltransferase			
	At3g51240	flavanone 3-hydroxylase			
	At4g14090	anthocyanidin 5-O-glucosyltransferase			
	At4g22880	leucoanthocyanidin dioxygenase			
	At5g07990	Required for flavonoid 3' hydroxylase activity			
	At5g08640	FLAVONOL SYNTHASE 1 (FLS1)			
	At5g13930	chalcone synthase (CHS)			
	At5g17220	GLUTATHIONE S-TRANSFERASE PHI 12 (GSTF12)			
	At5g42800	dihydroflavonol reductase			
	At5g48880	peroxisomal 3-keto-acyl-CoA thiolase 2 precursor			
	At5g62210	Embryo-specific protein 3			
Glucose:flavonoid transferase	At5g54060	anthocyanin 3-O-glucoside			
H2O2 scavenging	At2g37770	NADPH-dependent aldo-keto reductase			
	At3g03630	protein that possesses S-sulfocysteine synthase activity			
	At4g31870	GLUTATHIONE PEROXIDASE 7 (GPX7)			
Light harvesting	At2g34430	LIGHT-HARVESTING CHLOROPHYLL-PROTEIN COMPLEX II SUBUNIT B1 (LHB1B1)			
NAD metabolism	At5g14760	L-aspartate oxidase			
Nutrient mobilization	At5g24770	VEGETATIVE STORAGE PROTEIN 2 (VSP2)			
	At5g24780	VEGETATIVE STORAGE PROTEIN 1 (VSP1)			
Oxidation-reduction process	At4g19170	similar to nine-cis-epoxycarotenoid dioxygenase			
Photoprotection	At4g04020	Fibrillin precursor protein, involved in abscisic acid-mediated photoprotection			
Stomatal functioning	At5g22920	sequence similarity to RING, zinc finger proteins			
Transcription factor activity	At1g27730	SALT TOLERANCE ZINC FINGER 10 (ZAT10)			
	At1g28370	ERF/AP2, B-1 family			
	At1g74930	ERF/AP2, DREB A-5 family			
	At2g20880	ERF/AP2, ERF53			
	At2g47460	MYB12			
	At4g17500	ERF/AP2, B-3 family			
	At5g05410	ERF/AP2 FAMILY, DEHYDRATION RESPONSE ELEMENT BINDING 2A (DREB2A)			
	At5g11590	ERF/AP2, DREB A-4 family			
	At5g25190	ERF/AP2, B-6 family			
	At5g44190	GOLDEN2-LIKE2 (GLK2)			
	At5g59820	RESPONSIVE TO HIGHR LIGHT 41 (RHL41)			
Transferase activity	At1g54570	PHYTYL ESTER SYNTHASE 1 (PES1)			
Chaperone	At1g53540	HSP20-like chaperones superfamily protein			
	At1g54050	HSP20-like chaperones superfamily protein			
	At1g59860	HSP20-like chaperones superfamily protein			
	At1g71000	Chaperone DnaJ-domain superfamily protein			
	At2g46240	A member of Arabidopsis BAG (Bcl-2-associated athanogene) proteins			
	At5g48570	tetratricopeptide repeat (TPR) proteins with potential to interact wither Hsp90/Hsp70			
	At5g56030	A member of heat-shock protein 90 (HSP90) gene family			
Heat-shock factor	At2g26150	HEAT-SHOCK TRANSCRIPTION FACTOR A2 (HSFA2)			
	At3g51910	HEAT-SHOCK TRANSCRIPTION FACTOR A7A (HSFA7A)			
	At4g36990	HEAT-SHOCK FACTOR 4 (HSF4)			

Fold Change HL vs LL: 24, 22, 20, 18, 16, 14, 12, 10, 8, 6, 4, 2, 1, 0.9, 0.8, 0.7, 0.6, 0.5, 0.4, 0.3, 0.2, 0.1, 0

FIGURE 5 Gene ontology and functional annotation of Arabidopsis Col-0 rosette genes showing time-point-specific transcriptional response upon exposure to increased irradiance. (a) Gene ontology enrichment for biological process of Arabidopsis Col-0 rosette genes >2 fold differentially expressed after exposure to increased irradiance (HL) when compared to low irradiance (LL) plants (control), and differentially expressed on at least two of the three tested time-points, 1, 3.5, and 25 hr after irradiance increase; (b) Heat map presenting gene expression fold changes of genes differentially expressed in Col-0 after exposure to increased irradiance when compared to low irradiance, control, plants, and differentially expressed between at least two of the three tested time-points (1, 3.5, 25 hr) annotated to be involved in one of the enriched biological processes as indicated in (a): photosynthesis (yellow highlight), response to heat (gray highlight), RNA binding (olive highlight), carbohydrate metabolism (blue highlight), photoreceptor activity (pink highlight), lipid-remodeling (light-green highlight), and cell organization (orange highlight). For the complete list of differentially expressed genes, see Supporting Information Tables S5 and S6

Category	Gene	Description
Heat-shock protein	At1g74310	belongs to the Casein lytic proteinase/heat-shock protein 100 (Clp/Hsp100) family
	At2g19310	HSP20-like chaperones superfamily protein
	At2g20560	DNAJ heat-shock family protein
	At2g25140	belongs to the Casein lytic proteinase/heat-shock protein 100 (Clp/Hsp100)
	At2g29500	HSP20-like chaperones superfamily protein
	At2g32120	heat-shock protein 70T-2 (HSP70T-2)
	At3g08970	J domain protein localized in ER lumen; similarity to HSP40 proteins
	At3g09350	Arabidopsis ortholog of the human Hsp70-binding protein 1 (HspBP-1)
	At3g25230	a high molecular weight member of the FK506 binding protein (FKBP) family
	At3g46230	HEAT-SHOCK PROTEIN 17.4 (HSP17.4)
	At4g12400	Hop3, a tetratricopeptide repeat (TPR) protein, interacts with Hsp90/Hsp70
	At4g21320	heat-stress-associated 32-kD protein
	At4g25200	AtHSP23.6-mito mRNA
	At5g02490	Heat-shock protein 70 (Hsp 70) family protein (Hsp70-2)
	At5g09590	heat-shock protein 70 (Hsc70-5)
	At5g12030	HEAT-SHOCK PROTEIN 17.6A (HSP17.6A)
	At5g37670	HSP20-like chaperones superfamily protein
	At5g51440	HSP20-like chaperones superfamily protein
	At5g52640	cytosolic heat-shock protein AtHSP90.1
Unknown	At1g30070	SGS domain-containing protein
	At1g66080	Unknown protein
	At5g13200	unknown
	At5g64510	TUNICAMYCIN INDUCED 1 (TIN1), a plant-specific ER stress-inducible protein
Alternative splicing	At1g07350	serine/arginine rich-like protein, SR45a
	At1g09140	serine/arginine rich-like protein, SR30
Methyl transferase	At5g57280	Gene encodes a methyltransferase-like protein involved in pre-rRNA processing
Organellar RNA editing	At3g22310	PUTATIVE MITOCHONDRIAL RNA HELICASE 1 (PMH1)
Ribosome biogenesis	At1g15440	nucleolar protein that is a ribosome biogenesis co-factor
	At5g22100	RNA cyclase family protein
	At5g39850	Ribosomal protein S4
rRNA editing	At4g25630	encodes a fibrillarin directing 2'-O-ribose methylation of the rRNA
Beta galactosidase	At5g63800	Involved in mucilage formation
Beta-hexosaminidase activity	At1g65590	protein with beta-hexosaminidase activity
Bnzyme, pectin lyase	At1g48100	Pectin lyase-like superfamily protein
Bnzyme, alpha amylase	At1g76130	alpha-amylase, putative / 1,4-alpha-D-glucan glucanohydrolase
Lectin family protein	At5g03350	Legume lectin family protein
Nodulin family protein	At3g28007	SWEET4; a nodulin MtN3 family protein
Starch degradation	At4g09020	isoamylase-like protein
	At4g15210	cytosolic beta-amylase
	At4g17090	beta-amylase targeted to the chloroplast
Sucrose-proton symporter	At1g71880	Sucrose transporter
Transporter	At5g13170	member of the SWEET sucrose efflux transporter family
	At1g61800	glucose6-Phosphate/phosphate transporter 2 (GPT2)
Unknown	At4g23820	Pectin lyase-like superfamily protein
HY5 homologue	At3g17609	Encodes a homolog of HY5 (HYH). Involved in phyB signaling pathway.
Phytochrome signalling	At5g04190	phytochrome kinase substrate 4
Unknown	At1g18810	phytochrome kinase substrate-related
Glycosyl transferase	At1g32900	GRANULE BOUND STARCHR SYNTHASE 1 (GBSS1)
Hydrolase activity	At1g58520	RXW8; functions in hydrolase activity, acting on ester bonds, lipase activity
Lipid binding	At2g37870	Bifunctional inhibitor/lipid-transfer protein
	At2g42540	COLD-REGULATED 15A (COR15A)
	At2g45180	Bifunctional inhibitor/lipid-transfer protein
	At4g22490	Bifunctional inhibitor/lipid-transfer protein
	At5g48490	Bifunctional inhibitor/lipid-transfer protein
	At5g59320	LIPID TRANSFER PROTEIN 3 (LTP3)
Lipid transport	At3g23080	Polyketide cyclase/dehydrase and lipid transport superfamily protein
MGDG synthesis	At5g20410	MONOGALACTOSYLDIACYLGLYCEROL SYNTHASE 2 (MGD2)
SQDG synthesis	At4g33030	SULFOQUINOVOSYLDIACYLGLYCEROL 1 (SQD1)
	At5g01220	SULFOQUINOVOSYLDIACYLGLYCEROL 2 (SQD2)
Phospholipid catabolism	At3g02040	GLYCEROPHOSPHODIESTER PHOSPHODIESTERASE 1 (GDPD1)
Transferase activity	At1g09390	GDSL-motif esterase/acyltransferase/lipase
	At3g16370	GDSL-motif esterase/acyltransferase/lipase
Unknown	At5g20150	SPX DOMAIN GENE 1 (SPX1)
Acetylesterase	At1g57590	Pectinacetylesterase family protein
Alpha expansin	At1g69530	Alpha-Expansin Gene Family. Inv. in the form. of nematode-induced syncytia in roots
	At2g40610	Alpha-Expansin Gene Family. Inv. in the form. of nematode-induced syncytia in roots
Auxin transport	At2g01420	Encodes PIN-FORMED 4 (PIN4), a putative auxin efflux carrier
	At2g38120	Encodes an auxin influx transporter
Cellulose synthase	At4g23990	encodes a protein similar to cellulose synthase
Fatty acid biosynthetic process	At1g01120	3-ketoacyl-CoA synthase 1 (KCS1), inv. in fatty acid elongation process in wax biosynth.
Hydrolase activity	At4g32940	a vacuolar processing enzyme, cysteine proteinase associated with cell death
Cytoskeleton organization	At4g16520	autophagy 8f (ATG8F)
	At4g20260	Encodes a Ca2+ and Cu2+ binding protein.
Transcription factor activity	At4g00150	Belongs to one of the LOM (LOST MERISTEMS) genes
Transferase activity	At4g03210	encodes a member of xyloglucan endotransglucosylase/hydrolases (XTHs)

FIGURE 5 (Continued)

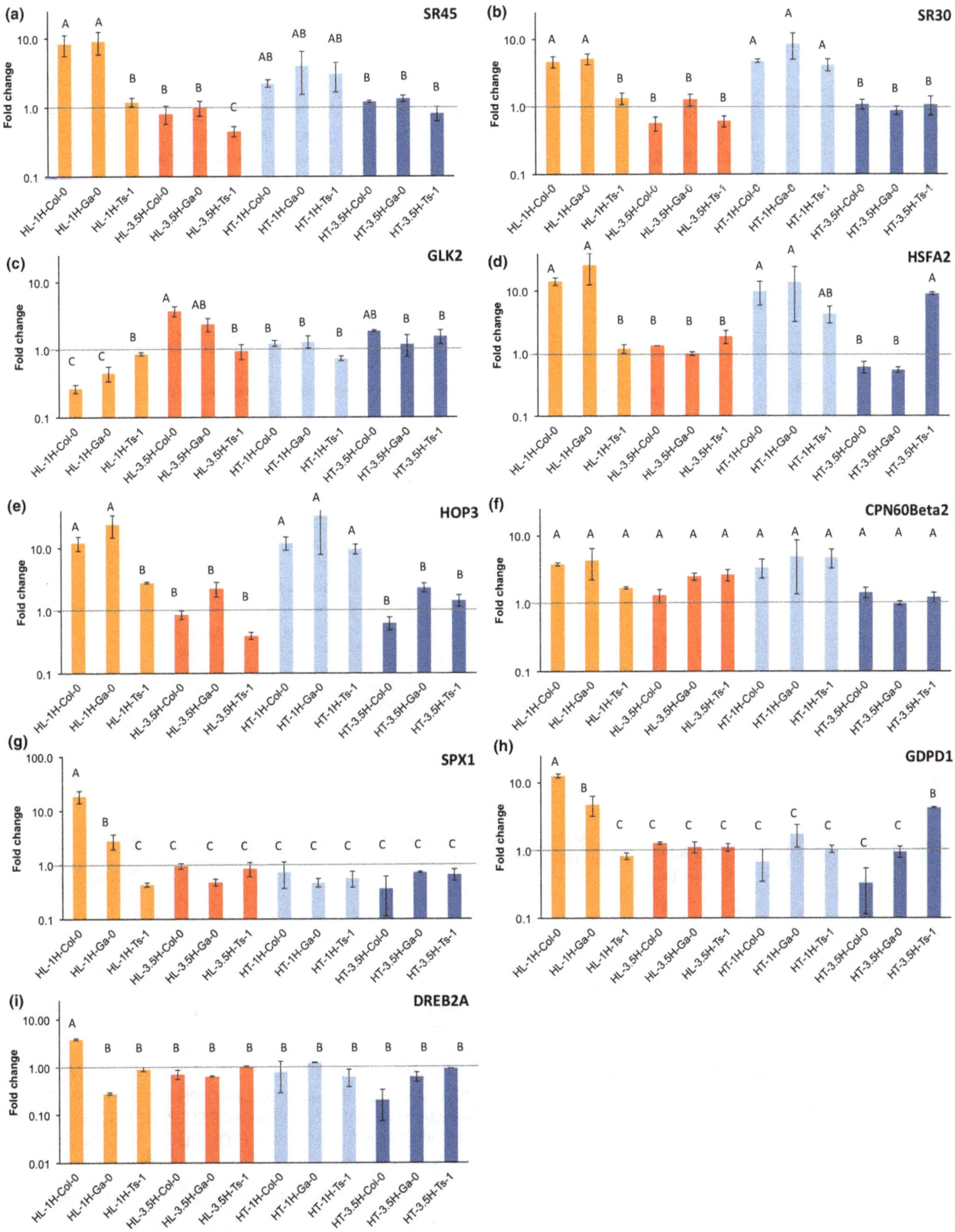

reduce the near-infrared irradiation, also noticed the induction of *SPX1*, providing additional evidence that this gene, and the Pi-response pathway it controls, is involved in the response to increase irradiation and not the parallel temperature increase. This is consistent with the proposed role of phospholipid-remodeling to provide additional Pi to accommodate increased photosynthesis and sugar

FIGURE 6 Gene expression changes in response to increased irradiance or in response to increased temperature, when compared to control conditions (no increased irradiance, no increased temperature). Gene expression differences, expressed as fold changes compared to control conditions (average ± *SE*), as determined by qRT-PCR analysis of nine representative genes selected from Figures 3, 4, and 5, in Arabidopsis rosettes at 1 hr (1H; orange) and 3.5 hr (3.5H; red) after an irradiance increase from 100 to 550 μmol m^{-2} s^{-1} (HL), as well as 1 hr (light blue) and 3.5 hr (dark blue) after a temperature increase from 20 to 30°C (HT) in accessions Col-0, Ga-0 and Ts-1. The control conditions were 100 μmol m^{-2} s^{-1} irradiance and 20°C temperature, with samples taken 1 hr after lights on. The following genes were analysed: (a) *SR45a* (At1 g07350); (b) *SR30* (At1 g09140); (c) *GLK2* (At5 g44190); (d) *HSFA2* (At2 g26150); (e) *HOP3* (At4 g12400); (f) *CPN60beta2* (At3 g13470); (g) *SPX1* (At5 g20150); (h) *GDPD1* (At3 g02040); (i) *DREB2A* (At5 g05410). Letters indicate statistically significant differences as determined by analysis of variance ($p < 0.05$)

(a)

Excess light →

Sensing Mechanism		Genetic Regulators	Physiological Response	Literature
Photoreceptor		CRY1/HY5	Flavonoid biosynthesis	Kleine et al, 2007
Redox	Reactive oxygen species (ROS)	Heat shock factors	ROS scavenging	Jung et al, 2013
Metabolites	Phosphate	PHR1/SPX1	Lipid remodeling	This study
	Sugar	ZAT10	Systemic resistance	Rossel et al, 2007
		ERF/AP2 family	Early response gene activation	Moore et al, 2014; Vogel et al 2014
	Reactive oxygen species (ROS)	HDS-MEcPP	Post-transcriptional gene silencing	Xiao et al, 2012
		SAL1-PAP	Post-transcriptional gene silencing	Estavillo et al, 2011
		mRNPs	RNA stability, stress granule (SG) formation	Uniacke et al, 2008
		TORC/SnRK1	Translational silencing	Robaglia et al, 2012

(b)

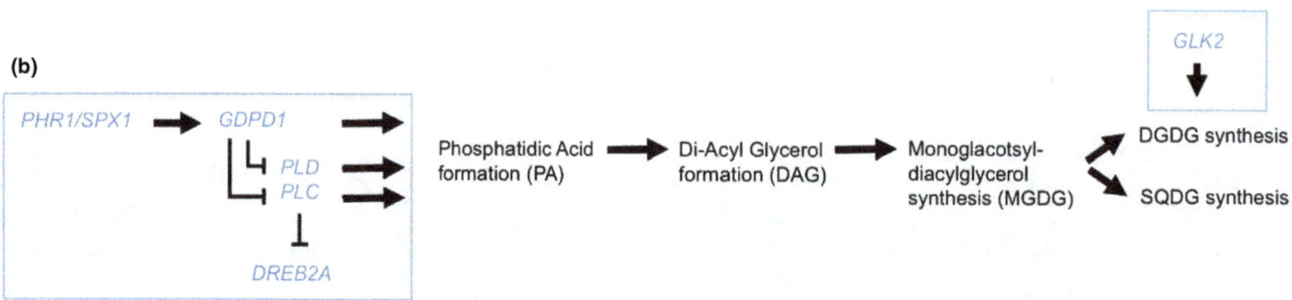

FIGURE 7 Summary table, incorporating lipid-remodeling as an additional response pathway to the known signal transduction pathways leading to transcriptional responses upon increased irradiance (a); and a schematic model (b) representing the newly proposed lipid-remodeling signal transduction pathway. Several genes of this pathway (*SPX1, GDPD1, DREB2A, GLK2*; in blue boxes) were found to be up-regulated in response to increased irradiance. The processes they influence are indicated in black. CRY1 = CRYPTOCHROME 1; HY5 = LONG HYPOCOTYL 5; PHR1 = PHOSPHATE STARVATION RESPONSE 1; ZAT10 = SALT TOLERANCE ZINC FINGER 10; ERF/AP2 = ETHYLENE RESPONSE FACTOR/APETALA2; SPX1 = SYG1, PHO81, AND XPR1 HOMOLOGUE 1; GDPD1 = GLYCEROPHOSPHO-DIESTER-PHOSPHO-DIESTERASE1; PLD = PHOSPHOLIPASE D; PLC = PHOSPHOLIPASE C; DREB2A = DEHYDRATION RESPONSE ELEMENT BINDING PROTEIN 2A; DGDG = DIGALACTOSYL DIACYLGLYCEROL; SQDG = SULFOQUINOVOSYL DIACYLGLYCEROL; GLK2 = GOLDEN-LIKE 2

assimilation synthesis. For a subset of the genotypes used in this study an irradiance of 550 μmol m^{-2} s^{-1} is just saturating (van Rooijen et al., 2015) and the experiments were conducted in normal air. Given that plants used were not cold-stressed or otherwise treated in a way that would lead to sink-limitation of photosynthesis, it is unexpected that they would develop phosphate limitation during the experiment – this is normally only encountered as triose-phosphate limited photosynthesis at high irradiances and carbon dioxide concentrations during, for example, A/Ci curves (Yang, Preiser, Li, Weise, & Sharkey, 2016). It seems possible, therefore, that the activation of metabolism that releases Pi might be more of a pre-emptive response for the prospective increase in assimilation that will result from acclimation to increased irradiance. Besides *SPX1* and *SPX2*, other lipid-remodeling genes found in this study to be responsive to increased irradiance have also been found by Puga et al. (2014), who compared the transcriptome of Pi-deficiency treated plants with that of control plants. Among these overlapping genes are *GDPD1*, *SQD2*, and *MGD2*, involved in galactolipid biosynthesis (Figure 4). This suggests that the increased irradiance causes an as yet unidentified Pi-limitation in the plant, that can be considered as a preparation for the increase in assimilation for what will develop over the coming days (van Rooijen et al., 2015). Considering the differential expression of these genes between the three accessions it is likely that this preparation contributes to the phenotypic variation we observed in the response to increased irradiance in Arabidopsis (Figure 1a). The response of lipid-remodeling genes to increased irradiance led us to propose a new sensing mechanism and signal transduction pathway leading to photosynthetic acclimation response, in addition to the already described pathways (Figure 7).

4.2 | The lipid-remodeling pathway

Three lipid-remodeling pathways are known to release of Pi from phospholipids during Pi deficiency (Ruelland et al., 2015). Two of these, phospholipid-remodeling into galactolipids through the generation of phosphatidic acid (PA) by phospholipase C (first route) or phospholipase D (second route), have been extensively studied in connection to phosphate deficiency responses (Gaude, Nakamura, Scheible, Ohta, & Dörmann, 2008; Nakamura et al., 2009). We do not find either of them to respond to increased irradiance, although they are known to repress expression of the transcription factor *DREB2A* (Figure 7; (Ruelland, Djafi, & Zachowski, 2013)). The activation of the GDPD pathway (third route), as a result of Pi limitation, probably outcompetes the PLC pathway, leading to reduced PLC-mediated repression of *DREB2A*. The third route is mediated through *GLYCEROPHOSPHO-DIESTER-PHOSPHO-DIESTERASE1* (*GDPD1*), a gene known to become expressed in response to phosphate limitation (Cheng et al., 2011). The promoters of all *GDPD* genes contain Pi deficiency response elements, recognized by the PHR1 transcription factor, to initiate transcription (Cheng et al., 2011). The *GDPD* family in Arabidopsis consists of 13 genes, of which *GDPD1* is the major one involved in hydrolyzing glycerophosphodiesters during Pi starvation (Cheng et al., 2011). In all accessions we found *GDPD1* to be up-regulated upon increased irradiance at 1 hr after the switch to an increased irradiance (Figures 4b and 5b). To release Pi from phospholipids, the phospholipids are first converted into glycerophophodiesters and then hydrolyzed into glycerol-3-phosphate (G-3-P) by glycerophospho-diester-phospho-diesterase GDPD1. It is unknown how Pi is released from G-3-P in Pi-starved plants, but this most likely occurs via the de novo synthesis of digalactosyl diacylglycerol (DGDG) and sulfoquinovosyl diacylglycerol (SQDG) (Cheng et al., 2011). In this pathway, G-3-P is first converted into phosphatidic acid (PA) and Pi is released in the subsequent conversion of PA into diacylglycerol (DAG) by phosphatidic acid phosphohydrolase (PAH), (Nakamura et al., 2009). DAG is the direct substrate for synthesis of either monogalactosyl diacylglycerol (MGDG; by MONOGALACTOSYLDIACYLGLYCEROL SYNTHASE 1 (MGD1), MGD2 and MGD3), digalactosyl diacylglycerol (DGDG; by DIGALACTOSYL DIACYLGLYCEROL SYNTHASE 1 (DGD1) and DGD2), or SULFOQUINOVOSYL DIACYLGLYCEROL (SQDG; by SULFOQUINOVOSYL DIACYLGLYCEROL SYNTHASE 1 (SQD1) and SQD2), where MGDG is the precursor of both DGDG and SQDG (Figure 7b). MGDG, DGDG, and SQDG are bilayer-forming galactolipids, important for the integrity of the photosynthetic membrane and of the chloroplast protein-import apparatus (Chen & Li, 1998; Kansy, Wilhelm, & Goss, 2014). DGDG is known to replace phospholipids in membranes in phosphate-limited conditions (Härtel, Dörmann, & Benning, 2000). Mutation of the *GDPD1* gene has no effect on the DGDG content, and the effect on the SQDG content is not known (Cheng et al., 2011). We do not observe induced transcription of *DGD1* and *DGD2* in response to the increase in irradiance, but transcription of both *SQD1* and *SQD2* is up-regulated in Col-0, comparable to the induction of *SPX1* and *GDPD1* (Figures 4b and 5b). This suggests that at least part of the release of Pi, which is needed to deal the incipient phosphate-limited conditions at increased irradiance, is mediated through SQDG synthesis.

The expression of lipid-remodeling genes, normally regulated by the SPX1/2-controlled PHR1 transcription factor, are induced within 1 hr after irradiance increase. We notice an interesting expression pattern for another transcription factor gene known to have a role in photosynthesis: *GOLDEN2-LIKE 2* (*GLK2*). Transcription of *GLK2* is down-regulated at 1 hr after the irradiance increase, but is induced after 3.5 hr (Figure 6). Golden2-like transcription factors (GLK) regulate nuclear genes involved in chlorophyll biosynthesis and are required for the development of photosynthetic proteins in chloroplasts (Chen et al., 2016; Wang et al., 2017; Waters et al., 2009). GLK2 is also a pivotal regulator of *DGD1* transcription, which encodes the key enzyme for DGDG synthesis (Kobayashi et al., 2014). The expression profile of *GLK2* (Figure 6c) implies that GLK2-induced DGDG replacement of phospholipids is activated only after 3.5 hr of excess light, following the initial induction of *SPX1/2* and *GDPD1*, after 1 hr. We therefore hypothesize that the preference for SQDG (over DGDG) for lipid-remodeling of photosynthetic membranes in response to irradiance increase is mediated by reduced DGDG synthesis as result of decreased *GLK2* expression.

4.3 | Expression variation in heat-shock factors associates with natural variation in photosynthesis response to increased irradiance

The core set of transcriptionally responsive genes is enriched for genes involved in photosynthesis, but also for heat-shock response and signaling through transcriptional response ("RNA binding"; Figure 3a). Leaf temperature is the result of the balance of different heat flows to and from the leaf (Nobel, 2005). An increase in irradiance represents an increase in the short-wave radiation flow to the leaf which tends to increase leaf temperature, but the increase in assimilation it produces results in an increase in stomatal conductance and thus transpiration, resulting in a latent heat flux from the leaf which tends to cool it. Whether the warming or the cooling effect associated with an increase in photosynthetically active irradiance will dominate will depend on the spectral distribution of the radiation, the boundary layer and stomatal conductances, and the vapor pressure deficit between the leaf air space and the mixed air around the leaf. In a controlled environment room with limited air movement (and thus a low boundary layer conductance) a temperature increase will nearly always parallel an irradiance increase, and temperature changes produce their own transcriptome changes. In the field, however, transpirative cooling can lower leaf temperature below air temperature (Larcher, 1995). The transcriptome changes provoked by an irradiance increase can actually be due to the temperature increase rather than the photochemical (photosynthetic and photoreceptor) changes resulting from a higher irradiance (Swindell et al., 2007), explaining the core responsiveness of heat-shock factors.

It is possible to separate the photosynthetic and photoreceptor-based effects of an irradiance increase from the temperature-based effect. This can be achieved by using heat filters to block the (near) infrared light that is responsible for part of the temperature increase (Rossel et al., 2002), or by running parallel experiments with an irradiance increase and an accompanying temperature increase alongside a temperature increase only. In this study, we use the latter approach and find that the induction of expression of several heat-shock response genes is mostly heat responsive and not specific for the irradiance increase, as they are also induced upon a temperature increase only. The expression of the *HSFA2* and *HOP3* genes is determined in more detail (Figure 6d,e). *HSF2A* encodes a Heat-shock transcription factor associated with photosynthesis acclimation through the ascorbate peroxidase pathway (Jung et al., 2013; Nishizawa et al., 2006). *HOP3* (*HSP70-HSP90 ORGANIZING PROTEIN3*), encodes a tetratricopeptide repeat protein that acts as a chaperone in the interaction with HSP70 and HSP90 to assist in the import of photosynthetic pre-proteins synthesized outside the chloroplast. It also regulates the stress response of the endoplasmic reticulum (Fellerer, Schweiger, Schöngruber, Soll, & Schwenkert, 2011; Fernández-Bautista, Fernández-Calvino, Muñoz, & Castellano, 2017). Both *HSFA2* and *HOP3* are induced in Ga-0 and Col-0, but only at 1 hr, and no longer at 3.5 hr, after the increase in irradiance and they are induced in all three accessions after the increase in temperature (Figure 6). Although out of all induced heat response genes (Figures 3, 4, 5) we only confirm the heat-induced transcriptional response for these two known heat response associated genes (Figure 6), it is likely to be the general pattern that their induction of expression is mainly in response to the increase in temperature that in our case accompanied the increase in irradiance (note that an increase in irradiance can result in leaf cooling due to the increase in transpiration). This does not mean that this response is unimportant for the acclimation to increased irradiance if this is accompanied by a temperature increase. In this respect it is important to note that the strong induction of heat-shock response genes in Col-0 and Ga-0 in response to increase in irradiance is not seen in Ts-1, which correlates well with the differences in photosynthesis efficiency between Ts-1, Col-0 and Ga-0 (Figure 1a). These observations lead us to conclude that the expression response of heat-shock genes might influence the degree in which photosynthetic light-use efficiency is affected in response to increased irradiance, and which, on longer time scale, might also affect the photosynthetic acclimation.

4.4 | Expression variation in ERF/AP2 transcription factors associates with natural variation in photosynthesis response to increased irradiance

Looking further among the core responsive genes, several members of the ERF/AP2 transcription factor family are found. The ERF/AP2 transcription factor family is known to sense and signal the response to changed metabolite levels resulting from increased irradiance levels; members of this family have been found before as regulators of early downstream targets in the acclimation response to high light (Alsharafa et al., 2014; Moore, Vogel, & Dietz, 2014). They are induced by the transcription factor ZAT10 (Alsharafa et al., 2014; Nguyen et al., 2012), which functions to initiate a systemic acclimation response to excess light (Munekage, Inoue, Yoneda, & Yokota, 2015; Rossel et al., 2007). *ZAT10* transcription is also up-regulated in our study (Figure 5b).

Among the irradiance responsive ERF/AP2 TFs, *DREB2A* stands out because its degree of up-regulation is accession-specific but not time-specific, and its expression correlates (negatively) with the variation in photosynthetic efficiency between the accessions (Supporting Information Tables S1 and S3). The expression of *DREB2A* is specifically responsive to increased irradiance and not to increased temperature (Figure 6). *DREB2A* encodes a transcription factor belonging to the ERF/AP2-type transcription factor family, binding to drought responsive elements (DREs) in the promoters of transcriptional target genes. *DREB2A* is known as a key regulator of drought response (Sakuma, Maruyama, Osakabe, et al., 2006), though it is also known to be activated in response to heat stress (Sakuma, Maruyama, Qin, et al., 2006), as well as increased irradiance (Rossel et al., 2002). *DREB2A* is transcriptionally up-regulated 1 hr after irradiance increase, only in Col-0 (Figure 6i), suggesting an accession-specific involvement of DREB2A in regulating the photosynthesis response to irradiance increase in Arabidopsis, probably through the SPX1/PHR1-mediated response pathway acting on lipid remodeling (Figure 7b). DREB2A is not the only ERF/AP2 transcription factor of which its expression correlates with photosynthetic efficiency response; also *ERF53* correlated, although in positive direction (Supporting Information Table S3). ERF53 is heat responsive and regulates stomatal movement (Hsieh, Cheng, & Lin, 2013).

4.5 | Expression variation in alternative splicing factors associates with natural variation in photosynthesis response to increased irradiance

Another core responsive gene for which the degree of up-regulation is accession-specific, but not time-specific, and for which the expression correlates (positively) with the variation in photosynthetic efficiency encodes the alternative splicing factor SR45 (Supporting Information Tables S1 and S3). SR45a is a protein involved in stress-induced alternative splicing (Sakuma, Maruyama, Qin, et al., 2006; Yoshimura et al., 2011). Expression of *SR45A* is responsive to both heat and light (Figure 6). Not only *SR45A*, but also another serine/arginine-rich protein (SR) gene, *SR30*, is found to be up-regulated in response to increased irradiance, and both respond most strongly 3.5 hr after the irradiance increase (Figure 5). Both genes were initially classified as members of the serine/arginine-rich protein family (Tanabe, Yoshimura, Kimura, Yabuta, & Shigeoka, 2007), however, a rational and consistent nomenclature for SR protein splicing factors excluded SR45a from the serine/arginine rich protein family, based on its sufficiently different protein sequence (Barta, Kalyna, & Reddy, 2010). Nevertheless, despite the sequence differences, later experimentation proved SR45a to function in stress-induced alternative splicing (Yoshimura et al., 2011). Both *SR30* and *SR45a* are up-regulated in response to high light stress (Rossel et al., 2002; Tanabe et al., 2007). No obvious aberrant phenotype has been observed for the *sr45a* mutant (Yoshimura et al., 2011). The *SR30* gene has so far not been investigated in great detail, but the similarity in gene expression profile (Figure 6b) and the similarity in subcellular localization (Mori et al., 2012) suggest that *SR45A* and *SR30* cooperate in the regulation of alternative splicing. The micro array we have used for transcript profiling does not allow the detection of alternative splicing, but expression of these genes suggests this may be affected by the increased irradiance.

5 | CONCLUSIONS

We identify a phospholipid-remodeling pathway as important for the acclimation to increased irradiance in Arabidopsis. This pathway starts with the SPX1-mediated activation of the transcription factor PHR1 leading to the transcription of *GDPD1*, enhancing galactolipid biosynthesis and eventually leading to DGDG and SQDG synthesis (Figure 7b), important for the lipid-remodeling of the photosynthetic membrane. Although the involvement of PHR1 in the response to high light had been noted (Nilsson, Lundmark, Jensen, & Nielsen, 2011), information on which subsequent steps might be involved was still lacking. We propose it involves the substitution of phospholipids by galactolipids through SQDG (preferred over DGDG) to release Pi needed for acclimating photosynthesis rates.

In addition to the up-regulated expression of phospholipid-remodeling pathway, we identify the up-regulation of *SR45A* and *SR30* transcription in response to irradiance increase. These genes are involved in mediation of alternative splicing, suggesting alternative transcripts are important for the acclimation to increased irradiance, but the targets of these regulators and their roles in the acclimation process remain to be elucidated.

It is known that retrograde signaling is light intensity-dependent, as being the prevalent sensory pathway in excess light providing a mechanism for protection against photo-oxidative damage by suppressing photomorphogenic development and minimizing tissue exposure to damaging irradiance (Martin et al., 2016; Szechynska-Hebda & Karpinski, 2013). We have shown this correlates with natural variation in photosynthetic response to high light as in this study the lower expression of cell growth regulatory genes positively correlated with Φ_{PSII} in the different accessions (Figure 4). In addition, we have shown the correlation of carbohydrate metabolism with natural variation as the gene ontology term enriched among the accessions specific up-regulated genes (Figure 4), shown before to act as a short-term response to high irradiance exposure (Schmitz et al., 2014).

In this study, one of the expression pathways that correlate with the observed phenotypic differences in acclimation between three Arabidopsis accessions is the up-regulation of *SR45A* and *SR30*. Second, stronger expression of heat responsive genes in the accession Ga-0 might explain its slightly higher photosynthesis efficiency compared to Col-0, while the absence of induced expression of heat responsive genes in Ts-1 could partly explain its lower photosynthetic efficiency. Third, the PHR-mediated gene activation pathway, that was found in this study to play a role in response to increased irradiance, was found to be active in both Ga-0 and Col-0, providing an additional explanation for the higher photosynthesis efficiencies in these accessions compared to Ts-1, in which the activation of both the heat-shock responsive pathway as well as the PHR-mediated pathway is absent. Fourth, we found a time-specific expression response of genes involved in flavonoid/anthocyanin biosynthesis in Col-0, correlating with a fast rise in anthocyanin content in Col-0, compared to Ts-1 and Ga-0, most likely involved in scavenging of ROS. As a last observation, transcription factors belonging to the ERF/AP2 family were found to be responsive to increased irradiance in all three accessions, but the specific genes of this family and the direction of the response differs between the three accessions. All in all, this study shows that natural variation for a complex physiological response such as photosynthetic efficiency response to increased irradiance, correlates to variation in transcriptional response of variable genes and processes. To identify the exact alleles that encode this variation will be the next challenge.

ACKNOWLEDGMENTS

We thank Wilco Ligterink and León̈ie Bentsink for their help with the analysis of the microarray data.

AUTHORS' CONTRIBUTIONS

All authors designed the study, RvR performed the experiments and analyzed the data, all authors wrote the manuscript.

ORCID

Roxanne van Rooijen http://orcid.org/0000-0002-7867-2670
Mark G. M. Aarts http://orcid.org/0000-0001-5257-0740

REFERENCES

Alsharafa, K., Vogel, M. O., Oelze, M. L., Moore, M., Stingl, N., Konig, K., ... Dietz, K. J. (2014). Kinetics of retrograde signalling initiation in the high light response of *Arabidopsis thaliana*. *Philosophical Transactions of the Royal Society of London. Series B, Biological Sciences, 369*, 20130424. https://doi.org/10.1098/rstb.2013.0424

Apel, K., & Hirt, H. (2004). Reactive oxygen species: Metabolism, oxidative stress, and signal transduction. *Annual Review of Plant Biology, 55*, 373–399. https://doi.org/10.1146/annurev.arplant.55.031903.141701

Asada, K. (2006). Production and scavenging of reactive oxygen species in chloroplasts and their functions. *Plant Physiology, 141*, 391–396. https://doi.org/10.1104/pp.106.082040

Badger, M. R., Sharkey, T. D., & von Caemmerer, S. (1984). The relationship between steady-state gas exchange of bean leaves and the levels of carbon-reduction-cycle intermediates. *Planta, 160*, 305–313. https://doi.org/10.1007/BF00393411

Barta, A., Kalyna, M., & Reddy, A. S. N. (2010). Implementing a rational and consistent nomenclature for serine/arginine-rich protein splicing factors (SR proteins) in plants. *Plant Cell, 22*, 2926–2929. https://doi.org/10.1105/tpc.110.078352

Benjamini, Y., & Hochberg, Y. (1995). Controlling the false discovery rate: A practical and powerful approach to multiple testing. *Journal of the Royal Statistical Society Series B (Methodological), 57*, 289–300.

Borisova-Mubarakshina, M. M., Ivanov, B. N., Vetoshkina, D. V., Lubimov, V. Y., Fedorchuk, T. P., Naydov, I. A., ... Bassi, R. (2015). Long-term acclimatory response to excess excitation energy: Evidence for a role of hydrogen peroxide in the regulation of photosystem II antenna size. *Journal of Experimental Botany, 66*, 7151–7164. https://doi.org/10.1093/jxb/erv410

Bustos, R., Castrillo, G., Linhares, F., Puga, M. I., Rubio, V., Pérez-Pérez, J., ... Paz-Ares, J. (2010). A central regulatory system largely controls transcriptional activation and repression responses to phosphate starvation in Arabidopsis. *Plos Genetics, 6*, e1001102. https://doi.org/10.1371/journal.pgen.1001102

Caemmerer, S., & Edmondson, D. (1986). Relationship between steady-state gas exchange, in vivo ribulose bisphosphate carboxylase activity and some carbon reduction cycle intermediates in raphanus sativus. *Functional Plant Biology, 13*, 669–688.

Chen, M., Ji, M., Wen, B., Liu, L., Li, S., Chen, X., & Li, L. (2016). GOLDEN 2-LIKE transcription factors of plants. *Frontiers in Plant Science, 7*, 1509.

Chen, L.-J., & Li, H.-M. (1998). A mutant deficient in the plastid lipid DGD is defective in protein import into chloroplasts. *The Plant Journal, 16*, 33–39. https://doi.org/10.1046/j.1365-313x.1998.00270.x

Cheng, Y., Zhou, W., El Sheery, N. I., Peters, C., Li, M., Wang, X., & Huang, J. (2011). Characterization of the Arabidopsis glycerophosphodiester phosphodiesterase (GDPD) family reveals a role of the plastid-localized AtGDPD1 in maintaining cellular phosphate homeostasis under phosphate starvation. *The Plant Journal, 66*, 781–795. https://doi.org/10.1111/j.1365-313X.2011.04538.x

Datt, B. (1999). A new reflectance index for remote sensing of chlorophyll content in higher plants: Tests using eucalyptus leaves. *Journal of Plant Physiology, 154*, 30–36. https://doi.org/10.1016/S0176-1617(99)80314-9

Demmig-Adams, B., & Adams, W. W. (1992). Photoprotection and other responses of plants to high light stress. *Annual Review of Plant Physiology and Plant Molecular Biology, 43*, 599–626. https://doi.org/10.1146/annurev.pp.43.060192.003123

Dietz, K. J., Wesemann, C., Wegener, M., & Seidel, T. (2018). Toward an integrated understanding of retrograde control of photosynthesis. *Antioxidants & Redox Signaling*. https://doi.org/10.1089/ars.2018.7519

Eberhard, S., Finazzi, G., & Wollman, F.-A. (2008). The dynamics of photosynthesis. *Annual Review of Genetics, 42*, 463–515. https://doi.org/10.1146/annurev.genet.42.110807.091452

Efron, B., & Tibshirani, R. (2002). Empirical Bayes methods and false discovery rates for microarrays. *Genetic Epidemiology, 23*, 70–86. https://doi.org/10.1002/(ISSN)1098-2272

Estavillo, G. M., Crisp, P. A., Pornsiriwong, W., Wirtz, M., Collinge, D., Carrie, C., ... Pogson, B. J. (2011). Evidence for a SAL1-PAP chloroplast retrograde pathway that functions in drought and high light signaling in Arabidopsis. *The Plant cell, 23*, 3992–4012. https://doi.org/10.1105/tpc.111.091033

Fellerer, C., Schweiger, R., Schöngruber, K., Soll, J., & Schwenkert, S. (2011). Cytosolic HSP90 cochaperones HOP and FKBP interact with freshly synthesized chloroplast preproteins of Arabidopsis. *Molecular Plant, 4*, 1133–1145. https://doi.org/10.1093/mp/ssr037

Fernández-Bautista, N., Fernández-Calvino, L., Muñoz, A., & Castellano, M. M. (2017). HOP3, a member of the HOP family in Arabidopsis, interacts with BiP and plays a major role in the ER stress response. *Plant, Cell and Environment, 40*(8), 1341–1355.https://doi.org/10.1111/pce.12927

Fey, V., Wagner, R., Brautigam, K., Wirtz, M., Hell, R., Dietzmann, A., ... Pfannschmidt, T. (2005). Retrograde plastid redox signals in the expression of nuclear genes for chloroplast proteins of *Arabidopsis thaliana*. *Journal of Biological Chemistry, 280*, 5318–5328. https://doi.org/10.1074/jbc.M406358200

Flood, P. J., Kruijer, W., Schnabel, S. K., van der Schoor, R., Jalink, H., Snel, J. F. H., ... Aarts, M. G. M. (2016). Phenomics for photosynthesis, growth and reflectance in *Arabidopsis thaliana* reveals circadian and long-term fluctuations in heritability. *Plant Methods, 12*, 14. https://doi.org/10.1186/s13007-016-0113-y

Foyer, C. H., Neukermans, J., Queval, G., Noctor, G., & Harbinson, J. (2012). Photosynthetic control of electron transport and the regulation of gene expression. *Journal of Experimental Botany, 63*, 1637–1661. https://doi.org/10.1093/jxb/ers013

Galvez-Valdivieso, G., Fryer, M. J., Lawson, T., Slattery, K., Truman, W., Smirnoff, N., ... Mullineaux, P. M. (2009). The high light response in Arabidopsis involves ABA signaling between vascular and bundle sheath cells. *The Plant cell, 21*, 2143–2162. https://doi.org/10.1105/tpc.108.061507

Gamon, J. A., Serrano, L., & Surfus, J. S. (1997). The photochemical reflectance index: An optical indicator of photosynthetic radiation use efficiency across species, functional types, and nutrient levels. *Oecologia, 112*, 492–501. https://doi.org/10.1007/s004420050337

Gaude, N., Nakamura, Y., Scheible, W. R., Ohta, H., & Dörmann, P. (2008). Phospholipase C5 (NPC5) is involved in galactolipid accumulation during phosphate limitation in leaves of Arabidopsis. *The Plant Journal, 56*, 28–39. https://doi.org/10.1111/j.1365-313X.2008.03582.x

Gentleman, R., Carey, V., Bates, D., Bolstad, B., Dettling, M., Dudoit, S., ... Zhang, J. (2004). Bioconductor: Open software development for computational biology and bioinformatics. *Genome Biology, 5*, R80. https://doi.org/10.1186/gb-2004-5-10-r80

Gitelson, A. A., Merzlyak, M. N., & Chivkunova, O. B. (2001). Optical properties and nondestructive estimation of anthocyanin content in plant leaves. *Photochemistry and Photobiology, 74*, 38. https://doi.org/10.1562/0031-8655(2001)074<0038:OPANEO>2.0.CO;2

Härtel, H., Dörmann, P., & Benning, C. (2000). DGD1-independent biosynthesis of extraplastidic galactolipids after phosphate deprivation in Arabidopsis. *PNAS, 97*, 10649–10654. https://doi.org/10.1073/pnas.180320497

Hruz, T., Wyss, M., Docquier, M., Pfaffl, M. W., Masanetz, S., Borghi, L., ... Zimmermann, P. (2011). RefGenes: Identification of reliable and condition specific reference genes for RT-qPCR data normalization. *BMC Genomics, 12*, 156. https://doi.org/10.1186/1471-2164-12-156

Hsieh, E. J., Cheng, M. C., & Lin, T. P. (2013). Functional characterization of an abiotic stress-inducible transcription factor AtERF53 in *Arabidopsis thaliana*. *Plant Molecular Biology, 82*, 223–237. https://doi.org/10.1007/s11103-013-0054-z

Irizarry, R. A., Hobbs, B., Collin, F., Beazer-Barclay, Y. D., Antonellis, K. J., Scherf, U., & Speed, T. P. (2003). Exploration, normalization, and summaries of high density oligonucleotide array probe level data. *Biostatistics*, *4*, 249–264. https://doi.org/10.1093/biostatistics/4.2.249

Jung, H.-S., Crisp, P. A., Estavillo, G. M., Cole, B., Hong, F., Mockler, T. C., ... Chory, J. (2013). Subset of heat-shock transcription factors required for the early response of Arabidopsis to excess light. *PNAS*, *110*, 14474–14479. https://doi.org/10.1073/pnas.1311632110

Kansy, M., Wilhelm, C., & Goss, R. (2014). Influence of thylakoid membrane lipids on the structure and function of the plant photosystem II core complex. *Planta*, *240*, 781–796. https://doi.org/10.1007/s00425-014-2130-2

Kleine, T., Kindgren, P., Benedict, C., Hendrickson, L., & Strand, A. (2007). Genome-wide gene expression analysis reveals a critical role for CRYPTOCHROME1 in the response of Arabidopsis to high irradiance. *Plant Physiology*, *144*, 1391–1406. https://doi.org/10.1104/pp.107.098293

Kleine, T., & Leister, D. (2016). Retrograde signaling: Organelles go networking. *Biochimica et Biophysica Acta*, *1857*, 1313–1325. https://doi.org/10.1016/j.bbabio.2016.03.017

Kobayashi, K., Fujii, S., Sasaki, D., Baba, S., Ohta, H., Masuda, T., & Wada, H. (2014). Transcriptional regulation of thylakoid galactolipid biosynthesis coordinated with chlorophyll biosynthesis during the development of chloroplasts in Arabidopsis. *Frontiers in Plant Science*, *5*, 272.

Larcher, W. (1995). *Physiological plant ecology*. Berlin: Springer Science & Business Media. https://doi.org/10.1007/978-3-642-87851-0

Lee, J., He, K., Stolc, V., Lee, H.-J., Figueroa, P., Gao, Y., ... Deng, X. W. (2007). Analysis of transcription factor HY5 genomic binding sites revealed its hierarchical role in light regulation of development. *Plant Cell*, *19*, 731–749. https://doi.org/10.1105/tpc.106.047688

Li, Z., Wakao, S., Fischer, B. B., & Niyogi, K. K. (2009). Sensing and responding to excess light. *Annual Review of Plant Biology*, *60*, 239–260. https://doi.org/10.1146/annurev.arplant.58.032806.103844

Long, S. P., Humphries, S. W., & Falkowski, P. G. (1994). Photoinhibition of photosynthesis in nature. *Annual Review of Plant Physiology and Plant Molecular Biology*, *45*, 633–662. https://doi.org/10.1146/annurev.pp.45.060194.003221

Martin, G., Leivar, P., Ludevid, D., Tepperman, J. M., Quail, P. H., & Monte, E. (2016). Phytochrome and retrograde signalling pathways converge to antagonistically regulate a light-induced transcriptional network. *Nature Communications*, *7*, 11431. https://doi.org/10.1038/ncomms11431

Moore, M., Vogel, M., & Dietz, K. (2014). The acclimation response to high light is initiated within seconds as indicated by upregulation of AP2/ERF transcription factor network in *Arabidopsis thaliana*. *Plant Signalling and Behavior*, *9*, 976479. https://doi.org/10.4161/15592324.2014.976479

Mori, T., Yoshimura, K., Nosaka, R., Sakuyama, H., Koike, Y., Tanabe, N., ... Shigeoka, S. (2012). Subcellular and subnuclear distribution of high-light responsive serine/arginine-rich proteins, atSR45a and atSR30, in *Arabidopsis thaliana*. *Bioscience Biotechnology Biochemistry*, *76*, 2075–2081. https://doi.org/10.1271/bbb.120425

Müller, R., Morant, M., Jarmer, H., Nilsson, L., & Hamborg Nielsen, T. (2007). Genome-wide analysis of the Arabidopsis leaf transcriptome reveals interaction of phosphate and sugar metabolism. *Plant Physiology*, *143*, 156–171.

Munekage, Y. N., Inoue, S., Yoneda, Y., & Yokota, A. (2015). Distinct palisade tissue development processes promoted by leaf autonomous signalling and long-distance signalling in *Arabidopsis thaliana*. *Plant, Cell and Environment*, *38*, 1116–1126. https://doi.org/10.1111/pce.12466

Murakawa, M., Shimojima, M., Shimomura, Y., Kobayashi, K., Awai, K., & Ohta, H. (2014). Monogalactosyldiacylglycerol synthesis in the outer envelope membrane of chloroplasts is required for enhanced growth under sucrose supplementation. *Frontiers in Plant Science*, *5*, 280.

Nakamura, Y., Koizumi, R., Shui, G., Shimojima, M., Wenk, M. R., Ito, T., & Ohta, H. (2009). Arabidopsis lipins mediate eukaryotic pathway of lipid metabolism and cope critically with phosphate starvation. *PNAS*, *106*, 20978–20983. https://doi.org/10.1073/pnas.0907173106

Nawkar, G. M., Kang, C. H., Maibam, P., Park, J. H., Jung, Y. J., Chae, H. B., ... Lee, S. Y. (2017). HY5, a positive regulator of light signaling, negatively controls the unfolded protein response in Arabidopsis. *Proceedings of the National Academy of Sciences of the United States of America*, *114*, 2084–2089. https://doi.org/10.1073/pnas.1609844114

Nguyen, X. C., Kim, S. H., Lee, K., Kim, K. E., Liu, X. M., Han, H. J., ... Chung, W. S. (2012). Identification of a C2H2-type zinc finger transcription factor (ZAT10) from Arabidopsis as a substrate of MAP kinase. *Plant Cell Reports*, *31*, 737–745. https://doi.org/10.1007/s00299-011-1192-x

Nilsson, L., Lundmark, M., Jensen, P. E., & Nielsen, T. H. (2011). The Arabidopsis transcription factor PHR1 is essential for adaptation to high light and retaining functional photosynthesis during phosphate starvation. *Physiologia Plantarum*, *144*, 35–47.

Nishizawa, A., Yabuta, Y., Yoshida, E., Maruta, T., Yoshimura, K., & Shigeoka, S. (2006). Arabidopsis heat shock transcription factor A2 as a key regulator in response to several types of environmental stress. *The Plant Journal*, *48*, 535–547. https://doi.org/10.1111/j.1365-313X.2006.02889.x

Niyogi, K. K. (1999). PHOTOPROTECTION REVISITED: Genetic and molecular approaches. *Annual Review of Plant Physiology and Plant Molecular Biology*, *50*, 333–359. https://doi.org/10.1146/annurev.arplant.50.1.333

Nobel, P. S. (2005). *Physicochemical & environmental plant physiology*, 3rd ed. Amsterdam: Elsevier.

Onate-Sánchez, L., & Vicente-Carbajosa, J. (2008). DNA-free RNA isolation protocols for *Arabidopsis thaliana*, including seeds and siliques. *BMC Research Notes*, *1*, 93. https://doi.org/10.1186/1756-0500-1-93

Paul, M. J., & Foyer, C. H. (2001). Sink regulation of photosynthesis. *Journal of Experimental Botany*, *52*, 1383–1400. https://doi.org/10.1093/jexbot/52.360.1383

Puga, M. I., Mateos, I., Charukesi, R., Wang, Z., Franco-Zorrilla, J. M., de Lorenzo, L., ... Paz-Ares, J. (2014). SPX1 is a phosphate-dependent inhibitor of Phosphate Starvation Response 1 in Arabidopsis. *PNAS*, *111*, 14947–14952. https://doi.org/10.1073/pnas.1404654111

Robaglia, C., Thomas, M., & Meyer, C. (2012). Sensing nutrient and energy status by SnRK1 and TOR kinases. *Current Opinion in Plant Biology*, *15*, 301–307. https://doi.org/10.1016/j.pbi.2012.01.012

Rossel, J. B., Wilson, P. B., Hussain, D., Woo, N. S., Gordon, M. J., Mewett, O. P., ... Pogson, B. J. (2007). Systemic and intracellular responses to photooxidative stress in Arabidopsis. *Plant Cell*, *19*, 4091–4110. https://doi.org/10.1105/tpc.106.045898

Rossel, J. B., Wilson, I. W., & Pogson, B. J. (2002). Global changes in gene expression in response to high light in Arabidopsis. *Plant Physiology*, *130*, 1109–1120. https://doi.org/10.1104/pp.005595

Ruelland, E., Djafi, N., & Zachowski, A. (2013). The phosphoinositide dependent-phospholipase C pathway differentially controls the basal expression of DREB1 and DREB2 genes. *Plant Signalling and Behavior*, *8*, e26895. https://doi.org/10.4161/psb.26895

Ruelland, E., Kravets, V., Derevyanchuk, M., Martinec, J., Zachowski, A., & Pokotylo, I. (2015). Role of phospholipid signalling in plant environmental responses. *Environmental and Experimental Botany*, *114*, 129–143. https://doi.org/10.1016/j.envexpbot.2014.08.009

Sakuma, Y., Maruyama, K., Osakabe, Y., Qin, F., Seki, M., Shinozaki, K., & Yamaguchi-Shinozaki, K. (2006). Functional analysis of an Arabidopsis transcription factor, DREB2A, involved in drought-responsive gene expression. *Plant Cell*, *18*, 1292–1309. https://doi.org/10.1105/tpc.105.035881

Sakuma, Y., Maruyama, K., Qin, F., Osakabe, Y., Shinozaki, K., & Yamaguchi-Shinozaki, K. (2006). Dual function of an Arabidopsis transcription factor DREB2A in water-stress-responsive and heat-stress-responsive gene expression. *PNAS*, *103*, 18822–18827. https://doi.org/10.1073/pnas.0605639103

Scheibe, R., Backhausen, J. E., Emmerlich, V., & Holtgrefe, S. (2005). Strategies to maintain redox homeostasis during photosynthesis under changing conditions. *Journal of Experimental Botany*, *56*, 1481–1489. https://doi.org/10.1093/jxb/eri181

Schmitz, J., Heinrichs, L., Scossa, F., Fernie, A. R., Oelze, M.-L., Dietz, K.-J., … Häusler, R. E. (2014). The essential role of sugar metabolism in the acclimation response of *Arabidopsis thaliana* to high light intensities. *Journal of Experimental Botany*, *65*, 1619–1636. https://doi.org/10.1093/jxb/eru027

Sinclair, T. R., & Muchow, R. C. (1999). Radiation use efficiency. *Advances in agronomy*, *65*, 215–265. https://doi.org/10.1016/S0065-2113(08)60914-1

Sivak, M. N., & Walker, D. A. (1988). Photosynhesis in vivo can be limited by phosphate supply. *New Phytologist*, *102*, 499–512.

Staneloni, R. J., Rodriguez-Batiller, M. J., & Casal, J. J. (2008). Abscisic acid, high-light, and oxidative stress down-regulate a photosynthetic gene via a promoter motif not involved in phytochrome-mediated transcriptional regulation. *Molecular Plant*, *1*, 75–83. https://doi.org/10.1093/mp/ssm007

Stitt, M. (1991). Rising CO2 levels and their potential significance for carbon flow in photosynthetic cells. *Plant, Cell and Environment*, *14*, 741–762. https://doi.org/10.1111/j.1365-3040.1991.tb01440.x

Suzuki, N., Koussevitzky, S., Mittler, R., & Miller, G. (2012). ROS and redox signalling in the response of plants to abiotic stress. *Plant, Cell and Environment*, *35*, 259–270. https://doi.org/10.1111/j.1365-3040.2011.02336.x

Swindell, W. R., Huebner, M., & Weber, A. P. (2007). Transcriptional profiling of Arabidopsis heat shock proteins and transcription factors reveals extensive overlap between heat and non-heat stress response pathways. *BMC Genomics*, *8*, 125. https://doi.org/10.1186/1471-2164-8-125

Szechynska-Hebda, M., & Karpinski, S. (2013). Light intensity-dependent retrograde signalling in higher plants. *Journal of Plant Physiology*, *170*, 1501–1516. https://doi.org/10.1016/j.jplph.2013.06.005

Tanabe, N., Yoshimura, K., Kimura, A., Yabuta, Y., & Shigeoka, S. (2007). Differential expression of alternatively spliced mRNAs of Arabidopsis SR protein homologs, atSR30 and atSR45a, in response to environmental stress. *Plant and Cell Physiology*, *48*, 1036–1049. https://doi.org/10.1093/pcp/pcm069

Uniacke, J., & Zerges, W. (2008). Stress induces the assembly of RNA granules in the chloroplast of Chlamydomonas reinhardtii. *The Journal of cell biology*, *182*, 641–646. https://doi.org/10.1083/jcb.200805125

van Rooijen, R., Aarts, M. G. M., & Harbinson, J. (2015). Natural genetic variation for acclimation of photosynthetic light use efficiency to growth irradiance in Arabidopsis. *Plant Physiology*, *167*, 1412–1429. https://doi.org/10.1104/pp.114.252239

van Rooijen, R., Kruijer, W., Boesten, R., van Eeuwijk, F. A., Harbinson, J., & Aarts, M. G. M. (2017). Natural variation of YELLOW SEEDLING1 affects photosynthetic acclimation of *Arabidopsis thaliana*. *Nature Communications*, *8*, 1421. https://doi.org/10.1038/s41467-017-01576-3

Vanderauwera, S., Zimmerman, P., Rombauts, S., Vandenabeele, S., Langebartels, C., Gruissem, W., … Van Breusegem, F. (2005). Genome-wide analysis of hydrogen peroxide-regulated gene expression in Arabidopsis reveals a high light-induced transcriptional cluster involved in anthocyanin biosynthesis. *Plant Physiology*, *139*, 806–821. https://doi.org/10.1104/pp.105.065896

Vass, I. (2012). Molecular mechanisms of photodamage in the Photosystem II complex. *Biochimica et Biophysica Acta*, *1817*, 209–217. https://doi.org/10.1016/j.bbabio.2011.04.014

Walker, D. A., & Sivak, M. N. (1986). Photosynthesis and phosphate: A cellular affair? *Trends in Biochemical Science*, *11*, 176–179. https://doi.org/10.1016/0968-0004(86)90136-2

Walters, R. G. (2005). Towards an understanding of photosynthetic acclimation. *Journal of Experimental Botany*, *56*, 435–447.

Wang, P., Khoshravesh, R., Karki, S., Tapia, R., Balahadia, C. P., Bandyopadhyay, A., … Langdale, J. A. (2017). Re-creation of a key step in the evolutionary switch from C3 to C4 leaf anatomy. *Current Biology : CB*, *27*, 3278–3287. e3276

Waters, M. T., Wang, P., Korkaric, M., Capper, R. G., Saunders, N. J., & Langdale, J. A. (2009). GLK transcription factors coordinate expression of the photosynthetic apparatus in Arabidopsis. *Plant Cell*, *21*, 1109–1128. https://doi.org/10.1105/tpc.108.065250

Williams, W. P. (1998). The physical properties of thylakoid membrane lipids and their relation to photosynthesis. In P.-A. S. Murata (Ed.), *Lipids in photosynthesis: Structure, function and genetics* (pp. 103–118). Dordrecht: Kluwer.

Wunder, T., Liu, Q., Aseeva, E., Bonardi, V., Leister, D., & Pribil, M. (2013). Control of STN7 transcript abundance and transient STN7 dimerisation are involved in the regulation of STN7 activity. *Planta*, *237*, 541–558. https://doi.org/10.1007/s00425-012-1775-y

Xiao, Y., Savchenko, T., Baidoo, E. E., Chehab, W. E., Hayden, D. M., Tolstikov, V., … Dehesh, K. (2012). Retrograde signaling by the plastidial metabolite MEcPP regulates expression of nuclear stress-response genes. *Cell*, *149*, 1525–1535. https://doi.org/10.1016/j.cell.2012.04.038

Yamamoto, Y., Shimada, Y., Kimura, M., Manabe, K., Sekine, Y., Matsui, M., … Yoshida, S. (2004). Global classification of transcriptional responses to light stress in *Arabidopsis thaliana*. *Endocytobiosis Cell Research*, *15*, 438–452.

Yang, J. T., Preiser, A. L., Li, Z., Weise, S. E., & Sharkey, T. D. (2016). Triose phosphate use limitation of photosynthesis: Short-term and long-term effects. *Planta*, *243*, 687–698. https://doi.org/10.1007/s00425-015-2436-8

Yoshimura, K., Mori, T., Yokoyama, K., Koike, Y., Tanabe, N., Sato, N., … Shigeoka, S. (2011). Identification of alternative splicing events regulated by an Arabidopsis serine/arginine-like protein, atSR45a, in response to high-light stress using a tiling array. *Plant and Cell Physiology*, *52*, 1786–1805. https://doi.org/10.1093/pcp/pcr115

Zhu, X., Long, S. P., & Ort, D. R. (2008). What is the maximum efficiency with which photosynthesis can convert solar energy into biomass? *Current Opinion in Biotechnology*, *19*, 153–159. https://doi.org/10.1016/j.copbio.2008.02.004

6

ShinyAIM: Shiny-based application of interactive Manhattan plots for longitudinal genome-wide association studies

Waseem Hussain[1,2] | **Malachy Campbell**[1,2] | **Harkamal Walia**[2] | **Gota Morota**[1,3]

[1]Department of Animal Science, University of Nebraska-Lincoln, Lincoln, Nebraska

[2]Department of Agronomy and Horticulture, University of Nebraska-Lincoln, Lincoln, Nebraska

[3]Department of Animal and Poultry Sciences, Virginia Polytechnic Institute and State University, Blacksburg, Virginia

Correspondence
Waseem Hussain, Department of Animal Science, Department of Agronomy and Horticulture, University of Nebraska-Lincoln, Lincoln, NE.
Email: waseem.hussain@unl.edu

Funding information
National Science Foundation, Grant/Award Number: 1736192

Abstract

Owning to advancements in sensor-based, non-destructive phenotyping platforms, researchers are increasingly collecting data with higher temporal resolution. These phenotypes collected over several time points are cataloged as longitudinal traits and used for genome-wide association studies (GWAS). Longitudinal GWAS typically yield a large number of output files, posing a significant challenge to data interpretation and visualization. Efficient, dynamic, and integrative data visualization tools are essential for the interpretation of longitudinal GWAS results for biologists; however, these tools are not widely available to the community. We have developed a flexible and user-friendly Shiny-based online application, ShinyAIM, to dynamically view and interpret temporal GWAS results. The main features of the application include (a) interactive Manhattan plots for single time points, (b) a grid plot to view Manhattan plots for all time points simultaneously, (c) dynamic scatter plots for *p*-value-filtered selected markers to investigate co-localized genomic regions across time points, (d) and interactive phenotypic data visualization to capture variation and trends in phenotypes. The application is written entirely in the R language and can be used with limited programming experience. ShinyAIM is deployed online as a Shiny web server application at https://chikudaisei.shinyapps.io/shinyaim/, enabling easy access for users without installation. The application can also be launched on a local machine in RStudio.

KEYWORDS
genome-wide association studies, interactive visualization, longitudinal traits, ShinyAIM

1 | INTRODUCTION

Due to the increased availability of high-throughput phenotyping platforms, there is growing interest in the quantitative genetics of longitudinally measured traits, i.e., traits that are measured over multiple time points by advanced imaging systems (Araus & Kefauver, 2018; Araus, Kefauver, Zaman-Allah, Olsen, & Cairns, 2018). For example, the application of GWAS to abiotic stress responses, such as drought, salinity, and temperature stress, measured at temporal resolution may provide insights into the mechanisms underlying plant physiological processes measured throughout the duration of stress or development (Busemeyer et al., 2013; Moore et al., 2013; Topp et al., 2013; Slovak et al., 2014; Würschum et al., 2014; Yang et al., 2014; Bac-Molenaar, Vreugdenhil, Granier, & Keurentjes, 2015; Campbell, Walia, & Morota, 2018).

Data visualization is a fundamental aspect of big data analysis in genetics. Manhattan plots are standard tools used to visualize GWAS results and to identify the genomic regions associated with a given phenotype. However, the static nature of these plots limits the information that can be displayed and extracted. Further, the number of Manhattan plots that can be viewed at one time is limited, making comparisons across phenotypes tedious. The situation becomes more challenging in the case of longitudinal GWAS, which is performed across multiple time points, with each time point producing a Manhattan plot. Furthermore, it is difficult to share GWAS outputs in an easy and convenient way, requiring novel applications for dynamic data visualization and sharing. Many browsers have been built to visualize GWAS outputs (e.g., Cuellar-Partida, Renteria, & MacGregor, 2015; Juliusdottir, Banasik, Robertson, Mott, & McCarthy, 2018; Khramtsova & Stranger, 2017; Ziegler, Hartsock, & Baxter, 2015). However, none of these are specifically tailored for longitudinal traits. Further, existing applications do not offer features for the dynamic visualization of Manhattan plots online and for comparisons across time points simultaneously.

To address these limitations, we have developed a Shiny-based application, ShinyAIM, for visualizing and interpreting longitudinal GWAS outputs in an interactive way. The application is distinct from previously developed GWAS visualization browsers because it is specifically designed for longitudinal traits, allowing the simultaneous visualization of all time points or phenotypes and comparisons of top associated markers across time points. The interactive and integrative GWAS and phenotypic data visualization features embedded in the application offers a new resource for users to readily extract extensive information from temporal GWAS results.

2 | OVERVIEW OF ShinyAIM

2.1 | Methods

ShinyAIM is entirely written in the R language (R Core Team 2018) with the underlying R code encapsulated by the shiny R package (Chang, Cheng, Allaire, Xie, & McPherson, 2018), which is a web application framework for R, offering an interactive graphical user interface. Shiny has been making inroads into plant breeding and quantitative genetics for research and teaching purposes, such as Be-Breeder (Fritsche-Neto & Matias, 2016) and ShinyGPAS (Morota, 2017). ShinyAIM leverages the cumulative utility of the R packages manhattanly (Bhatnagar, 2016), plotly (Sievert et al., 2017) and ggplot2 (Wickham, 2016) to create a cohesive web browser-based application. The ShinyAIM application does not require any working knowledge of R and is intuitively operated through graphical user interface. ShinyAIM is hosted by a Shiny web server (https://chikudaisei.shinyapps.io/shinyaim/) for online use or can be run locally within RStudio by running the code *shiny::runGitHub("ShinyAIM", "whussain2")*. Alternatively, the ShinyAIM source code and sample files can be directly downloaded from the GitHub repository (https://github.com/whussain2/ShinyAIM). From the downloaded directory, the source file named app.R in RStudio can be run by clicking the *Run App* button. The ShinyAIM application is open source and is distributed under Artistic License 2.0.

2.2 | Usage

The starting page of the ShinyAIM application includes the Information tab with detailed information on how to format and upload the data. The video demonstration illustrating the application usage is also available (https://youtu.be/5-JLMpSiwv4). ShinyAIM is aimed for visualization of GWAS outputs and does not perform GWAS analysis. There are five required columns in the user data file labeled as "timepoint" (time point), "marker" (marker name), "chrom" (chromosome number), "pos" (marker position), and "P" (marker *p*-value) for Manhattan plot visualizations. For phenotypic data visualization, the data file must have two columns including "timepoint" (time point), and "Value" (phenotypic value). Further detailed instructions regarding the data formatting and column naming can be found in the main Information tab. In addition, the sample data files can be directly downloaded by clicking the "Download Sample File" button shown on top of the sidebar panel in the main tab.

The ShinyAIM application hosted on the server can handle 200–330 k markers for the visualization of interactive Manhattan plots. However, we suggest to launch the application locally by running the code *shiny::runGitHub("ShinyAIM", "whussain2")* in RStudio for datasets with millions of markers. Alternatively, filtering can be done based on *p*-values by removing markers with large *p*-values prior to uploading the input file for visualization.

3 | MAIN FEATURES AND FUNCTIONALITY

The application has four main features to explore GWAS results: (a) interactive Manhattan plots for single time points, (b) Manhattan grid plot to compare results across all time points simultaneously, (c) dynamic views of *p*-value filtered top associated markers in a scatter plot to identify co-localized markers over time, and (d) visualization of phenotypic data used for GWAS (Figure 1). These features are supported by user-defined data filtering criteria in ShinyAIM to smoothly navigate the application. Each feature is briefly described in the following sections.

3.1 | Interactive Manhattan plots

In the interactive Manhattan plots panel, users can interactively view the Manhattan plot for each time point (Figure 1A). After the correct file format has been selected and the file has been uploaded, the available time points will be automatically updated in the "Choose Time Point or Phenotypes" menu. An interactive Manhattan plot is automatically generated on the right-hand panel after selecting a target time point. Users can move the mouse over the points in the plot to display detailed information, including the marker name, position, chromosome location, and −log10 *p*-value. Furthermore, it is possible to zoom in on potential candidate regions to obtain additional detail. ShinyAIM offers the flexibility to choose the significance level by moving the slider input bar. In addition, users have a choice to display a list of markers arranged in decreasing order of *p*-values in the table below the Manhattan plot panel. The display also

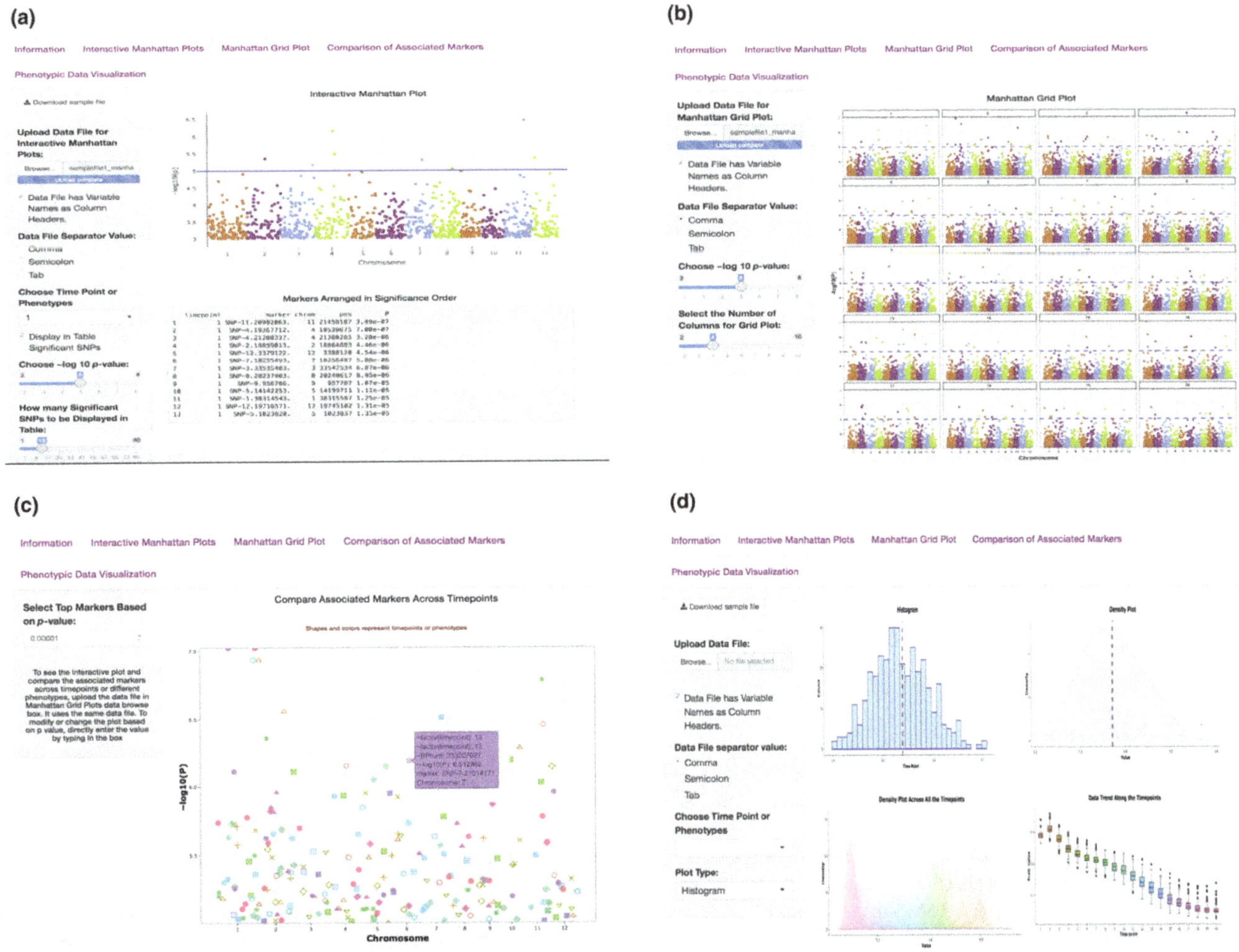

FIGURE 1 Main interface of the ShinyAIM application. Screenshots of panels for the main tabs are shown. (a) The "Interactive Manhattan Plots" tab allows users to display interactive Manhattan plots for a selected time point. Users have the flexibility to choose the significance level and can display the top associated markers in tabular format. (b) The "Manhattan Grid Plot" tab allows users to visualize Manhattan plots for all time points simultaneously. Users have the flexibility to choose the significance level and the number of columns in the grid plot. (c) The "Comparison of Associated Markers" tab allows users to filter markers based on *p*-values, display a scatter plot for comparisons across all time points, and search for co-localized markers. (d) The "Phenotypic Data Visualization" tab generates histogram and density plots and summarizes trends in temporal phenotypic data in the form of box plots

includes marker information in the input data file. The slider input bar controls the number of markers shown in the table.

3.2 | Manhattan grid plot

The Manhattan grid plot tab allows users to visualize the Manhattan plots combined for all time points and can be used to explore how GWAS peaks change over time to facilitate data interpretation (Figure 1B). The significance threshold for markers can be modified by moving the slider input bar. Moreover, ShinyAIM enables users to choose the number of columns and rows in the grid plot by moving the slider input bar "Select the Number of Columns in Grid Plot."

3.3 | Comparison of associated markers

Users are able to dynamically view only the top associated markers in a scatter plot (Figure 1C). This feature is implemented in ShinyAIM to enable users to focus only on the topmost associated markers and compare these markers across time points to identify co-localized regions. Users can select the number of markers displayed in a scatter plot by filtering the markers based on *p*-values. This is achieved by directly typing or selecting the option "Select Top Markers Based on *p*-value." The scatter plot is interactive, and users can move the mouse over a point to display information, including the time point, chromosome name, position of the marker, name of the marker, and −log10 *p*-value (Figure 1C).

3.4 | Phenotypic data visualization

Phenotypic data visualization helps users to view phenotypes used for GWAS in the form of dynamic histograms and density plots (Figure 1D). The trends and variability in phenotypic values at each time point can be visualized using box plots. All plot types are interactive, and users can move the mouse over a particular point to obtain detailed information.

4 | CONCLUSION

We have developed a user-friendly integrative Shiny-based application to dynamically visualize and interpret longitudinal GWAS results, providing an easy-to-use online tool to the community.

ACKNOWLEDGMENT

This work was funded and supported by National Science Foundation (Grant No. 1736192).

CONFLICT OF INTEREST

The authors declare there are no competing interests.

AUTHOR CONTRIBUTIONS

Study was conceived by W.H. and G.M. W.H. wrote the code and drafted the manuscript. M.C., H.W., and G.M. revised the manuscript.

REFERENCES

Araus, J. L., & Kefauver, S. C. (2018). Breeding to adapt agriculture to climate change: Affordable phenotyping solutions. *Current Opinion in Plant Biology*. https://doi.org/10.1016/j.pbi.2018.05.003

Araus, J. L., Kefauver, S. C., Zaman-Allah, M., Olsen, M. S., & Cairns, J. E. (2018). Translating high-throughput phenotyping into genetic gain. *Trends in Plant Science*, *23*(5), 451–466. https://doi.org/10.1016/j.tplants.2018.02.001

Bac-Molenaar, J. A., Vreugdenhil, D., Granier, C., & Keurentjes, J. J. B. (2015). Genome-wide association mapping of growth dynamics detects time-specific and general quantitative trait loci. *Journal of Experimental Botany*, *66*(18), 5567–5580. https://doi.org/10.1093/jxb/erv176

Bhatnagar, S. (2016). *Manhattanly: Interactive Q–Q and Manhattan plots using "Plotly.Js."* https://CRAN.R-project.org/package=manhattanly

Busemeyer, L., Ruckelshausen, A., Möller, K., Melchinger, A. E., Alheit, K. V., Maurer, H. P., … Würschum, T. (2013). Precision phenotyping of biomass accumulation in triticale reveals temporal genetic patterns of regulation. *Scientific Reports*, *3*, 2442. https://www.nature.com/articles/srep02442 https://doi.org/10.1038/srep02442

Campbell, M. T., Walia, H., & Morota, G. (2018). Utilizing random regression models for genomic prediction of a longitudinal trait derived from high-throughput phenotyping. *BioRxiv*, 319897. https://doi.org/10.1101/319897

Chang, W., Cheng, J., Allaire, J. J., Xie, Y., & McPherson, J. (2018). *Shiny: Web application framework for R*. Retrieved from https://CRAN.R-project.org/package=shiny

Cuellar-Partida, G., Renteria, M. E., & MacGregor, S. (2015). LocusTrack: Integrated visualization of GWAS results and genomic annotation. *Source Code for Biology and Medicine*, *10*, 1. https://doi.org/10.1186/s13029-015-0032-8

Fritsche-Neto, R., & Matias, F. I. (2016). Be-Breeder – Learning: A new tool for teaching and learning plant breeding principles. *Crop Breeding and Applied Biotechnology*, *16*(3), 240–245. https://doi.org/10.1590/1984-70332016v16n3n36

Juliusdottir, T., Banasik, K., Robertson, N. R., Mott, R., & McCarthy, M. I. (2018). Toppar: An interactive browser for viewing association study results. *Bioinformatics*, *34*(11), 1922–1924. https://doi.org/10.1093/bioinformatics/btx840

Khramtsova, E. A., & Stranger, B. E. (2017). Assocplots: A Python package for static and interactive visualization of multiple-group GWAS results. *Bioinformatics*, *33*(3), 432–434. https://doi.org/10.1093/bioinformatics/btw641

Moore, C. R., Johnson, L. S., Kwak, I. Y., Livny, M., Broman, K. W., & Spalding, E. P. (2013). High-throughput computer vision introduces the time axis to a quantitative trait map of a plant growth response. *Genetics*, *195*, 1077–1086. https://doi.org/10.1534/genetics.113.153346

Morota, G. (2017). ShinyGPAS: Interactive Genomic Prediction Accuracy Simulator based on deterministic formulas. *Genetics, Selection, Evolution*, *49*, 91. https://doi.org/10.1186/s12711-017-0368-4

R Core Team (2018). *R: The R project for statistical computing*. Vienna, Austria. Retrieved from https://www.r-project.org/

Sievert, C., Parmer, C., Hocking, T., Chamberlain, S., Ram, K., Corvellec, M., & Despouy, P. (2017). *Plotly: Create interactive web graphics via "Plotly.Js."* Retrieved from https://CRAN.R-project.org/package=plotly

Slovak, R., Göschl, C., Su, X., Shimotani, K., Shiina, T., & Busch, W. (2014). A scalable open-source pipeline for large-scale root phenotyping of Arabidopsis. *Plant Cell*, *26*, 2390–2403. https://doi.org/10.1105/tpc.114.124032

Topp, C. N., Iyer-Pascuzzi, A. S., Anderson, J. T., Lee, C. R., Zurek, P. R., Symonova, O., … Benfey, P. N. (2013). 3D phenotyping and quantitative trait locus mapping identify core regions of the rice genome controlling root architecture. *Proceedings of National Academy of Science of the United States of America*, *110*, E1695–E1704. https://doi.org/10.1073/pnas.1304354110

Wickham, H. (2016). *ggplot2: Elegant graphics for data analysis*. New York, NY: Springer-Verlag. ISBN 978-3-319-24277-4. http://ggplot2.org. https://doi.org/10.1007/978-3-319-24277-4

Würschum, T., Liu, W., Busemeyer, L., Tucker, M. R., Reif, J. C., Weissmann, E. A., … Maurer, H. P. (2014). Mapping dynamic QTL for plant height in triticale. *BMC Genetics*, *15*, 59. https://doi.org/10.1186/1471-2156-15-59

Yang, W., Guo, Z., Huang, C., Duan, L., Chen, G., Jiang, N., … Xiong, L. (2014). Combining high-throughput phenotyping and genome-wide association studies to reveal natural genetic variation in rice. *Nature Communications*, *5*, 5087. https://www.nature.com/articles/ncomms6087

Ziegler, G. R., Hartsock, R. H., & Baxter, I. (2015). Zbrowse: An interactive GWAS results browser. *PeerJ Computer Science*, *1*, e3. https://peerj.com/articles/cs-3/

7

A chromosome-scale assembly of the model desiccation tolerant grass *Oropetium thomaeum*

Robert VanBuren[1,2] | **Ching Man Wai**[1] | **Jens Keilwagen**[3] | **Jeremy Pardo**[4]

[1]Department of Horticulture, Michigan State University, East Lansing, Michigan

[2]Plant Resilience Institute, Michigan State University, East Lansing, Michigan

[3]Institute for Biosafety in Plant Biotechnology, Julius Kühn-Institut (JKI) – Federal Research Centre for Cultivated Plants, Quedlinburg, Germany

[4]Department of Plant Biology, Michigan State University, East Lansing, Michigan

Correspondence
Robert VanBuren, Department of Horticulture, Michigan State University, East Lansing, MI.
Email: bobvanburen@gmail.com

Funding information
National Science Foundation, Grant/Award Number: MCB-1817347

Abstract

Oropetium thomaeum is an emerging model for desiccation tolerance and genome size evolution in grasses. A draft genome of Oropetium was recently sequenced, but the lack of a chromosome-scale assembly has hindered comparative analyses and downstream functional genomics. Here, we reassembled Oropetium, and anchored the genome into 10 chromosomes using high-throughput chromatin conformation capture (Hi-C) based chromatin interactions. A combination of high-resolution RNAseq data and homology-based gene prediction identified thousands of new, conserved gene models that were absent from the V1 assembly. This includes thousands of new genes with high expression across a desiccation timecourse. Comparison between the Sorghum and Oropetium genomes revealed a surprising degree of chromosome-level collinearity, and several chromosome pairs have near perfect synteny. Other chromosomes are collinear in the gene rich chromosome arms but have experienced pericentric translocations. Together, these resources will be useful for the grass-comparative genomic community and further establish Oropetium as a model resurrection plant.

KEYWORDS
chromosome-scale, comparative genomics, desiccation tolerance, grasses, Hi-C

1 | INTRODUCTION

Desiccation tolerance evolved as an adaptation to extreme and prolonged drying, and resurrection plants are among the most resilient plants on the planet. The molecular basis of desiccation tolerance is still largely unknown, but a number of models have emerged to dissect the genetic control of this trait (Hoekstra, Golovina, & Buitink, 2001; Zhang & Bartels, 2018). The genomes of several model resurrection plants have been sequenced including *Boea hygrometrica* (Xiao et al., 2015), *Oropetium thomaeum* (VanBuren et al., 2015), *Xerophyta viscosa* (Costa et al., 2017), *Lindernia brevidens* (VanBuren, Wai, Pardo, et al., 2018), *Selaginella lepidophylla* (VanBuren, Wai, Ou, et al., 2018), and *Selaginella tamariscina* (Xu et al., 2018). To date, no chromosome scale assemblies are available for these species, limiting large-scale quantitative genetics and comparative genomics-based approaches. Many resurrection plants are polyploid or have prohibitively large genomes including those in the genera *Boea*, *Xerophyta*, *Eragrostis*, *Sporobolus*, and *Craterostigma*. This complexity complicates genome assembly and gene redundancy in the polyploid species hinders downstream functional genomics work.

Oropetium thomaeum (hereon referred to as Oropetium), is a diploid resurrection plant with the smallest genome among the grasses (245 Mb) (Bartels & Mattar, 2002). Oropetium plants are similar in size to Arabidopsis, but significantly smaller than the model grasses *Setaria italica* (Li & Brutnell, 2011) and *Brachypodium distachyon* (Brkljacic et al., 2011), with a short generation time of ~4 months. Oropetium is in the Chloridoideae subfamily of grasses and is closely related to the orphan cereal crops tef (*Eragrostis* tef) and finger millet (*Eleusine coracana*). Desiccation tolerance evolved independently several times within Chloridoideae (Gaff, 1977, 1987; Gaff & Latz, 1978) making it a useful system for studying convergent evolution. Together, these traits make Oropetium an attractive model for exploring the origin and molecular basis of desiccation tolerance. Oropetium was one of the first plants to be sequenced using the long reads of PacBio technology, and the assembly quality was comparable to early Sanger sequencing-based plant genomes such as rice and Arabidopsis (VanBuren et al., 2015). Despite the high contiguity of Oropetium V1, the assembly has 625 contigs and the BioNano based genome map was unable to produce chromosome-scale scaffolds. Furthermore, the V1 annotation was based on limited transcript evidence, and a high proportion of conserved plant genes were missing (VanBuren et al., 2015). Here, we reassembled the Oropetium genome using a more refined algorithm and generated a chromosome-scale assembly using Hi-C based chromatin interactions. The annotation quality was improved using high-resolution RNAseq data and protein homology, facilitating detailed comparative genomics with other grasses.

2 | METHODS

2.1 | Genome reassembly

The raw PacBio reads from the Oropetium V1 release (VanBuren et al., 2015) were reassembled with improved algorithms to better resolve highly complex and repetitive regions. PacBio data were error corrected and assembled using Canu (V1.4) (Koren et al., 2017) with the following modifications: minReadLength = 1,500, GenomeSize = 245 Mb, and minOverlapLength = 1,000. Other parameters were left as default. The resulting assembly graph was visualized in Bandage (Wick, Schultz, Zobel, & Holt, 2015). The assembly graph was free of heterozygosity related bubbles, but many nodes (contigs) were interconnected by a high copy number retrotransposon. The Canu based contigs (assembly V1.2) were first polished using Quiver (Chin et al., 2013) with the raw PacBio data and default parameters. Contigs were further polished with Pilon (V1.22) (Walker et al., 2014) using ~120× coverage of paired-end 150 bp Illumina data. Quality-trimmed Illumina reads were aligned to the draft contigs using bowtie2 (V2.3.0) (Langmead & Salzberg, 2012) with default parameters. The overall alignment rate was 95.5%, which was slightly higher than alignment against the HGAP V1 assembly (94.5%). The following parameters for Pilon were modified: –flank 7, –K 49, and –mindepth 25. Other parameters were left as default. Pilon was run four times with an updated reference and realignment of Illumina data after each iteration. Indel corrections plateaued after the third iteration, suggesting polishing removed most residual assembly errors.

2.2 | Hi-C library construction analysis and genome anchoring

Oropetium descended from the original plants used for PacBio sequencing was collected for Hi-C library construction and RNAseq. Oropetium is highly selfing with low heterozygosity, and we expect minimal differences to be introduced in the new version. Oropetium seeds are available upon request. Oropetium plants were maintained under day/night temperatures of 26 and 22°C, respectively, with a light intensity of 200 E m^{-2} s^{-1} and 16/8 hr photoperiod. Young leaf tissue was used for Hi-C library construction with the Proximo™ Hi-C Plant kit (Phase Genomics) following the manufacture's protocol. Briefly, 0.2 g of fresh, young leaf tissue was finely chopped and the chromatin was immediately crosslinked. The chromatin was fragmented and proximity ligated, followed by library construction. The final library was size selected for 300–600 bp and sequenced on the Illumina HiSeq 4000 under paired-end 150 bp mode. Adapters were trimmed and low-quality sequences were removed using Trimmomatic (V0.36) (Bolger, Lohse, & Usadel, 2014). Read pairs were aligned to the Oropetium contigs using bwa (V0.7.16) (Li, 2013) with strict parameters (-n 0) to prevent mismatches and non-specific alignments in duplicated and repetitive regions. SAM files from bwa were used as input in the Juicer pipeline, and PCR duplicates with the same genome coordinates were filtered prior to constructing the interaction based distance matrix. In total, 101 filtered read pairs were used as input for the Juicer and 3d-DNA Hi-C analysis and scaffolding pipelines (Dudchenko et al., 2017; Durand et al., 2016). Contig ordering, orientation, and chimera splitting were done using the 3d-DNA pipeline (Dudchenko et al., 2017) under default parameters. Contig misassemblies and scaffold misjoins were manually detected and corrected based on interaction densities from visualization in Juicebox. In total, six chimeric contigs were identified and split at the junction with closest interaction data. The manually validated assembly was used as input to build the 10 scaffolds (chromosomes) using the finalize-output.sh script from 3d-DNA. Chromosomes and unanchored contigs were renamed by size, producing the V2 assembly.

2.3 | Genome annotation

The Oropetum V2 assembly was reannotated using the homology-based gene prediction program Gene Model Mapper (GeMoMa: V 1.5.2) (Keilwagen et al., 2016, 2018). GeMoMa uses protein homology and RNAseq evidence to predict gene models. Genome assemblies and gene annotation for the following 11 species were downloaded from Phytozome (V12) and used as homology-based evidence: *Arabidopsis thaliana, Brachypodium distachyon, Glycine max, Oryza sativa, Panicum hallii, Populus trichocarpa, Prunus persica, Setaria italica, Solanum lycopersicum, Sorghum bicolor*, and *Theobroma cacao*. Translated coding exons and proteins from the reference gene annotations and genome assemblies were extracted using the module Extractor function of GeMoMa (module Extractor: Ambiguity = AMBIGUOUS, r = true). RNAseq data from Oropetium desiccation and rehydration timecourses (VanBuren et al., 2017) were aligned to the V2 Oropetium genome using HISAT2 (Kim, Langmead, & Salzberg, 2015) with default parameters. The resulting BAM files were used to extract intron and exon boundaries using the module ERE (module ERE: s = FR_FIRST_STRAND, c = true). Translated coding exons from other species were aligned to the Oropetium genome using tblastn and transcripts were predicted based on each reference species independently using the extracted introns and coverage (module GeMoMa). Finally, the predictions based on the 11 reference species were combined to obtain a final prediction using the module GAF. Gene models containing transposases were filtered, resulting in a final annotation of 28,835 gene models. The annotation completeness was assessed using the plant specific Benchmarking Universal Single Copy Ortholog (BUSCO) dataset (version 3.0.2, embryophyta_odb9) (Simão, Waterhouse, Ioannidis, Kriventseva, & Zdobnov, 2015). The following report was obtained from BUSCO: 98.9% overall, 95.4% single copy, 3.5% duplicated, 0.6% fragmented, and 0.5% missing. Gene model names from V1 were conserved where possible, and new gene models received new names.

2.4 | Expression analysis

Oropetium RNAseq data from desiccation and rehydration time-courses were reanalyzed using the updated gene model annotations (VanBuren et al., 2017). Four timepoints during dehydration (days 7, 14, 21, and 30), two during rehydration (24 and 48 hr), and one well-watered sample were analyzed. Based on principle component analysis, replicate two of the "well-watered" and "D21" samples were excluded from the analysis. Each other timepoint had three replicates. Gene expression was quantified on a transcript level using salmon (v 0.9.1) in quasi-mapping mode (Patro, Duggal, Love, Irizarry, & Kingsford, 2017). Default parameters were used with the internal GC bias correction in salmon. The R package tximport (v 1.2.0) was used to map transcript level quantifications to gene level counts (R Core Team, 2013; Soneson, Love, & Robinson, 2015). We conducted differential expression analysis with the remaining samples using the R package DESeq2 (v 1.14.1) set to default parameters [3,4].

2.5 | Identification of LTR-RTs

A preliminary list of candidate long terminal repeat retrotransposons (LTR-RTs) from Oropetium was identified using LTR_Finder (V1.02) (Xu & Wang, 2007) and LTRharvest (Ellinghaus, Kurtz, & Willhoeft, 2008). The following parameters for LTRharvest were modified: -similar 90 –vic 10 –seed 20 –minlenltr 100 –maxlenltr 7000 –mintsd 4 –maxtsd 6 –motif TGCA –motifmis 1. LTR_Finder parameters were: -D 15000 –d 1000 –L 7000 –l 100 –p 20 –C –M 0.9. LTR_retriever (Ou & Jiang, 2017) was used to filter out false LTR retrotransposons using the target site duplications, terminal motifs, and Pfam domains. Default parameters were used for LTR_retriever. LTR_retriever produced a list of full length, high-quality LTRs. LTRs were annotated across the genome using RepeatMasker (http://www.repeatmasker.org/) (Smit, Hubley, & Green, 1996) and the non-redundant LTR-RT library constructed by LTR_retriever. The insertion time of intact LTRs was calculated in LTR_retriever using the formula $T = K/2\mu$ with a neutral mutation rate of $\mu = 1 \times 10^{-8}$ mutations per bp per year.

2.6 | Comparative genomics

Syntenic gene pairs between the Oropetium and Sorghum genomes were identified using the MCSCAN toolkit (V1.1) (Wang et al., 2012) implemented in python (https://github.com/tanghaibao/jcvi/wiki/MCscan-(Python-version)). Default parameters were used. Gene models were aligned using LAST and hits were filtered to find the best 1:1 syntenic blocks. Macrosyntenic dotplots were constructed in MCScan.

3 | RESULTS AND DISCUSSION

The first version of the Oropetium genome (V1) was sequenced with high coverage PacBio data (~72×) followed by error correction and assembly using the hierarchical genome assembly process (HGAP) (VanBuren et al., 2015). We reassembled this PacBio data using the Canu assembler (Koren et al., 2017), which can more accurately assemble and phase complex repetitive regions. The resulting Canu based assembly (hereon referred to as V1.2) had fewer contigs than the V1 HGAP assembly, but had otherwise similar assembly metrics (Table 1). Draft contigs were polished using a two-step process to remove residual insertion/deletion (Indel) and single nucleotide errors. Contigs were first polished using the raw PacBio data with Quiver (Chin et al., 2013), followed by four rounds of reiterative polishing with Pilon (Walker et al., 2014) using high coverage Illumina paired-end data. The final V1.2 assembly contains 436 contigs with an N50 of 2.0 Mb and total assembly size of 236 Mb. This is six megabases smaller than the V1 assembly, with slightly lower contiguity. More intact LTR-RTs and centromere specific repeat arrays were identified in Oropetium V1.2 compared to V1, suggesting the Canu assembler resolved these repetitive elements more accurately. Thus, V1.2 was used for pseudomolecule construction.

The Oropetium V1.2 contigs were ordered and oriented into chromosome-scale pseudomolecules using Hi-C. Hi-C leverages long-range interactions across distal regions of chromosomes to order and orient contigs. This approach is similar to genetic map-based anchoring, but with higher resolution. Illumina data generated from the Hi-C library were mapped to the V1.2 Oropetium genome using bwa (Li, 2013) and the proximity-based clustering matrix was generated using the Juicer and 3d-DNA pipelines (Dudchenko et al.,

TABLE 1 Comparison of the Oropetium V1 and V2 assembly and annotation statistics

Statistics	V1	V2
# of contigs	625	436
Contig N50	2.38 Mb	2.02 Mb
Scaffold N50	NA	20.5 Mb
Total assembly size	243 Mb	236 Mb
Gene models	28,446	28,835
BUSCO	72.1%	98.9%

2017; Durand et al., 2016). After filtering and manual curation, 10 high-confidence clusters were identified (Figure 1). These 10 clusters correspond to the haploid chromosome number of Oropetium. Regions with low density interactions highlight the centromeric and pericentromeric regions, and regions with higher than expected interactions represent topologically associated domains. After splitting six chimeric PacBio contigs, 239 contigs were anchored and oriented into 10 chromosomes spanning 226.5 Mb or 95.8% of the total assembled genome (Table 1). Chromosomes range in size from 11.0 to 34.7 Mb with an average size of 22.6 Mb. Most of the unanchored contigs are small (average size 42 kb), or are entirely composed of rRNA, centromeric repeat arrays, or centromere specific LTR-RTs. Telomeres were identified at both ends of Chromosomes 1, 2, 3, 4, 5, 7, and 9 and on one end of Chromosomes 6, 8, and 10. Three unanchored contigs contain the remaining telomeres. This supports the completeness and accuracy of the pseudomolecule construction.

The chromosome scale Oropetium genome (hereon referred to as V2) was reannotated using the homology-based gene prediction program GeMoMa (Keilwagen, Hartung, Paulini, Twardziok, & Grau, 2018; Keilwagen et al., 2016). Protein coding sequences from 11 angiosperm genomes and RNAseq data from Oropetium (VanBuren et al., 2017) were used as evidence. After filtering gene models derived from transposases, the final annotation consists of 28,835 high-confidence gene models. The annotation completeness was assessed using the Benchmarking Universal Single-Copy Ortholog (BUSCO) embryophyta dataset. The V2 gene models have a BUSCO score of 98.9%, suggesting that the updated annotation is high-quality. In comparison, the Oropetium V1 annotation has a BUSCO score of 72%, and many conserved gene models were likely missing or misannotated. Nearly 40% (11,227) of the gene models in V2 are new and were unannotated in V1. In addition, 10,837 gene models from V1 were removed or substantially improved in the V2 annotation. These discarded gene models either had little support based on protein homology to other species and transcript evidence from Oropetium, or they were misannotated transposable elements. In total, 94.3% of the gene models (27,216) were anchored to the 10 chromosomes. Among the newly annotated gene models are 3,525 tandem gene duplicates (Figure 2a). Tandem duplicates span 3,062 arrays with 7,760 total genes. Of the arrays containing three or more genes, only 49 are new to V2, and the majority contain genes previously identified in V1. The boundaries of tandem duplicates are difficult to correctly annotate, resulting in fusions of two or more gene copies. The homology-based annotation used in V2 was able to parse the previously fused gene models.

The expression patterns of newly annotated genes were surveyed using high-resolution RNAseq expression data (VanBuren et al., 2017). This dataset consists of seven leaf samples collected during desiccation and rehydration timecourses. Timepoints include well-watered, 7, 14, 21, and 30 days desiccated as well as 24 and

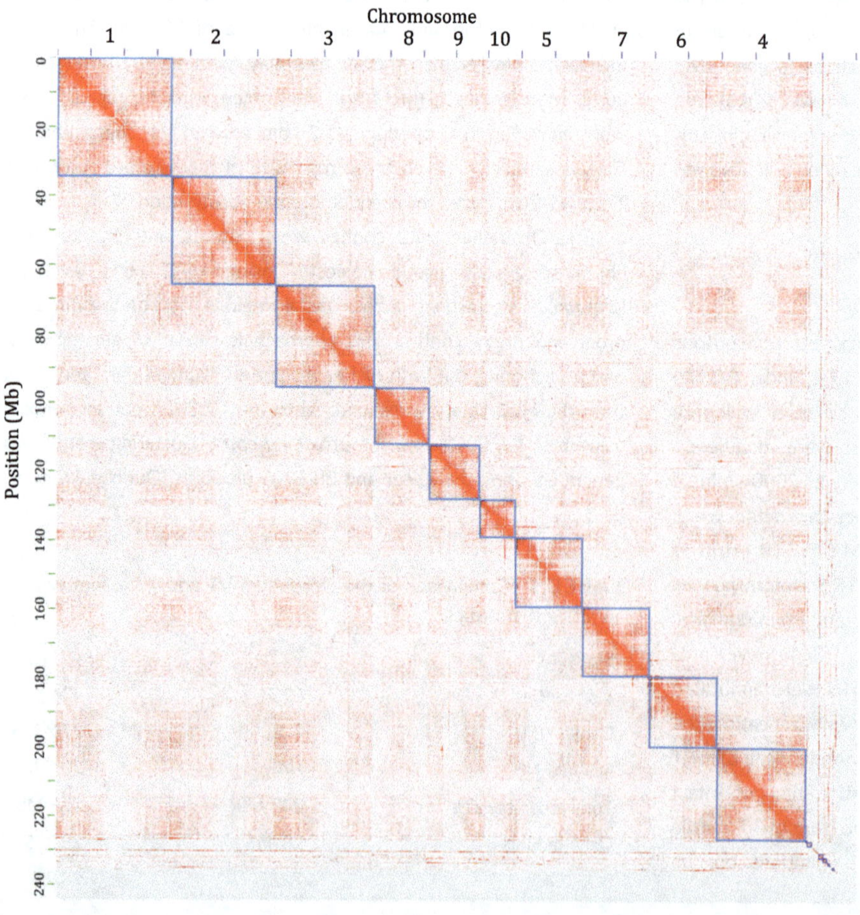

FIGURE 1 Hi-C based contig anchoring. Post-clustering heat map showing density of Hi-C interactions between contigs from the Juicer and 3d-DNA pipeline. The 10 Oropetium chromosomes are highlighted by blue squares

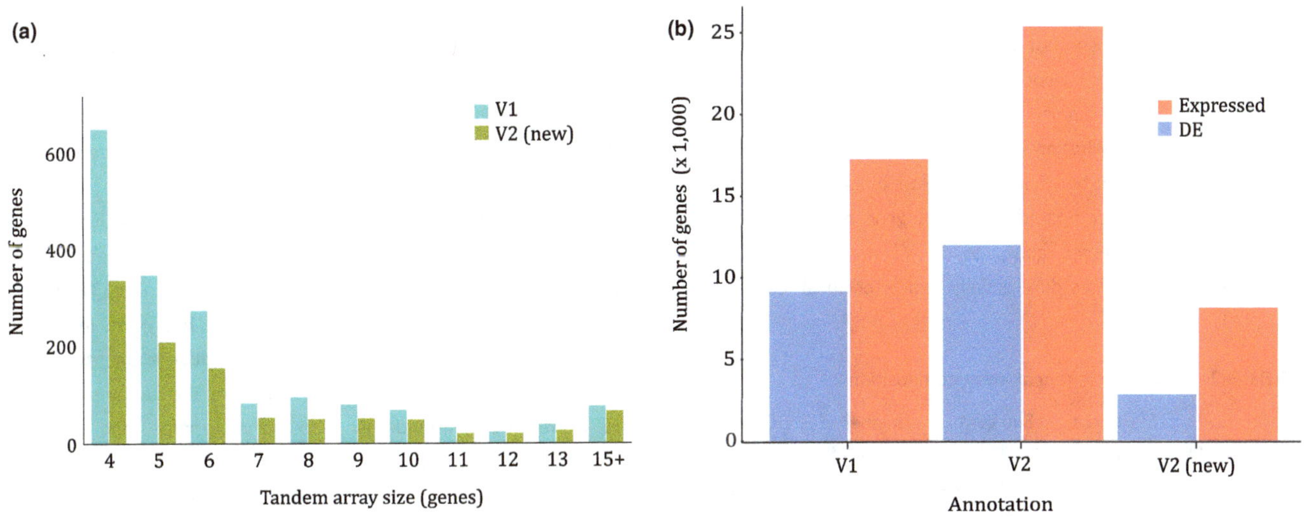

FIGURE 2 Characterization of the updated V2 Oropetium annotation. (a) Tandem gene array size comparison of the V1 and V2 annotation. Tandem genes identified in V1 are shown in blue and tandem genes newly annotated in V2 are shown in gold. (b) Comparison of expression patterns from the V1 and V2 annotation. The total number of genes with detectable expression and differential expression (DE) in the Oropetium desiccation/rehydration timecourse are plotted

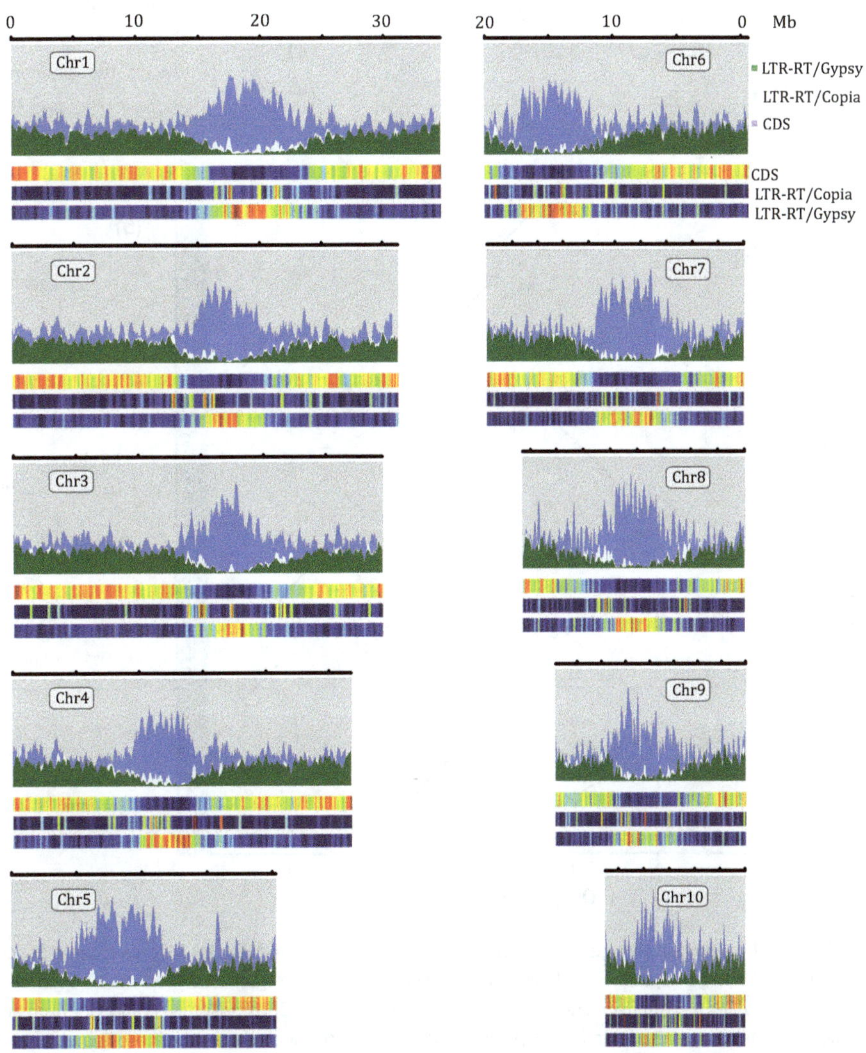

FIGURE 3 Landscape of the Oropetium genome. *Gypsy* and *Copia* long terminal repeat retrotransposons (LTR-RT) and CDS density are plotted for the 10 Oropetium chromosomes. Features are plotted in sliding windows of 50 kb with 25 kb step size. The location of centromere specific tandem arrays is highlighted by red bars. The heat maps below each landscape show relative density with red indicating high density and blue indicating low density for each feature

48 hr post rehydration. Differentially expressed genes were identified based on comparisons of well-watered leaves with each dehydration or rehydration timepoint. In addition, each timepoint was compared with the timepoint immediately following it in the time-course (i.e., day 7 dehydration vs. day 14). In total, 17,204 gene models from the V1 annotation had detectable expression (count > 0 in at least one sample) compared to 25,314 gene models in V2 (Figure 2b). Of the expressed genes, 9,149 V1 and 11,948 V2 gene models were classified as differentially expressed in at least one of the comparisons. Most newly annotated genes (8,110) have detectable expression in at least one of the seven timepoints, and the majority are expressed in all timepoints. In total, 2,799 new V2 gene models were differentially expressed, suggesting the newly annotated genes have important and previously uncharacterized roles in desiccation tolerance.

TABLE 2 Centromeric repeat array composition

Chromosome	Start cent. array (bp)	End cent. array (bp)	Number of cent. repeats	Cent. size (bp)
Chr_1	18,899,082	19,114,162	154	215,080
Chr_2	18,277,215	18,463,229	786	186,014
Chr_3	18,882,303	18,993,598	308	111,295
Chr_4	11,739,636	13,338,554	176	1,598,918
Chr_5	10,361,368	10,828,355	800	466,987
Chr_6	3,649,010	3,746,417	513	97,407
Chr_7	12,434,273	12,559,564	272	125,291
Chr_8	8,288,262	9,010,114	306	721,852
Chr_9	6,142,739	7,433,209	1,044	1,290,470
Chr_10	3,147,692	3,209,432	155	61,740
Unanchored			4,258	982,774

We used the chromosome-scale assembly of Oropetium to survey patterns of genome organization and evolution related to maintaining a small genome size. The proportion of LTR-RTs in Oropetium V1 and V2 is similar, though V2 has more intact elements. LTR-RTs are the most abundant repetitive elements in Oropetium and collectively span 27% (62 Mb) of the genome. LTR-RTs are distributed non-randomly across the genome, and peaks of Gypsy LTR-RTs are observed in each of the 10 chromosomes (Figure 3). These peaks of Gypsy LTR-RTs correspond to the pericentromeric regions. The pericentromeric regions show reduced intra-chromosomal interactions in the Hi-C matrix and contain arrays of centromeric repeats. The Oropetium V2 genome contains 8,965 155 bp monomeric centromeric repeats; considerably more than the 4,315 identified in the V1 assembly. The centromeric array sizes vary from 61 kb in chromosome 10 to 1,598 kb in Chromosome 4 (Figure 3; Table 2). Array sizes are likely underestimated, as only 52% of centromeric arrays were anchored to chromosomes, and 23 unanchored contigs contain centromeric repeat arrays. Gene density is low in the pericentromeric regions, consistent with the rice, sorghum, maize, and Brachypodium genomes (Du et al., 2017; Initiative,

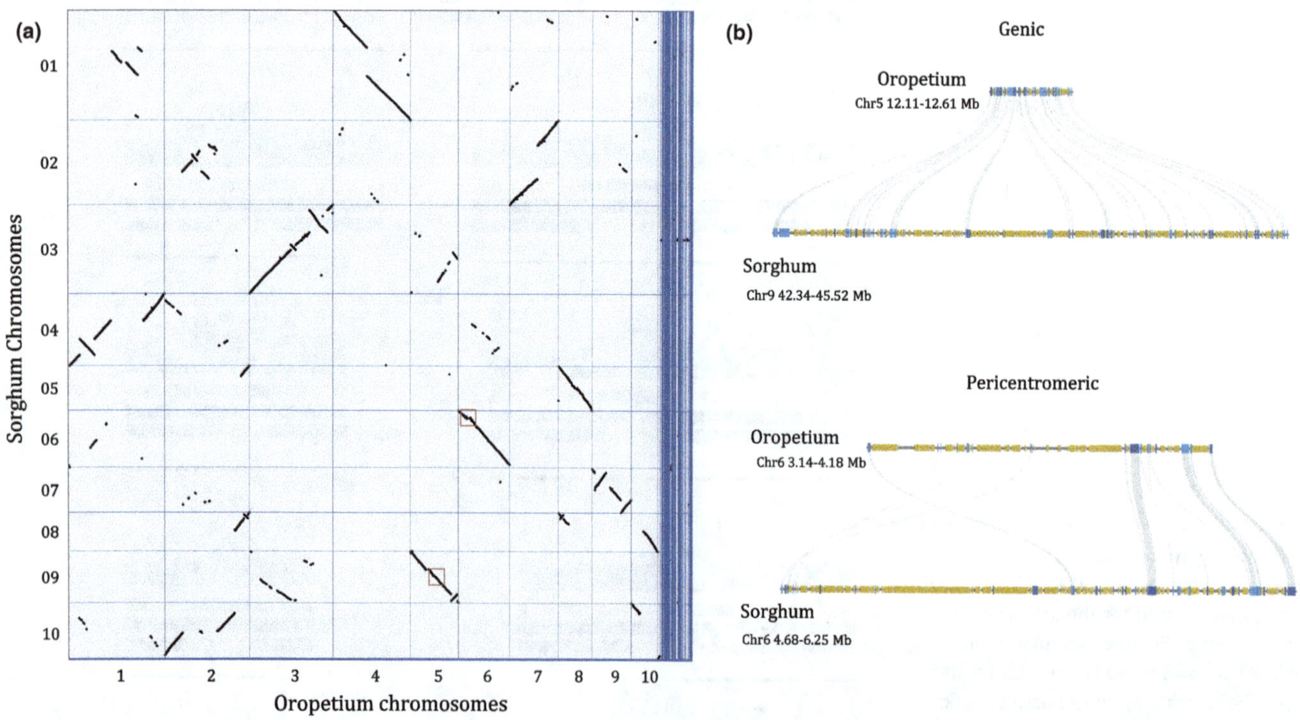

FIGURE 4 Comparative genomics between Oropetium and Sorghum. (a) Macrosyntenic dotplot of the Oropetium and Sorghum chromosomes based on 18,889 gene pairs. Each black dot represents a syntenic region between the two genomes. (b) Microsynteny of a typical genic region of Sorghum and Oropetium (top) and the pericentromeric region of Chromosome 6 of Oropetium and Sorghum (bottom). LTR-RTs are shown in yellow and genes are shown in blue. Syntenic orthologs are connected by gray lines. The centromeric repeat array in Oropetium is shown in red

2010; Jiao et al., 2017; Paterson et al., 2009). Collectively, pericentromeric regions span 67.5 Mb or 29% of the genome, a much smaller proportion than Sorghum (62%; 460 Mb) (Paterson et al., 2009), but higher than rice (15%; 63 Mb) (Goff et al., 2002). The majority of intact LTRs (86%; 628) have an insertion time of less than one million years ago, with a steep drop off of insertion time after 0.4 MYA. This suggests that LTRs are highly active in Oropetium but rapidly fragmented and purged to maintain its small genome size.

Previous comparative genomic analyses supported a high degree of collinearity between Oropetium and other grass genomes, but the draft assembly prevented detailed chromosome-level comparisons. To date, no chromosome scale assemblies are available for other Chloridoideae grasses, though draft genomes are available for the orphan grain crops tef (*Eragrostis tef*) (Cannarozzi et al., 2014) and finger millet (*Eleusine coracana*) (Hittalmani et al., 2017). We compared the V2 Oropetium chromosomes to the high-quality BTX 623 Sorghum genome (McCormick et al., 2018). Sorghum is in the Panicoideae subfamily of grasses which diverged from the ancestors of Chloridoideae ~31 MYA (Cotton et al., 2015). Despite this divergence, the 10 chromosomes in Oropetium are largely collinear to the corresponding 10 chromosomes in Sorghum, though large-scale inversions and translocations were identified (Figure 4a). Oropetium chromosomes 5, 6, and 8 are collinear along their length to Sorghum chromosomes 9, 6, and 5, respectively. Oropetium chromosomes 1, 2, 4, and 7, are collinear to the arms of Sorghum chromosomes 4, 10, 1, and 2, but the pericentric regions have translocated to other chromosomes. Oropetium chromosome 9 and Sorghum chromosome 7 are syntenic but have two large-scale inversions, and Oropetium and Sorghum chromosome 3 are syntenic with one inversion.

The Sorghum genome is roughly threefold larger than Oropetium, and genome size dynamics in grasses are driven by purge and accumulation of retrotransposons (Wicker, Buchmann, & Keller, 2010). Gene rich regions of Oropetium are 2–3× more compact than orthologous regions in sorghum, and much of this expansion in Sorghum is caused by intergenic blocks of LTR-RTs (Figure 4b), consistent with patterns observed in the V1 assembly (VanBuren et al., 2015). The chromosome-scale nature of Oropetium V2 allowed us to survey patterns of collinearity in the pericentromeric regions. These regions have a lower degree of synteny with Sorghum compared to gene rich euchromatin, consistent with retrotransposon-mediated rearrangements (Figure 4b). Pericentromeres are greatly expanded in Oropetium compared to the gene rich euchromatic blocks, similar to patterns observed in sorghum.

The read lengths of third generation sequencing technologies enable near gapless assemblies with high contiguity for virtually any plant genome. PacBio and Nanopore based genomes have a better representation of gene and regulatory sequences, but often lack the chromosome-scale scaffolding required for comparative genomics and quantitative genetics. The pseudomolecules in Oropetium V2 allowed us to more accurately identify syntenic orthologs in other grasses and make detailed comparisons of chromosome evolution. The V2 chromosome-scale assembly will serve as a reference for future population genomics work and positional cloning of desiccation related genes. Together, this highlights the need to improve and scaffold existing high-quality reference genomes.

ACKNOWLEDGMENTS

This work is supported by funding from the National Science Foundation (MCB-1817347 to R.V.).

AUTHOR CONTRIBUTIONS

R.V. designed research; R.V., C.M.W, J.K. and J.P. performed research and/or analyzed data; and R.V. wrote the paper. All authors reviewed the manuscript.

REFERENCES

Bartels, D., & Mattar, M. (2002). *Oropetium thomaeum*: A resurrection grass with a diploid genome. *Maydica*, *47*, 185–192.

Bolger, A. M., Lohse, M., & Usadel, B. (2014). Trimmomatic: A flexible trimmer for Illumina sequence data. *Bioinformatics*, *30*, 2114–2120. https://doi.org/10.1093/bioinformatics/btu170

Brkljacic, J., Grotewold, E., Scholl, R., Mockler, T., Garvin, D. F., Vain, P., … Budak, H. (2011). Brachypodium as a model for the grasses: Today and the future. *Plant Physiology*, *157*, 3–13. https://doi.org/10.1104/pp.111.179531

Cannarozzi, G., Plaza-Wüthrich, S., Esfeld, K., Larti, S., Wilson, Y. S., Girma, D., … Farinelli, L. (2014). Genome and transcriptome sequencing identifies breeding targets in the orphan crop tef (*Eragrostis tef*). *BMC Genomics*, *15*, 581. https://doi.org/10.1186/1471-2164-15-581

Chin, C.-S., Alexander, D. H., Marks, P., Klammer, A. A., Drake, J., Heiner, C., … Eichler, E. E. (2013). Nonhybrid, finished microbial genome assemblies from long-read SMRT sequencing data. *Nature Methods*, *10*, 563–569. https://doi.org/10.1038/nmeth.2474

Costa, M., Artur, M., Maia, J., Jonkheer, E., Derks, M., Nijveen, H., … Hesselink, T. (2017). A footprint of desiccation tolerance in the genome of *Xerophyta viscosa*. *Nature Plants*, *3*, 17038. https://doi.org/10.1038/nplants.2017.38

Cotton, J. L., Wysocki, W. P., Clark, L. G., Kelchner, S. A., Pires, J. C., Edger, P. P., … Duvall, M. R. (2015). Resolving deep relationships of PACMAD grasses: A phylogenomic approach. *BMC Plant Biology*, *15*, 178. https://doi.org/10.1186/s12870-015-0563-9

Du, H., Yu, Y., Ma, Y., Gao, Q., Cao, Y., Chen, Z., … Zhao, X. (2017). Sequencing and de novo assembly of a near complete indica rice genome. *Nature Communications*, *8*, 15324. https://doi.org/10.1038/ncomms15324

Dudchenko, O., Batra, S. S., Omer, A. D., Nyquist, S. K., Hoeger, M., Durand, N. C., … Aiden, A. P. (2017). De novo assembly of the *Aedes aegypti* genome using Hi-C yields chromosome-length scaffolds. *Science*, *356*, 92–95. https://doi.org/10.1126/science.aal3327

Durand, N. C., Shamim, M. S., Machol, I., Rao, S. S., Huntley, M. H., Lander, E. S., & Aiden, E. L. (2016). Juicer provides a one-click system for analyzing loop-resolution Hi-C experiments. *Cell Systems*, *3*, 95–98. https://doi.org/10.1016/j.cels.2016.07.002

Ellinghaus, D., Kurtz, S., & Willhoeft, U. (2008). LTRharvest, an efficient and flexible software for de novo detection of LTR retrotransposons. *BMC Bioinformatics*, *9*, 18. https://doi.org/10.1186/1471-2105-9-18

Gaff, D. (1977). Desiccation tolerant vascular plants of Southern Africa. *Oecologia*, *31*, 95–109. https://doi.org/10.1007/BF00348713

Gaff, D. (1987). Desiccation tolerant plants in South America. *Oecologia*, *74*, 133–136. https://doi.org/10.1007/BF00377357

Gaff, D., & Latz, P. (1978). The occurrence of resurrection plants in the Australian flora. *Australian Journal of Botany*, *26*, 485–492. https://doi.org/10.1071/BT9780485

Goff, S. A., Ricke, D., Lan, T.-H., Presting, G., Wang, R., Dunn, M., … Varma, H. (2002). A draft sequence of the rice genome (*Oryza sativa* L. ssp. *japonica*). *Science*, *296*, 92–100. https://doi.org/10.1126/science.1068275

Hittalmani, S., Mahesh, H., Shirke, M. D., Biradar, H., Uday, G., Aruna, Y., … Mohanrao, A. (2017). Genome and Transcriptome sequence of Finger millet (*Eleusine coracana* (L.) Gaertn.) provides insights into drought tolerance and nutraceutical properties. *BMC Genomics*, *18*, 465. https://doi.org/10.1186/s12864-017-3850-z

Hoekstra, F. A., Golovina, E. A., & Buitink, J. (2001). Mechanisms of plant desiccation tolerance. *Trends in Plant Science*, *6*, 431–438. https://doi.org/10.1016/S1360-1385(01)02052-0

Initiative, I. B. (2010). Genome sequencing and analysis of the model grass *Brachypodium distachyon*. *Nature*, *463*, 763.

Jiao, Y., Peluso, P., Shi, J., Liang, T., Stitzer, M. C., Wang, B., … Chin, C.-S. (2017). Improved maize reference genome with single-molecule technologies. *Nature*, *546*, 524–527.

Keilwagen, J., Hartung, F., Paulini, M., Twardziok, S. O., & Grau, J. (2018). Combining RNA-seq data and homology-based gene prediction for plants, animals and fungi. *BMC Bioinformatics*, *19*, 189. https://doi.org/10.1186/s12859-018-2203-5

Keilwagen, J., Wenk, M., Erickson, J. L., Schattat, M. H., Grau, J., & Hartung, F. (2016). Using intron position conservation for homology-based gene prediction. *Nucleic Acids Research*, *44*, e89. https://doi.org/10.1093/nar/gkw092

Kim, D., Langmead, B., & Salzberg, S. L. (2015). HISAT: A fast spliced aligner with low memory requirements. *Nature Methods*, *12*, 357. https://doi.org/10.1038/nmeth.3317

Koren, S., Walenz, B. P., Berlin, K., Miller, J. R., Bergman, N. H., & Phillippy, A. M. (2017). Canu: Scalable and accurate long-read assembly via adaptive k-mer weighting and repeat separation. *Genome Research*, *27*, 722–736. https://doi.org/10.1101/gr.215087.116

Langmead, B., & Salzberg, S. L. (2012). Fast gapped-read alignment with Bowtie 2. *Nature Methods*, *9*, 357–359. https://doi.org/10.1038/nmeth.1923

Li, H. (2013). *Aligning sequence reads, clone sequences and assembly contigs with BWA-MEM*. arXiv preprint arXiv:1303.3997.

Li, P., & Brutnell, T. P. (2011). *Setaria viridis* and *Setaria italica*, model genetic systems for the Panicoid grasses. *Journal of Experimental Botany*, *62*, 3031–3037. https://doi.org/10.1093/jxb/err096

McCormick, R. F., Truong, S. K., Sreedasyam, A., Jenkins, J., Shu, S., Sims, D., … McKinley, B. (2018). The *Sorghum bicolor* reference genome: Improved assembly, gene annotations, a transcriptome atlas, and signatures of genome organization. *The Plant Journal*, *93*, 338–354. https://doi.org/10.1111/tpj.13781

Ou, S., & Jiang, N. (2017). LTR_retriever: A highly accurate and sensitive program for identification of LTR retrotransposons. *Plant Physiology*, *176*, 1410–1422.

Paterson, A. H., Bowers, J. E., Bruggmann, R., Dubchak, I., Grimwood, J., Gundlach, H., … Poliakov, A. (2009). The *Sorghum bicolor* genome and the diversification of grasses. *Nature*, *457*, 551–556. https://doi.org/10.1038/nature07723

Patro, R., Duggal, G., Love, M. I., Irizarry, R. A., & Kingsford, C. (2017). Salmon provides fast and bias-aware quantification of transcript expression. *Nature Methods*, *14*, 417. https://doi.org/10.1038/nmeth.4197

R Core Team (2013). *R: A language and environment for statistical computing*. Vienna, Austria: R Foundation for Statistical Computing.

Simão, F. A., Waterhouse, R. M., Ioannidis, P., Kriventseva, E. V., & Zdobnov, E. M. (2015). BUSCO: Assessing genome assembly and annotation completeness with single-copy orthologs. *Bioinformatics*, *31*, 3210–3212. https://doi.org/10.1093/bioinformatics/btv351

Smit, A. F., Hubley, R., & Green, P. (1996). *RepeatMasker Open-3.0*. Retrieved from http://www.repeatmasker.org/

Soneson, C., Love, M. I., & Robinson, M. D. (2015). Differential analyses for RNA-seq: Transcript-level estimates improve gene-level inferences. *F1000Research*, *4*, 1521. https://doi.org/10.12688/f1000research

VanBuren, R., Bryant, D., Edger, P. P., Tang, H., Burgess, D., Challabathula, D., … Lyons, E. (2015). Single-molecule sequencing of the desiccation-tolerant grass *Oropetium thomaeum*. *Nature*, *527*, 508–511. https://doi.org/10.1038/nature15714

VanBuren, R., Wai, C. M., Ou, S., Pardo, J., Bryant, D., Jiang, N., … Michael, T. P. (2018). Extreme haplotype variation in the desiccation-tolerant clubmoss *Selaginella lepidophylla*. *Nature Communications*, *9*, 13. https://doi.org/10.1038/s41467-017-02546-5

VanBuren, R., Wai, C. M., Pardo, J., Giarola, V., Ambrosini, S., Song, X., & Bartels, D. (2018). Desiccation tolerance evolved through gene duplication and network rewiring in Lindernia. *Plant Cell*, https://doi.org/10.1105/tpc.18.00517.

VanBuren, R., Wai, J., Zhang, Q., Song, X., Edger, P. P., Bryant, D., … Bartels, D. (2017). Seed desiccation mechanisms co-opted for vegetative desiccation in the resurrection grass *Oropetium thomeaum*. *Plant, Cell & Environment*, *40*, 2292–2306. https://doi.org/10.1111/pce.13027

Walker, B. J., Abeel, T., Shea, T., Priest, M., Abouelliel, A., Sakthikumar, S., … Young, S. K. (2014). Pilon: An integrated tool for comprehensive microbial variant detection and genome assembly improvement. *PLoS ONE*, *9*, e112963. https://doi.org/10.1371/journal.pone.0112963

Wang, Y., Tang, H., DeBarry, J. D., Tan, X., Li, J., Wang, X., … Guo, H. (2012). MCScanX: A toolkit for detection and evolutionary analysis of gene synteny and collinearity. *Nucleic Acids Research*, *40*, e49. https://doi.org/10.1093/nar/gkr1293

Wick, R. R., Schultz, M. B., Zobel, J., & Holt, K. E. (2015). Bandage: Interactive visualization of de novo genome assemblies. *Bioinformatics*, *31*, 3350–3352. https://doi.org/10.1093/bioinformatics/btv383

Wicker, T., Buchmann, J. P., & Keller, B. (2010). Patching gaps in plant genomes results in gene movement and erosion of colinearity. *Genome Research*, *gr.*, *20*, 1229–1237. https://doi.org/10.1101/gr.107284.110

Xiao, L., Yang, G., Zhang, L., Yang, X., Zhao, S., Ji, Z., … Chen, M. (2015). The resurrection genome of Boea hygrometrica: A blueprint for survival of dehydration. *Proceedings of the National Academy of Sciences of the United States of America*, *112*, 5833–5837. https://doi.org/10.1073/pnas.1505811112

Xu, Z., & Wang, H. (2007). LTR_FINDER: An efficient tool for the prediction of full-length LTR retrotransposons. *Nucleic Acids Research*, *35*, W265–W268. https://doi.org/10.1093/nar/gkm286

Xu, Z., Xin, T., Bartels, D., Li, Y., Gu, W., Yao, H., … Zhou, J. (2018). Genome analysis of the ancient tracheophyte *Selaginella tamariscina* reveals evolutionary features relevant to the acquisition of desiccation tolerance. *Molecular Plant*, *11*, 983–994. https://doi.org/10.1016/j.molp.2018.05.003

8

Growth- and stress-related defects associated with wall hypoacetylation are strigolactone-dependent

Vicente Ramírez[1,2] | **Guangyan Xiong**[1,3] | **Kiyoshi Mashiguchi**[4] | **Shinjiro Yamaguchi**[4] | **Markus Pauly**[1,2]

[1]Department of Plant & Microbial Biology, Energy Biosciences Institute, University of California, Berkeley, California

[2]Institute for Plant Cell Biology and Biotechnology and Cluster of Excellence on Plant Sciences (CEPLAS), Heinrich Heine University, Düsseldorf, Germany

[3]Department of Anatomical Sciences and Neurobiology, School of Medicine, University of Louisvile, Louisville, Kentucky

[4]Department of Biomolecular Sciences, Graduate School of Life Sciences, Tohoku University, Sendai, Japan

Correspondence
Markus Pauly, Institute for Plant Cell Biology and Biotechnology and Cluster of Excellence on Plant Sciences (CEPLAS), Heinrich Heine University, Düsseldorf, 40225, Germany.
Email: m.pauly@hhu.de

Funding information
This research was supported by CEPLAS (Cluster of Excellence on Plant Sciences—deutsche Forschungsgemeinschaft EXC1028) and Marie Curie PIOF-GA-2013-623553 to V.R.

Abstract

Mutants affected in the Arabidopsis TBL29/ESK1 xylan O-acetyltransferase display a strong reduction in total wall O-acetylation accompanied by a dwarfed plant stature, collapsed xylem morphology, and enhanced freezing tolerance. A newly identified *tbl29/esk1* suppressor mutation reduces the expression of the *MAX4* gene, affecting the biosynthesis of methyl carlactonoate (MeCLA), an active strigolactone (SL). Genetic and biochemical evidence suggests that blocking the biosynthesis of this SL is sufficient to recover all developmental and stress-related defects associated with the TBL29/ESK1 loss of function without affecting its direct effect—reduced wall O-acetylation. Altered levels of the *MAX4* SL biosynthetic gene, reduced branch number, and higher levels of MeCLA, were also found in *tbl29/esk1* plants consistent with a constitutive activation of the SL pathway. These results suggest that the reduction in O-acetyl substituents in xylan is not directly responsible for the observed *tbl29/esk1* phenotypes. Alternatively, plants may perceive defects in the structure of wall polymers and/or wall architecture activating the SL hormonal pathway as a compensatory mechanism.

1 | INTRODUCTION

The success of higher plants to colonize terrestrial habitats rests greatly upon the evolution of the vascular system, the xylem, a highly specialized tissue for conducting water, minerals, and nutrients from soil to leaves over long distances (Lucas et al., 2013). Xylem tissue is characterized by the presence of specialized cells with tubular-shape structures formed by reinforcement of their cell walls, followed by a controlled cell death allowing a continuous column of water (Fukuda, 1997; Greenberg, 1996). During differentiation, these xylem cells modify the composition and structure of their secondary wall to increase the mechanical strength preventing the collapse of the vascular system under the pressure of water transpiration and extensive upright growth (reviewed in Schuetz, Smith, & Ellis, 2012). So far, forward and reverse genetic approaches have indicated that a correct synthesis and/or coordinate deposition of the secondary wall components may be necessary for the proper formation of xylem cells and the functionality of the plant vascular system. Several *irregular xylem* (*irx*) mutants have been identified in multiple plant species including poplar (Xi, Song, Sun, Shen, & Li, 2016), rice (Chiniquy et al., 2013; Gao et al., 2017), maize (Sindhu et al., 2007; Vermerris,

Sherman, & McIntyre, 2010) and Arabidopsis (Jones, Ennos, & Turner, 2001; Pauly et al., 2013; Scheller & Ulvskov, 2010; Taylor, Howells, Huttly, Vickers, & Turner, 2003; Xiong, Cheng, & Pauly, 2013). Most of the *irx* mutants display structural alterations in the main polymer components of xylem secondary walls (i.e., cellulose, hemicelluloses, and/or lignin) and a concomitant collapse of the xylem cells resulting in a stunted growth phenotype and activation of stress responses.

The identification of three protein families involved in the O-acetylation of hemicellulosic polysaccharides [Reduced Wall Acetylation (RWA), Trichome Birefringence-Like (TBL) and Altered Xyloglucan 9 (AXY9)] brought to light the crucial function of xylan O-acetyl substituents in xylem development as the corresponding mutants with reduced xylan acetylation displayed also collapsed xylem morphology, dwarfism, and stress-related phenotypes (Manabe et al., 2013; Schultink, Naylor, Dama, & Pauly, 2015; Xiong et al., 2013; Yuan, Teng, Zhong, Haghighat, et al., 2016; Yuan, Teng, Zhong, & Ye, 2016a,b). One of the most extensively studied mutants in this category is *tbl29/eskimo1*. Chemical analyses of mutant walls revealed that mutations in this gene caused a reduction in 2-O- and 3-O-monoacetylation of xylan:xylosyl-residues without causing apparent alterations in the xylan chain length and the abundance of the reducing end sequence. *tbl29* mutant plants show collapsed xylem and reduced plant growth correlated with a constitutive expression of stress-related genes and enhanced tolerance to salt stress and freezing (Xin & Browse, 1998; Xiong et al., 2013; Yuan, Teng, Zhong, & Ye, 2013). Replacing the acetyl-substitutents in the *tbl29* mutant with glucuronic acid moieties through the expression of a glucuronic acid transferase lead to the abolishment of the irregular xylem phenotype and thus normal plant growth (Xiong, Dama, & Pauly, 2015; Mortimer et al., 2010). These data suggested that a properly acetylated xylan structure is required for proper xylem function.

However, in this study, we describe a detailed characterization of a *tbl29* suppressor mutant identified through a genetic suppressor screen uncoupling reduced xylan acetylation from the irregular xylem phenotype and providing evidence of an hitherto unreported role of the phytohormone strigolactone (SL) in the perception of secondary cell wall defects.

2 | MATERIALS AND METHODS

2.1 | Growth conditions

Unless otherwise indicated, seeds were stratified in sterile water for 4 days in the dark and 4°C, sown on soil and grown in environmentally controlled growth chambers under long day conditions (10,000 luxes, 16 hr light, 22°C/8 hr dark, 19°C).

For the hydroponic experiment, seeds were surface-sterilized incubating them for 7 min in a 70% ethanol (Sigma-Aldrich), 0.1% Triton X-100 (Panreac Applichem) solution followed by five washes with sterile water. Seeds were then transferred to the Araponics seed holders (Araponics) filled with half-strength pH5.7 Murashige and Skoog media (Duchefa Biochemie) supplemented with 0.65% Plant Agar (Duchefa Biochemie) and incubated for four days in the dark at 4°C before being transferred to the growth chamber (7,000 luxes, continuous day, 22°C). New Araponics nutrient solution (0.5 ml/L) was added every week supplemented with rac-GR24 (Strigolab) or acetone as a control. After 6 weeks, plant height was recorded, pictures were taken, and stems were collected for toluidine-O-blue staining and cell wall analyses.

2.2 | Identification of *max4-7* (*tbl29S*)

The *tbl29S* suppressor was found in a population of homozygous *tbl29-1* mutant plants transformed with the *pORE-E4* vector (Malik et al., 2002). A bulk segregant population was generated by crossing the *tbl29S* suppressor to the *tbl29* parental and 80 individuals in the F2 progeny were screened for the presence/absence of the suppressed (dwarf) phenotype. Genomic DNA from the different individuals was prepared using the DNAeasy plant mini kit according to the manufacturer's instructions (Qiagen) and equimolar amounts of DNA from the two groups were used for whole-genome sequencing. The 250PER Illumina sequencing was performed by the Vincent J. Coates Genomics Sequencing Laboratory in Berkeley using a Hi-Seq 2500 rapid instrument (Illumina). The obtained sequence reads were assembled using CLC Genomics Workbench 7.5 and aligned to the TAIR10 version of the Col-0 reference genome. Differences found in the suppressed and non-suppressed DNA pools were compared to identify the causative *tbl29S* mutation.

2.3 | Mutant genotyping

The different mutant lines were obtained from the NASC (http://arabidopsis.info) (Scholl, May, & Ware, 2000) and ABRC (https://abrc.osu.edu) Arabidopsis stock centers. For the genotyping of the *tbl29* mutant (Stock number: SAIL_856_G11) and *max4-1* mutant (Stock number: CS9568), two PCR reactions were performed. Reaction A was performed using primers binding-specific regions for each gene flanking the T-DNA insertion site (LP and RP primers). Reaction B was performed using a T-DNA-specific primer (LB) and the RP primer used in reaction A. Similarly, the insertion in the *MAX4* (At4 g32810) promoter region was confirmed using the LP and RP MAX4-specific primers flanking the insertion in reaction A, and in reaction B, the LP primer was combined with the insertion-specific primer LB. Primer sequences used for genotyping of the mutants described in this work are presented in Supporting Information Table S1.

2.4 | Plant measurements

The length of the primary steam (Plant height) and the number of shoot branches arising from axillary meristems (number of stems) were recorded 6 weeks after sowing the seeds and individual pictures were taken using a Lumix DMC-FZ35 camera (Panasonic).

2.5 | Xylem visualization

Plants were grown as indicated and the stems carefully collected. For comparative reasons, the middle of the first stem internode was selected for analysis. Hand-cut stem sections of the indicated genotypes were incubated one minute in 1 ml 0.02% Toluidine Blue O solution (Sigma-Aldrich) and then rinsed three times with 10 ml sterile water. Sections were then gently pipetted onto a microscope slide, covered with a coverslip and observed under bright-field lighting microscope (Leica DM2000 LED).

2.6 | Cell wall composition

Individual primary stems were collected from 6-week-old plants, lyophilized using a ScanVac CoolSafe Freeze-dryer (Labogene) and homogenized to a fine powder using a MM400 mixer mill (Retsch Technology) at 30 Hz for two minutes. Total destarched alcohol insoluble residue (AIR) was extracted and one milligram per assay was used for the determination of cellulose content as previously described (Foster, Martin, & Pauly, 2010).

For the determination of matrix polysaccharide composition, 1 mg of destarched cell wall material was hydrolyzed in 2M TFA for 90 min and 121°C and washed twice with 2-propanol before resuspension in H_2O. A 940 Professional IC Vario HPAEC system with a pulse amperometric detector (Metrohm) was used to measure the different monosaccharides in the hydrolysate. Samples were separated with a Dionex CarboPac PA20 analytical column (3 × 150 mm) and a Dionex CarboPac PA20 guard column (3 × 30 mm) (Thermo Fischer Scientific), using a 0.4 ml/min flow rate at 40°C. The eluents used were: A-Double distilled H_2O; B-10 mM NaOH (Sigma-Aldrich); and C-700 mM NaOH with the following elution profile: 20 min 80%A + 20%B, 5 min 30%A + 70%C, 3 min 100%C and 30 min 80%A + 20%B.

Total wall acetate was determined using the 96-well-adapted format of the Acetic Acid Kit (Megazyme) as described previously (Gille et al., 2011).

2.7 | Analysis of methyl carlactonoate (MeCLA)

The analysis of MeCLA in Arabidopsis seedlings was performed as previously described (Abe et al., 2014). 13-day-old plants vertically grown on half-strength Murashige and Skoog medium were homogenized in 5 mL of acetone containing [10-$^2H_1$]-MeCLA (D_1-MeCLA) as an internal standard, using POLYTORON PT3100D (Kinematica). The filtrates were evaporated under N_2 gas and extracted twice with 4 ml of ethyl acetate after the addition of 3 ml of water and 1 ml of saturated sodium chloride solution. The organic phase was evaporated to dryness under N_2 gas, dissolved in 0.4 ml of hexane, and loaded onto a Sep-Pak Silica cartridge column (1 cc; Waters). The column was washed with 2 ml of 5% (vol/vol) ethyl acetate/hexane and eluted with 2 ml of 15% (vol/vol) ethyl acetate/hexane. The eluate was then evaporated to dryness under N_2 gas, dissolved in 0.3 ml of hexane, and loaded onto a Sep-Pak cyanopropyl cartridge column (1 cc; Waters), washed with 1 ml of hexane, and isocratically eluted with 3 ml of hexane. The eluate was evaporated to dryness, dissolved in acetonitrile, and then subjected to LC-MS/MS. LC-MS/MS analysis was carried out using a system consisting of TripleTOF 5600 (AB SCIEX) and Nexera ultra high performance liquid chromatography (Shimadzu) equipped with a CORTECS UPLC C18 + column (1.6 μm, ϕ2.1 × 100 mm; Waters). Elution of the samples was carried out with 0.05% (vol/vol) acetic acid/water (solvent A) and 0.05% (vol/vol) acetic acid/acetonitrile (solvent B) using a gradient from 30% to 95% solvent B (0 min to 6.5 min) at a flow rate of 0.3 ml/min. The temperature of the column was set to 35°C. The MS/MS analysis was operated under the enhanced mass mode as following parameters: Declustering potential, 55 V; collision energy, 25 V; GS1, 35 psi; GS2, 35 psi; temperature, 400°C; precursor ion (*m/z*), 347.2 for MeCLA and 348.2 for D_1-MeCLA. A fragment ion with *m/z* 97.03 was used to identify both MeCLA and D_1-MeCLA. A standard curve for quantification was generated using MeCLA and D_1-MeCLA.

2.8 | Freezing tolerance assay

4-week-old plants grown in soil under the described conditions were incubated at −5°C in a Coolfreeze CF-110 cooling box (Waeco) for 20 hr and then transferred back to the growth chamber. After three days pictures were taken and the number of surviving plants recorded.

2.9 | Gene expression analysis

Primary stems from 6-week-old plants were collected and immediately frozen in liquid nitrogen. Tissue was homogenized to a fine powder using a MM400mixer mill (Retsch Technology) at 30 Hz for one minute and total RNA was extracted DNAse-treated using the Plant RNeasyMini Kit according to the manufacturer's instructions (Qiagen). One microgram of RNA was then used for the synthesis of complementary DNA using the Tetro cDNA Synthesis Kit (Bioline) and one microliter was used as template to perform real-time quantitative PCR with SsoAdvanced Universal SYBR Green Super Mix (BioRad) in a CFX96 System (Biorad). Relative MAX4 transcript abundance was calculated using the ACTIN2 (At3 g18780) gene as an internal control using the following primers: MAX4-F: 5′-TGTGGAGTAGCCGTCGAAGAG-3′, MAX4-R: 5′- GAAAGATACCCACTTGGCTGAATG-3′, ACT2-F: 5′-TCTTCCGCTCTTTCTTTCCAAGC-3′ and ACT2-R: 5′- ACCATTGTCACACACGATTGGTTG-3′.

3 | RESULTS

3.1 | Identification and characterization of Arabidopsis *tbl29* suppressors

A recessive mutation (*tbl29S*) was found to rescue *tbl29*-associated growth defects such as reduction in plant size, altered plant architecture and a dark-green appearance (Figure 1a–b). To identify the causative mutation we used bulk segregant analysis combined with whole-genome sequencing. This approach allowed us to identify an insertion co-segregating with the *tbl29*-suppressed phenotype around position -1269 in the promoter region of the *MORE AXILLARY BRANCHES 4* (*MAX4*) gene. This insertion leads to a 95% reduction in *MAX4* gene expression in a *tbl29S* background compared to the *tbl29* parental line (Supportin information Figure S1). To confirm that the reduced expression of *MAX4* is responsible for the suppression of the *tbl29* developmental phenotypes, we crossed *tbl29* with a previously characterized *max4-1* knockdown mutant (Sorefan et al., 2003). The *max4-1* mutation was also able to suppress the *tbl29* phenotypes (Figure 1), confirming that the *tbl29* suppression in *tbl29S* is due to a mutation in *MAX4*. Accordingly, we changed the name of *tbl29S* to an additional allele of *max4*, *max4-7*.

A detailed characterization of the two suppressor alleles, *tbl29 max4-7* and *tbl29 max4-1*, was performed (Figure 1). Both double mutant plants lost the characteristic small size and dark-green color of the leaves (Figure 1a). The 62% reduction in stem height observed in the *tbl29* mutant is almost completely rescued in both *tbl29 max4-7* and *tbl29 max4-1* (Figure 1b). In addition, the altered shoot system architecture in *tbl29* is also modified (Figure 1c). *tbl29* plants have on average half the number of shoot branches arising from axillary meristems compared to Col-0. This reduced number of

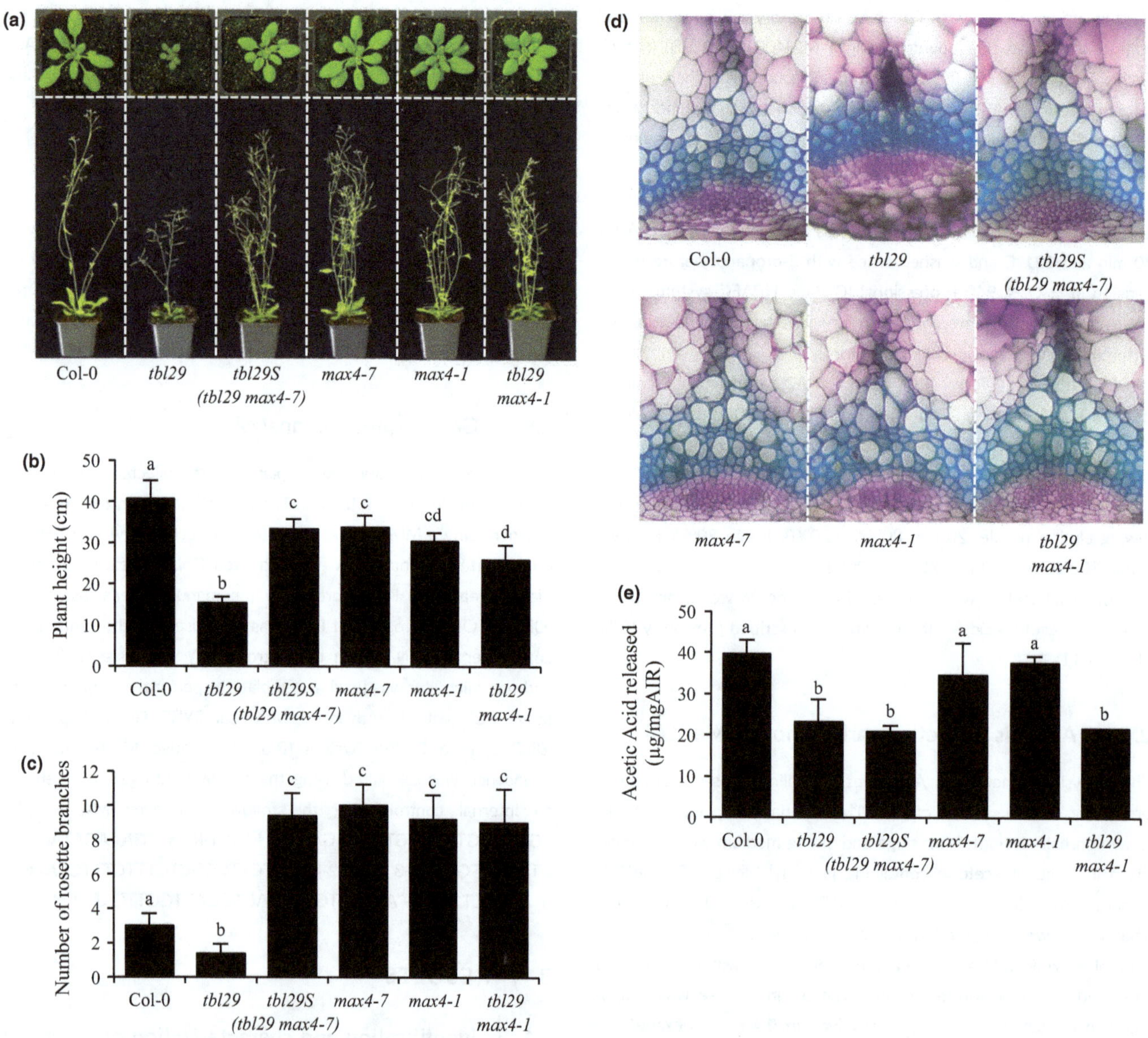

FIGURE 1 *tbl29S* mutation suppresses *tbl29*-associated dwarfism, altered plant architecture and collapsed xylem. (a) Growth phenotypes of 4-week-old (upper panel) and 6-week-old (lower panel) plants. (b) Primary inflorescence heights (cm). (c) Number of rosette branches. (d) Toluidine-O-Blue stained cross section of inflorescence stems. (e) Alkali-released acetate content of wall material (AIR) from stems. Data are represented as the mean of biological replicates (≥15 plants) with standard deviation. Means with different letters are significantly different (Tukey's HSD, $p<0.05$)

branches in *tbl29* is not only reverted to wildtype levels, but significantly increased in both double mutant alleles as previously described for other *max4* mutants (Sorefan et al., 2003).

Analysis of stem cross sections revealed that *max4* alleles also suppress the *tbl29*-associated irregular xylem morphology. In toluidine blue-stained sections from *tbl29* mutant plants, the vascular tissue contained irregular, collapsed-looking xylem elements. In contrast, in the equivalent stem sections from all plants containing *max4* mutations alone or in combination with *tbl29* showed a wild-type xylem morphology, where xylem elements can be easily identified as large round empty-looking cells with thick uniform cell walls (Figure 1d).

Cell wall compositional analyses on stem tissue showed that the reduced cellulose content in *tbl29* walls, which is typical among *irregular xylem* mutants, was fully restored in the *tbl29 max4-7* and *tbl29 max4-1*. The increased levels of non-cellulosic monosaccharides such as arabinose or xylose in *tbl29* stem tissue are also restored (Supporting Information Figure S2). In contrast, the characteristic low wall acetylation content in *tbl29* (42% reduction compared to wild-type) was retained in *tbl29 max4-7* and *tbl29 max4-1* plants (47% and 46% reduction, respectively), implying that xylan acetylation remains low in the *tbl29*-suppressed plants. Taken together, these results suggest that blocking the expression of *MAX4* is able to suppress the characteristic dwarf phenotype and the collapsed xylem in a *tbl29* mutant background. Concomitantly, the indirectly altered cellulose and matrix polysaccharide levels and composition are restored. But this reversion does not involve a restoration of the direct effect of *TBL29* (xylan O-acetyltransferase) loss of function, reduced wall acetylation.

3.2 | Effect of exogenous SL application

MAX4 represents a carotenoid cleavage dioxygenase involved in the synthesis of SL (reviewed in Al-Babili & Bouwmeester, 2015). SL are a group of structurally similar terpenoid lactones controlling shoot branching. *max4* plants are affected in the production of carlactone (CL), an endogenous biosynthetic precursor of SL (Seto et al., 2014). Thus, mutant plants are highly branched, a phenotype that can be complemented by application of the synthetic SL GR24 (Gomez-Roldan et al., 2008; Umehara et al., 2008). To test if SL-deficiency was responsible for the suppression of *tbl29*-associated growth defects in the *tbl29 max4-7* double mutant, we examined the effect of exogenous applications of GR24 on plant architecture and cell wall composition of selected genotypes (Figure 2a–c). Supplementing a hydroponic culture media with 1 μM GR24 did not result in any changes in WT plants (normal growth) or *tbl29* plants (dwarfed growth). However, when adding GR24 to the *tbl29 max4-7* double mutant, plant growth reverted back to dwarfism as found in the *tbl29* single mutant and the suppression through the *max4-7* mutation was abolished. Regarding plant architecture, GR24 treatment inhibited the axillary shoot growth in all genotypes tested except in *tbl29*, probably due to the already low branch number in this mutant under the growth conditions used (Figure 2c).

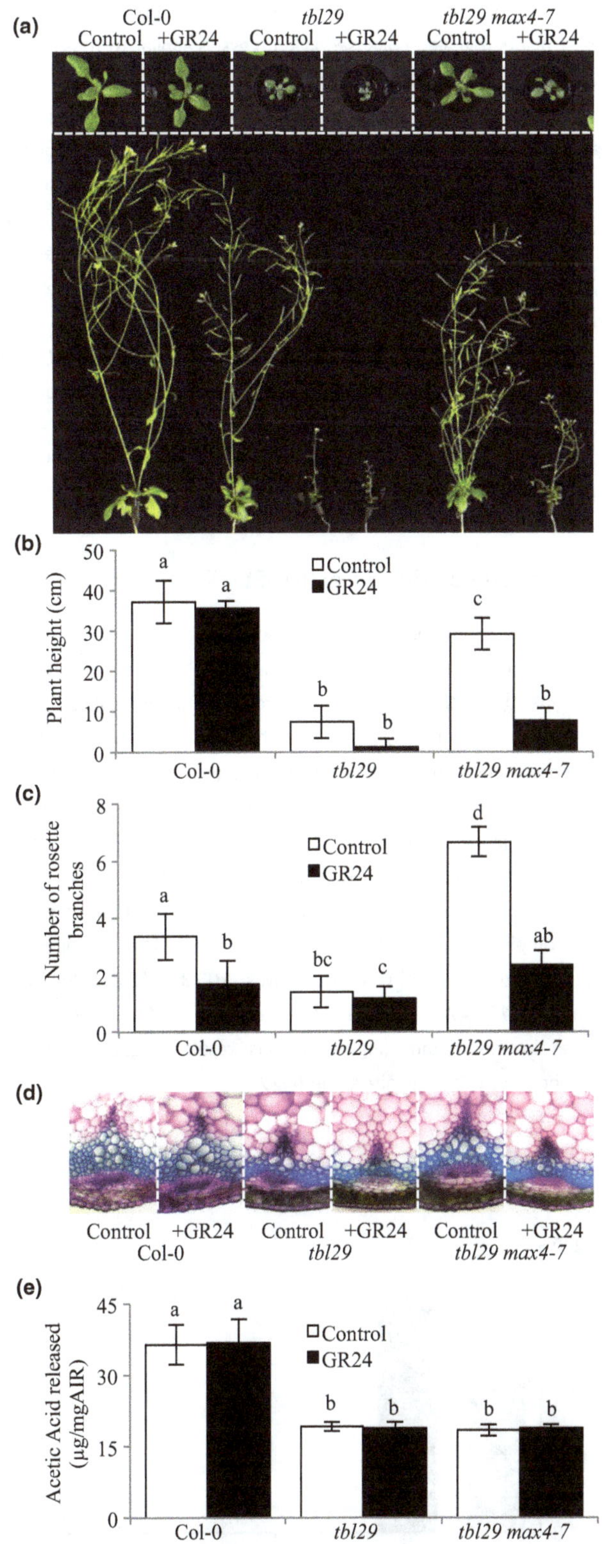

FIGURE 2 Effect of strigolactone application. (a) Growth phenotypes of 3-week-old (upper panel) and 6-week-old (lower panel) plants. (b) Primary inflorescence heights (cm). (c) Number of rosette branches. (d) Toluidine-O-Blue stained cross section of inflorescence stems. (e) Alkali-released acetate content of wall material (AIR) from stems. Data are represented as the mean of biological replicates (N=6 plants) with standard deviation. Means with different letters are significantly different (Tukey's HSD, $p<0.05$)

Examinations of stems sections from GR24-supplemented *tbl29 max4-7* plants revealed collapsed xylem vessels similar to those characteristics of *tbl29* (Figure 2d).

Wall compositional analyses indicate that *tbl29 max4-7* plants treated with GR24 are similar to those observed for *tbl29* (Supporting Information Figure S3). In particular, xylose levels increased in GR24-treated *tbl29 max4-7* plants to levels comparable to non-treated *tbl29* plants. No differences in the total wall acetate content in response to GR24 treatment were observed in any of the analyzed genotypes (Figure 2e).

Hence, the *max4-7* mutation is not able to complement *tbl29* in the presence of exogenously applied SL, indicating that a reduction in SL content suppresses the developmental defects including some wall structural defects caused by xylan hypoacetylation, but not hypoacetylation itself.

3.3 | Accumulation of active SL

Based on the results presented, a proper endogenous production of SL may be important for the expression of the highly pleiotropic phenotype of the *tbl29* mutant. To investigate if the reduced xylan acetylation in *tbl29* influences the synthesis of SL we measured the endogenous levels of methyl carlactonoate (MeCLA) in *tbl29* plants by LC-MS/MS. MeCLA is a non-canonical SL shown to be active in Arabidopsis plants (Abe et al., 2014). Endogenous MeCLA levels were not detectable in the two *max4* mutant alleles and the corresponding *tbl29 max4* double mutants confirming that *max4* mutations completely block MeCLA accumulation. The endogenous MeCLA content in Col-0 extracts was 103.3 ± 12.3 pg/g fresh weight and 156.47 ± 28.9 in *tbl29*, denoting a 50% steady state increase in the mutant (Figure 3). This result indicates that the biosynthesis of this SL is altered in *tbl29*.

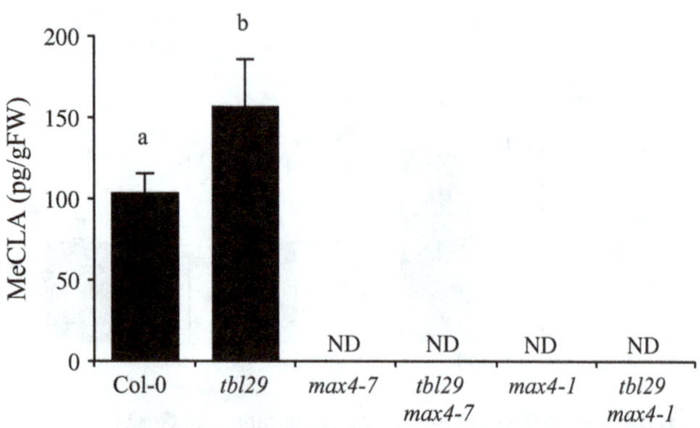

FIGURE 3 Quantification of endogenous MeCLA. Endogenous MeCLA was quan1fied by LC-MS/MS in Col-0, *tbl29*, *max4-7*, *tbl29 max4-7*, *max4-1* and *tbl29 max4-1* plants using D1-MeCLA as an internal standard. Data are represented as the mean of biological replicates (3-4) with standard devia1on. Means with different leLers are significantly different (Tukey's HSD, p<0.05). FW, fresh weight. ND, not detectable

3.4 | Suppression of *tbl29*-associated freezing tolerance

tbl29 was first identified as *eskimo 1* (*esk1*) exhibiting an increased resistance to freezing temperatures (Xin & Browse, 1998; Xin, Mandaokar, Chen, Last, & Browse, 2007). The survival of plants when subjected to freezing temperatures (−5°C) during a 20 hr period was tested. Under these conditions wildtype plants were reproducibly killed while most of the *tbl29* plants did not show any visible injury and were capable of continued growth after freezing and completion of their life cycle. To obtain a quantitative measure of freezing tolerance in the different genotypes, we determined the percentage of plants that are alive after the freezing treatment (Figure 4). The survival rate of *tbl29* in these freezing assays exceeded 90%, in contrast to Col-0 (3.3%). Interestingly, *tbl29 max4-7* plants showed the lethal wildtype-like phenotype with a survival rate of only 10%. These results indicate that blocking SL synthesis also suppresses the *tbl29/esk1*-associated freezing tolerance.

4 | DISCUSSION

4.1 | Reduced wall acetylation can be uncoupled from dwarfism and collapsed xylem phenotypes

The functionality of the plant vascular system is dependent on the correct deposition and reinforcement of a secondary cell wall in

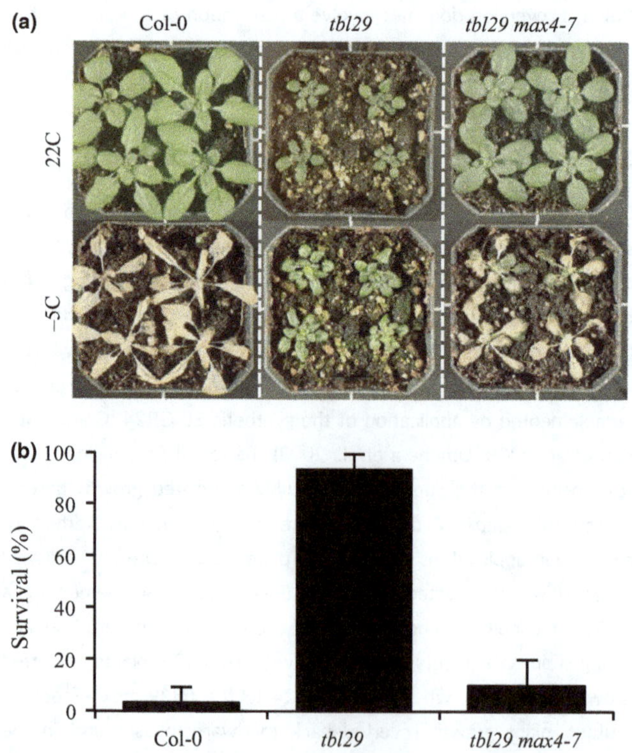

FIGURE 4 *tbl29*-associated freezing tolerance is suppressed by *max4-7*. (a) 4-week-old plants before (upper panel) and 3 days after freezing treatment (lower panel). (b) Survival rate calculated as the percentage of plants alive after treatment. Error bars represent standard deviation (N=3 experiments, 20 plants per experiment)

xylem cells allowing the plant to resist the negative forces generated during water transport. The physiochemical properties of the secondary wall depend on the structure of the polymers (mainly cellulose, xylan and lignin) forming these walls. The degree of O-acetylation of xylan, the major acetylated polymer in secondary walls, seems to have a strong impact on these properties affecting the xylan-cellulose interaction (Ebringerová, Hromádková, & Heinze, 2005; Grantham et al., 2017). Although the biological significance of this modification is not clear, single and multiple mutants characterized in several plant species have shown a strong correlation between the level of reduction in xylan acetylation and a stunted plant growth and xylem collapse. The strongest reduction in wall O-acetylation has been reported so far for the *Arabidopsis axy9* mutant, with a 70% overall reduction affecting multiple polymers including xylan and another hemicellulose, xyloglucan (Schultink et al., 2015). No growth phenotypes have been reported for mutants lacking xyloglucan O-acetylation, so the severe growth defects, including dwarfed organs, a dark-green leaf color and an extreme collapsed xylem, exhibited by the mutant has been attributed to the 80% reduction in xylan O-acetylation in stem tissue of *axy9*. The *Arabidopsis rwa2* single mutant exhibits a moderate 20% reduction in overall wall O-acetylation with no apparent plant development and/or xylem morphology defects, but an additive effect was observed by simultaneous mutation of multiple *RWA* genes (Lee, Teng, Zhong, & Ye, 2011). The quadruple *rwa* mutant shows a 42% reduction in xylan acetylation associated with a reduction in secondary wall thickening and collapsed xylem morphology. In the case of TBL family, several genes have been involved in xylan acetylation, with additive degrees of impact on O-acetylation and developmental defects. In rice, the *OsTBL1* and *OsTBL2* genes have recently been shown to be involved in xylan monoacetylation. The reduced growth and collapsed xylem in the double mutant is accompanied by a 55% reduction on total wall acetylation (Gao et al., 2017). In *Arabidopsis*, mutants affecting *TBL3*, *TBL31*, *TBL32*, *TBL33*, *TBL34* and *TBL35* genes show a minor decrease on xylan acetylation when analyzed individually but several double mutant combinations have been reported to cause moderate reductions ranging from 7% to 20% (Yuan, Teng, Zhong, Haghighat, et al., 2016; Yuan, Teng, Zhong, & Ye, 2016a,b). Although a side-by-side comparison of all different mutant combinations has not been reported, there seems to be a direct correlation between the reduction in xylan acetylation and growth/irregular xylem phenotypes. On the other hand, single loss-of-function alleles of the TBL29/ESK1 xylan acetyltransferase show a strong reduction on xylan O-acetylation (~ 40% depending on the report) accompanied by a strong arrest in plant growth and collapse of the xylem vessels (Figure 1; Supporting Information Figure S2; Xin & Browse, 1998; Xin et al., 2007; Yuan et al., 2013; Xiong et al., 2013). *tbl29* mutant plants exhibit a reduction on the rosette size and overall plant height and a characteristic dark-green color of the leaves. Interestingly, these growth defects can be complemented by replacing the missing acetyl-substitutents in the *tbl29* mutant with functionally equivalent glucuronic acid moieties through the expression of the AtGUX1 glycosyltransferase in the vascular tissue, suggesting that appropriately substituted xylan is required for proper xylem function (Mortimer et al., 2010; Xiong et al., 2015).

However, we found a suppressor mutation able to complement all these *tbl29*-associated defects without affecting xylan hypoacetylation (Figure 1). This demonstrates that in *tbl29* reduced xylan O-acetylation is not directly responsible for the collapse of the xylem vessels and the corresponding developmental alterations. The structural defect of xylan is hence not essential for the xylem functionality and normal development. This notion of uncoupling xylan hypoacetylation and growth/morphological defects is also supported by a previously identified *tbl29/esk1* suppressor mutation, *kaktus*, which was found to partially complement dwarfism and collapsed xylem also without recovering wall acetylation or other wall modifications (Bensussan et al., 2015). The *KAKTUS* (*KAK*) gene encodes a protein belonging to the E3-ubiquitin protein ligase family involved in endoreduplication and trichome architecture (Downes, Stupar, Gingerich, & Vierstra, 2003; El Refy et al., 2003). It is thought that *kak* is able to partially suppress *tbl29* dwarfism and collapsed xylem due to a hyperactivation of vascular development. However, the mechanism is unknown (Bensussan et al., 2015). A possible relationship between *KAK1* and SL requires further investigations.

4.2 | Irregular xylem and constitutive freezing tolerance phenotypes associated with wall hypoacetylation are SL-dependent

The *tbl29* suppressor mutation identified here is a new allele of *max4* (*max4-7*) representing an insertion in the promoter region of *MAX4*, causing a 95% reduction in the expression of the gene (Supporting Information Figure S1). This insertion could be disrupting a transcriptional regulatory element in the *MAX4* promoter, preventing *MAX4* from specifically responding to a *tbl29*-specific signal. However, a similar *tbl29* suppression effect was observed when crossing the mutant to a previously described *max4* allele (i.e., *max4-1*), where a T-DNA insertion in the 1st intron of the *MAX4* gene significantly impairs its expression (Sorefan et al., 2003). This result indicates that in both cases, the reduced expression of *MAX4* is the cause of the suppression of the *tbl29*-associated defects. MAX4/CCD8 is a carotenoid cleaving deoxygenase catalyzing the biosynthesis of carlactone, the precursor of biologically active SL such as MeCLA in *Arabidopsis*. Consequently, *max4* mutants show a drastic reduction in SL accumulation (Figure 3; Seto et al., 2014; Abe et al., 2014). Genetic evidence demonstrates that blocking the endogenous production of SL is sufficient to suppress all *tbl29*-associated phenotypes as in the *tbl29 max4* double mutants (Figure 1; Supporting Information Figure S2). Moreover, complementing SL levels by exogenous applications of GR24, a synthetic SL, block the suppression effect of the *max4-7* mutation. While GR24 has a weak effect on wildtype plants, *tbl29 max4-7* plants treated with GR24 mimic the dwarf phenotype, collapsed xylem and most of the wall alteration attributes of *tbl29* control plants (Figure 2; Supporting Information Figure S3), suggesting that the reduction in SL synthesis is responsible for the *tbl29* suppression in the *tbl29 max4-7* double mutant. How blocking the SL biosynthesis in a secondary cell wall deficient mutant recovers the aberrant xylem development and prevents the activation of a general stress response requires further investigation, but several lines of evidence indicate that the SL pathway might be involved. Multiple indications in this report suggest that *tbl29* plants have a conspicuous alteration in the SL biosynthesis/perception. First, *tbl29* plants present a reduced branching phenotype (Figure 1c), consistent with the main role of SL as repressors of the outgrowth of the axillary bud (Gomez-Roldan et al., 2008; Umehara et al., 2008). The branch number has been used as typical phenotypic effect of SL signaling. While SL-deficient or SL-insensitive mutants show highly branched phenotype, activation of SL pathway has been associated with a reduction in branching both endogenously, through nutrient deficiency-triggered SL activation, or exogenously upon application of GR24 (Ito et al., 2016; Mashiguchi et al., 2009). Second, *tbl29* shows a 50% increased accumulation of active SLs (i.e., MeCLA) suggesting that SL biosynthesis is altered (Figure 3). Third, *MAX4* gene expression is down-regulated in *tbl29* stems (Supporting Information Figure S1), mimicking what has been reported for GR24-treated plants and opposed to the *MAX4* up-regulation observed in SL-deficient mutants due to a negative feedback regulation of SL biosynthetic genes observed in several plant species such as *Arabidopsis*, pea or rice (Arite et al., 2007; Johnson et al., 2006; Mashiguchi et al., 2009). Finally, an altered SL homeostasis in the *tbl29* mutant could also explain its general abiotic stress tolerance. SL are carotenoid-derived plant hormones controlling multiple processes involved in plant development and adaptation to environmental cues (reviewed in Waters, Gutjahr, Bennett, & Nelson, 2017). Although there is some controversy, it has been suggested that SL positively regulates plant's adaptation to abiotic stresses such as drought or salinity. In several plant species such as *Arabidopsis*, *Lotus japonica* or corn, GR24 applications increase the tolerance to drought, while conversely, SL-deficient mutants show an increased sensitivity (Van Ha et al., 2014; Liu et al., 2015; Davidson, Bayer, Windram, & Hleba, 2015). It is therefore likely that an increased endogenous SL content or perception in *tbl29* plants is part of the increased tolerance to multiple abiotic factors such as freezing, drought or salt stress (Figure 4; Xin et al., 2007; Bouchabke-Coussa et al., 2008; Xu et al., 2014). Accordingly, in the *tbl29 max4-7* double mutant, the *tbl29*-associated freezing tolerance is completely abolished (Figure 4), reinforcing the hypothesis that proper SL biosynthesis is necessary for the freezing tolerance presented by *tbl29/esk1* plants.

4.3 | Relationship between SL and xylem development

TBL29/ESK1 was first proposed to be a specific regulator of freezing tolerance (Xin & Browse, 1998), but several subsequent studies have reported an increased tolerance of *tbl29/esk1* mutants to abiotic factors including drought, salinity, osmotic stress or dehydration as well as biotic factors, indicating a more general role. Several metabolic alterations have been used to explain the constitutive tolerance of the *tbl29/esk1* mutant plants to these stresses such as increased accumulation of osmoprotectant metabolites (e.g., proline or soluble sugars), up-regulation of stress-specific subset of genes, or constitutive activation of the abscisic acid (ABA) signaling pathway (Escudero et al., 2017; Lugan et al., 2009). Recent reports favor the hypothesis that all these are indirect effects of the alteration in the water transport so the increased stress tolerance shown by the *tbl29/esk1* mutants is likely a consequence of the irregular xylem phenotype displayed by these plants (Bensussan et al., 2015; Bouchabke-Coussa et al., 2008; Lefebvre et al., 2011; Lugan et al., 2009).

The results presented in this manuscript suggest that proper endogenous SL production or sensitivity is necessary for the development of the irregular xylem morphology in this mutant. How SL might regulate xylem morphology is still unknown. It has been previously described that high-humidity conditions can partially rescue vessel collapse and consequently the severity of the developmental and stress-related phenotypes in other irregular xylem mutants (Lao et al., 2003). Detailed analyses of the suppression effect of *max4* mutations on the *tbl29*-associated phenotypes under high-humidity regimes or drought conditions could contribute to address this possibility. One could also speculate that the change in plant architecture (i.e., increased branching) caused by the block of the SL biosynthetic pathway could be alleviating the transpiration and vascular stress in the *tbl29* mutant, thus preventing the xylem collapse. Although the current data cannot exclude this possibility, it seems unlikely as the *max4*-suppression effect is already noticeable at the rosette stage rescuing *tbl29*-associated reduced rosette size and increased freezing tolerance (Figures 1, 2 and 4). As an alternative, we propose a model in which a defect in xylan acetylation in the secondary wall is somehow perceived by the cell by a hitherto unknown mechanism triggering the activation of a SL-dependent response. In the *tbl29* mutant, the continuous perception of a defective wall would lead to a constitutive activation of this signaling, leading to xylem collapse with the corresponding growth/morphological/stress response defects (Figure 5). Whether this mechanism is specific to xylan hypoacetylation or represents a common feature of other cell wall defects in the vasculature remains to be addressed. Future experiments to determine the SL biosynthesis/perception in other irregular xylem mutants affected not only in hemicellulose structure [e.g., *irx9* or *parvus*; Lee, Zhong, et al., 2007; Lee, O'Neill, Tsumuraya, Darvill, & Ye, 2007), but also cellulose (e.g., *irx1*, *irx3* or *irx5*; Taylor et al., 2003), and lignin content (e.g., *irx4*; Jones et al., 2001) will provide valuable data to address this possibility.

Future work is also needed to understand this SL-dependent cell wall sensing mechanism. The SL field is making great advances to identify and measure the SL compounds which are active in different plant species, and to unravel how they are perceived and the signal transduced in some SL-regulated processes (Reviewed in Lumba, Holbrook-Smith, & McCourt, 2017; and Waters et al., 2017). SL

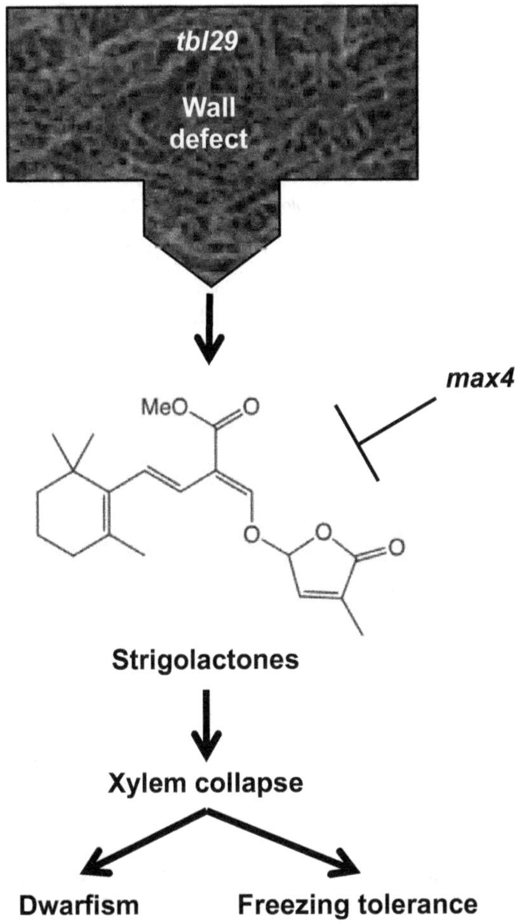

FIGURE 5 Proposed model. Defects in xylan acetylation in the secondary wall are perceived by an unknown mechanism triggering the activation on of a SL-dependent response. In the *tbl29* mutant, the continuous perception of a defective wall would lead to a constitutive activation of this signaling, leading to xylem collapse with the corresponding growth/morphological/ stress response defects. Altering the production of SL (as in the SL-deficient *max4* mutant) blocks the signaling preventing the developmental- and stress-related phenotypes

signal transduction controlling shoot branching seems to be based upon hormone-activated proteolysis of SL target proteins. There are a few SL targets proposed to be involved, including members of the SUPPRESSOR OF MAX2 1 (SMAX1) family (Stanga, Smith, Briggs, & Nelson, 2013). The function of these proteins remains poorly understood, but some reports support the hypothesis that they are transcriptional regulators (Jiang et al., 2015; Soundappan et al., 2015). This is the case of other putative SL targets such as the BRI1-EMS SUPPRESSOR1 (BES1) and the DELLA family of GRAS transcriptional regulators, known targets of the brassinosteroid and gibberellin signaling, respectively (Nakamura et al., 2013; Wang et al., 2013). In that regard, some xylem-specific components of the wall biosynthesis/remodeling machinery are regulated by the SL pathway. For example, in hybrid aspen, a SL-dependent up-regulation of 1,3-β-glucanases (GH17 family, α-clade) controls shoot elongation via callose hydrolysis at sieve plates and plasmodesmata (Rinne, Paul, Vahala, Kangasjärvi, & van der Schoot, 2016). In addition, SL have been suggested as positive regulators of cambium activity in *Arabidopsis*, pea and in the woody plant *Eucalyptus globulus*. Although the molecular basis is unknown, local applications of GR24 in stems promote secondary growth by increasing the production of secondary vascular tissues and the lateral extension of the cambium zone (Agusti et al., 2011).

ACKNOWLEDGMENTS

We thank Ulrike Klauss for excellent technical assistance.

AUTHORS' CONTRIBUTIONS

M.P. and V.R. designed the research; G.X. performed the identification of *tbl29S* and initial characterization; K.M. and S.Y performed the MeCLA measurements; V.R. conducted all other experiments; M.P. and V.R. wrote the paper; M.P., V.R., G.Y., K.M., and S.Y. revised the article.

ACCESSION NUMBERS

Sequence data from this article can be found in the EMBL/GenBank data libraries under accession numbers: *TBL29/ESK1*, At3 g55990; *MAX4/CCD8*, At4 g32810.

REFERENCES

Abe, S., Sado, A., Tanaka, K., Kisugi, T., Asami, K., Ota, S., ... Nomura, T. (2014). Carlactone is converted to carlactonoic acid by MAX1 in Arabidopsis and its methyl ester can directly interact with AtD14 in vitro. *Proceedings of the National Academy of Sciences of the United States of America*, *111*, 18084–18089. https://doi.org/10.1073/pnas.1410801111

Agusti, J., Herold, S., Schwarz, M., Sanchez, P., Ljung, K., Dun, E. A., ... Greb, T. (2011). Strigolactone signaling is required for auxin-dependent stimulation of secondary growth in plants. *Proceedings of the National Academy of Sciences of the United States of America*, *108*, 20242–20247. https://doi.org/10.1073/pnas.1111902108

Al-Babili, S., & Bouwmeester, H. J. (2015). Strigolactones, a novel carotenoid-derived plant hormone. *Annual Review of Plant Biology*, *66*, 161–186. https://doi.org/10.1146/annurev-arplant-043014-114759

Arite, T., Iwata, H., Ohshima, K., Maekawa, M., Nakajima, M., Kojima, M., ... Kyozuka, J. (2007). DWARF10, an RMS1/MAX4/DAD1 ortholog, controls lateral bud outgrowth in rice. *The Plant Journal*, *51*, 1019–1029. https://doi.org/10.1111/j.1365-313X.2007.03210.x

Bensussan, M., Lefebvre, V., Ducamp, A., Trouverie, J., Gineau, E., Fortabat, M. N., ... Durand-Tardif, M. (2015). Suppression of dwarf and irregular xylem phenotypes generates low-acetylated biomass lines in arabidopsis. *Plant Physiology*, *168*, 452–463. https://doi.org/10.1104/pp.15.00122

Bouchabke-Coussa, O., Quashie, M. L., Seoane-Redondo, J., Fortabat, M. N., Gery, C., Yu, A., ... Durand-Tardif, M. (2008). ESKIMO1 is a key gene involved in water economy as well as cold acclimation and salt tolerance. *BMC Plant Biology*, *8*, 125–148. https://doi.org/10.1186/1471-2229-8-125

Chiniquy, D., Varanasi, P., Oh, T., Harholt, J., Katnelson, J., Singh, S., ... Ronald, P. C. (2013). Three novel rice genes closely related to the Arabidopsis IRX9, IRX9L, and IRX14 genes and their roles in xylan biosynthesis. *Frontiers in Plant Science*, *4*, 83.

Davidson, E. A., Bayer, T. S., Windram, O., & Hleba, Y. (2015). U.S. Patent No. 20,150,274,690. U.S. Patent and Trademark Office, Washington, DC.

Downes, B. P., Stupar, R. M., Gingerich, D. J., & Vierstra, R. D. (2003). The HECT ubiquitin-protein ligase (UPL) family in Arabidopsis: UPL3 has a specific role in trichome development. *The Plant Journal.*, *35*, 729–742. https://doi.org/10.1046/j.1365-313X.2003.01844.x

Ebringerová, A., Hromádková, Z., & Heinze, T. (2005). Hemicellulose. *Advances in Polymer Science*, *186*, 1–67.

El Refy, A., Perazza, D., Zekraoui, L., Valay, J. G., Bechtold, N., Brown, S., … Bonneville, J. M. (2003). The Arabidopsis KAKTUS gene encodes a HECT protein and controls the number of endoreduplication cycles. *Molecular Genetics and Genomics*, *270*, 403–414.

Escudero, V., Jordá, L., Sopeña-Torres, S., Mélida, H., Miedes, E., Muñoz-Barrios, A., … Molina, A. (2017). Alteration of cell wall xylan acetylation triggers defense responses that counterbalance the immune deficiencies of plants impaired in the β-subunit of the heterotrimeric G-protein. *The Plant Journal.*, *92*, 386–399. https://doi.org/10.1111/tpj.13660

Foster, C. E., Martin, T. M., & Pauly, M. (2010). Comprehensive compositional analysis of plant cell walls (lignocellulosic biomass) Part II: Carbohydrates. *Journal of Visualized Experiments*, *37*, e1745.

Fukuda, H. (1997). Tracheary element differentiation. *Plant Cell*, *9*, 1147–1156. https://doi.org/10.1105/tpc.9.7.1147

Gao, Y., He, C., Zhang, D., Liu, X., Xu, Z., Tian, Y., … Zhang, B. (2017). Two trichome birefringence-like proteins mediate xylan acetylation, which is essential for leaf blight resistance in rice. *Plant Physiology*, *173*, 470–481. https://doi.org/10.1104/pp.16.01618

Gille, S., de Souza, A., Xiong, G., Benz, M., Cheng, K., Schultink, A., … Pauly, M. (2011). O-acetylation of Arabidopsis hemicellulose xyloglucan requires AXY4 or AXY4L, proteins with a TBL and DUF231 domain. *Plant Cell*, *23*, 4014–4053.

Gomez-Roldan, V., Fermas, S., Brewer, P. B., Puech-Pages, V., Dun, E. A., Pillot, J. P., … Rochange, S. F. (2008). Strigolactone inhibition of shoot branching. *Nature*, *455*, 189–194. https://doi.org/10.1038/nature07271

Grantham, N. J., Wurman-Rodrich, J., Terrett, O. M., Lyczakowski, J. J., Stott, K., Iuga, D., … Dupree, P. (2017). An even pattern of xylan substitution is critical for interaction with cellulose in plant cell walls. *Nature Plants*, *3*, 859–865. https://doi.org/10.1038/s41477-017-0030-8

Greenberg, J. T. (1996). Programmed cell death: A way of life for plants. *Proceedings of the National Academy of Sciences of the United States of America*, *93*, 12094–12097. https://doi.org/10.1073/pnas.93.22.12094

Ha, C. V., Leyva-González, M. A., Osakabe, Y., Tran, U. T., Nishiyama, R., Watanabe, Y., … Tran, L. S. (2014). Positive regulatory role of strigolactone in plant responses to drought and salt stress. *Proceedings of the National Academy of Sciences of the United States of America*, *111*, 851–856. https://doi.org/10.1073/pnas.1322135111

Ito, S., Ito, K., Abeta, N., Takahashi, R., Sasaki, Y., & Yajima, S. (2016). Effects of strigolactone signaling on Arabidopsis growth under nitrogen deficient stress condition. *Plant Signaling & Behavior*, *11*, e1126031. https://doi.org/10.1080/15592324.2015.1126031

Jiang, L., Liu, X., Xiong, G., Liu, H., Chen, F., Wang, L., … Li, J. (2015). DWARF 53 acts as a repressor of strigolactone signalling in rice. *Nature*, *504*, 401–405.

Johnson, X., Brcich, T., Dun, E. A., Goussot, M., Haurogne, K., Beveridge, C. A., & Rameau, C. (2006). Branching genes are conserved across species. Genes controlling a novel signal in pea are coregulated by other long- distance signals. *Plant Physiology*, *142*, 1014–1026. https://doi.org/10.1104/pp.106.087676

Jones, L., Ennos, A. R., & Turner, S. R. (2001). Cloning and characterization of irregular xylem4 (irx4): A severe lignin-deficient mutant of Arabidopsis. *The Plant Journal*, *26*, 205–216. https://doi.org/10.1046/j.1365-313x.2001.01021.x

Lao, N. T., Long, D., Kiang, S., Coupland, G., Shoue, D. A., Carpita, N. C., & Kavanagh, T. A. (2003). Mutation of a family 8 glycosyltransferase gene alters cell wall carbohydrate composition and causes a humidity-sensitive semi-sterile dwarf phenotype in Arabidopsis. *Plant Molecular Biology*, *53*, 647–661.

Lee, C. H., O'Neill, M. A., Tsumuraya, Y., Darvill, A. G., & Ye, Z. H. (2007). The irregular xylem9 mutant is deficient in xylan xylosyltransferase activity. *Plant and Cell Physiology*, *48*, 1624–1634. https://doi.org/10.1093/pcp/pcm135

Lee, C., Teng, Q., Zhong, R., & Ye, Z. H. (2011). The four Arabidopsis REDUCED WALL ACETYLATION genes are expressed in secondary wall-containing cells and required for the acetylation of xylan. *Plant and Cell Physiology*, *52*, 1289–1301. https://doi.org/10.1093/pcp/pcr075

Lee, C. H., Zhong, R. Q., Richardson, E. A., Himmelsbach, D. S., McPhail, B. T., & Ye, Z. H. (2007). The PARVUS gene is expressed in cells undergoing secondary wall thickening and is essential for glucuronoxylan biosynthesis. *Plant and Cell Physiology*, *48*, 1659–1672. https://doi.org/10.1093/pcp/pcm155

Lefebvre, V., Fortabat, M. N., Ducamp, A., North, H. M., Maia-Grondard, A., Trouverie, J., … Durand-Tardif, M. (2011). ESKIMO1 disruption in Arabidopsis alters vascular tissue and impairs water transport. *PLoS ONE*, *6*, e16645. https://doi.org/10.1371/journal.pone.0016645

Liu, J., He, H., Vitali, M., Visentin, I., Charnikhova, T., Haider, I., … Cardinale, F. (2015). Osmotic stress represses strigolactone biosynthesis in Lotus japonicus roots: Exploring the interaction between strigolactones and ABA under abiotic stress. *Planta*, *241*, 1435–1451. https://doi.org/10.1007/s00425-015-2266-8

Lucas, W. J., Groover, A., Lichtenberger, R., Furuta, K., Yadav, S. R., Helariutta, Y., … Kachroo, P. (2013). The plant vascular system: Evolution, development and functions. *Journal of Integrative Plant Biology*, *55*, 294–388. https://doi.org/10.1111/jipb.12041

Lugan, R., Niogret, M. F., Kervazo, L., Larher, F. R., Kopka, J., & Bouchereau, A. (2009). Metabolome and water status phenotyping of Arabidopsis under abiotic stress cues reveals new insight into ESK1 function. *Plant, Cell and Environment*, *32*, 95–108. https://doi.org/10.1111/j.1365-3040.2008.01898.x

Lumba, S., Holbrook-Smith, D., & McCourt, P. (2017). The perception of strigolactones in vascular plants. *Nature Chemical Biology*, *13*, 599–606. https://doi.org/10.1038/nchembio.2340

Malik, K., Wu, K., Li, X. Q., Martin-Heller, T., Hu, M., Foster, E., … Miki, B. (2002). A constitutive gene expression system derived from the tCUP cryptic promoter elements. *TAG. Theoretical and Applied Genetics.*, *105*, 505–514.

Manabe, Y., Verhertbruggen, Y., Gille, S., Harholt, J., Chong, S. L., Pawar, P. M., … Scheller, H. V. (2013). Reduced wall acetylation proteins play vital and distinct roles in cell wall O-acetylation in Arabidopsis. *Plant Physiology*, *163*, 1107–1117. https://doi.org/10.1104/pp.113.225193

Mashiguchi, K., Sasaki, E., Shimada, Y., Nagae, M., Ueno, K., Nakano, T., … Asami, T. (2009). Feedback-regulation of strigolactone biosynthetic genes and strigolactone-regulated genes in Arabidopsis. *Bioscience, Biotechnology, and Biochemistry*, *73*, 2460–2465. https://doi.org/10.1271/bbb.90443

Mortimer, J. C., Miles, G. P., Brown, D. M., Zhang, Z., Segura, M. P., Weimar, T., … Dupree, P. (2010). Absence of branches from xylan in Arabidopsis gux mutants reveals potential for simplification of lignocellulosic biomass. *Proceedings of the National Academy of Sciences of the United States of America*, *107*, 17409–17414. https://doi.org/10.1073/pnas.1005456107

Nakamura, H., Xue, Y. L., Miyakawa, T., Hou, F., Qin, H. M., Fukui, K., … Asami, T. (2013). Molecular mechanism of strigolactone perception by DWARF14. *Nature Communications*, *4*, 2613. https://doi.org/10.1038/ncomms3613

Pauly, M., Gille, S., Liu, L. F., Mansoori, N., de Souza, A., Schultink, A., & Xiong, G. Y. (2013). Hemicellulose biosynthesis. *Planta*, *238*, 627–642. https://doi.org/10.1007/s00425-013-1921-1

Rinne, P. L., Paul, L. K., Vahala, J., Kangasjärvi, J., & van der Schoot, C. J. (2016). Axillary buds are dwarfed shoots that tightly regulate GA pathway and GA-inducible 1,3-β-glucanase genes during branching in hybrid aspen. *Journal of Experimental Botany*, *67*, 5975–5991. https://doi.org/10.1093/jxb/erw352

Scheller, H. V., & Ulvskov, P. (2010). Hemicelluloses. *Annual Review of Plant Biology*, *61*, 263–289. https://doi.org/10.1146/annurev-arplant-042809-112315

Scholl, R. L., May, S. T., & Ware, D. H. (2000). Seed and molecular resources for Arabidopsis. *Plant Physiology*, *124*, 1477–1480. https://doi.org/10.1104/pp.124.4.1477

Schuetz, M., Smith, R., & Ellis, B. (2012). Xylem tissue specification, patterning, and differentiation mechanisms. *Journal of Experimental Botany*, *64*, 11–31.

Schultink, A., Naylor, D., Dama, M., & Pauly, M. (2015). The role of the plant-specific ALTERED XYLOGLUCAN9 protein in Arabidopsis cell wall polysaccharide O-acetylation. *Plant Physiology*, *167*, 1271–1283. https://doi.org/10.1104/pp.114.256479

Seto, Y., Sado, A., Asami, K., Hanada, A., Umehara, M., Akiyama, K., & Yamaguchi, S. (2014). Carlactone is an endogenous biosynthetic precursor for strigolactones. *Proceedings of the National Academy of Sciences of the United States of America*, *111*, 1640–1645. https://doi.org/10.1073/pnas.1314805111

Sindhu, A., Langewisch, T., Olek, A., Multani, D. S., McCann, M. C., Vermerris, W., … Johal, G. (2007). Maize brittle stalk2 encodes a COBRA-like protein expressed in early organ development but required for tissue flexibility at maturity. *Plant Physiology*, *145*, 1444–1459. https://doi.org/10.1104/pp.107.102582

Sorefan, K., Booker, J., Haurogné, K., Goussot, M., Bainbridge, K., Foo, E., … Leyser, O. (2003). MAX4 and RMS1 are orthologous dioxygenase-like genes that regulate shoot branching in Arabidopsis and pea. *Genes & Development*, *17*, 1469–1474. https://doi.org/10.1101/gad.256603

Soundappan, I., Bennett, T., Morffy, N., Liang, Y., Stanga, J. P., Abbas, A., … Nelson, D. C. (2015). SMAX1-LIKE/D53 family members enable distinct MAX2-dependent responses to strigolactones and karrikins in Arabidopsis. *Plant Cell*, *27*, 3143–3159. https://doi.org/10.1105/tpc.15.00562

Stanga, J. P., Smith, S. M., Briggs, W. R., & Nelson, D. C. (2013). SUPPRESSOR OF MORE AXILLARY GROWTH2 1 controls seed germination and seedling development in Arabidopsis. *Plant Physiology*, *163*, 318–330. https://doi.org/10.1104/pp.113.221259

Taylor, N. G., Howells, R. M., Huttly, A. K., Vickers, K., & Turner, S. R. (2003). Interactions among three distinct CesA proteins essential for cellulose synthesis. *Proceedings of the National Academy of Sciences of the United States of America*, *100*, 1450–1455. https://doi.org/10.1073/pnas.0337628100

Umehara, M., Hanada, A., Yoshida, S., Akiyama, K., Arite, T., Takeda-Kamiya, N., … Yamaguchi, S. (2008). Inhibition of shoot branching by new terpenoid plant hormones. *Nature*, *455*, 195–200. https://doi.org/10.1038/nature07272

Vermerris, W., Sherman, D. M., & McIntyre, L. M. (2010). Phenotypic plasticity in cell walls of maize brown midrib mutants is limited by lignin composition. *Journal of Experimental Botany*, *61*, 2479–2490. https://doi.org/10.1093/jxb/erq093

Wang, Y., Sun, S., Zhu, W., Jia, K., Yang, H., & Wang, X. (2013). Strigolactone/MAX2-induced degradation of brassinosteroid transcriptional effector BES1 regulates shoot branching. *Developmental Cell*, *27*, 681–688. https://doi.org/10.1016/j.devcel.2013.11.010

Waters, M. T., Gutjahr, C., Bennett, T., & Nelson, D. C. (2017). Strigolactone signaling and evolution. *Annual Review of Plant Biology*, *68*, 291–322. https://doi.org/10.1146/annurev-arplant-042916-040925

Xi, W., Song, D., Sun, J., Shen, J., & Li, L. (2016). Formation of wood secondary cell wall may involve two type cellulose synthase complexes in Populus. *Plant Molecular Biology*, *93*, 419–429.

Xin, Z., & Browse, J. (1998). Eskimo1 mutants of Arabidopsis are constitutively freezing-tolerant. *Proceedings of the National Academy of Sciences of the United States of America*, *95*, 7799–7804. https://doi.org/10.1073/pnas.95.13.7799

Xin, Z., Mandaokar, A., Chen, J., Last, R. L., & Browse, J. (2007). Arabidopsis ESK1 encodes a novel regulator of freezing toler- ance. *The Plant Journal.*, *49*, 786–799. https://doi.org/10.1111/j.1365-313X.2006.02994.x

Xiong, G., Cheng, K., & Pauly, M. (2013). Xylan O-acetylation impacts xylem development and enzymatic recalcitrance as indicated by the Arabidopsis mutant tbl29. *Molecular Plant*, *6*, 1373–1375. https://doi.org/10.1093/mp/sst014

Xiong, G. Y., Dama, M., & Pauly, M. (2015). Glucuronic acid moieties on xylan are functionally equivalent to O-acetyl-substituents. *Molecular Plant*, *8*, 1119–1121. https://doi.org/10.1016/j.molp.2015.02.013

Xu, F., Liu, Z., Xie, H., Zhu, J., Zhang, J., Kraus, J., … Wang, H. (2014). Increased drought tolerance through the suppression of ESKMO1 gene and overexpression of CBF-related genes in Arabidopsis. *PLoS ONE*, *9*, e106509. https://doi.org/10.1371/journal.pone.0106509

Yuan, Y., Teng, Q., Zhong, R., Haghighat, M., Richardson, E. A., & Ye, Z. H. (2016). Mutations of Arabidopsis TBL32 and TBL33 affect xylan acetylation and secondary wall deposition. *PLoS ONE*, *11*, e0146460. https://doi.org/10.1371/journal.pone.0146460

Yuan, Y., Teng, Q., Zhong, R., & Ye, Z. H. (2013). The Arabidopsis DUF231 domain-containing protein ESK1 mediates 2-O- and 3-O-acetylation of xylosyl residues in xylan. *Plant and Cell Physiology*, *54*, 1186–1199. https://doi.org/10.1093/pcp/pct070

Yuan, Y., Teng, Q., Zhong, R., & Ye, Z. H. (2016a). Roles of Arabidopsis TBL34 and TBL35 in xylan acetylation and plant growth. *Plant Science*, *243*, 120–130. https://doi.org/10.1016/j.plantsci.2015.12.007

Yuan, Y., Teng, Q., Zhong, R., & Ye, Z. H. (2016b). TBL3 and TBL31, two Arabidopsis DUF231 domain proteins, are required for 3-O-monoacetylation of xylan. *Plant and Cell Physiology*, *57*, 35–45. https://doi.org/10.1093/pcp/pcv172

Maize GO Annotation—Methods, Evaluation, and Review (maize-GAMER)

Kokulapalan Wimalanathan[1,2] | Iddo Friedberg[1,3] | Carson M. Andorf[4,5] | Carolyn J. Lawrence-Dill[1,2,6]

[1]Bioinformatics and Computational Biology, Iowa State University, Ames, IA, USA

[2]Department of Genetics Development and Cell Biology, Iowa State University, Ames, IA, USA

[3]Department of Veterinary Microbiology and Preventive Medicine, Iowa State University, Ames, IA, USA

[4]USDA-ARS Corn Insects and Crop Genetics Research Unit, Iowa State University, Ames, IA, USA

[5]Department of Computer Science, Iowa State University, Ames, IA, USA

[6]Department of Agronomy, Iowa State University, Ames, IA, USA

Correspondence
Carolyn J. Lawrence-Dill, Bioinformatics and Computational Biology, Iowa State University, Ames, IA, USA.
Email: triffid@iastate.edu

Funding information
Iowa State University Plant Sciences Institute Faculty Scholars Program; National Science Foundation, Grant/Award Number: IOS #1027527, ABI-1458359; United States Department of Agriculture—Agricultural Research Services

Abstract

We created a new high-coverage, robust, and reproducible functional annotation of maize protein-coding genes based on Gene Ontology (GO) term assignments. Whereas the existing Phytozome and Gramene maize GO annotation sets only cover 41% and 56% of maize protein-coding genes, respectively, this study provides annotations for 100% of the genes. We also compared the quality of our newly derived annotations with the existing Gramene and Phytozome functional annotation sets by comparing all three to a manually annotated gold standard set of 1,619 genes where annotations were primarily inferred from direct assay or mutant phenotype. Evaluations based on the gold standard indicate that our new annotation set is measurably more accurate than those from Phytozome and Gramene. To derive this new high-coverage, high-confidence annotation set, we used sequence similarity and protein domain presence methods as well as mixed-method pipelines that were developed for the Critical Assessment of Function Annotation (CAFA) challenge. Our project to improve maize annotations is called maize-GAMER (GO Annotation Method, Evaluation, and Review), and the newly derived annotations are accessible via MaizeGDB (http://download.maizegdb.org/maize-GAMER) and CyVerse (B73 RefGen_v3 5b+ at doi.org/10.7946/P2S62P and B73 RefGen_v4 Zm00001d.2 at doi.org/10.7946/P2M925).

KEYWORDS
functional annotation, gene ontology, genomics, GO, maize

1 | INTRODUCTION

Maize is an agriculturally important crop species and model organism for genetics and genomics research (Lawrence, Dong, Polacco, Seigfried, & Brendel, 2004). Not only is maize historically important for genetics research, along with other model species, significant efforts have been made to transition existing datasets into a more sequence-centric paradigm (Sen et al., 2009), thus enabling genomics approaches to be brought to bear on both basic research problems and applied breeding (Lawrence et al., 2008). In 2009, the maize genome's reference sequence was made available to the research community (Schnable et al., 2009). Since then, much work has gone into improving the utility of the genome sequence to scientists with a focus on sequence annotation.

In practice, making a genome sequence useful involves three basic steps: assembling the genome sequence, assigning gene structures, and assigning functions to genes. The quality of data generated at each step influences downstream inferences, with high-quality sequence, assembly, and gene structure assignments generally resulting in better functional annotations overall. Functional predictions serve as the basis for formulating hypotheses that are subsequently tested in the laboratory. As such, experimentalists have a great interest in high-quality functional annotation sets that cover all or most of the genes in their species of interest.

The Gene Ontology (GO) is a controlled vocabulary of semi-hierarchically related terms that describe gene product function. It consists three categories: Biological Process (BP), Cellular Component (CC), and Molecular Function (MF). In the context of GO, functional annotation of a gene consists of the assignment of one or more GO terms from one or more of the GO categories to a given gene or gene model (here, we will refer to genes and gene models simply as "genes" for simplicity).

For individual GO term associations to genes, Evidence Codes (ECs) are assigned to assert how the association of term to gene was made (Harris et al., 2004). GO Evidence Codes are aggregated into five general categories: Experimental, Computational Analysis, Curator Statement, Author Statement, and Automatically Assigned. See Table 1 and http://www.geneontology.org/page/guide-go-evidence-codes for a detailed explanation of the GO Evidence Codes.

TABLE 1 Evidence Codes used in gene ontology annotations

Type	Evidence code	Description
Experimental	EXP	Inferred from Experiment
	IDA	Inferred from Direct Assay
	IPI	Inferred from Physical Interaction
	IMP	Inferred from Mutant Phenotype
	IGI	Inferred from Genetic Interaction
	IEP	Inferred from Expression Pattern
Computational Analysis	ISS	Inferred from Sequence or structural Similarity
	ISO	Inferred from Sequence Orthology
	ISA	Inferred from Sequence Alignment
	ISM	Inferred from Sequence Model
	IGC	Inferred from Genomic Context
	IBA	Inferred from Biological aspect of Ancestor
	IBD	Inferred from Biological aspect of Descendant
	IKR	Inferred from Key Residues
	IRD	Inferred from Rapid Divergence
	RCA	Inferred from Reviewed Computational Analysis
Author statement	TAS	Traceable Author Statement
	NAS[a]	Non-traceable Author Statement
Curatorial Statement	IC	Inferred by Curator
	ND[a]	No biological Data available
Automatically Assigned	IEA[a]	Inferred from Electronic Annotation

[a]Evidence Codes without either curation or biological data supporting them.

The use of experimental ECs asserts that the assignment results from a physical characterization of the protein's function as described in a publication. Computational approaches are based on in silico analyses. One of the simplest and most commonly conducted computational approaches involves matching similar genes between an existing, well-annotated genome and an unannotated genome. Once the matches are assigned, annotations are inferred to genes in the unannotated genome. Such assignments receive the ISS (Inferred from Sequence or Structural Similarity) EC. The ISS EC is also assigned if an uncharacterized sequence contains a characterized domain. In such instances, the presence of the domain itself can be used to predict function for the uncharacterized sequence. For Curator and Author Statements, included EC types are based on judgment by curators and scientists in their expert opinion. As such, they are considered to be reviewed annotation types, although these do include two ECs based on little data: NAS (Non-traceable Author Statement) and ND (No biological Data available). The Automatically Assigned EC type contains only one EC: Inferred from Electronic Annotation (IEA). IEA is unique in that no reviewed analysis of the assignment is required. Put another way, no curatorial judgment is applied, making it the least supported EC of the group.

Sequence-based approaches to automated functional annotation generally fall into three basic categories: sequence similarity, domain-based methods, and mixed methods. Sequence similarity-based gene matching most often relies on BLAST (e.g., BLAST2GO) followed by limiting the number of accepted matches based on e-value or a reciprocal best hit (RBH) strategy (Altschul, Gish, Miller, Myers, & Lipman, 1990; Conesa & Götz, 2008; Moreno-Hagelsieb & Latimer, 2008). Domain-based methods score sequences for the presence of well-described protein domain such as those included in Pfam, PANTHER, and ProSite (Finn et al., 2017). InterProScan is a commonly used domain-based GO annotation pipeline (Jones et al., 2014). Mixed methods combine sequence similarity, domain-based approaches, and other evidence such as inferred orthology through phylogenetics to assign GO terms systematically (Clark & Radivojac, 2011; Falda et al., 2012; Koskinen, Törönen, Nokso-Koivisto, & Holm, 2015). For more of the latest methods, see Jiang et al. (2016).

For maize, two genome-scale GO annotation sets exist for the B73 reference assembly and gene set (i.e., B73 RefGen_v3 and 5b+, respectively). These functional annotations are generated by and accessible from the Gramene (www.gramene.org (Tello-Ruiz et al., 2016)) and Phytozome (phytozome.jgi.doe.gov (Goodstein et al., 2012)) projects and websites, respectively. Gramene annotations are based on the Ensembl annotation pipeline (http://ensemblgenomes.org/info/data/cross_references), which is a mixed-method approach. The primary sources of the Ensembl annotations are from UniProtKB, community-based annotations from MaizeGDB (Andorf et al., 2016), InterPro2GO, and projections from orthologs inferred from phylogenetic analyses. Phytozome has a two-step process for GO annotation. First, Pfam domains are assigned to proteins. Second, GO annotations are determined based on the Pfam2GO mapping (Hunter et al., 2009).

Given the wealth of functional descriptions derived from mutational analyses, many researchers rely on the available maize GO-based functional annotations from large-scale, high-profile community resources such as Gramene and Phytozome for formulating experimental hypotheses and also as input datasets to transitively annotate predicted functions to newly sequenced grass species and crop genomes (e.g., Hirsch et al., 2016). However, if we compare the EC types for GO assignments between the model species *Arabidopsis thaliana* and the Gramene and Phytozome functional annotations of the maize reference line B73, it is clear that the evidence supporting GO term assignments for these maize datasets is comparatively lacking (see Figure 1). Both the Gramene and Phytozome maize annotations have few annotations beyond those Inferred from Electronic Annotation (IEA). This situation is not intuitive to researchers given that maize has a wealth of functional descriptions in the literature.

Exacerbating this problem, transfers of predicted function often are based on sequence similarity alone with no restriction of input data to associations based on well-documented EC types. Furthermore, although mixed-method pipelines such as the Ensemble COMPARA pipeline used by Gramene and the Phytozome Pfam2GO (Goodstein et al., 2012; Herrero et al., 2016) mappings may seem reproducible in principle given that they are based on the use of specific systems and software, details including input files and parameters often are unavailable or incomplete, making it impossible for research groups outside the group that generated those annotation resources to reproduce the annotation sets. In addition, because many computational pipelines inherit functional annotations that were also purely computationally derived, a single errant annotation can be propagated to many genomes (Andorf, Dobbs, & Honavar, 2007), making it mistakenly appear that many genomes agree on the errant function. For these reasons, existing computational functional annotations of maize (and many other plant genomes) should be approached with skepticism.

Given these issues with the maize functional annotation, we endeavored to create an improved annotation set. This task requires both application of robust and reproducible methods and a gold standard set of maize GO annotations to compare generated result sets to each other as well as to the Gramene and Phytozome maize functional annotations. One small dataset of well-curated GO-based functional annotations does exist for maize. It was initially created by curators at MaizeGDB for the purpose of enriching the MaizeCyc metabolic pathway database (Monaco et al., 2013) and expanded through manual literature curation. This dataset constitutes 1,621 genes and 2,002 GO terms.

We annotated the maize B73 RefGen_v3 annotation set 5b+ using only experimentally based annotations by filtering out GO assignments with IEA, NAS, and ND ECs from the input data and assigned GO terms using multiple input datasets then compared the performance of sequence similarity, domain presence, and mixed methods based on how well the methods predicted function for genes included in the MaizeGDB gold standard dataset. For mixed methods, we used pipelines developed for the Critical Assessment of Functional Annotation (CAFA) challenge, a competition designed to evaluate the latest computational functional annotation methods and

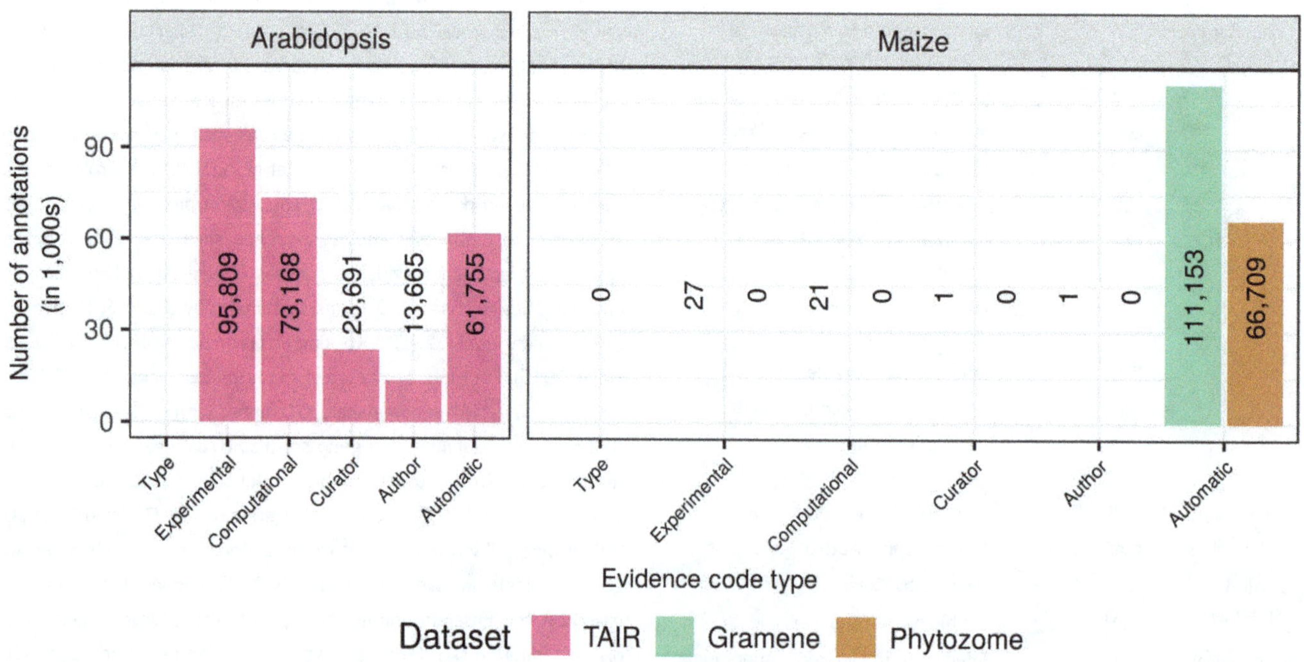

FIGURE 1 Numbers of annotations by EC category. Arabidopsis (TAIR10) shown in magenta and maize in green and orange for Gramene and Phytozome, respectively. Annotation counts on the y-axis are shown in thousands. Each bar in the histogram is labeled with the actual count to show where counts are so small that no bar is visible

to promote improvement of methods for functional annotation (Jiang et al., 2016; Radivojac et al., 2013). Groups competing in the CAFA challenge create tools that are applied to a set of specified target sequences. GO assignments are subsequently evaluated based on accumulation of functional data in the literature for the target sequence set. Some CAFA tools use preprocessing steps combined with a number of different computational and statistical approaches to reduce the number of false-positive and false-negative annotations (Clark & Radivojac, 2011; Falda et al., 2012; Koskinen et al., 2015). Some mixed-method pipelines performed better on average than other methods in the first iteration of the CAFA competition (Radivojac et al., 2013), indicating that the use of mixed-method pipelines for large-scale GO annotations could potentially improve the overall quality of the annotation sets.

The project to evaluate and improve maize GO annotations is called GAMER: GO Annotation Method, Evaluation, and Review. We compared GAMER annotations to annotations based on sequence similarity, domain, and three CAFA mixed methods. Next, we combined GAMER outputs to generate an aggregate maize-GAMER GO annotation set and compared it to the existing Phytozome and Gramene GO annotations based on the hF_1 score. The GAMER annotations had three major advantages compared to the Gramene and Phytozome annotations: (i) an increased number of maize genes annotated with GO terms; (ii) more than twice the number of annotations (GO terms assigned) for maize protein-coding genes; and (iii) similar or better quality scores relative to existing annotations sets based on hF_1 score. The B73 RefGen_v3 5b+ maize-GAMER functional annotation dataset described here is accessible via MaizeGDB (http://download.maizegdb.org/maize-GAMER) and CyVerse (doi.org/10.7946/P2S62P). Scripts used to generate the annotation are available via GitHub at https://github.com/Dill-PICL/maize-GAMER.

2 | MATERIALS AND METHODS

2.1 | Functional annotation of maize genes

Three sequence-based approaches were used to annotate function to genes in the maize reference genome: sequence similarity, domain-based, and mixed-method pipelines (see Figure 2; also described in Sections 2.1.1, 2.1.2, and 2.1.3, respectively). The scripts (bash, R, and Python) used to generate the annotations for maize B73 RefGen_v3 are available at https://github.com/Dill-PICL/maize-GAMER. These scripts run free and open-source tools on different inputs required for these tools to generate annotation datasets. Please refer to the reproducibility supplemental file for details on versions of software, version of input datasets used, and commands and parameters used to run these tools.

The B73 genome and protein sequences for gene models included in the filtered gene set (FGS) were downloaded from Gramene Release 42 (Tello-Ruiz et al., 2016). The downloaded protein

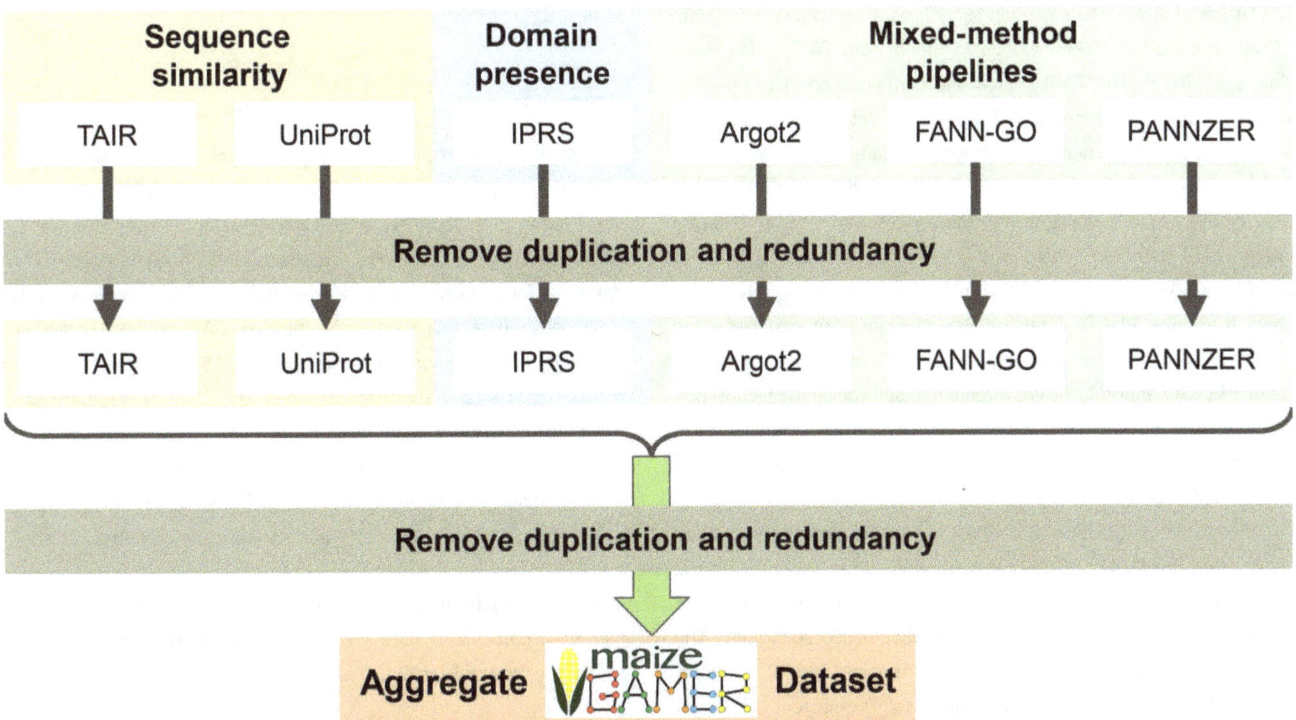

FIGURE 2 Overview of steps to produce the maize-GAMER datasets. Three types of methods are used: sequence similarity (yellow), domain presence (blue), and CAFA mixed-method pipelines (green). Within sequence similarity, two input datasets were subjected to reciprocal best hit against maize: TAIR10 (Arabidopsis) and UniProt (the ten most well-annotated plant species). For domain presence, InterPro signatures were applied to maize using InterProScan (IPRS). From the CAFA mixed-method pipelines, Argot2, FANN-GO, and PANNZER were applied to maize. For each individual output, duplications and redundancies were removed, and then, the datasets were combined. A second round of duplication and redundancy removal was carried out to produce the maize-GAMER Aggregate dataset

FASTA file contained sequences for all FGS transcripts (e.g., the gene model X has transcript models X_T01, X_T02, and X_T03). For each gene model, only the longest translated protein sequence derived from the transcripts was analyzed. The gold standard annotations used for evaluations were obtained from MaizeGDB, and they encompass GO annotations for 1,619 gene models from RefGen_v3. The number of annotations for Cellular Component (CC), Molecular Function (MF), and Biological Process (BP) was 1,584, 88, and 323, respectively.

2.1.1 | Sequence similarity

The sequence similarity-based annotation method has three main steps: (i) calculation of sequence similarity, (ii) valid hit detection, and (iii) inheritance of high-confidence GO annotations. BLASTP was used (Altschul et al., 1990) with default parameters to calculate sequence similarity between maize protein sequences and two other datasets: the "Arabidopsis" dataset from TAIR, The Arabidopsis Information Resource (Berardini et al., 2015), and the "Plant" dataset from UniProt (UniProt Consortium, 2015). Valid hits were detected using the RBH method from BLASTP results. GO terms with nonreviewed ECs (i.e., IEA, NAS, and ND described in the introduction) were removed from input datasets. All others were inherited between the RBH pairs of maize and the other plant.

Arabidopsis has the largest number of reviewed (human curated) EC GO annotations among plant model organisms (see Table S1). A FASTA file of Arabidopsis protein sequences along with the cognate GO Annotation File (GAF) was downloaded from TAIR v.10 (Berardini et al., 2015). The TAIR protein file contained predicted protein sequences from all transcripts. This file was filtered to retain only the protein sequence derived from longest transcript for each gene. Retained protein sequences from TAIR were used to create the TAIR BLAST database, and maize protein sequences were used to create a maize BLAST database. Maize protein sequences were used to query the TAIR BLAST database. Likewise, TAIR sequences were used to query the maize BLAST database. Results from both searches were used to detect RBH pairs between Arabidopsis and maize. All nonreviewed EC GO annotations were removed, and remaining GO associations to Arabidopsis genes were inherited to maize genes for each RBH pair. This maize/Arabidopsis RBH ortholog dataset is called "maize-TAIR GO annotations".

All reviewed EC GO annotations and protein sequences for all plants from the UniProt-GOA database were downloaded using the QuickGO tool hosted at EBI (Binns et al., 2009). Protein sequences and reviewed EC GO annotations were downloaded separately. The UniProt plant GO annotation dataset containing 304,426 annotations from 75,537 unique protein sequences. The protein sequences downloaded spanned 292 taxa. Only ten species had more than 1,000 annotations (see Table S1). Annotations from the top 10 species (in terms of number of reviewed GO annotations) were retained for our analyses. The process to annotate maize genes using UniProt plant data was similar to that for Arabidopsis. Maize protein sequences were matched against protein sequences from each species separately using BLASTP. Putative orthologs were determined using RBH for each maize-plant pair. Terms annotated to the other plant protein were inherited to the maize protein sequence for each putative ortholog pair. GO annotations inherited from each plant species were concatenated together. The derived dataset is called the "maize-UniProt GO annotations".

2.1.2 | Domain presence

InterProScan5 (IPRS) version 5.16-55.0 was used to create domain-based GO annotation of maize protein-coding genes (Jones et al., 2014). IPRS was used to annotate GO terms to maize genes to produce the "maize-IPRS GO annotations".

2.1.3 | Mixed-method pipelines

At the beginning of this project, the first iteration of the CAFA challenge (CAFA1; described in the Introduction) had been completed. The results from the challenge indicated that CAFA1 mixed-method pipelines performed as well or better than standard methods (Radivojac et al., 2013). To determine their predictive power for functional annotation in plants, the top-performing mixed-method pipelines from CAFA1 were reviewed to identify a group that could be implemented based upon availability of code and sufficient documentation. Three tools were selected as follows: Argot2, FANN-GO, and PANNZER (Clark & Radivojac, 2011; Falda et al., 2012; Koskinen et al., 2015).

Argot2

Argot2 has a batch processing tool that can annotate up to 5,000 preprocessed input sequences. There are two different preprocessing steps for Argot2: (i) querying the UniProt database for sequence similarity matches to the input sequences and (ii) querying the Pfam database for putative domains present in the input sequences. The maize sequences were split into multiple FASTA files containing a maximum of 5,000 sequences. The eight FASTA files resulting from the previous step were used to query the UniProt database using BLASTP for matches, and the output was saved. HMMER was used to search a local Pfam database for potential hits for all the input protein sequences (Finn, Clements, & Eddy, 2011). Preprocessing each input FASTA resulted in a pair of input files for Argot2: BLAST and HMMER files. Each pair of preprocessed files was compressed and submitted to Argot2 batch processing tool. Results from each pair of preprocessed data were downloaded and concatenated to create the "maize-Argot2 GO annotations."

FANN-GO

The file containing maize protein sequences was imported into MATLAB using a built-in function (MATLAB:2017). The MAIN function from FANN-GO was used to preprocess and annotate maize protein sequences. FANN-GO uses BLASTP to query FANN-GO training sequence dataset (derived from UniProt) for potential matches for the input sequences and converts the results to input feature vectors. The FANN-GO predictor built from the training dataset is then used to process the input feature vectors and calculate the probability that a particular protein is associated with a particular GO term. These probabilities are represented in a matrix where rows represent sequences and columns represent GO terms. The matrix was converted to a GAF (GO Annotation File Format) file to be used for subsequent evaluations. This dataset is referred to as the "maize-FANN-GO annotations".

PANNZER

Maize protein sequences were preprocessed using BLASTP to query a local UniProt protein BLAST database, and the output was saved in XML format as required by PANNZER. PANNZER was run on the xml file output from the previous step, and the output was converted into a GAF file. This dataset from PANNZER is referred to as the "maize-PANNZER GO annotations."

2.2 | Metrics used in maize-GAMER

A number of metrics defined and described by the AIGO (Analysis and the Inter-comparison of GO functional annotations) library were used to select high-confidence annotations, clean, and evaluate the maize-GAMER-derived annotation sets (Defoin-Platel et al., 2011). AIGO has defined two types of metrics: analysis metrics and comparison metrics (see Table 2).

2.2.1 | Analysis metrics

Analysis metrics defined by AIGO measure features of a given annotation set. Four AIGO analysis metrics were used for maize-GAMER: Duplication, Redundancy, Coverage, and Specificity (see Table 2). With respect to calculating AIGO metrics, an annotation is defined as a single gene–GO term pair. Duplication is the proportion of annotations that are not unique to a given annotation set. Duplication is calculated for each annotation set as described in Table 2. The collection of more general GO terms that can be inherited from a specific GO term is called "ancestral". Redundancy occurs when a GO term and one or more of its ancestral terms are annotated to the same gene in a dataset. Redundancy metric was calculated by obtaining the mean of the proportion of ancestral terms annotated for each gene in a particular dataset. Coverage is the proportion of genes that have at least one GO term assigned. Specificity is measured by counting the number of ancestral terms that exist for a given annotation, then averaging those counts across all annotations in the set. An ideally cleaned annotation dataset would have no Duplication, no Redundancy, high Coverage, and high Specificity.

2.2.2 | Comparison metrics

Comparison metrics defined by AIGO measure how well a given set of annotations match with another set of annotations. The AIGO comparison metrics hierarchical Precision (*hPr*) and hierarchical Recall (*hRc*) were used to evaluate annotation sets against gold standard

TABLE 2 Metrics used to analyze and evaluate the annotation datasets

Metric	Description	Formula
Analysis metrics[a]		
Duplication (%)	Proportion of duplicate annotations in a dataset	$\frac{\#\text{of Total Annotations} - \#\text{of Unique Annotations}}{\#\text{of Total Annotations}}$
Redundancy (%)	Proportion of redundant terms in a dataset	$\frac{\sum_{y=1}^{M} \frac{\#\text{of Ancestral terms annotated to Gene}_{(y)}}{\#\text{of total Annotations to Gene}_{(y)}} \times 100}{M}$
Coverage (%)	Proportion of Genes annotated in a dataset	$\frac{\#\text{of Genes with} \geq \text{one GO annotation}}{\#\text{of Genes}}$
Specificity	Specificity of the Annotation for a given gene	$\frac{\sum_{x=1}^{N} \#\text{of Ancestral terms inferred for Annotation}_{(x)}}{\#\text{of total Annotations}}$
Comparison metrics[b]		
hPr	Hierarchical Precision calculated by evaluating against the gold standard	$\frac{\text{GO terms predicted in AS} \cap \text{GO terms annotated in GS}}{\text{GO terms predicted in AS}}$
hRc	Hierarchical Recall calculated by evaluating against the gold standard	$\frac{\text{GO terms predicted in AS} \cap \text{GO terms annotated in GS}}{\text{GO terms annotated in GS}}$
hF_1	Harmonic mean of *hPr* and *hRc*	$2 \times \frac{hPr \times hRc}{hPr + hRc}$

N, Total # of Annotations, *M*, Total # of Genes, AS, Annotation Set, GS, Gold Standard.
[a]Comparison Metrics used here have been described in detail in (Defoin-Platel et al., 2011).
[b]See Supplementary document for precise steps to calculate *hPr* and *hRc*.

annotations from MaizeGDB (see Table 2 and Supplementary Materials). Different metrics have been defined for the evaluation of GO annotations against a gold standard (Clark & Radivojac, 2013; Defoin-Platel et al., 2011; Jiang et al., 2016; Radivojac et al., 2013). AIGO provided a set of well-described evaluation metrics which were adapted by maize-GAMER and was utilized for large number of annotations produced by mixed-method pipelines. Both *hPr* and *hRc* evaluations start with propagating the GO terms in the annotations to the root. *hPr* is the proportion of the GO terms (directly annotated and inferred by propagation) in an annotation set which is shared with the GO terms (directly annotated and inferred by propagation) in the gold standard. *hRc* is the proportion of the GO terms in the gold standard, which are found in the annotation set. *hPr* and *hRc* were calculated for the genes in the gold standard dataset and were calculated independently for each GO category. If a gene was annotated in the gold standard, but was not annotated in the annotations set then both *hPr* and *hRc* were set to 0. See Supplementary Materials for precise steps used to calculate *hPr* and *hRc*. Harmonic mean (hF_1) of *hPr* and *hRc* was calculated for each annotation set for each GO category to use a single number to compare different annotation methods.

2.3 | Cleaning and combining component datasets

2.3.1 | Score threshold selection for mixed methods

Mixed-method pipelines used in the maize-GAMER project provide a confidence score for each GO annotation. The confidence score ranges from 0.0 to 1.0, where a higher score indicates more confidence for a given annotation. A score threshold which maximizes hF_1 (*hFmax*) will select the optimal set of annotations which reduces the total number of false positives and false negatives (see Section 2.2 for description of the metrics). The range of annotation scores from mixed-method pipelines did not span the whole 0.0–1.0 range, so the scores were normalized to fall between 0.0 and 1.0 independently for each annotation set. A set of thresholds (every 0.05 from 0.0 to 1.0; i.e., 0.00, 0.05, 0.1, 0.1,…., 0.95, 1.00) were selected. hF_1 score for each GO category, and each threshold was calculated by selecting the annotations with a normalized score that was ≥ to the threshold and evaluating against the gold standard annotations. *hFmax* for each GO category was determined by getting the highest hF_1 obtained from the previous step. The score thresholds which resulted in *hFmax* were used to select a subset of annotations from each mixed-method pipeline (see Table S2). The maize-Argot2, maize-FANN-GO, and maize-PANNZER GO annotations described in subsequent sections refer to the subset of annotations selected via this selection step.

2.3.2 | Removing redundancy and duplication

Duplication is the presence of two or more instances of the same gene–GO term pair in a single annotation set. Redundancy is the presence of an ancestral GO term in the annotations of a gene which also contains a specific annotation from which the ancestral GO term can be inferred by propagation. Component annotation sets from all methods described above were cleaned by removing redundancy and duplication for each annotation set across all three GO categories. Duplication was cleaned by replacing multiple instances of a gene–GO term pair with a single instance for a given annotation set. Duplicate annotations from all six raw annotation sets were removed, and files with nonduplicate annotations were created for each annotation set. Redundancy was cleaned by removing annotations containing GO terms that could be inferred from other terms based on the GO hierarchy, and only retaining the annotations with GO terms that cannot be inferred.

2.3.3 | The maize-GAMER Aggregate dataset

Clean (nonredundant and nonduplicated) annotation sets from all component methods were merged to generate the maize-GAMER Aggregate annotation set. Redundancy and duplication introduced by concatenating multiple datasets were removed.

A new genome assembly (B73 RefGen_v4) and annotation set (Zm00001.2) for maize inbred line B73 were recently released (Jiao et al., 2017). Because this dataset has not been available for long, only few published analyses are available and the research community is only now in the process of transitioning to general use of RefGen_v4 for large-scale analyses. As such, analyses and results described here derive from the well-annotated v3 assembly and annotation set. To extend outcomes of the work described here for future v4 efforts, maize-GAMER Aggregate annotations have also been created for the maize B73 RefGen_v4, which can be accessed at MaizeGDB (http://download.maizegdb.org/maize-GAMER) and via CyVerse (doi.org/10.7946/P2M925).

2.4 | Evaluation of GAMER-derived annotation sets

Component and aggregate annotation sets were compared at two levels; a general comparison and a GO category-specific comparison.

Comparison metrics mentioned in Section 2.2 were calculated for the general comparison (see Table 2). All metrics were calculated independently for each annotation set and compared among component annotation sets as well as the aggregate annotation set. Coverage and the number of annotations were calculated directly for each annotation set. Specificity was calculated for each annotation, and the mean across all annotations is reported.

The annotations from component annotation sets and the aggregate annotation set were divided into specific GO categories, and category-specific annotations were evaluated separately. Three different metrics were used for GO category-specific evaluations: coverage, number of annotations, and hF_1 (see Section 2.2 for more details). Coverage and the number of annotations were calculated individually for each GO category for each dataset. hF_1 score was calculated for each annotation set for each GO category.

2.5 | Comparisons among the maize-GAMER Aggregate, gramene, and phytozome annotation sets

The existing Gramene, Phytozome, and maize-GAMER annotations were compared to each other. Redundancy and duplication were removed from the Gramene and Phytozome annotation sets before evaluations were performed. Evaluation and comparisons were identical to the analyses performed in the previous Section 2.4. General evaluations for the maize-GAMER, Gramene, and Phytozome annotation sets were based on coverage, number of annotations, and specificity. These metrics were calculated as described in the previous Section 2.4. The Gramene, Phytozome, and maize-GAMER annotation sets were also compared in a GO category-specific manner to account for biases in performance among different categories (i.e., CC, BP, and MF). Comparisons were made based on coverage, number of annotations, and mean hF_1 score.

2.6 | Case study of the gene *nana plant1 (na1)*

The gene *na1* (GRMZM2G449033) had the most terms (7 GO terms) associated with it in the gold standard dataset and all the terms were from BP GO category. Annotations for *na1* from the three maize annotation sets were obtained. The ancestral nodes were inferred from the leaf nodes for each annotation set, and a subgraph for the BP ontology was generated (see Figure 5). The nodes in the subgraph were compared to gold standard, and nodes shared between a given annotation set and the gold standard dataset were identified. Nodes exclusively found only in a given annotation set or the gold standard were also identified. Illustrations of the subgraphs without node labels were drawn to compare among the three different GO annotation sets.

3 | RESULTS

3.1 | Evaluation of maize-GAMER-derived component annotation sets

The maize-GAMER-derived component annotation sets (i.e., the TAIR, UniProt, IPRS, Argot2, FANN-GO, and PANNZER) and the maize-GAMER Aggregate annotation set were evaluated across GO categories as well as within each GO category using metrics described in Table 2.

3.1.1 | General evaluation of maize-GAMER component annotation sets

Initial evaluations and comparisons of datasets created by the maize-GAMER pipeline were assessed based on coverage, number of annotations, and specificity among all clean component annotation sets as well as the aggregate annotation set (see Tables 2 and 3). The TAIR and UniProt annotation sets had the lowest coverage and number of annotations among all maize-GAMER component annotation sets (Table 3). The Argot2, FANN-GO, and PANNZER annotation sets had the highest number of annotations compared to other annotation sets, as well as higher coverage compared to other annotation sets. Notably, FANN-GO had the highest coverage at 100% of genes, and Argot2 had annotations for more than 90% of the genes. The IPRS annotation set had a lower number of annotations compared to the CAFA mixed-method pipelines, but covered more genes than sequence similarity methods. Although sequence similarity methods and IPRS covered a lower number of genes, they had higher specificity compared to mixed-method pipelines in general. Of the three mixed-method pipelines, only PANNZER had comparable specificity to the methods, but had lower coverage than both Argot2 and FANN-GO. Both Argot2 and FANN-GO had lower average specificity, but had higher coverage than other methods. The maize-GAMER Aggregate annotation (made up of all component annotation sets) covered all maize genes with at least one annotation (as expected given that the FANN-GO component annotation set also covers 100% of genes). In addition, the aggregate annotation set contains more than double the number of annotations that occur in any component annotation set. This indicates that different component methods assign different GO terms to genes. Therefore, combining annotations from different methods results in increased diversity of GO term assignments. Moreover, the aggregate annotation set has higher specificity than the mixed-method pipelines which have higher coverage, but has lower specificity than all other component annotation sets.

Genes that are annotated with at least one GO term from each component annotation set were compared among the three different method types (i.e., sequence similarity, domain-based, and mixed methods; see Figure 3a). This comparison revealed that less than a quarter of genes had been annotated by all three methods, but more than half were annotated by two different methods. The remainder were only annotated by mixed-method pipelines. Sequence similarity

TABLE 3 Results from the general evaluation of maize-GAMER-derived annotation sets

Method type	Sequence similarity		Domain	CAFA mixed methods			Union
Annotation set	TAIR-RBH	Plant-RBH	IPRS	Argot2	FANN-GO	PANNZER	Aggregate
Raw annotations	61,528	106,053	200,324	450,013	618,312	1,091,123	
Duplication (%)	9.11	65.89	72.18	0.33	1.07	0.97	
Redundancy (%)	14.18	9.17	10.91	48.25	59.84	73.09	
Clean annotations	35,791	32,085	46,599	224,827	187,850	219,984	515,059
Coverage (%)	23.70	20.10	48.48	92.64	100.00	47.86	100.00
Specificity	11.54	12.20	10.45	8.54	5.30	11.58	9.56

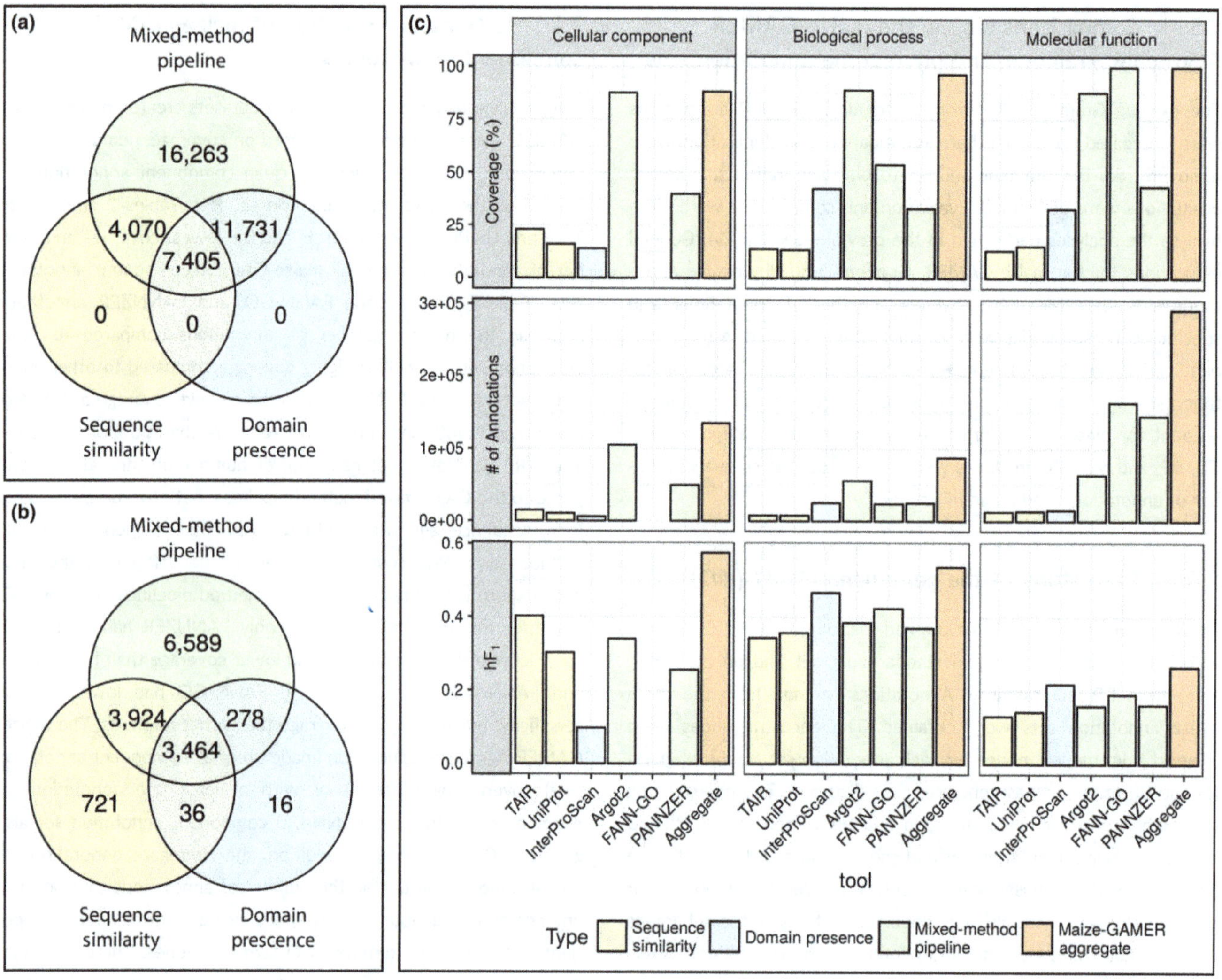

FIGURE 3 GO assignment metrics for each method type. Sequence similarity in yellow, domain presence in blue, and mixed-method pipeline in green. (a) Number of genes with at least one GO term annotated. (b) Number of GO terms with at least one gene annotated. (c) Percent coverage, number of annotations, and average *hF*1 score for each annotation set across the three GO graphs (i.e., Cellular Component, Molecular Function, and Biological Process). Color codes as used in (a) and (b), with the aggregate dataset shown in orange

and domain-based methods resulted in annotations to genes that were also annotated by mixed-method pipelines. The number of genes annotated by domain-based methods and mixed-method pipelines and not annotated by sequence similarity methods is higher than genes annotated by all three methods. In contrast, sequence similarity methods shared more genes with both other annotation sets than only with mixed-method pipelines. Moreover, only mixed-method pipelines annotate at least one GO term to all genes in the maize FGS.

Although the mixed-method pipelines annotated all genes, they did not capture all GO terms annotated to genes by the other methods. GO term assignments were compared to evaluate the diversity of the GO terms present in the three types of GO annotation methods (see Figure 3b). GO terms annotated directly and ancestral terms inferred from the direct terms annotated to genes were compared among the three GO annotation methods used in maize-GAMER. The number of GO terms annotated by sequence similarity, domain-based, and mixed-method pipelines was 3,794, 8,145, and 14,225, respectively. The number of GO terms annotated by the mixed-method pipelines is significantly higher than both other methods; however, there are a small number of GO terms that are only annotated by sequence similarity (721) and domain presence (16) methods. Only a small proportion (23.05%) of the total (15,028) GO terms are annotated by all three methods.

3.1.2 | GO category-specific evaluations of maize-GAMER component annotation sets

CAFA1 indicated that annotations for some GO categories are easier to predict than others (Radivojac et al., 2013). This indicated that the GO category-specific evaluations could provide a more accurate comparison between component methods. This would also allow unbiased comparison of tools which do not predict certain categories (e.g., FANN-GO does not predict the CC category). Therefore, maize-GAMER-derived annotation sets were divided into specific GO categories (i.e., CC, BP, and MF) and each category was evaluated separately based on coverage, number of annotations, and hF_1.

Mixed-method pipelines had higher coverage across all three GO categories. Argot2 covered more than 80% genes across all three categories (see Figure 3c). FANN-GO does not annotate GO terms for CC category, but had 100% coverage in BP category and covered about 50% genes in MF category. PANNZER had the lowest coverage compared to the other mixed-method pipelines, covered only 30%–50% of genes across different categories, and had highest coverage in BP. Sequence similarity methods consistently had lowest coverage compared to other methods in BP and MF, but IPRS had the lowest coverage in CC. IPRS covered higher number of genes than sequence similarity methods in BP and MF, but had lower coverage than mixed-method pipelines. When comparing IPRS coverage across three GO categories, the coverage was highest in MF. Aggregate annotation set covered slightly more genes than the component annotation sets with highest coverage in each category and covered more than 88% of the FGS genes in all categories. In the BP category, the aggregate annotation set annotated all genes from maize FGS with at least one annotation.

Mixed-method pipelines produce a higher number of annotations than other methods in all three GO categories. Moreover, the number of annotations from mixed-method pipelines loosely correlates with coverage in different GO categories. The only exception was PANNZER, which annotated more GO terms per gene in BP category (data not shown), than any other component annotation set. The number of annotations from sequence similarity methods and IPRS was consistently lower than mixed-method pipelines. The variation in the number of annotations was proportional to the number of genes annotated in sequence similarity and IPRS methods. The lowest number of annotations was seen in the CC category from IPRS, and sequence similarity methods in other GO categories. As the union of all component method annotations, the aggregate annotation set had a higher number of annotations in all three GO categories. The highest number of annotations for the aggregate annotation set was from the BP GO category, followed by CC and MF.

We used hF_1 scores as a representation of the quality of annotations and annotation sets. As described in Section 2, the hF_1 was calculated individually for all genes in the gold standard dataset and then averaged across all genes within an annotation set. There are clear differences in hF_1 across different GO categories. The highest performance was seen in the MF category, and the lowest performance is seen in the BP category. This fits the observation from CAFA1 (Radivojac et al., 2013). Mixed-method pipelines outperformed other methods in all three GO categories. PANNZER produced the highest hF_1 within the MF category, but Argot2 had the highest hF_1 scores in CC and BP. IPRS outperformed sequence similarity methods in both MF and BP categories, but was the lowest performing method in the CC category. Comparison between two sequence similarity methods indicated that maize-UniProt method performs better than the maize-TAIR method in MF and BP categories. On the other hand, maize-TAIR method performs better than maize-UniProt method in the CC category. Aggregating the component annotations from maize-GAMER increased the performance in the CC category. In contrast, aggregating the component annotation sets did not increase the performance compared to the top-performing tool in other categories.

3.2 | Evaluation of existing maize GO annotation sets and comparison with the maize-GAMER Aggregate annotation set

Two existing maize GO annotation sets, Gramene and Phytozome, were downloaded, evaluated, cleaned (i.e., redundancies and duplicates were removed), and compared with maize-GAMER Aggregate annotations (referred to as the "maize-GAMER annotation set"). The same metrics used for the evaluation of maize-GAMER-derived annotation sets were used for the comparison among the existing maize GO annotation sets and maize-GAMER Aggregate annotation set.

3.2.1 | General evaluation of public maize GO annotation sets

The maize-GAMER Aggregate annotations covered all genes in the maize FGS with at least one GO term, but Gramene and Phytozome covered only about half the genes (see Table 4). Phytozome covered the fewest genes (less than half of the genes), and Gramene covered slightly more than half of the genes (see Table 4). The maize-GAMER annotation set had more annotations than both Gramene and Phytozome. Gramene had twofold more annotations than Phytozome, and the maize-GAMER had severalfold more annotations than Gramene. While the maize-GAMER has higher coverage and a higher number of annotations, it has lower average specificity than Gramene and Phytozome. Gramene has the highest average specificity of all three annotation sets.

Genes with annotations from each set were compared to see the distribution of annotated genes among different annotations (see Figure 4a). Genes from Gramene and Phytozome annotations were a subset of the maize-GAMER annotations. Less than half of the genes were annotated in all three sets, and slightly more than half of the genes were annotated in at least two sets. Comparison of Gramene and Phytozome annotations shows that most of the genes that were annotated were shared. Both Gramene and Phytozome had genes

TABLE 4 Overall results from maize-GAMER and other existing maize datasets

	Existing		maize-GAMER
Analysis metric	Gramene	Phytozome	Aggregate
Raw annotations	111,203	66,709	
Duplication (%)	0.00	37.90	
Redundancy (%)	23.25	7.24	
Clean annotations	81,315	36,987	515,059
Coverage (%)	55.55	40.87	100.00
Specificity	10.90	10.41	9.56

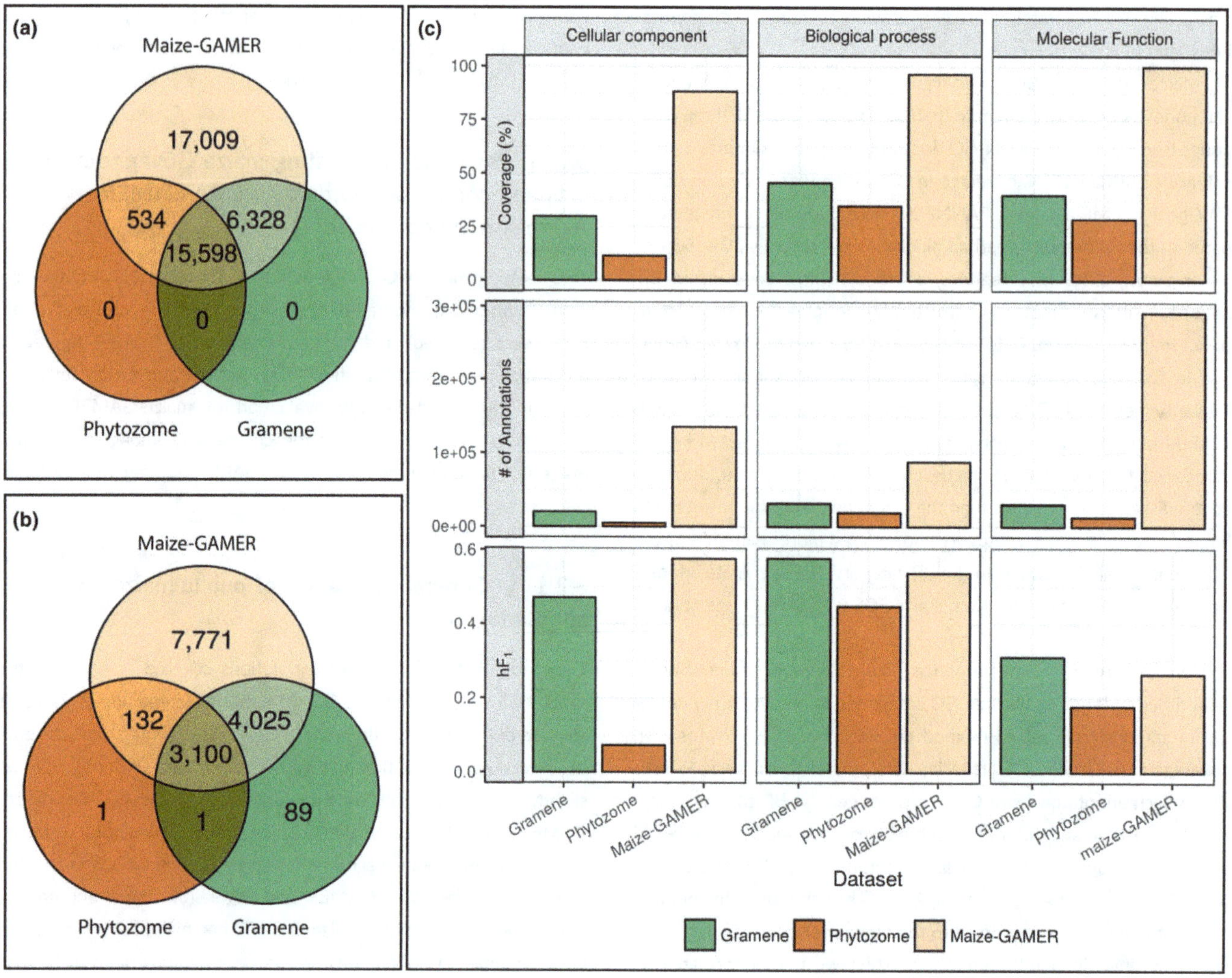

FIGURE 4 GO assignment metrics for Gramene, Phytozome, and maize-GAMER. Gramene in green, Phytozome in rust, and maize-GAMER in tan. (a) Number of genes with at least one GO term annotated. (b) Number of GO terms with at least one gene annotated. (c) Percent coverage, number of annotations, and *hF*1 score for each annotation dataset across the three GO graphs (i.e., Cellular Component, Molecular Function, and Biological Process)

that were annotated in only one of the two (i.e., Gramene or Phytozome but not both; See Figure 4a).

GO terms annotated directly to genes by different methods and ancestral GO terms propagated from these annotations were compared among the three annotation sets. The number of GO terms annotated in each set varied greatly. The least diverse set in terms of number of GO terms annotated was Phytozome, which was annotated with only 3,234 GO terms (approximately 7% of total GO terms). Gramene has annotated 7,215 GO terms (approximately 16%) and was more diverse than Phytozome, but had lower diversity than maize-GAMER. maize-GAMER had the highest diversity and contained 15,028 GO terms (approximately 33%). A small number of GO terms were used by all three annotation sets, and the majority of terms from Phytozome were shared across all three annotation sets. Only a single GO term was exclusive to the Phytozome annotations, and small number of terms was found to be exclusive to Gramene annotations. Approximately 50% of the GO terms from maize-GAMER were unique. The maize-GAMER Aggregate annotations shared a higher number of GO terms with Gramene than Phytozome.

3.2.2 | GO category-specific evaluations of maize-GAMER and existing maize GO annotation sets

Annotations from the three maize GO annotation sets were analyzed in a GO category-specific manner to identify differences in performance among the different categories (see Figure 4c). As was true for the component annotation sets, three different metrics were used for evaluation and comparison: coverage, number of annotations, and hF_1 score.

Comparison of coverage across GO categories indicated that all annotation sets had lower coverage in CC category, compared to other categories. Both Gramene and Phytozome had lower coverage in BP than MF, but maize-GAMER had higher coverage in BP than MF. Lowest coverage for all annotation sets and categories was seen in the CC category for the Phytozome annotation set, and the highest coverage was seen in the maize-GAMER Aggregate annotation set in the BP category. Comparison among the three maize annotation sets indicates that the maize-GAMER annotation set had the highest coverage in all three categories by a large margin. Coverage from maize-GAMER was almost twofold that of Gramene, which had the next highest coverage in all GO categories. Gramene had higher coverage than Phytozome in all three categories.

When the number of annotations was compared across different GO categories, the lowest number of annotations for Gramene and Phytozome annotation sets was seen in the CC category. In contrast, maize-GAMER had the lowest number of annotations in the MF category. Moreover, both Gramene and Phytozome both had a higher number of annotations in the MF category, whereas maize-GAMER had the highest number of annotations in the BP category. Comparison among the annotations sets illustrated that the maize-GAMER annotation set has the highest number of annotations in all three categories. Phytozome had the fewest annotations in all three GO categories. Number of annotations loosely correlated with coverage in different GO categories for both Gramene and Phytozome. Furthermore, maize-GAMER had the highest number of annotations in the BP category. The number of annotations from the maize-GAMER annotations for the BP was severalfold higher than other annotation sets (approximately 9× that of Gramene and approximately 28× that of Phytozome).

The hF_1 score reflects the overall quality of annotations produced by different pipelines used by the three maize annotation projects. Comparing performance of different pipelines across the three GO categories revealed a similar trend that was seen in the previous section. All pipelines had higher hF_1 scores in the MF category and had lower hF_1 scores in the BP category. The only pipeline that did not fit this trend was Phytozome, which had lowest performance in the CC category. maize-GAMER had a higher hF_1 score than other pipelines in the CC category. maize-GAMER also had higher performance than Phytozome in other categories, but performed slightly lower than Gramene in those categories. Gramene performed better than other pipelines in the MF and BP categories. Phytozome consistently had lower performance than other pipelines across all three GO categories. Phytozome's performance was especially low in the CC category, which was the lowest hF_1 score seen for any annotation set in any GO category.

To visualize how GO annotation methods perform, comparatively, the distribution of metrics *hPr*, *hRc*, and hF_1 can be calculated across all genes included in the gold standard dataset. In Figure S1, lower coverage by sequence similarity and domain presence methods is illustrated by the high number of genes with a value of zero. In general, mixed-method pipelines are shown to perform better than other methods we used. They cover more genes, and two of the three mixed-method pipelines have higher *hPr* and *hRc* than all other methods. The FANN-GO distribution is different from both other mixed methods: In general, it has lower performance than other methods. This can be attributed to that fact FANN-GO annotations have lower specificity than other methods, and lower specificity results in lower values for *hPr* and *hRc*. In addition, the maize-GAMER has fewer genes with a value of 0 than both Phytozome and Gramene.

3.3 | Example annotations from *nana plant1* (*na1*)

The gene *nana plant1* (*na1*; GRMZM2G449033) has more annotations than any other gene in the gold standard dataset. A classical maize mutant with a dwarf phenotype (Hartwig et al., 2011), the *na1* recessive mutant results from a loss-of-function mutation in the gene that affects the brassinosteroid (BR) biosynthetic pathway where BR is a plant hormone that is required for normal plant growth (Hartwig et al., 2011). In the gold standard dataset, *na1* had seven biological process GO terms annotated. Annotations for *na1* from different maize annotation sets were compared to the gold standard, and a subgraph for each annotation set and gold standard dataset was plotted (see Figure 5). Phytozome did not annotate any GO terms to *na1* (see Figure 5a), but both Gramene (see Figure 5b) and maize-GAMER (see Figure 5c) have annotated BP GO terms for *na1*. Gramene annotates three GO terms *na1*, while maize-GAMER has annotated 13 GO terms to *na1*. Two GO terms from the gold standard are known to be related to *na1* dwarf phenotype from previous studies, "brassinosteroid biosynthetic process" (GO:0016132) and "unidimensional cell growth" (GO:0009826). While both of these were annotated correctly by maize-GAMER (see Figure 5c), only one of them was correctly annotated by Gramene (see Figure 5b). Comparison of overlapping nodes indicates that the maize-GAMER Aggregate annotation set also contains a number of less specific nonleaf terms which overlap with nodes inferred from gold standard dataset. Overall, the maize-GAMER has larger proportion of overlapping nodes with the gold standard than the Gramene for the BP GO category.

The different approaches taken by the pipelines from Gramene and maize-GAMER result in different annotations for the example case study of the maize *na1* gene. Gramene has a lower number of GO terms annotated to *na1* than maize-GAMER. The average specificity of GO terms annotated in the BP category for *na1* (see Figure 5) is not significantly different between GAMER (mean = 12.154) and Gramene (mean = 12.667) pipelines (two-sided two-group Wilcoxon rank-sum test; $p = .89$). This example from *na1* indicates that the specificity of the annotations is not significantly different for specific instances, but is different when compared overall. We further compared the GAMER component and aggregate dataset creation methodologies annotate gene function to that of Gramene and Phytozome by comparing the hPr, hRc, and hF_1 metrics for the *na1* gene (see Figure S2). Phytozome's method stands out because it has no annotations for *na1*; thus, all the metrics have a value of 0. The GAMER aggregate dataset has higher hF_1 score compared to Gramene and has higher *hPr* and *hRc* as well. Among the component datasets, PANNZER has the highest hF_1 score and highest *hRc*. InterProScan has the highest *hPr*, but its lower *hRc* reduces the hF_1. FANN-GO and Argot2 have lower metrics due to lower specificity of annotations compared to other methods.

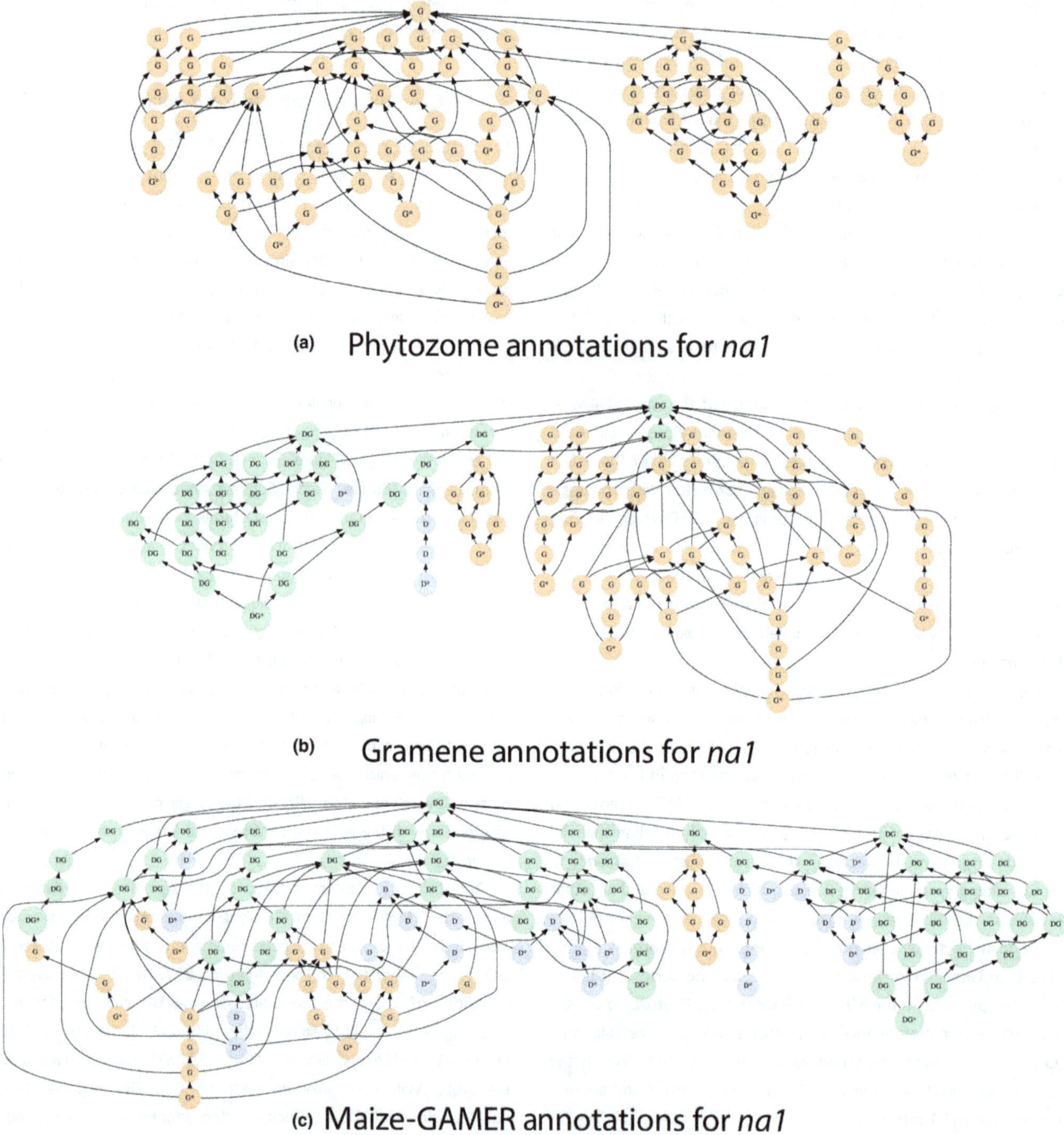

FIGURE 5 Biological Process GO graph for maize *na1*. Leaf terms are toward the bottom, and root terms are toward the top. Terms covered only by the gold standard are shown in orange (labeled G), those in the dataset but absent from the gold standard are shown in blue (labeled D), and those that appear in both are shown in green (labeled DG). Leaf terms in each subgraph have an * next to them. Phytozome graph is shown at the top (a), Gramene graph is shown in the middle (b), and maize-GAMER Aggregate graph is shown at the bottom (c)

4 | DISCUSSION

In keeping with our goal, through the maize-GAMER project, we were able to improve the GO annotation dataset for maize and to document inputs, methods, and results at a level that enables both reproducibility and reuse of the pipeline for future genome versions.

To determine how best to create an improved maize annotation dataset, we tried out multiple different methods and compared the resulting datasets based on a gold standard set of gene functions. This also enabled us to better understand differences in term assignments among the methods we used. Using the same gold standard, we also were able to compare resulting datasets to those produced by and available from Gramene and Phytozome.

We used the *hFmax* metric to select high-confidence annotations from mixed-method pipelines and to evaluate annotation sets resulting from all methods under evaluation. We found that mixed-method pipelines developed for the CAFA1 challenge outperformed RBH and domain presence methods for GO annotation (Radivojac et al., 2013). They covered more genes with annotations, produced higher number of annotations, and had higher hF_1 score than both sequence similarity and domain-based methods. The higher performance from mixed-method pipelines is the outcome of advanced statistical (Falda et al., 2012; Koskinen et al., 2015) and machine learning approaches (Clark & Radivojac, 2011) used to reduce the false-positive and false-negative annotations. Mixed-method pipelines do have a limitation: They have higher coverage, but annotations are less specific in general when compared with datasets produced using other approaches. This could be due to the dearth of training dataset for the more specific GO terms, which is required for training machine learning methods.

When we aggregated the predictions from RBH, domain-based methods, and three tools from CAFA1, we produced the maize-GAMER Aggregate dataset, which covers more gene space than the datasets produced by Gramene and Phytozome, and with similar or better accuracy. While the higher coverage could cause concern, evaluating the annotations using the gold standard has shown that the performance is similar or better than existing datasets. This also indicates that maize-GAMER annotations are no less reliable than Gramene and are in fact are better than those for Phytozome. Removing less specific GO terms annotated by some methods in cases where there were more specific GO terms annotated to the same genes was important for aggregating different datasets. In certain cases, more than one annotation with lower specificity was replaced by a single annotation with higher specificity. As with any computational approach to annotate GO terms, the current maize-GAMER dataset should be considered as an initial step in improving the GO annotations in maize. As future iterations of the CAFA competition evaluate new tools and methods for GO annotations, we anticipate that the quality of computational maize GO annotations could be iteratively improved in a reproducible manner by continuing to apply the newest, best performing methods.

To enable better reproducibility, we have generated a supplementary document with exact parameters and commands used to generate the maize dataset. We are currently in the process of formalizing the code used to generate the maize GO annotation set into a reusable pipeline called GO-MAP. Once completed, the GO-MAP pipeline can be used for GO annotation of newly sequenced plant genomes as well as existing plant genomes. The pipeline will be made freely available and will utilize the same methods and datasets used for maize.

The set of manually reviewed gene function annotations for maize that we call the gold standard is both incomplete and sparse. This situation does not reflect the amount of published literature describing gene function for maize. Instead, this situation is due to limited curation of gene function into GO terms. While tools exist at MaizeGDB that enable researchers to assign GO terms to genes directly, these tools remain poorly utilized. In an effort to improve community engagement and to upgrade the Evidence Codes for GO assignments, our next step for maize-GAMER will be to develop and deploy a tool to enable experts in the maize community to review existing GO annotations. By enabling GO annotation review through expert crowdsourcing, term assignments produced by computational pipelines including GAMER can be upgraded from IEA (Inferred from Electronic Annotation) to RCA (reviewed computational analysis). In this way, we will enable the transfer of collective knowledge members of the maize community have generated over the years to produce higher-quality functional annotation datasets for maize with clear extension of this practice for other species.

ACKNOWLEDGMENTS

We thank Darwin Campbell and Ramona Walls for efforts to create the DOIs for datasets available via CyVerse and John Portwood and Ethalinda Cannon for their efforts to populate resulting datasets for MaizeGDB. Thanks to members of the Dill Plant Informatics and Computation Lab (dill-picl.org) for critical review and helpful suggestions, with particular thanks to Darwin Campbell for help with figures. This work was supported by funding from the Iowa State University Plant Sciences Institute Faculty Scholars Program to C.J.L.D.; the National Science Foundation [IOS #1027527] to C.J.L.D.; and the United States Department of Agriculture—Agricultural Research Services to C.M.A. IF was supported, in part, by National Science Foundation award ABI-1458359.

CONFLICT OF INTERESTS

None declared.

AUTHORS' CONTRIBUTIONS

K.W. carried out all of the research and development efforts. K.W., C.M.A., I.F., and C.J.L.D. designed the experiments and wrote the manuscript.

REFERENCES

Altschul, S. F., Gish, W., Miller, W., Myers, E. W., & Lipman, D. J. (1990). Basic local alignment search tool. *Journal of Molecular Biology*, *215*, 403–410. https://doi.org/10.1016/S0022-2836(05)80360-2

Andorf, C. M., Cannon, E. K., Portwood, J. L., Gardiner, J. M., Harper, L. C., Schaeffer, M. L., … Lawrence-Dill, C. J. (2016). MaizeGDB update: New tools, data and interface for the maize model organism database. *Nucleic Acids Research*, *44*, D1195–D1201. https://doi.org/10.1093/nar/gkv1007

Andorf, C., Dobbs, D., & Honavar, V. (2007). Exploring inconsistencies in genome-wide protein function annotations: A machine learning approach. *BMC Bioinformatics*, *8*, 284. https://doi.org/10.1186/1471-2105-8-284

Berardini, T. Z., Reiser, L., Li, D., Mezheritsky, Y., Muller, R., Strait, E., & Huala, E. (2015). The Arabidopsis information resource: Making and mining the "gold standard" annotated reference plant genome. *Genesis*, *53*, 474–485. https://doi.org/10.1002/dvg.22877

Binns, D., Dimmer, E., Huntley, R., Barrell, D., O'Donovan, C., & Apweiler, R. (2009). QuickGO: A web-based tool for Gene Ontology searching. *Bioinformatics*, *25*, 3045–3046. https://doi.org/10.1093/bioinformatics/btp536

Clark, W. T., & Radivojac, P. (2011). Analysis of protein function and its prediction from amino acid sequence. *Proteins, 79*, 2086–2096. https://doi.org/10.1002/prot.23029

Clark, W. T., & Radivojac, P. (2013). Information-theoretic evaluation of predicted ontological annotations. *Bioinformatics, 29*, i53–i61. https://doi.org/10.1093/bioinformatics/btt228

Conesa, A., & Götz, S. (2008). Blast2GO: A comprehensive suite for functional analysis in plant genomics. *International Journal of Plant Genomics, 2008*, 619832.

Defoin-Platel, M., Hindle, M. M., Lysenko, A., Powers, S. J., Habash, D. Z., Rawlings, C. J., & Saqi, M. (2011). AIGO: Towards a unified framework for the analysis and the inter-comparison of GO functional annotations. *BMC Bioinformatics, 12*, 431. https://doi.org/10.1186/1471-2105-12-431

Falda, M., Toppo, S., Pescarolo, A., Lavezzo, E., Di Camillo, B., Facchinetti, A., … Fontana, P. (2012). Argot2: a large scale function prediction tool relying on semantic similarity of weighted Gene Ontology terms. *BMC Bioinformatics, 13*(Suppl 4), S14. https://doi.org/10.1186/1471-2105-13-S4-S14

Finn, R. D., Attwood, T. K., Babbitt, P. C., Bateman, A., Bork, P., Bridge, A. J., … Mitchell, A. L. (2017). InterPro in 2017-beyond protein family and domain annotations. *Nucleic Acids Research, 45*, D190–D199. https://doi.org/10.1093/nar/gkw1107

Finn, R. D., Clements, J., & Eddy, S. R. (2011). HMMER web server: Interactive sequence similarity searching. *Nucleic Acids Research, 39*, W29–W37. https://doi.org/10.1093/nar/gkr367

Goodstein, D. M., Shu, S., Howson, R., Neupane, R., Hayes, R. D., Fazo, J., … Rokhsar, D. S. (2012). Phytozome: A comparative platform for green plant genomics. *Nucleic Acids Research, 40*, D1178–D1186. https://doi.org/10.1093/nar/gkr944

Harris, M. A., Clark, J., Ireland, A., Lomax, J., Ashburner, M., Foulger, R., … Gene Ontology Consortium (2004). The Gene Ontology (GO) database and informatics resource. *Nucleic Acids Research, 32*, D258–D261.

Hartwig, T., Chuck, G. S., Fujioka, S., Klempien, A., Weizbauer, R., Potluri, D. P. V., … Schulz, B. (2011). Brassinosteroid control of sex determination in maize. *Proceedings of the National Academy of Sciences of the United States of America, 108*, 19814–19819. https://doi.org/10.1073/pnas.1108359108

Herrero, J., Muffato, M., Beal, K., Fitzgerald, S., Gordon, L., Pignatelli, M., … Flicek, P. (2016). Ensembl comparative genomics resources. Database (Oxford) *2016*, bav096.

Hirsch, C. N., Hirsch, C. D., Brohammer, A. B., Bowman, M. J., Soifer, I., Barad, O., … Mikel, M. A. (2016). Draft Assembly of Elite Inbred Line PH207 provides insights into genomic and transcriptome diversity in maize. *Plant Cell, 28*, 2700–2714. https://doi.org/10.1105/tpc.16.00353

Hunter, S., Apweiler, R., Attwood, T. K., Bairoch, A., Bateman, A., Binns, D., … Yeats, C. (2009). InterPro: The integrative protein signature database. *Nucleic Acids Research, 37*, D211–D215. https://doi.org/10.1093/nar/gkn785

Jiang, Y., Oron, T. R., Clark, W. T., Bankapur, A. R., D'Andrea, D., Lepore, R., … Radivojac, P. (2016). An expanded evaluation of protein function prediction methods shows an improvement in accuracy. *Genome Biology, 17*, 184. https://doi.org/10.1186/s13059-016-1037-6

Jiao, Y., Peluso, P., Shi, J., Liang, T., Stitzer, M. C., Wang, B., … Ware, D. (2017). Improved maize reference genome with single-molecule technologies. *Nature, 546*, 524–527.

Jones, P., Binns, D., Chang, H.-Y., Fraser, M., Li, W., McAnulla, C., … Hunter, S. (2014). InterProScan 5: Genome-scale protein function classification. *Bioinformatics, 30*, 1236–1240. https://doi.org/10.1093/bioinformatics/btu031

Koskinen, P., Törönen, P., Nokso-Koivisto, J., & Holm, L. (2015). PANNZER: High-throughput functional annotation of uncharacterized proteins in an error-prone environment. *Bioinformatics, 31*, 1544–1552. https://doi.org/10.1093/bioinformatics/btu851

Lawrence, C. J., Dong, Q., Polacco, M. L., Seigfried, T. E., & Brendel, V. (2004). MaizeGDB, the community database for maize genetics and genomics. *Nucleic Acids Research, 32*, D393–D397. https://doi.org/10.1093/nar/gkh011

Lawrence, C. J., Harper, L. C., Schaeffer, M. L., Sen, T. Z., Seigfried, T. E., & Campbell, D. A. (2008). MaizeGDB: The maize model organism database for basic, translational, and applied research. *International Journal of Plant Genomics, 2008*, 496957.

Monaco, M. K., Sen, T. Z., Dharmawardhana, P. D., Ren, L., Schaeffer, M., Naithani, S., … Jaiswal, P. (2013). Maize metabolic network construction and transcriptome analysis. *The Plant Genome, 6*.

Moreno-Hagelsieb, G., & Latimer, K. (2008). Choosing BLAST options for better detection of orthologs as reciprocal best hits. *Bioinformatics, 24*, 319–324. https://doi.org/10.1093/bioinformatics/btm585

Radivojac, P., Clark, W. T., Oron, T. R., Schnoes, A. M., Wittkop, T., Sokolov, A., … Friedberg, I. (2013). A large-scale evaluation of computational protein function prediction. *Nature Methods, 10*, 221–227. https://doi.org/10.1038/nmeth.2340

Schnable, P. S., Ware, D., Fulton, R. S., Stein, J. C., Wei, F., Pasternak, S., … Wilson, R. K. (2009). The B73 maize genome: Complexity, diversity, and dynamics. *Science, 326*, 1112–1115. https://doi.org/10.1126/science.1178534

Sen, T. Z., Andorf, C. M., Schaeffer, M. L., Harper, L. C., Sparks, M. E., Duvick, J., … Lawrence, C. J. (2009). MaizeGDB becomes "sequence-centric". Database (Oxford) *2009*, bap020.

Tello-Ruiz, M. K., Stein, J., Wei, S., Preece, J., Olson, A., Naithani, S., … Ware, D. (2016). Gramene 2016: Comparative plant genomics and pathway resources. *Nucleic Acids Research, 44*, D1133–D1140. https://doi.org/10.1093/nar/gkv1179

UniProt Consortium (2015). UniProt: A hub for protein information. *Nucleic Acids Research, 43*, D204–D212.

10

Nonpolar residues in the presumptive pore-lining helix of mechanosensitive channel MSL10 influence channel behavior and establish a nonconducting function

Grigory Maksaev | Jennette M. Shoots | Simran Ohri | Elizabeth S. Haswell

Department of Biology and Center for Engineering MechanoBiology, Washington University in Saint Louis, Saint Louis, Missouri

Correspondence
Elizabeth S. Haswell, Department of Biology and Center for Engineering MechanoBiology, Washington University in Saint Louis, One Brookings Drive, Saint Louis, MO 63130.
Email: ehaswell@wustl.edu

Present address
Grigory Maksaev, Department of Cell Biology and Physiology and Center for the Investigation of Membrane Excitability Diseases, Washington University School of Medicine, Saint Louis, MO.

Funding information
National Institutes of Health Grant, Grant/Award Number: R01GM084211; National Science Foundation Grant, Grant/Award Number: MCB1253103; National Science Foundation Graduate Research Fellowship, Grant/Award Number: DGE-1745038

Abstract

Mechanosensitive (MS) ion channels provide a universal mechanism for sensing and responding to increased membrane tension. MscS-like (MSL) 10 is a relatively well-studied MS ion channel from *Arabidopsis thaliana* that is implicated in cell death signaling. The relationship between the amino acid sequence of MSL10 and its conductance, gating tension, and opening and closing kinetics remains unstudied. Here, we identify several nonpolar residues in the presumptive pore-lining transmembrane helix of MSL10 (TM6) that contribute to these basic channel properties. F553 and I554 are essential for wild type channel conductance and the stability of the open state. G556, a glycine residue located at a predicted kink in TM6, is essential for channel conductance. The increased tension sensitivity of MSL10 compared to close homolog MSL8 may be attributed to F563, but other channel characteristics appear to be dictated by more global differences in structure. Finally, MSL10 F553V and MSL10 G556V provided the necessary tools to establish that MSL10's ability to trigger cell death is independent of its ion channel function.

KEYWORDS
Arabidopsis thaliana, electrophysiology, mechanosensitive ion channel, MscS-Like10

1 | INTRODUCTION

The ability to respond to mechanical stimuli is an ancient and intrinsic property of cells (Anishkin, Loukin, Teng, & Kung, 2014; Booth, Miller, Müller, & Lehtovirta-Morley, 2015). One of the most universal mechanisms for mechanotransduction is the use of mechanosensitive (MS) ion channels. MS channels are oligomeric protein structures embedded in the lipid bilayer, and their primary function is to form a conductive pore in response to increased lateral membrane tension or force transduced from cytoskeletal filaments (Bavi et al., 2017). MS channels mediate the perception of external mechanical stimuli (touch, gravity, vibration) and internal mechanical stresses in plants, animals, and bacteria (Basu & Haswell, 2017; Martinac et al., 2014; Ranade, Syeda, & Patapoutian, 2015). Mammalian MS channel dysfunctions are associated with numerous pathologies (Gu & Gu, 2014) and both mammalian and bacterial MS channels are under investigation as drug targets (Boulos, 2013; Gottlieb, Suchyna, Ostrow, & Sachs, 2004; Iscla et al., 2015). It is therefore important for both basic and applied reasons to determine the molecular mechanisms of MS ion channel function, including the relationship between channel structure and its conductance, gating tension, and opening and closing kinetics.

We already have considerable insight into these questions, in part due to decades of research into the structure and function of a MS ion channel from *Escherichia coli*, the Mechanosensitive ion channel of Small conductance (*Ec*MscS). *Ec*MscS is directly opened by membrane tension (Sukharev, 2002), has a unitary conductance of 1.2 nS in giant *E. coli* spheroplasts (Levina et al., 1999; Martinac, Buechner, Delcour, Adler, & Kung, 1987), and demonstrates a slight preference for anions (1.2-fold to threefold, summarized in [Maksaev & Haswell, 2013; Cox, Wann, & Martinac, 2014)]. The primary physiological function of *Ec*MscS is to promote bacterial survival when subjected to hypoosmotic shock (Boer, Anishkin, & Sukharev, 2011; Levina et al., 1999). *Ec*MscS also shows inactivation behavior whereby sustained tension leads to a nonconductive state of the channel that cannot be opened again until a period of recovery (Akitake, Anishkin, Liu, & Sukharev, 2007; Belyy, Kamaraju, Akitake, Anishkin, & Sukharev, 2010; Edwards, Bartlett, & Booth, 2008; Kamaraju, Belyy, Rowe, Anishkin, & Sukharev, 2011; Vásquez, 2013).

A wealth of structural information on MscS is available, derived from several members of the MscS family from different bacterial species. Crystal structures thought to represent the conducting state of *Ec*MscS or the nonconducting states of MscS from *E. coli*, *Thermoanaerobacter tengcongensis*, and *Helicobacter pylori* (Bass, Strop, Barclay, & Rees, 2002; Lai, Poon, Kaiser, & Rees, 2013; Zhang et al., 2012) have been determined. A cryo-electron micrograph structure of a MscS homolog from *E. coli*, Ynal, has also been reported (Böttcher, Prazak, Rasmussen, Black, & Rasmussen, 2015). These structures reveal that MscS forms a homoheptamer with a transmembrane (TM) domain localized to the inner *E. coli* membrane and a cytoplasmic "vestibule." Each subunit contains an N-terminal domain comprised of three TM helices and a soluble C-terminal domain. The most C-terminal of the TM helices, TM3, lines the permeation pore. It comprises two regions, TM3a and TM3b, which are separated by a distinctive kink at residue G113. Other key residues include L105 and L109, which form the narrowest constriction of the closed or nonconducting pore, and G121, which is thought to be critical to formation of the closed state (Akitake et al., 2007; Bass et al., 2002).

A comparison of the open-state versus closed-state structures suggests that gating involves swinging a tension-sensitive paddle made up of the TM1/TM2 helices and twisting TM3a about G113. This motion allows L105 and L109 to move out of the pore. Mutational analyses support important roles for L105 and L109 (Miller et al., 2003; Rasmussen et al., 2010) and have shown that G104, A106, and G108 play critical roles in channel gating (Anishkin, Akitake, & Sukharev, 2008; Edwards et al., 2005). A single mutation in this region, A106V, locks the channel in an open conformation and was used to obtain the first open-state crystal structure of *Ec*MscS (Wang et al., 2008). Mutating other residues in TM3, such as S114, L118, A120, L123, F127 (Malcolm & Blount, 2015), Q112, and A120 (Boer et al., 2011) has less dramatic impact on channel properties, only modulating its gating and inactivation kinetics. However, changing the kink-forming residue G113 to alanine prevents inactivation and in combination with G121A severely alters channel opening and closing (Akitake et al., 2007). Residues F68 and L111 may form a force-transmitting clutch between TM2 and TM3, transmitting membrane tension sensed by the TM1/TM2 paddle to the pore-forming TM3 (Belyy, Anishkin, Kamaraju, Liu, & Sukharev, 2010). Surprisingly, the channel's weak ion selectivity is governed by its cytoplasmic cage, rather than the pore-lining TM3 (Cox et al., 2013; Gamini, Sotomayor, Chipot, & Schulten, 2011). In summary, a combination of modeling and functional assays now provide a general understanding of the conformational changes that ultimately result in channel gating for *Ec*MscS (Booth & Blount, 2012; Martinac et al., 2014; Naismith & Booth, 2012).

These insights into the structural basis of *Ec*MscS mechanosensitivity provide a strong foundation for studying homologs of *Ec*MscS, which are found in all kingdoms of life (Booth et al., 2015; Cox, Nakayama, Nomura, & Martinac, 2015; Haswell, 2007; Martinac & Kloda, 2003; Pivetti et al., 2003). The domain conserved among all these MscS family members is limited to the pore-lining helix and about 100 amino acids of the following soluble domain. The number of predicted transmembrane domains and the structure of N and C termini are highly variable. The channel behavior and physiological function of multiple *Ec*MscS homologs from other bacterial species, archaea, fission yeast, green algae, and land plants have been reported (Çetiner et al., 2017; Hamilton et al., 2015; Kloda & Martinac, 2001; Lee et al., 2016; Maksaev & Haswell, 2012; Malcolm et al., 2012; Nakayama, Fujiu, Sokabe, & Yoshimura, 2007; Nakayama, Yoshimura, & Iida, 2012, 2013). These channels are all mechanically gated and generally function in hypoosmotic stress relief, but have a range of conductances and ion channel selectivities and play different physiological and developmental roles.

The ten *Ec*MscS homologs encoded in the genome of the model land plant *Arabidopsis thaliana* have been named MscS-like or MSL channels (Haswell, 2007). MSL proteins exhibit diverse tissue expression patterns, subcellular localizations, and domain structures (Basu & Haswell, 2017). To date, MSL1, MSL8, and MSL10 are the best-characterized MSLs in terms of ion channel physiology. All three provide tension-gated ion channel activities in native plant cells (Haswell, Peyronnet, Barbier-Brygoo, Meyerowitz, & Frachisse, 2008) and/or when expressed in heterologous systems (Hamilton et al., 2015; Lee et al., 2016; Maksaev & Haswell, 2012). MSL1 is localized to the mitochondrial inner membrane (Lee et al., 2016), while MSL8 and MSL10 are primarily localized to the plasma membrane (Hamilton et al., 2015; Haswell et al., 2008). The unitary conductances of MSL8 and MSL10 expressed in *Xenopus* oocytes are approximately 60 pS and 105 pS, respectively (compare to 340 pS for *Ec*MscS expressed in oocytes) (Hamilton & Haswell, 2017; Hamilton et al., 2015; Maksaev & Haswell, 2011, 2012). MSL8 and MSL10 have a slightly higher preference for anions ($P_{Cl}:P_{Na}$ = 5.6–6.3) than *Ec*MscS. Another intrinsic feature of an MS ion channel is its tension sensitivity, defined as the amount of tension applied to the membrane required for channel opening. The tension at which a channel opens may or may not be the same as the tension at which it closes. Different opening and closing tensions lead to an asymmetric gating profile, and this phenomenon is referred to as hysteresis. Both MSL10 and MSL8 have lower tension sensitivity than *Ec*MscS, and both exhibit strong hysteresis, with much higher opening than closing pressures (Hamilton et al., 2015; Maksaev & Haswell, 2012). We interpret this to mean that, once opened, MSL8 and MSL10 are very stable and do not close until most of the tension is relieved (Maksaev & Haswell, 2012).

In terms of physiological function, MSL8 appears to serve in a role analogous to that of EcMscS, as it protects pollen from multiple hypoosmotic challenges associated with pollen development and function (Hamilton et al., 2015). MSL8 is primarily localized to the plasma membrane of pollen grains, where it is required for full survival of rehydration, germination, and tube growth. Two mutations in the presumptive pore-lining helix of MSL8, I711S and F720L, alter channel behavior and fail to complement these mutant phenotypes (Hamilton & Haswell, 2017). These observations link ion flux through the MSL8 channel to protection from osmotic stresses during pollen development.

Much less clear is the functional role of MSL10. MSL10 is required for the predominant MS ion channel activity in root cells (Haswell et al., 2008), but to date no other loss-of-function phenotype has been established. Fortunately, gain-of-function phenotypes have been revealing; overexpression of MSL10 leads to cell death, as does a single ethyl methanesulfonate-induced point mutation in the MSL10 C terminus, S640L (*rea1*) (Veley et al., 2014; Zou et al., 2016). Overexpression of the soluble N terminus of MSL10 is sufficient to induce cell death in tobacco epidermal cells (Veley et al., 2014). Thus, all existing data suggest that MSL10 is a multifunctional MS ion channel that is capable of mediating adaptation to hypoosmotic shock in the short term (reducing pressure by releasing osmolytes) and of signaling to change cellular state in the long-term (inducing cell death in response to biotic or abiotic stress). However, more structural information about MSL proteins and their pore-forming region is required to fully and directly test the possibility that MSL10 has a nonconducting function.

To gain additional information about the structure of the channel pore, the mechanism of gating, and how ion flux through the channel is related to its genetic functions, we used mutational analysis and single-channel patch-clamp electrophysiology to identify residues in the presumptive pore-lining domain of MSL10 that are important for channel behavior, including tension sensitivity, conductivity, and stability of the open state. We then tested two tension-insensitive mutants for the ability to induce cell death in a previously established transient expression assay. These data provide critical information about the structural component of tension-sensitive ion transport and a useful comparison to *Ec*MscS and other MS channels in animals and bacteria. In addition, these mutant MSL10 channels provide tools for studying the relationship between tension sensitivity, open-state stability, ion flux, and cell death signaling.

2 | METHODS

2.1 | Molecular biology

All construct were based on pOO2-MSL10 (Maksaev & Haswell, 2012). Site-directed mutagenesis was used to introduce point mutations into the *MSL10* sequence and confirmed by sequencing. Capped RNA was transcribed in vitro with SP6 polymerase using the mMessenger mMachine kit (Ambion, Thermo Fisher Scientific) and stored at −80°C at approximately 1000 ng/μl.

2.2 | Oocyte preparation

Xenopus laevis oocytes (Dumont stage V or VI) were purchased (Ecocyte Bioscience US LLC, Austin, Texas) and handled as described (Maksaev & Haswell, 2015). Oocytes were injected with 50 nl of 1000 ng/μl of RNA the day after isolation. Fluorescent imaging of the oocytes was carried out 48–72 h after injection. Oocytes were mounted on concave slides and covered with coverslips. Confocal imaging of the periphery of the oocytes was performed using Olympus Fluoview 1000 with BX61 microscope and the Olympus FV10-ASW software suite.

2.3 | Electrophysiology

The buffer used was 60 mM $MgCl_2$, 5 mM HEPES, adjusted to pH 7.38 with TEA-OH. All the traces presented were obtained from excised inside-out patches. Data were acquired using an Axopatch 200B amplifier and a Digidata 1440A digitizer (Molecular Devices) at 20 kHz and low-pass filtered at 5 kHz. Channels were activated by symmetric 5-s pressure ramps. Pressure was applied and monitored with a HSPC-1 high speed pressure clamp system (ALA Scientific Instruments), and traces analyzed with Clampfit 10.6 (Molecular Devices). Patch pipettes of approximately the same resistance (3.00 ± 0.25 MOhm) were used in all experiments. The gating threshold of a channel variant was defined as the pressure at which the second channel of the population in a patch opened. For each patch, several pressure ramps of −30 mmHg were run to accommodate for patch creep. Only after that, measurements at −20 mV membrane were performed. For each patch, the results of 7–12 consecutive pressure ramps were averaged. The number of channels per patch was estimated from the peak current at patch rupturing pressure. In cases when the number of open channels was 100 or more, the correction for series pipette resistance was introduced. For closing pressures, the average was taken only in cases when closing pressure was not zero. In case when at least one pull for a patch resulted in at least one open channel after the pressure was released, the closing pressure was considered to be zero.

2.4 | Software

The putative structure of *At*MSL10 TM6 (Figure 1b) was obtained from the I-TASSER prediction server (Zhang, 2008) using the *Ec*MscS closed-state crystal structure [2OAU:A, (Bass et al., 2002; Steinbacher, Bass, Strop, & Rees, 2007)] as a template. Sequence alignments and analyses were made using the Unipro UGENE bioinformatics toolkit (Okonechnikov, Golosova, & Fursov, 2012). Visualization of crystal structures and imaging of the putative MSL10 pore region were performed in the VMD suite (Humphrey, Dalke, & Schulten, 1996). Secondary structure of MSL10 was predicted using ARAMEMNON plant membrane protein database (Schwacke et al., 2003).

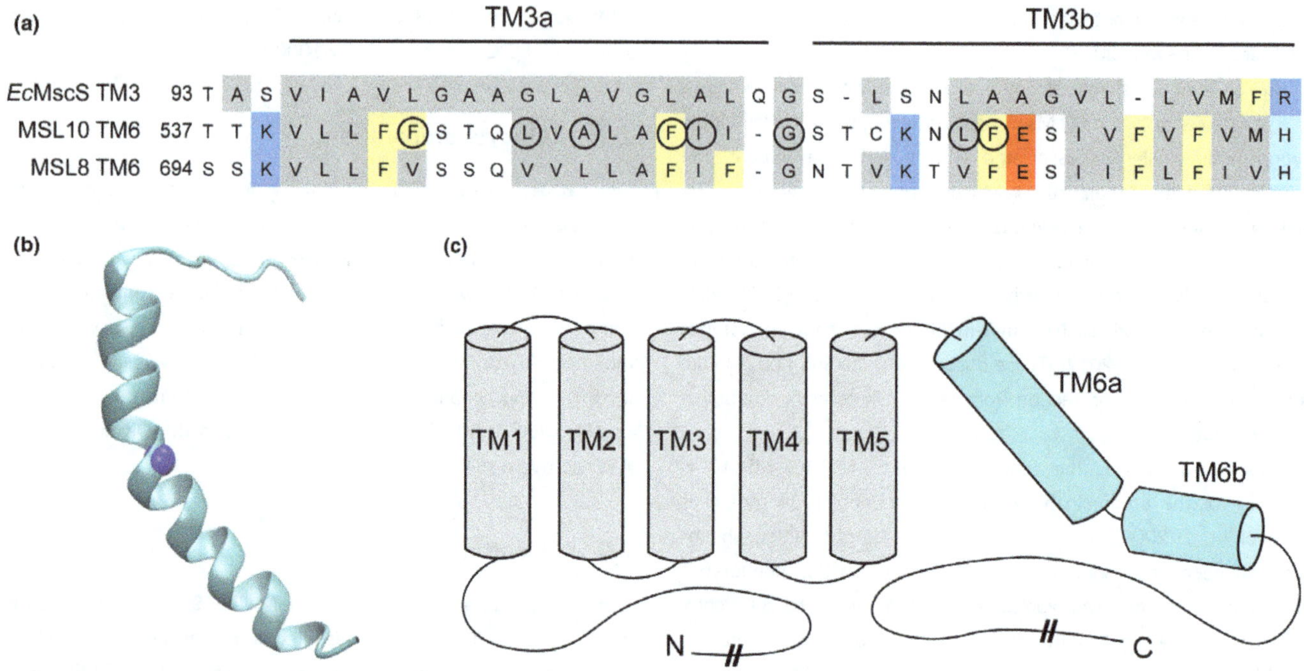

FIGURE 1 Identification of potential pore-disrupting residues in MSL10. (a) Alignment of the pore-lining domain of *Escherichia coli* MscS and corresponding regions of *Arabidopsis thaliana* MSL10 and MSL8. Acidic residues are indicated in red; basic residues in blue; and nonpolar in gray. Phe residues are marked yellow. Circles indicate residues that were analyzed in this report. (b) Side view of the predicted structure of the MSL10 TM6, created with I-TASSER. The side chain at G556, predicted to form a kink, is indicated with a blue sphere. (c) Predicted topology of MSL10, indicating soluble N- and C-termini and six membrane-spanning helices. The lengths of the N- and C-termini are not to scale

2.5 | Cell death assays

The coding sequences of wild-type MSL10, MSL10 F553V, and MSL10 G556V were amplified from the pOO2 vectors described above and cloned into the pK7FWG2 vector for C-terminal GFP tagging and transient expression in *Nicotiana benthamiana* leaves under the control of the 35S promoter. To quantify the amount of dead versus viable epidermal pavement cells, leaves were dual stained with fluorescein diacetate (FDA) and propidium iodide (PI) 5 days after infiltration as described in (Garabagi, Gilbert, Loos, McLean, & Hall, 2012) and visualized via confocal microscopy. As previously outlined, cells were considered dead (i) if their nucleus was stained by PI and/or (ii) their vacuole disappeared, which is apparent when cytoplasmic FDA/GFP signal fills the entire body of the pavement cell (Veley et al., 2014). In this study, we added a third criterion whereby cells were also considered dead if their vacuole had mostly, though not completely, disappeared. The percentage of cell death reported was the average of four separate infiltration experiments, each consisting of three to four leaves per construct, with $n \sim 60$ cells imaged per leaf.

3 | RESULTS

3.1 | Selection of potential pore-disrupting residues

To identify residues likely to be important for MSL10 ion channel activity, we performed a sequence alignment between the region of highest homology between *Ec*MscS, MSL10, and MSL8 (Figure 1a). In the *Ec*MscS crystal structure, this sequence forms transmembrane (TM) helix 3, the domain that lines the channel pore (Bass et al., 2002). *Ec*MscS TM3 is 33 amino acids long and has a pronounced kink at G113 that splits it into TM3a and TM3b. The analogous sequence in MSL10 is its most C-terminal TM helix, TM6. We therefore generated a hypothetical structure of TM6 using the I-TASSER prediction server (Zhang, 2008) and the *Ec*MscS closed-state crystal structure (2OAU:A (Bass et al., 2002; Steinbacher et al., 2007)), as a template (Figure 1b). These data, along with a topology prediction from the ARAMEMNON server (Schwacke et al., 2003), support the overall topology for MSL10 shown in Figure 1c. Based on this alignment, we identified four key classes of residues within the TM6 sequence that were likely to contribute to MSL10 channel function: (i) multiple phenylalanine residues, (ii) the putative glycine kink, (iii) several nonpolar TM6 residues that differ between MSL10 and MSL8, and (iv) an isoleucine residue known to play a role in MSL8 function. Those residues selected for study are indicated with circles in Figure 1a.

3.2 | Phenylalanine 553 maintains channel conductance and the stability of the open state

One distinction between the amino acid sequences of *Ec*MscS TM3 and MSL10 TM6 is the presence or absence of multiple phenylalanine residues. *Ec*MscS TM3 contains only one Phe residue, and it is located at very end of TM3b. MSL10 has six Phe residues scattered through the pore-lining domain. MSL8 has six Phe residues in its

TM6; five of these are conserved with MSL10 (indicated in yellow, Figure 1c), and F720 is essential for channel function (Hamilton & Haswell, 2017). To determine whether Phe residues in TM6 are also critical for MSL10 channel function, we used site-directed mutagenesis to change F544, F553, and F563 to smaller nonpolar residues, and to introduce a series of increasingly disruptive changes at F553. We introduced F544V, F553W, F553L, F553V, and F563L mutations into the MSL10 coding sequence of pOO2-MSL10-GFP, and pOO2-MSL10 for in vitro capped RNA (cRNA) production (Maksaev & Haswell, 2012). cRNA for all variants was injected into Xenopus oocytes for expression and characterization as previously described (Maksaev & Haswell, 2015).

To determine whether these mutations affected protein folding or stability, we monitored the trafficking of MSL10 variants fused to GFP by confocal microscopy. All variants produced similar signal at the oocyte periphery 5 days after injection, while no signal was present in water-injected oocytes (Figure 2a). Single-channel patch-clamp electrophysiology was then used to assess the channel behavior (single channel conductance, open-state stability, and tension sensitivity) of each of these MSL10 variants, as in (Maksaev & Haswell, 2015). Current traces from excised inside-out patches derived from oocytes expressing wild type MSL10 showed the expected stable single-channel openings (119.4 ± 4.0 pS at −100 mV, Figure 2b, top). MSL10 F544V did not appear different from the wild type channel. However, MSL10 F553W, MSL10 F553L, and MSL10 F563L exhibited "flickery" channel activity, defined here as rapid increases and decreases in conductance without a clear time spent in the open state (middle and bottom, Figure 2b). We were unable to detect any channel activity in oocytes expressing MSL10 F553V.

We measured the unitary currents produced by MSL10 and MSL10 Phe variant channels under transmembrane potential from 0 to −200 mV. The unitary conductances of MSL10 F544V (114.8 ± 4.0 pS), F553W (121.8 ± 4.3 pS) and F563L (117.7 ± 5.4 pS), calculated at −100 mV membrane potential, were indistinguishable from wild type MSL10 (Figure 2c, Table 1). However, MSL10 F553L produced single-channel currents that were significantly lower than that of the wild type at every voltage tested (Figure 2c), and its conductance at −100 mV was 78.0 ± 4.4 pS, 0.6-fold that of the wild type (Table 1). As noted above, MSL10 F553V did not produce any activity. Thus, successively reducing the bulkiness of residues at position 553 resulted in successively lower unitary channel conductance (W≈F>L>V). We conclude that F553 and F563 both play important roles in maintaining the stability of the open state of MSL10, while F553 also controls channel conductance.

3.3 | Glycine 556 is a key residue for MSL10 channel function

The next residue in MSL10 we chose to investigate was G556. MSL10 G556 aligns with the kink-forming residue G113 in *Ec*MscS (Figure 1a)

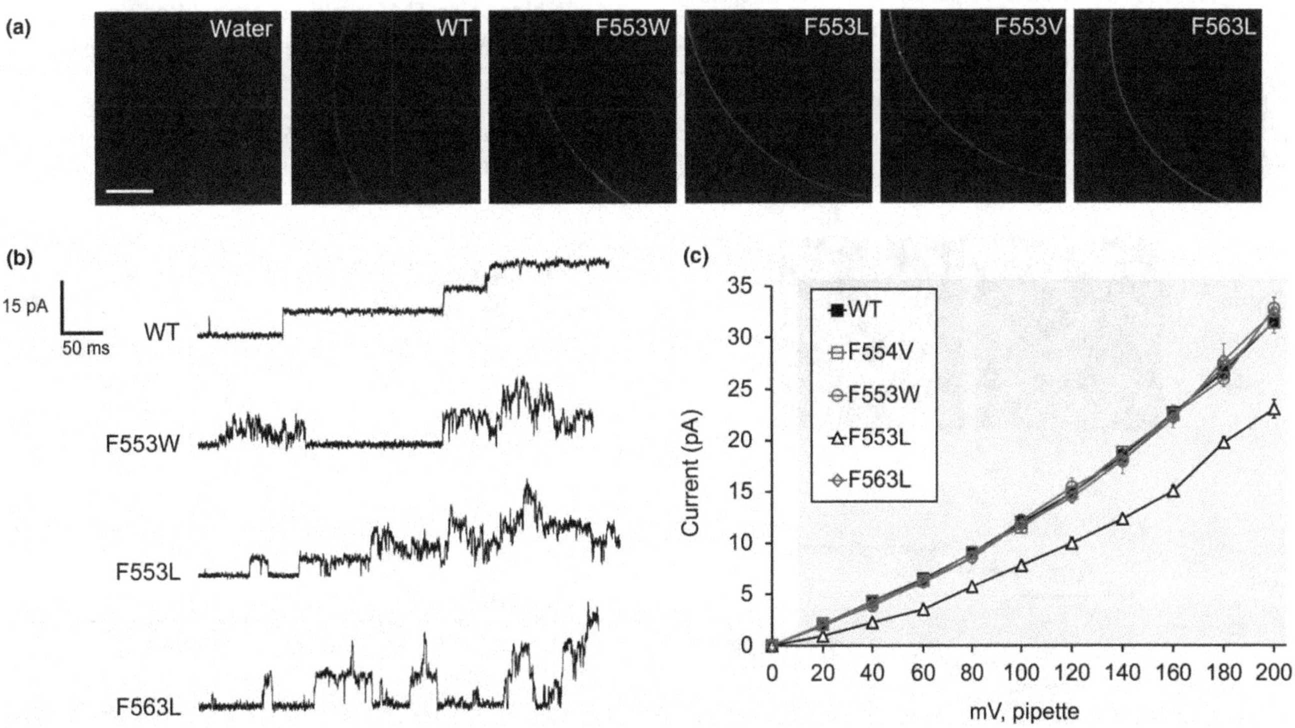

FIGURE 2 Mutagenesis of two TM6 phenylalanines reveals a role for bulky nonpolar residues in maintaining channel conductance and the stability of the open state. (a) Portions of the oocyte periphery 5 days after injection with the indicated MSL10-GFP variant cRNA. Scale bar is 100 μm. (b) Examples of traces at −100 mV membrane potential. 0.5-s fragments of 5-s records of channel activation by symmetric pressure ramps are shown. (c) Current/voltage curves for the indicated MSL10 variants. Each data point is the average current from 3 to 9 patches in 60 mM $MgCl_2$, 4 mM HEPES. Error bars indicate standard deviation but are obscured by the symbols. Gray background indicates voltages where the current produced by MSL10 F553L differed significantly from wild-type MSL10, $p < 0.001$ (Student's t test)

TABLE 1 Properties of the *At*MSL10 TM6 mutants. Conductance at −100 mV

	Mutation	Conductance at −100 mV	Conductance compared to WT	Open state	Gating characteristics	Cell death
MSL10 WT	None	119.4 ± 4.0 pS	1.0 WT	Stable	Strong hysteresis	Strong
Phe residues	F544V	114.8 ± 4.0 pS	1.0 WT	Stable		
	F553W	121.8 ± 4.3 pS	1.0 WT	Slight flicker		
	F553L	78.0 ± 4.4 pS	0.65 WT	Slight flicker	No hysteresis	
	F553V	N/A	N/A	N/A	N/A	Strong
	F563L	117.7 ± 5.4 pS	1.0 WT	Stable	High gating threshold No hysteresis	
Putative Gly kink	G556A	106.0 ± 4.4 pS	0.9 WT	Stable		
	G556V	N/A	N/A	N/A	N/A	Strong
Change to MSL8	L548V	114.3 ± 5.0 pS	1.0 WT	Stable		
	A550L	112.7 ± 4.6 pS	1.0 WT	Stable		
	L562V	119.2 ± 6.1 pS	1.0 WT	Stable	High gating threshold	
	I554V	121.8 ± 8.1 pS	1.0 WT	Stable		
	I554S	61.1 ± 4.9 pS	0.5 WT	Flickery		Strong

and is predicted to sit at a similar kink in TM6 (Figure 1b). To determine whether G556 plays a role in MSL10 channel function, we changed this residue to the larger nonpolar residues alanine (G556A) or valine (G556V). While both MSL10 variants were expressed in Xenopus oocytes and trafficked normally to the plasma membrane (Figure 3a), neither functioned like the wild type. The MSL10 G556A mutant was active and produced a relatively stable channel opening (Figure 3b). However, it had a unitary conductance of 106.0 ± 4.4 pS, 0.9-fold that of the wild-type channel (Figure 3c). MSL10 G556V did not produce any activity even under extreme membrane tensions and high membrane potentials. Thus, the G556A substitution produced a modest effect, and G556V completely ablated function, presumably because of increasing size of the side chains. These results establish that G556 plays a key role in the function of MSL10 and are consistent with the prediction that there is bending at this residue within the helix and that mobility at this site is important for the conformational changes associated with channel opening.

3.4 | Nonpolar TM6 residues that differ between MSL10 and MSL8 are not required for wild-type conductance and open-state stability

Nonpolar residues in MSL10 TM6 that are not conserved in MSL8 include F544, L548, A550, I555, and L562. To determine

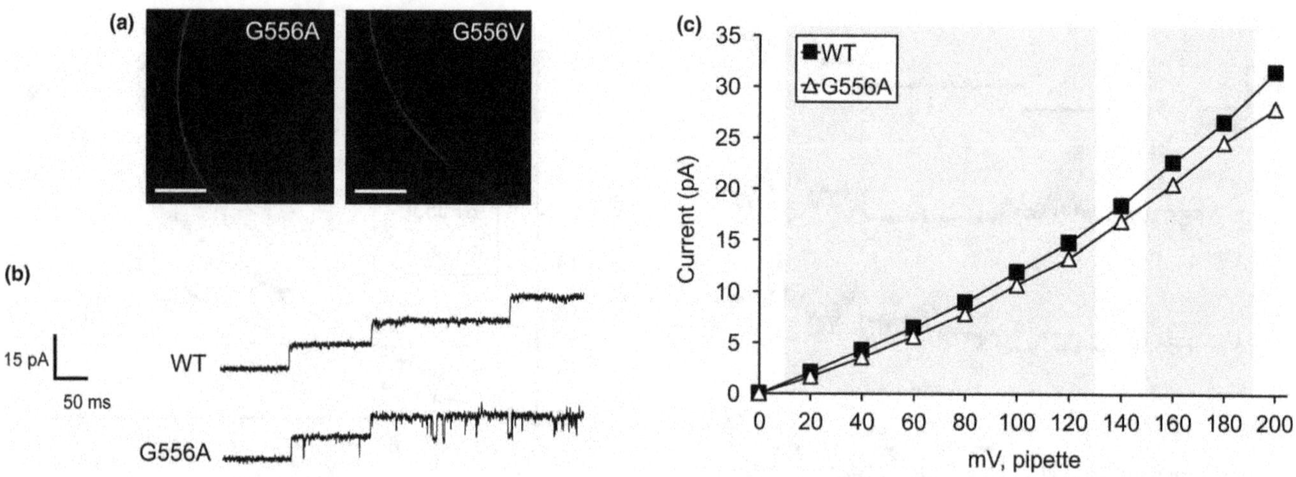

FIGURE 3 Replacing G556 with larger nonpolar residues reduces or ablates channel conductance. (a) Portions of the oocyte periphery 5 days after injection with the indicated MSL10-GFP variant cRNA. Scale bar is 100 μm. (b) Examples of traces at −100 mV membrane potential. 0.5-s fragments of 5-s records of channel activation by symmetric pressure ramps are shown. (c) Current/voltage curves for wild type and MSL10 G556A. Each data point is the average current from 3 to 9 patches in 60 mM $MgCl_2$, 4 mM HEPES. Error bars indicating standard deviation are present but are obscured by the symbols. Wild-type MSL10 curve is from the same data as in Figure 2c. Gray background indicates voltages where the current produced by MSL10 G556A differed significantly from wild-type MSL10, $p < 0.001$ (Student's *t* test)

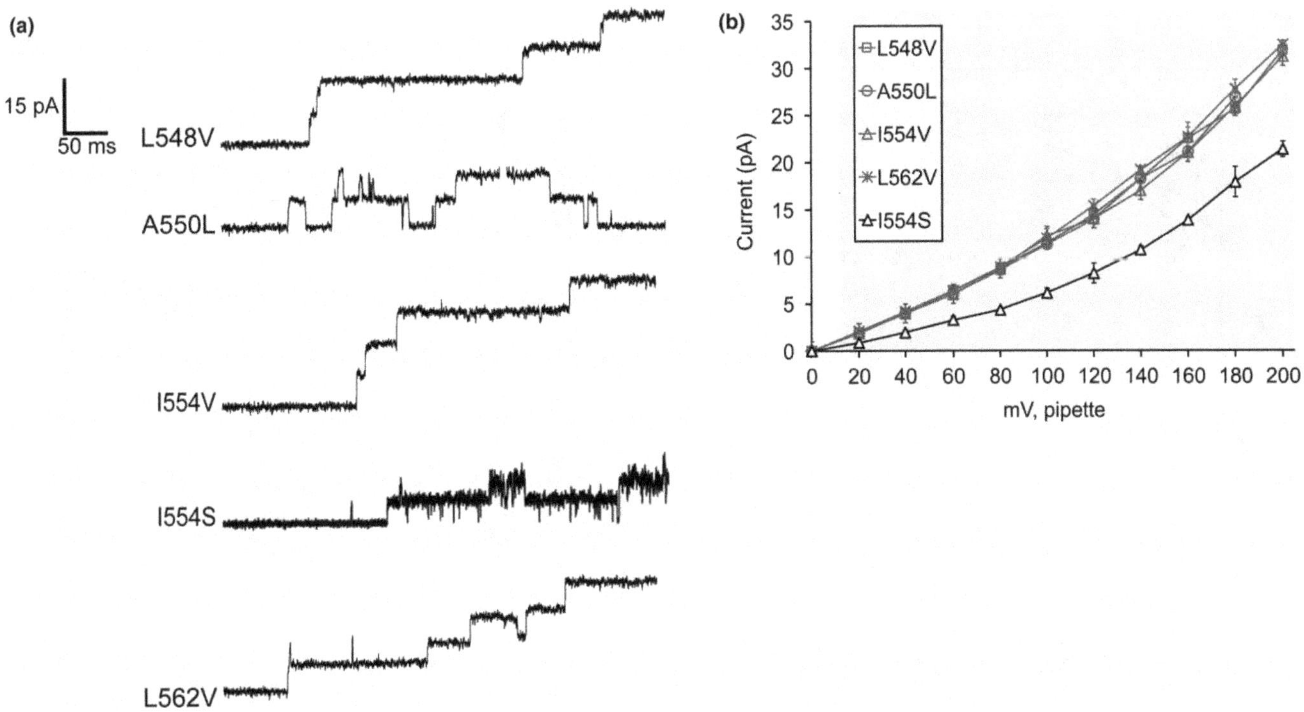

FIGURE 4 Effects of mutating nonpolar residues in TM6 that differ between MSL8 and MSL10. (a) Examples of traces at −100 mV membrane potential. 0.5-s fragments of 5-s records of channel activation by symmetric pressure ramps are shown. (b) Current/voltage curves for the indicated MSL10 variants. L548V, A550L, I554V, and L562V do not differ from the WT. Each data point is the average current from 3 to 9 patches in 60 mM $MgCl_2$, 4 mM HEPES. Error bars indicate standard deviation. Gray background indicates voltages where the current produced by MSL10 I554S differed significantly from wild-type MSL10, $p < 0.001$ (Student's *t* test)

whether these residues confer the differences in conductance and tension sensitivity between MSL10 and MSL8, we changed several of these residues in MSL10 to the corresponding residue found in MSL8. MSL10 F544V was analyzed with other Phe substitutions in Figure 2 and did not appreciably affect MSL10 channel behavior. MSL10 L548V (114.3 ± 5.0 pS), MSL10 A550L (112.7 ± 4.6 pS), and MSL10 L562V (119.2 ± 6.1 pS) were also indistinguishable from the wild type (119.4 ± 4.0 pS) with respect to unitary conductance and open-state stability (Figure 4a and b, Table 1).

3.5 | Changing isoleucine 554 to serine disrupts channel conductance and the stability of the open state

We previously showed that MSL8 I711S produced a channel with wild-type conductance but an increased tension threshold (Hamilton & Haswell, 2017). We mutated the corresponding residue in MSL10, I554, to valine or serine. While MSL10 I554V exhibited a wild-type unitary conductance (121.8 ± 8.1 pS) and gating pattern, MSL10 I554S was very flickery (Figure 4a), and had a significantly lower unitary conductance than WT (61.1 ± 4.9 pS compared to 119.4 ± 4.0 pS of WT at −100 mV membrane potential, Figure 4b). These unexpected results suggest that MSL10 I554S has normal tension sensitivity, but, has an unstable open state. Taken together, the data shown in Figure 4 suggest that the differences in MSL8 and MSL10 channel behavior cannot be easily attributed to individual amino acid residues within TM6.

3.6 | Opening and/or closing tension sensitivities were altered in MSL10 F553L, L562V, and F563L

The established approach for measuring tension sensitivity of *Ec*MscS, by calculating the activation midpoint [e.g., refer (Blount, Sukharev, Schroeder, Nagle, & Kung, 1996; Edwards et al., 2005; Maksaev & Haswell, 2011)], is not possible with MSL10 expressed in oocytes, as one typically cannot reach current saturation before the patch collapses (Maksaev & Haswell, 2012). Instead, we measured the gating threshold, or the amount of tension required to open the second channel in the patch. We used pipettes with the same resistance (3.00 ± 0.25 MOhm) to reduce variability in patch size and geometry.

We first analyzed the gating threshold for wild-type MSL10. We observed that the gating threshold depended on the number of channels we observed in a patch prior to saturation or patch rupture and that patches with more channels had a lower opening threshold (Figure 5a). Assuming low open probability and minimal spontaneous gating at zero tension, these data can be fit to a line with slope 0.049 ± 0.02 channels/mm Hg (Figure 5a, circles). The same analyses for MSL10 F553L (squares) and MSL10 L562V (triangles) produced slopes similar to that of the wild type channel, 0.047 ± 0.04 and 0.036 ± 0.01 channels/mm Hg, respectively (Figure 5a).

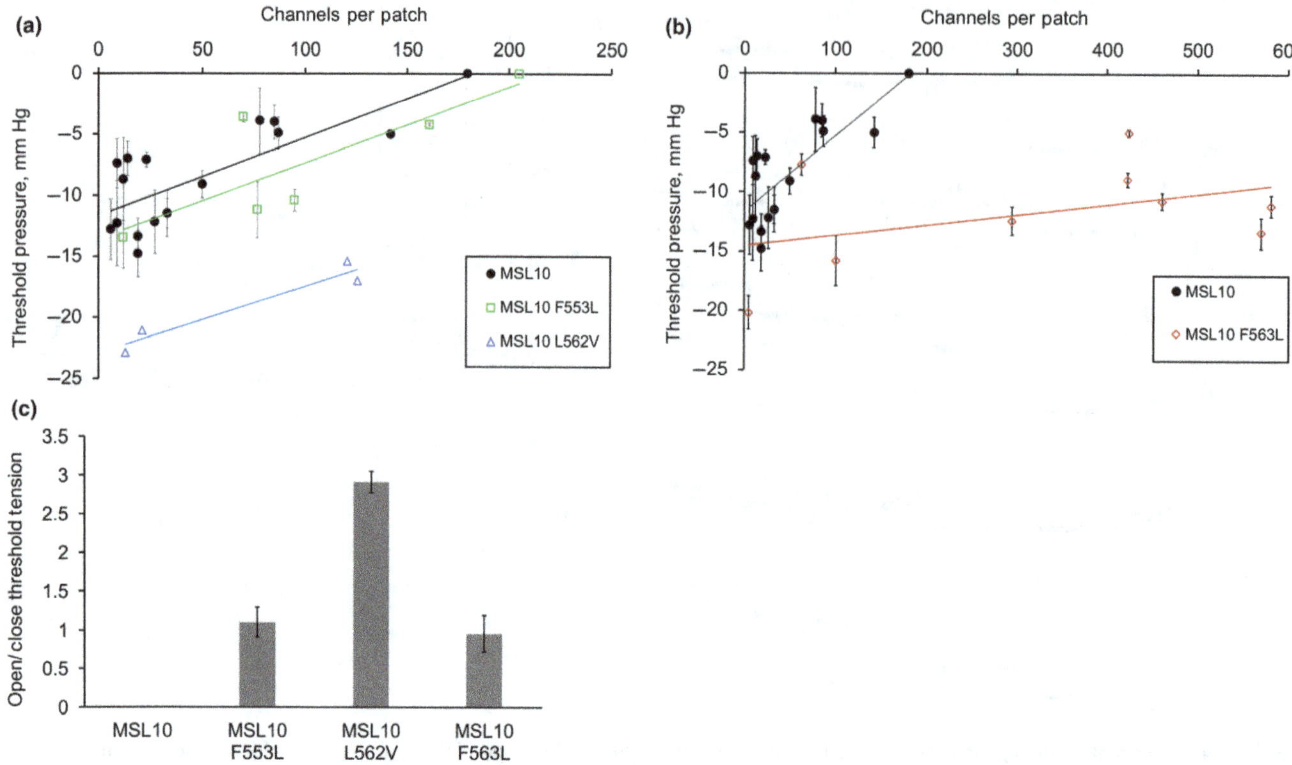

FIGURE 5 Three bulky nonpolar residues affect the threshold tensions of opening and closing. (a, b) Plot of channel number versus threshold tension for wild type MSL10 and the indicated variants. Data were corrected for access resistance. The data for wild-type MSL10 are the same in both plots. Error bars indicate standard deviation between the threshold pressure measured from 5 to 10 trials from a single patch containing the indicated number of channels. (c) The average open/close membrane tension ratios for MSL10 and variants including data from all patches shown in panels a and b. One sample for MSL10 F553L was left out as outlier. Error bars indicate standard deviation between the average open/close ratio for each patch

However, MSL10 F563L did not show a strong dependence of channel number on threshold pressure, with patches containing a range of channel numbers all opening at essentially the same tension. The linear regression line slope for MSL10 F563L was only 0.009 ± 0.007 channels/mm Hg (Figure 5b, diamonds). As shown in Figure 5b, we routinely observed very high channel numbers per patch for MSL10 F563L; in some cases as many as 600 channels per patch.

It is not possible to calculate an open/closed ratio for WT MSL10. In almost every case, one or more channels are still open when all applied tension is released. As a result, the closing threshold tension cannot be calculated. However, the fact that all three MSL10 variants did not show this behavior made it possible to calculate their open/close ratio, using patches with 5–600 channels (Figure 5c). For MSL10 F563L, the open/close ratio was close to one (0.95 ± 0.23). MSL10 L562V had an open/close ratio of 2.9 ± 0.13. For MSL10 F553L, the open/close ratio was close to one (1.10 ± 0.19) when only the four closest data points were considered. A fifth strongly outlying data point was not included, as its value exceeded the upper fence as defined by the quartile method. To summarize, wild-type MSL10 showed threshold opening pressure of around −15 mm Hg and exhibited hysteresis. MSL10 L562V had a high tension threshold for opening but maintained hysteresis. MSL10 F563L had a high tension threshold for both opening and closing and did not exhibit hysteresis. Due to the small sample size, we can only tentatively conclude that MSL10 F553L lost hysteresis.

3.7 | MSL10 TM6 mutations that ablate channel function do not alter the ability of MSL10 to trigger cell death signaling

These mutations that alter MSL10 channel behavior provide tools to test the structural requirements for MSL10's cell death signaling function. The two MSL10 mutations that produced no ion channel activity (F553V and G556V), along with wild-type MSL10, were fused to GFP and transiently expressed under the strong constitutive cauliflower mosaic virus 35S promoter (35Sp) in tobacco leaf epidermal cells as described previously (Veley et al., 2014). Constructs were coinfiltrated with a plasmid expressing P19 to suppress gene silencing. As shown in Figure 6A, all MSL10 variants were expressed and trafficked normally in tobacco cells. Five days after Agrobacterium infiltration, leaf samples were dual stained for FDA and PI to assess cell death (Figure 6B). As expected, 35% of the cells in leaf samples expressing wild type full-length MSL10 were dead, compared to 14% when P19 was infiltrated alone. Neither MSL10-F553V nor MSL10 G556V produced levels of cell death that were statistically distinguishable from wild type MSL10 (Figure 6C),

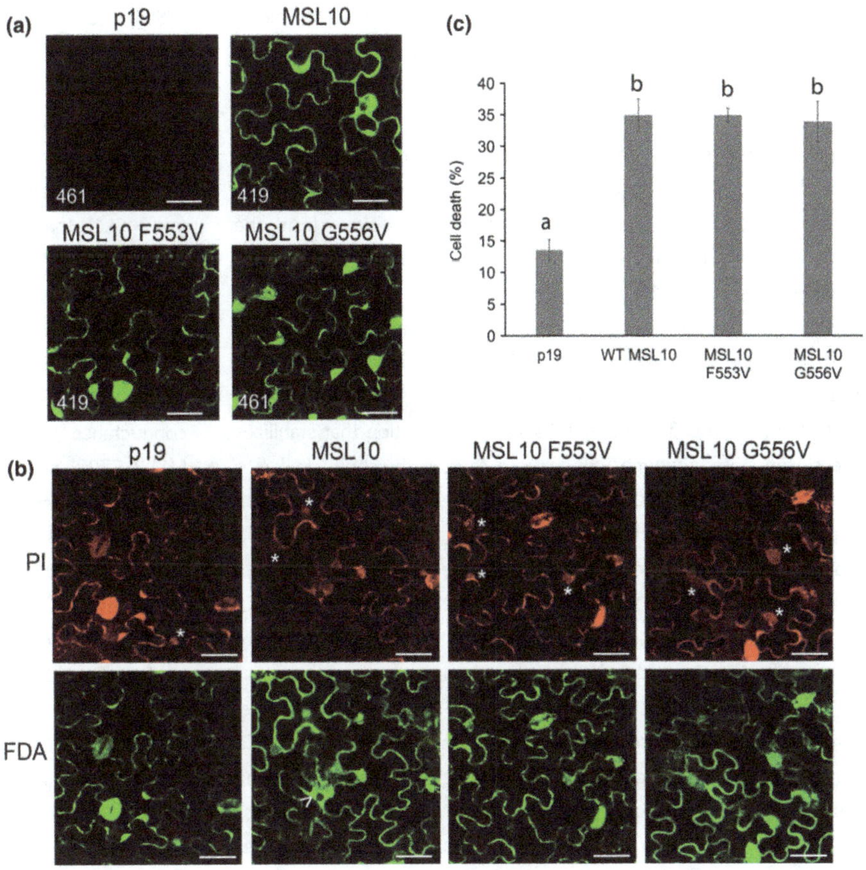

FIGURE 6 MSL10 mechanosensitive channel activity is not required for death signaling *in planta*. (a) Localization of MSL10-GFP variants transiently expressed in tobacco epidermal pavement cells. Images were taken 5 days after infiltration. Infiltrations with P19 alone were used as a negative control. Scale bar is 40 μm. Numbers in the lower left-hand corner indicate the voltage setting of the PMT detector when imaging.
(b) Examples of cell viability assays. Tobacco leaves expressing MSL10-GFP variants were dual stained with FDA and PI; cells were scored as dead if they had a PI-stained nucleus (indicated by asterisks) or a disappearing vacuole, evidenced by spreading of cytoplasmic signal (arrowhead). Scale bar is 50 μm. (c) Percentage of dead cells quantified from dual staining of 20 leaves (P19 and MSL10) or 10 leaves (MSL10 variants) from multiple infiltration experiments. Error bars indicate standard error. Statistical differences were assessed by one-way ANOVA and Scheffe's test; groups with the same letter did not significantly differ from each other ($p > 0.05$)

indicating that ion channel function is not required for the ability of full-length MSL10 to trigger cell death in this transient assay.

4 | DISCUSSION

Here, we report the functional effect of twelve different point mutations in the mechanosensitive channel MscS-like (MSL)10 from *Arabidopsis thaliana*. All substitutions were made in a set of eight nonpolar amino acids located in the putative pore-lining transmembrane (TM) helix 6. We observed that four of these mutations (F544V, L548V, A550L, and I554V) did not detectably alter channel behavior in our assays, four primarily affected unitary conductance and/or open-state stability (F553W, F553L, G556A, and I554S), three affected tension sensitivity and/or hysteresis (F553L, L562V, and F563L), and two produced no mechanosensitive channel activity at all (F553V and G556V). Finally, we observed that the latter two variants were as capable as wild type MSL10 of inducing cell death in an *in planta* expression assay.

As Phe residues are found in TM6 of MSL8 and MSL10 but not in TM3 of *Ec*MscS, they were reasonable candidates for residues that underlie the differing channel properties of MSLs and *Ec*MscS. MSL10 has a smaller conductance than *Ec*MscS, and we originally hypothesized that these bigger hydrophobic residues found in the MSL10 TM6 could form a "vapor lock," like *Ec*MscS L105 and L109 (Anishkin & Sukharev, 2004; Bass et al., 2002), or partially obstruct the channel pore as suggested for F450 in the Arabidopsis SLAC1 channel (Chen et al., 2010). If this were the case, replacing Phe with smaller side chains such as Leu or Val would lead to a deregulated or constitutively open channel. Alternatively, bulky Phe residues could be involved in the slow closing kinetics observed in MSL10 but not in MscS if they participated in a hydrophobic residue-mediated force transduction pathway from the periphery of the channel to its pore, in a manner analogous to the proposed force-transmitting clutch of *Ec*MscS (Belyy et al., 2010). However, none of these hypotheses were correct. Instead, our data suggest that the large hydrophobic nature of the Phe residues at 553 and 563 facilitate the unitary conductance and the stability of the open state of MSL10.

While mutation of F553 to tryptophan had no significant effect on conductance, mutation to leucine led to a decrease in unitary conductance and mutation to valine produced a channel without any apparent mechanosensitive response (Figure 2). Because MSL10 F553L produced a partially functional channel, and because MSL10 F553V was expressed at wild-type levels, MSL10 F553V is likely to produce a stable oligomeric channel that lacks a conductive pore. We thus speculate that the pore size of MSL10 is altered by the size of the side chain at F553; as the side chains at F553 decrease in size, pore-lining domains grow more and more closely packed, thus successively reducing conductance.

While gating pressures for the Phe mutants did not significantly differ from those of WT, both MSL10 F553L and MSL10 F563L showed almost no hysteresis (Figure 5c). MSL10 typically shows delayed closing relative to opening kinetics; with MSL10 F553L and MSL10 F563L, the ratio of opening and closing tensions was close to 1. This suggested that these mutant channels have an open state that transitions back to the closed state more easily than the wild-type channel. We speculate that the open state is stabilized by interaction between F553 and another hydrophobic residue from the adjacent pore-lining domain of the channel. In this model, the flickery behavior of MSL10 F553L is explained by a steric mismatch between the mutated residue and its interaction partner from the adjacent pore-lining domain. The F553V mutant would have no channel activity as the nonpolar side chain could not create even a partially open pore. There may be a similar interaction that stabilizes the closed state, although F553L has a tension threshold similar to that of the WT MSL10 channel, suggesting that this explanation might be too simple. If reducing the size of the side chain at F553 stabilizes the closed pore, one would expect the flickery F553L channel to be harder to activate. Taken together, the data shown in Figure 2 show that multiple Phe residues in MSL10 TM6 are required for the size and stability of the open pore.

We also found that G556 plays a critical role in MSL10 channel function. We first targeted G556 for mutation because a glycine at a similar position in *Ec*MscS TM3 (G113) has been implicated in gating movements as suggested by crystal structures and molecular dynamics. The *Ec*MscS G113A mutation prevents channel inactivation, presumably by preventing the helix kinking thought to accompany MscS entry from the open state into the inactivated state (Akitake et al., 2007; Boer et al., 2011). While MSL10 does not show inactivation in oocytes [although it does in plant cells (Peyronnet, Tran, Girault, & Frachisse, 2014)], the model for MSL10 TM6 shown in Figure 1 suggests a similar geometry and is consistent with a similar gating movement in the pore-lining helix between MscS and MSL10. We therefore hypothesized that mutations at this site might affect the conformational changes required for channel gating.

While MSL10 G556A showed only a subtle difference from the wild type—a small, but statistically significant, decrease in conductance—MSL10 G556V did not produce any mechanosensitive activity at any applied tensions, though the mutant was expressed at WT levels (Figure 3). We speculate here that the MSL10 G556 side chain faces the pore lumen in the open state and therefore introduction of alanine at position 556 resulted in a decrease in unitary currents (Figure 3b, c). Introduction of the even bigger hydrophobic residue valine at this site might then either fully occlude the channel pore, or prevent the conformational changes associated with channel gating. Whether the same movements are made during the gating transitions of *Ec*MscS and MSL10 is not clear and will certainly require additional experimentation, but the fact that the G556 is a key structural feature of MSL10 TM6 is established.

We addressed the similarities and differences between MSL10 and its close homolog MSL8. We first attempted to link the difference in MSL8 and MSL10 channel characteristics to the differences in the sequence of their putative pore-lining domains. We mutated 3 residues of MSL10 into their counterparts in MSL8, generating MSL10 L548V, A550L, and L562V (Figure 4). Only the latter had any effect; MSL10 L562V had a high gating threshold (Figure 5a), similar to that previously observed with MSL8 (Hamilton et al., 2015). Thus, L562/V719 may in part be responsible for the difference between MSL8 and MSL10 with respect to tension sensitivity.

We also made mutations in residues conserved between MSL8 and MSL10 that have known effects on MSL8 function. MSL8 I711S exhibited normal conductance but an increased gating threshold and MSL8 F720L was a completely disrupted channel (Hamilton & Haswell, 2017). Analagous mutations altered MSL10 function, but they did not produce the same effects they did in MSL8. MSL10 I554S (aligns with MSL8 I711) produced a flickery channel with half the conductance of the wild type, while MSL10 F563L (aligns with MSL8 F720) produced a channel with normal conductance but an increased gating threshold (Figures 4 and 5). These unexpected results suggest that local or global differences in structure dictate the characteristics of these two channels, so trying to identify individual residues responsible for particular channel characteristics may not be successful.

Residues L562 and F563 form a hydrophobic patch flanked by polar and charged residues (Figure 1a) and our experiments indicate that they are essential for normal tension sensitivity and gating kinetics of the MSL10 channel (Figure 5). Hysteresis is the strongest in WT MSL10, wherein the threshold tension for closing is much lower than for opening. The difference between opening and closing tensions was decreased in MSL10 F562V and completely abolished in MSL10 F563L (Figure 5c). We speculate that these two residues may function as a "force transmitters," similar to L111 of *Ec*MscS TM3. It has been proposed that L111 interacts with F68 from TM2, enabling transduction of the membrane tension into a gating force (Belyy et al., 2010). Similarly, L562 and F563 residues may be a part of an intra-transmembrane helix system that serves to control the closing kinetics of MSL10 by stabilizing the open state.

The concept that ion channels have functions separable from their ability to mediate ion flux (nonconducting functions) is not new and has been established for sodium and potassium channels in animal systems (Feinshreiber et al., 2010; Kruger & Isom, 2016). However, nonconducting functions have not been previously demonstrated for MS ion channels. We previously established that MSL10 is capable of inducing cell death when overexpressed in stable Arabidopsis lines and in transient tobacco expression experiments (Veley et al., 2014). This effect was modulated by seven phosphorylation sites in the soluble N terminus, but these sites did not alter ion channel function, suggesting that the two functions might be separable. Transient overexpression of the soluble N terminus of MSL10 was capable triggering programmed cell death on its own. However, these results left open the possibility that MSL10 ion channel function is required indirectly to activate the cell death signaling mediated by the N terminus and that this effect could be simulated by expressing the truncated, soluble N terminus.

Two of the twelve mutations we tested, MSL10 F553V and MSL10 G556V, were normally expressed and trafficked in Xenopus oocytes but did not exhibit any mechanosensitive ion channel activity

(Figures 2 and 3). These mutants provided the opportunity to test the idea of a nonconducting function for MSL10 more directly and without the caveats of the truncation experiment. The effect of over-expressing these mutant channels did not differ in any way from the wild type, indicating that MS ion channel activity is not required for MSL10's ability to induce cell death in tobacco epidermal cells. The results presented here now firmly establish that MSL10 induces programmed cell death through a mechanism independent of mechanically-induced ion flux. A key future experiment will be to determine whether the nonconducting function of MSL10 is regulated by membrane tension. It will also be worth testing whether nonconducting functions are a conserved feature of proteins in the MscS family, as suggested by a report that the soluble C terminus of MscS interacts with the bacterial fission protein FtsZ (Koprowski et al., 2015).

ACKNOWLEDGMENTS

We acknowledge Debarati Basu for generating the mutant MSL10 constructs used in Figure 6. These experiments were funded by the National Institutes of Health Grant R01GM084211 (to ESH), National Science Foundation Grant MCB1253103 (to ESH), and NSF Graduate Research Fellowship DGE-1745038 (to JMS).

AUTHOR CONTRIBUTIONS

GM, JMS, and ESH designed experiments and wrote the manuscript; GM, JMS, and SO performed experiments.

ORCID

Elizabeth S. Haswell http://orcid.org/0000-0002-4246-065X

REFERENCES

Akitake, B., Anishkin, A., Liu, N., & Sukharev, S. (2007). Straightening and sequential buckling of the pore-lining helices define the gating cycle of MscS. *Nature Structural & Molecular Biology*, *14*(12), 1141–1149. https://doi.org/10.1038/nsmb1341

Anishkin, A., Akitake, B., & Sukharev, S. (2008). Characterization of the resting MscS: Modeling and analysis of the closed bacterial mechanosensitive channel of small conductance. *Biophysical Journal*, *94*(4), 1252–1266. https://doi.org/10.1529/biophysj.107.110171

Anishkin, A., Loukin, S. H., Teng, J., & Kung, C. (2014). Feeling the hidden mechanical forces in lipid bilayer is an original sense. *Proceedings of the National Academy of Sciences of the United States of America*, *111* (22), 7898–7905. https://doi.org/10.1073/pnas.1313364111

Anishkin, A., & Sukharev, S. (2004). Water dynamics and dewetting transitions in the small mechanosensitive channel MscS. *Biophysics*, *86*(5), 2883–2895. https://doi.org/10.1016/S0006-3495(04)74340-4

Bass, R. B., Strop, P., Barclay, M., & Rees, D. C. (2002). Crystal structure of *Escherichia coli* MscS, a voltage-modulated and mechanosensitive channel. *Science*, *298*(5598), 1582–1587. https://doi.org/10.1126/science.1077945

Basu, D., & Haswell, E. S. (2017). Plant mechanosensitive ion channels: An ocean of possibilities. *Current Opinion in Plant Biology*, *40*, 43–48. https://doi.org/10.1016/j.pbi.2017.07.002

Bavi, N., Nikolaev, Y. A., Bavi, O., Cox, C. D., Qin, Q. H., Martinac, B., & Battle, A. R. (2017). Principles of mechanosensing at the membrane interface. In: R. M. Epand & J.-M. Ruysschaert (Eds.), *The biophysics of cell membranes. Vol 19. Springer series in biophysics* (pp. 85–119). Singapore: Springer. https://doi.org/10.1007/978-981-10-6244-5_4

Belyy, V., Anishkin, A., Kamaraju, K., Liu, N., & Sukharev, S. (2010). The tension-transmitting "clutch" in the mechanosensitive channel MscS. *Nature Structural & Molecular Biology*, *17*(4), 451–458. https://doi.org/10.1038/nsmb.1775

Belyy, V., Kamaraju, K., Akitake, B., Anishkin, A., & Sukharev, S. (2010). Adaptive behavior of bacterial mechanosensitive channels is coupled to membrane mechanics. *Journal of General Physiology*, *135*(6), 641–652. https://doi.org/10.1085/jgp.200910371

Blount, P., Sukharev, S. I., Schroeder, M. J., Nagle, S. K., & Kung, C. (1996). Single residue substitutions that change the gating properties of a mechanosensitive channel in *Escherichia coli*. *Proceedings of the National Academy of Sciences of the United States of America*, *93*(21), 11652–11657.

Boer, M., Anishkin, A., & Sukharev, S. (2011). Adaptive MscS gating in the osmotic permeability response in *E. coli*: The question of time. *Biochemistry*, *50*(19), 4087–4096. https://doi.org/10.1021/bi1019435

Booth, I. R., & Blount, P. (2012). The MscS and MscL families of mechanosensitive channels act as microbial emergency release valves. *Journal of Bacteriology*, *194*(18), 4802–4809. https://doi.org/10.1128/JB.00576-12

Booth, I. R., Miller, S., Müller, A., & Lehtovirta-Morley, L. (2015). The evolution of bacterial mechanosensitive channels. *Cell Calcium*, *57*(3), 140–150. https://doi.org/10.1016/j.ceca.2014.12.011

Böttcher, B., Prazak, V., Rasmussen, A., Black, S. S., & Rasmussen, T. (2015). The structure of YnaI implies structural and mechanistic conservation in the MscS family of mechanosensitive channels. *Structure*, *23*(9), 1705–1714. https://doi.org/10.1016/j.str.2015.06.023

Boulos, R. A. (2013). Antimicrobial dyes and mechanosensitive channels. *Antonie van Leeuwenhoek*, *104*(2), 155–167. https://doi.org/10.1007/s10482-013-9937-x

Çetiner, U., Rowe, I., Schams, A., Mayhew, C., Rubin, D., Anishkin, A., & Sukharev, S. (2017). Tension-activated channels in the mechanism of osmotic fitness in *Pseudomonas aeruginosa*. *Journal of General Physiology*, *149*(5), 595–609. https://doi.org/10.1085/jgp.201611699

Chen, Y.-H., Hu, L., Punta, M., Bruni, R., Hillerich, B., Kloss, B., ... Hendrickson, W. A. (2010). Homologue structure of the SLAC1 anion channel for closing stomata in leaves. *Nature*, *467*(7319), 1074–1080. https://doi.org/10.1038/nature09487

Cox, C. D., Nakayama, Y., Nomura, T., & Martinac, B. (2015). The evolutionary "tinkering" of MscS-like channels: Generation of structural and functional diversity. *Pflugers Archiv-European Journal of Physiology*, *467*(1), 3–13. https://doi.org/10.1007/s00424-014-1522-2

Cox, C. D., Nomura, T., Ziegler, C. S., Campbell, A. K., Wann, K. T., & Martinac, B. (2013). Selectivity mechanism of the mechanosensitive channel MscS revealed by probing channel subconducting states. *Nature Communications*, *4*, 2137. https://doi.org/10.1038/ncomms3137

Cox, C. D., Wann, K. T., & Martinac, B. (2014). Selectivity mechanisms in MscS-like channels: From structure to function. *Channels (Austin)*, *8* (1), 5–12. https://doi.org/10.4161/chan.27107

Edwards, M. D., Bartlett, W., & Booth, I. R. (2008). Pore mutations of the *Escherichia coli* MscS channel affect desensitization but not ionic preference. *Biophysical Journal*, *94*(8), 3003–3013. https://doi.org/10.1529/biophysj.107.123448

Edwards, M. D., Li, Y., Kim, S., Miller, S., Bartlett, W., Black, S., ... Booth, I. R. (2005). Pivotal role of the glycine-rich TM3 helix in gating the MscS mechanosensitive channel. *Nature Structural & Molecular Biology*, *12*(2), 113–119. https://doi.org/10.1038/nsmb895

Feinshreiber, L., Singer-Lahat, D., Friedrich, R., Matti, U., Sheinin, A., Yizhar, O., ... Lotan, I. (2010). Non-conducting function of the Kv2.1

channel enables it to recruit vesicles for release in neuroendocrine and nerve cells. *Journal of Cell Science*, *123*(Pt 11), 1940–1947. https://doi.org/10.1242/jcs.063719

Gamini, R., Sotomayor, M., Chipot, C., & Schulten, K. (2011). Cytoplasmic domain filter function in the mechanosensitive channel of small conductance. *Biophysical Journal*, *101*(1), 80–89. https://doi.org/10.1016/j.bpj.2011.05.042

Garabagi, F., Gilbert, E., Loos, A., McLean, M. D., & Hall, J. C. (2012). Utility of the P19 suppressor of gene-silencing protein for production of therapeutic antibodies in Nicotiana expression hosts. *Plant Biotechnology Journal*, *10*(9), 1118–1128. https://doi.org/10.1111/j.1467-7652.2012.00742.x

Gottlieb, P. A., Suchyna, T. M., Ostrow, L. W., & Sachs, F. (2004). Mechanosensitive ion channels as drug targets. *Current Drug Targets: CNS & Neurological Disorders*, *3*(4), 287–295.

Gu, Y., & Gu, C. (2014). Physiological and pathological functions of mechanosensitive ion channels. *Molecular Neurobiology*, *50*(2), 339–347. https://doi.org/10.1007/s12035-014-8654-4

Hamilton, E. S., & Haswell, E. S. (2017). The tension-sensitive ion transport activity of MSL8 is critical for its function in pollen hydration and germination. *Plant and Cell Physiology*, *58*, 1–16. https://doi.org/10.1093/pcp/pcw230

Hamilton, E. S., Jensen, G. S., Maksaev, G., Katims, A., Sherp, A. M., & Haswell, E. S. (2015). Mechanosensitive channel MSL8 regulates osmotic forces during pollen hydration and germination. *Science*, *350* (6259), 438–441. https://doi.org/10.1126/science.aac6014

Haswell, E. S. (2007). MscS-like proteins in plants. In O. P. Hamill (Ed.), *Mechanosensitive ion channels, part A. Vol 58. Current topics in membranes* (pp. 329–359). London: Academic Press.

Haswell, E. S., Peyronnet, R., Barbier-Brygoo, H., Meyerowitz, E. M., & Frachisse, J.-M. (2008). Two MscS homologs provide mechanosensitive channel activities in the Arabidopsis root. *Current Biology*, *18*(10), 730–734. https://doi.org/10.1016/j.cub.2008.04.039

Humphrey, W., Dalke, A., & Schulten, K. (1996). VMD: Visual molecular dynamics. *Journal of Molecular Graphics*, *14*(1), 33–38, 27–8.

Iscla, I., Wray, R., Blount, P., Larkins-Ford, J., Conery, A. L., Ausubel, F. M., … Boulos, R. A. (2015). A new antibiotic with potent activity targets MscL. *Journal of Antibiotics*, *68*(7), 453–462. https://doi.org/10.1038/ja.2015.4

Kamaraju, K., Belyy, V., Rowe, I., Anishkin, A., & Sukharev, S. (2011). The pathway and spatial scale for MscS inactivation. *Journal of General Physiology*, *138*(1), 49–57. https://doi.org/10.1085/jgp.201110606

Kloda, A., & Martinac, B. (2001). Molecular identification of a mechanosensitive channel in archaea. *Biophysics*, *80*(1), 229–240. https://doi.org/10.1016/S0006-3495(01)76009-2

Koprowski, P., Grajkowski, W., Balcerzak, M., Filipiuk, I., Fabczak, H., & Kubalski, A. (2015). Cytoplasmic domain of MscS interacts with cell division protein FtsZ: A possible non-channel function of the mechanosensitive channel in *Escherichia coli*. *PLoS ONE*, *10*(5), e0127029. https://doi.org/10.1371/journal.pone.0127029

Kruger, L. C., & Isom, L. L. (2016). Voltage-gated Na+ channels: Not just for conduction. *Cold Spring Harbor Perspectives Biology*, *8*(6), a029264. https://doi.org/10.1101/cshperspect.a029264

Lai, J. Y., Poon, Y. S., Kaiser, J. T., & Rees, D. C. (2013). Open and shut: Crystal structures of the dodecylmaltoside solubilized mechanosensitive channel of small conductance from *Escherichia coli* and *Helicobacter pylori* at 4.4 Å and 4.1 Å resolutions. *Protein Science*, *22*(4), 502–509. https://doi.org/10.1002/pro.2222

Lee, C. P., Maksaev, G., Jensen, G. S., Murcha, M. W., Wilson, M. E., Fricker, M., … Sweetlove, L. J. (2016). MSL1 is a mechanosensitive ion channel that dissipates mitochondrial membrane potential and maintains redox homeostasis in mitochondria during abiotic stress. *Plant Journal*, *88*(5), 809–825. https://doi.org/10.1111/tpj.13301

Levina, N., Tötemeyer, S., Stokes, N. R., Louis, P., Jones, M. A., & Booth, I. R. (1999). Protection of *Escherichia coli* cells against extreme turgor by activation of MscS and MscL mechanosensitive channels: Identification of genes required for MscS activity. *EMBO Journal*, *18*(7), 1730–1737. https://doi.org/10.1093/emboj/18.7.1730

Maksaev, G., & Haswell, E. S. (2011). Expression and characterization of the bacterial mechanosensitive channel MscS in *Xenopus laevis* oocytes. *Journal of General Physiology*, *138*(6), 641–649. https://doi.org/10.1085/jgp.201110723

Maksaev, G., & Haswell, E. S. (2012). MscS-Like10 is a stretch-activated ion channel from *Arabidopsis thaliana* with a preference for anions. *Proceedings of the National Academy of Sciences of the United States of America*, *109*(46), 19015–19020. https://doi.org/10.1073/pnas.1213931109

Maksaev, G., & Haswell, E. S. (2013). Recent characterizations of MscS and its homologs provide insight into the basis of ion selectivity in mechanosensitive channels. *Channels (Austin)*, *7*(3), 215–220. https://doi.org/10.4161/chan.24505

Maksaev, G., & Haswell, E. S. (2015). Expressing and characterizing mechanosensitive channels in Xenopus oocytes. *Methods in Molecular Biology*, *1309*(Chapter 13), 151–169. https://doi.org/10.1007/978-1-4939-2697-8_13

Malcolm, H. R., & Blount, P. (2015). Mutations in a conserved domain of *E. coli* MscS to the most conserved superfamily residue leads to kinetic changes. *PLoS ONE*, *10*(9), e0136756. https://doi.org/10.1371/journal.pone.0136756

Malcolm, H. R., Heo, Y.-Y., Caldwell, D. B., McConnell, J. K., Hawkins, J. F., Guayasamin, R. C., … Maurer, J. A. (2012). Ss-bCNGa: A unique member of the bacterial cyclic nucleotide gated (bCNG) channel family that gates in response to mechanical tension. *European Biophysics Journal*, *41*(12), 1003–1013. https://doi.org/10.1007/s00249-012-0855-z

Martinac, B., Buechner, M., Delcour, A. H., Adler, J., & Kung, C. (1987). Pressure-sensitive ion channel in *Escherichia coli*. *Proceedings of the National Academy of Sciences of the United States of America*, *84*(8), 2297–2301.

Martinac, B., & Kloda, A. (2003). Evolutionary origins of mechanosensitive ion channels. *Progress in Biophysics and Molecular Biology*, *82*(1–3), 11–24.

Martinac, B., Nomura, T., Chi, G., Petrov, E., Rohde, P. R., Battle, A. R., … Landsberg, M. J. (2014). Bacterial mechanosensitive channels: Models for studying mechanosensory transduction. *Antioxidants & Redox Signaling*, *20*(6), 952–969. https://doi.org/10.1089/ars.2013.5471

Miller, S., Bartlett, W., Chandrasekaran, S., Simpson, S., Edwards, M., & Booth, I. R. (2003). Domain organization of the MscS mechanosensitive channel of *Escherichia coli*. *EMBO Journal*, *22*(1), 36–46. https://doi.org/10.1093/emboj/cdg011

Naismith, J. H., & Booth, I. R. (2012). Bacterial mechanosensitive channels–MscS: Evolution's solution to creating sensitivity in function. *Annual Review of Biophysics*, *41*(1), 157–177. https://doi.org/10.1146/annurev-biophys-101211-113227

Nakayama, Y., Fujiu, K., Sokabe, M., & Yoshimura, K. (2007). Molecular and electrophysiological characterization of a mechanosensitive channel expressed in the chloroplasts of Chlamydomonas. *Proceedings of the National Academy of Sciences of the United States of America*, *104* (14), 5883–5888. https://doi.org/10.1073/pnas.0609996104

Nakayama, Y., Yoshimura, K., & Iida, H. (2012). Organellar mechanosensitive channels in fission yeast regulate the hypo-osmotic shock response. *Nature Communications*, *3*, 1020. https://doi.org/10.1038/ncomms2014

Nakayama, Y., Yoshimura, K., & Iida, H. (2013). Electrophysiological characterization of the mechanosensitive channel MscCG in *Corynebacterium glutamicum*. *Biophysical Journal*, *105*(6), 1366–1375. https://doi.org/10.1016/j.bpj.2013.06.054

Okonechnikov, K., Golosova, O., & Fursov, M. (2012). UGENE team. Unipro UGENE: A unified bioinformatics toolkit. *Bioinformatics*, *28*(8), 1166–1167. https://doi.org/10.1093/bioinformatics/bts091

Peyronnet, R., Tran, D., Girault, T., & Frachisse, J.-M. (2014). Mechanosensitive channels: Feeling tension in a world under pressure. *Front Plant Science*, *5*(204), 558. https://doi.org/10.3389/fpls.2014.00558

Pivetti, C. D., Yen, M. R., Miller, S., Busch, W., Tseng, Y. H., Booth, I. R., & Saier, M. H. Jr (2003). Two families of mechanosensitive channel proteins. *Microbiology and Molecular Biology Reviews*, *67*(1), 66–85. https://doi.org/10.1128/mmbr.67.1.66-85.2003

Ranade, S. S., Syeda, R., & Patapoutian, A. (2015). Mechanically activated ion channels. *Neuron*, *87*(6), 1162–1179. https://doi.org/10.1016/j.neuron.2015.08.032

Rasmussen, T., Edwards, M. D., Black, S. S., Rasmussen, A., Miller, S., & Booth, I. R. (2010). Tryptophan in the pore of the mechanosensitive channel MscS: Assessment of pore conformations by fluorescence spectroscopy. *Journal of Biological Chemistry*, *285*(8), 5377–5384. https://doi.org/10.1074/jbc.M109.071472

Schwacke, R., Schneider, A., van der Graaff, E., Fischer, K., Catoni, E., Desimone, M., … Kunze, R. (2003). ARAMEMNON, a novel database for Arabidopsis integral membrane proteins. *Plant Physiology*, *131*(1), 16–26. https://doi.org/10.1104/pp.011577

Steinbacher, S., Bass, R., Strop, P., & Rees, D. C. (2007). Structures of the prokaryotic mechanosensitive channels MscL and MscS. In: O. P. Hamill (Ed.), *Mechanosensitive ion channels, part A. Vol 58. Current topics in membranes* (pp. 1–24). London: Academic Press.

Sukharev, S. (2002). Purification of the small mechanosensitive channel of *Escherichia coli* (MscS): The subunit structure, conduction, and gating characteristics in liposomes. *Biophysical Journal*, *83*(1), 290–298. https://doi.org/10.1016/S0006-3495(02)75169-2

Vásquez, V. (2013). MscS inactivation: An exception rather than the rule. An extremophilic MscS reveals diversity within the family. *Biophysical Journal*, *104*(7), 1391–1393. https://doi.org/10.1016/j.bpj.2013.02.010

Veley, K. M., Maksaev, G., Frick, E. M., January, E., Kloepper, S. C., & Haswell, E. S. (2014). Arabidopsis MSL10 has a regulated cell death signaling activity that is separable from its mechanosensitive ion channel activity. *Plant Cell*, *26*(7), 3115–3131. https://doi.org/10.1105/tpc.114.128082

Wang, W., Black, S. S., Edwards, M. D., Miller, S., Morrison, E. L., Bartlett, W., … Booth, I. R. (2008). The structure of an open form of an *E. coli* mechanosensitive channel at 3.45 A resolution. *Science*, *321*(5893), 1179–1183. https://doi.org/10.1126/science.1159262

Zhang, Y. (2008). I-TASSER server for protein 3D structure prediction. *BMC Bioinformatics*, *9*(1), 40. https://doi.org/10.1186/1471-2105-9-40

Zhang, X., Wang, J., Feng, Y., Ge, J., Li, W., Sun, W., … Yang, M. (2012). Structure and molecular mechanism of an anion-selective mechanosensitive channel of small conductance. *Proceedings of the National Academy of Sciences of the United States of America*, *109*(44), 18180–18185. https://doi.org/10.1073/pnas.1207977109

Zou, Y., Chintamanani, S., He, P., Fukushige, H., Yu, L., Shao, M., … Zhou, J. M. (2016). A gain-of-function mutation in Msl10 triggers cell death and wound-induced hyperaccumulation of jasmonic acid in Arabidopsis. *Journal of Integrative Plant Biology*, *58*(6), 600–609. https://doi.org/10.1111/jipb.12427

11

Organ-specific phytohormone synthesis in two *Geranium* species with antithetical responses to far-red light enrichment

Charlotte M. M. Gommers[1,2] | Sara Buti[1] | Danuše Tarkowská[3] | Aleš Pěnčík[3] | Jason P. Banda[1] | Vincent Arricastres[1] | Ronald Pierik[1]

[1]Plant Ecophysiology, Institute of Environmental Biology, Utrecht University, Utrecht, The Netherlands

[2]Plant Development and Signal Transduction Program, Center for Research in Agricultural Genomics (CRAG), Barcelona, Spain

[3]Laboratory of Growth Regulators, Centre of the Region Haná for Biotechnological and Agricultural Research, Institute of Experimental Botany ASCR, Faculty of Science, Palacký University, Olomouc, Czechia

Correspondence
Ronald Pierik, Plant Ecophysiology, Institute of Environmental Biology, Utrecht University, Utrecht, The Netherlands.
Email: r.pierik@uu.nl

Present address
Jason P. Banda, Centre for Plant Integrative Biology, School of Biosciences, University of Nottingham, Sutton Bonington, UK.

Funding information
Netherlands Organisation for Scientific Research (NWO), Grant/Award Number: 022.001.018, 864.12.003; the Ministry of Education, Youth and Sports of the Czech Republic, Grant/Award Number: LO 1204

Abstract

Plants growing in high densities experience a reduced red (R) to far-red (FR) light ratio and shade-intolerant species respond with accelerated elongation growth to reach the top of the canopy: the shade avoidance syndrome (SAS). FR-enriched light inactivates phytochrome photoreceptors, which results in subsequent action of several plant hormones regulating growth. SAS is adaptive for shade-intolerant plants, but is suppressed in shade-tolerant plant species. Inspired by a previously published transcriptome analysis, we use two species of the genus *Geranium* here to study the involvement of auxin, brassinosteroids (BRs), and gibberellins (GAs) in supplemental FR-induced elongation growth. *G. pyrenaicum*, a shade-avoiding species, strongly induces auxin and gibberellin levels, but not BR, in elongating petioles. We show that, in this species, FR light perception, hormone synthesis, and growth are local and restricted to the petiole, and not the leaf lamina. Using chemical hormone inhibitors, we confirm the essential role of auxin and GAs in supplemental FR-induced elongation growth. Shade-tolerant *G. robertianum* does not display the change in hormone levels upon FR light enrichment, resulting in the lack of a shade avoidance response.

KEYWORDS

auxin, brassinosteroids, *Geranium*, gibberellins, phytochrome signaling, shade avoidance

1 | INTRODUCTION

Shade-intolerant plants growing in dense stands, for example, a grassland or agricultural field, compete with neighbors to secure light interception. To do so, they enhance the elongation of stems and leaves, thus growing taller than surrounding plants. This adaptive trait is referred to as the shade avoidance syndrome (SAS) and largely depends on the modulation of the homeostasis of several plant hormones (reviewed in: Ballaré & Pierik, 2017; de Wit, Costa Galvão, & Fankhauser, 2016; de Wit, Keuskamp, et al., 2016). SAS is a favorable strategy for shade-intolerant species, but it is suppressed by shade-tolerant plants in the forest understory that are unable to outgrow neighboring trees (Gommers, Visser, St Onge, Voesenek, & Pierik, 2013; Gommers et al., 2017).

The reflection of far-red (FR) and absorbance of red (R) light by green tissues causes a decline of the R to FR (R:FR) light ratio, which is perceived by the R and FR light-sensitive phytochrome (Phy) photoreceptors of surrounding plants. Inactivation of Phy in low R:FR light releases the suppression of a group of basic helix–loop–helix (bHLH) transcription factors, the PHYTOCHROME INTERACTING FACTORS (PIFs). PIFs trigger the transcriptional cascade required for the elongation response, at least in part by regulating biosynthesis and signaling of several plant hormones (reviewed in Casal, 2013).

Auxins are often considered the major regulatory signals for phytochrome-mediated shade avoidance. In *Arabidopsis thaliana*, exposure to low R:FR conditions induces auxin production in cotyledons or leaf laminas via transcriptional induction of *YUCCA* genes by PIF4, PIF5, and PIF7 (de Wit, Ljung, & Fankhauser, 2015; Hornitschek et al., 2012; Pantazopoulou et al., 2017). In addition, PIF4 and PIF5 induce the expression of several genes contributing to auxin responsiveness during low R:FR light (Hornitschek, Lorrain, Zoete, Michielin, & Fankhauser, 2009; Roig-Villanova et al., 2007). Therefore, *Arabidopsis* mutants deficient in auxin biosynthesis, transport, or signaling are unable to induce low R:FR-mediated hypocotyl and petiole elongation, leaf movement, and phototropism (Goyal et al., 2016; Keuskamp, Pollmann, Voesenek, Peeters, & Pierik, 2010; Michaud, Fiorucci, Xenarios, & Fankhauser, 2017; Nozue et al., 2015; Pantazopoulou et al., 2017; Tao et al., 2008).

Auxins coact with the steroidal plant hormones brassinosteroids (BRs) to induce elongation responses to FR enrichment (Kozuka et al., 2010) or blue light depletion (Keuskamp et al., 2011) in *Arabidopsis*. Although BRs biosynthesis is necessary for the response to end of day FR enrichment light (Kozuka et al., 2010), other studies report a reduction in BRs in shade-grown *Arabidopsis* (Bou-Torrent et al., 2014) or dark-grown pea plants (Symons et al., 2002). BR sensitivity in low R:FR light conditions is enhanced via the transcription factors BR-ENHANCED EXPRESSION (BEE) and BES1-INTERACTING MYC-LIKE (BIM) (Cifuentes-Esquivel et al., 2013), and the PIF4–BRASSINAZOLE RESISTANT1 (BZR1) transcription factor complex (Oh, Zhu, & Wang, 2012), which induce elongation-promoting pathways. In addition, BRASSINOSTEROID INSENSITIVE2 (BIN2), a BR signaling kinase, phosphorylates PIF4, which make BRs direct regulators of part of the light signaling cascade (Bernardo-García et al., 2014).

Another group of PIF interactors are DELLA proteins, such as REPRESSOR OF GA1-3 (RGA), GIBBERELLIN INSENSITIVE (GAI), and RGA-LIKE1 (RGL1). DELLAs bind PIF4 and PIF5, inhibit their promoter binding ability, and thus suppress SAS (de Lucas et al., 2008; Feng et al., 2008). DELLA proteins are degraded under FR-enriched light conditions due to enhanced action of bioactive gibberellins (GAs) (Djakovic-Petrovic, de Wit, Voesenek, & Pierik, 2007). The GA biosynthetic genes *GA 20-OXIDASE-1* (*GA20ox1*), *GA20ox2*, and *GA3ox1* are transcriptionally induced by FR light enrichment, which leads to higher levels of GA_1 and GA_4, biologically active GAs (García-Martinez & Gil, 2001; Reed, Foster, Morgan, & Chory, 1996).

As major regulators of plant development and growth, these phytohormones play a key role in regulation of SAS. Nevertheless, SAS is suppressed in plants that are unable to outgrow a shaded environment (Gommers et al., 2013). To date, it remains unknown how SAS is suppressed in such species and if auxins, BRs, and GAs play a role in this process. Our previous work has shown that two *Geranium* species from contrasting natural habitats (*G. pyrenaicum* and *G. robertianum*) display contrasting growth and gene expression patterns when exposed to FR-enriched light conditions (Gommers et al., 2017). Here, we show that differential regulation of hormone levels correlate with enhanced elongation growth under FR enrichment in *G. pyrenaicum* and its absence in *G. robertianum*. In addition, we show that the site of FR light perception, hormone accumulation, and elongation is restricted to the petiole, and not to the lamina, of *G. pyrenaicum* leaves.

2 | MATERIALS AND METHODS

2.1 | Gene ontology enrichment analysis

For gene ontology analysis of previously published RNA sequencing data from Gommers et al. (2017) (Array Express E-MTAB-5371), significantly up- and downregulated *Geranium* OMCL groups upon FR light enrichment, with a BLAST E-value $<10^{-10}$ with *A. thaliana* genes, were clustered using the R package GOseq (Young, Wakefield, Smyth, & Oshlack, 2010), with correction for the total length of all transcripts in the *Geranium* OMCL group.

2.2 | Plant material and growth conditions

For rosette experiments, *G. pyrenaicum* and *G. robertianum* seeds were sown and grown in long day conditions as described before (Gommers et al., 2017). Treatments started 2 weeks after transplanting.

For seedling experiments, seeds were surface sterilized in 70% EtOH followed by a 5% (*G. pyrenaicum*) or 10% (*G. robertianum*) bleach solution and rinsed with sterile water for four times. Seeds were sown on sterile agar plates (0.5 strength Murashige and Skoog medium, with 1 g/L MES buffer, 0.8% Plant Agar w/v) and stratified in the dark at 4°C for five (*G. pyrenaicum*) or seven (*G. robertianum*) days. Afterward, plates were transferred to long day light conditions, placed vertically in an angle of 70° to germinate for five (*G. pyrenaicum*) or eight (*G. robertianum*) days. Thereafter, seedlings were transferred to new agar plates, either or not with a pharmacological treatment. 24 hr later, light treatments would start.

2.3 | Light treatments

FR-enriched light conditions were obtained by supplementing standard growth chamber white light (WL, R:FR 1.8, 180 μmol m^{-2} s^{-1} PAR, Philips HPI 400 W) with far-red LEDs (± 730 nm, Philips), to obtain a R:FR of 0.2 (WL + FR) without changing PAR (Gommers et al., 2017). Seedlings were exposed to a far-red treatment for 5 (PEO-IAA and NPA) or 4 days (PAC and BRZ), rosettes for 24 hr.

Local FR light treatments were applied using custom built light bundles, each containing four LED lights (diodes from Shinkoh Electronics) (Pantazopoulou et al., 2017). For $FR_{petiole}$, the apical half of the petiole was enriched with FR light, and for FR_{lamina}, the whole lamina was enriched with FR from below, to prevent illumination of other plant parts.

2.4 | RNA isolation and gene expression

RNA extraction, cDNA synthesis, and RT-qPCR were performed as described before (Gommers et al., 2017). The apical 1 cm of the petioles of the second leaf of three independent *Geranium* plants, were pooled as one biological replicate. RNA was extracted using the RNeasy kit (Qiagen) with on-column DNAseI treatment, followed by cDNA synthesis using the Superscript III reverse transcriptase (Invitrogen) with RNAse inhibitors and random primers. Real-time quantitative PCR (RT-qPCR) was performed using Sybr Green Supermix (Bio-Rad) in a Viia7 PCR. A list of the used primers is provided in Supporting Information Table S1. The *Geranium* orthologue for *AT3G57890* was used as a reference gene. Relative gene expression was calculated as $2^{-\Delta\Delta CT}$.

2.5 | Hormone analysis

For auxins, BRs, and GAs measurements, *Geranium* leaf lamina and petiole samples were harvested after 2 and 11.5 hr (12:00 and 21:30, respectively) of FR-enriched (WL+FR) or control (WL) light treatment, as three biological replicates. These time points are identical to the ones used for transcriptomics in our previous study (Gommers et al., 2017).

For BRs content, these samples were analyzed as previously described (Tarkowská, Novák, Oklestkova, & Strnad, 2016) with a few modifications. In brief, fresh *Geranium* tissue samples of 50 mg were homogenized to a fine consistency using 3-mm zirconium oxide beads (Retsch GmbH & Co. KG, Haan, Germany) and a MM 301 vibration mill at a frequency of 30 Hz for 3 min (Retsch GmbH & Co. KG, Haan, Germany). The samples were then extracted overnight with stirring at 4°C using a benchtop laboratory rotator Stuart SB3 (Bibby Scientific Ltd., Staffordshire, UK) after adding with 1 mL ice-cold 60% acetonitrile and 10 pmol of [$^2H_3$]brassinolide, [$^2H_3$]castasterone, [$^2H_3$]24-*epi*brassinolide, [$^2H_3$]24-*epi*castasterone, [$^2H_3$]28-norbrassinolide, [$^2H_3$]28-norcastasterone, and [$^2H_3$]typhasterol as internal standards (OlChemIm Ltd., Olomouc, Czech Republic). The samples were further centrifuged, purified on polyamide SPE columns (Supelco, Bellefonte, PA, USA), and then analyzed by UHPLC-MS/MS (Micromass, Manchester, UK). The data were analyzed using Masslynx 4.1 software (Waters, Milford, MA, USA), and BR content was finally quantified by the standard isotope dilution method (Rittenberg & Foster, 1940). Each sample was analyzed three times, and the average of these technical replicates was used as one biological replicate.

Auxins content in these samples was determined by slightly modified method described by Pěnčík et al. (2009). In brief, 10 mg of tissue was extracted with 1 ml cold phosphate buffer (50 mM; pH 7.0) containing 0.02% sodium diethyldithiocarbamate with internal standards: [$^{13}C_6$]IAA, [$^{13}C_6$]oxIAA, [$^{15}N,^2H_5$]IAAsp, and [$^{15}N,^2H_5$]IAGlu. Samples were then centrifuged at 36.000 *g* for 10 min, acidified with 1 M HCl to pH 2.7, and purified by solid phase extraction (SPE) using C8 columns (Bond Elut, 500 mg, 3 ml; Varian). After evaporation under reduced pressure, samples were diluted in 30 µl of 10% methanol and analyzed for auxin content using Acquity UHPLC™ (Waters, USA) linked to a triple quadrupole mass detector (Xevo TQ-S™; Waters, USA). Each sample was analyzed three times, and the average of these technical replicates was used as one biological replicate. Deionized (Milli-Q) water obtained from a Simplicity185 water system (Millipore, Bedford, MA, USA) was used to prepare all aqueous solutions. All other chemicals (analytical grade or higher purity) were from Sigma-Aldrich Chemie (Steinheim, Germany).

GAs content in these samples was analyzed as described before by Urbanová, Tarkowská, Novák, Hedden, & Strnad, 2013; Shahnejat-Bushehri, Tarkowska, Sakuraba, & Balazadeh, 2016.

2.6 | Pharmacological treatments

For rosette experiments, different concentrations of 1-naphthaleneacetic acid (1-NAA (Duchefa Haarlem); 0.1, 1, 10, 50, 100, or 500 µM, in 0.1% EtOH, 0.1% Tween-20), 24-*epi*brassinolide (*epi*BL (Sigma-Aldrich); 10, 50, or 100 µM, in 0.01% DMSO, 0.1% Tween-20), and GA_3 ((Sigma-Aldrich) 10, 50, or 100 µM, in 0.01% EtOH, 0.1% Tween-20) were sprayed on one leaf (petiole and lamina) of each plant (total 250 µL/leaf). Mock controls were either 0.01% DMSO, 0.01%, or 0.1% EtOH.

Different concentrations of auxin biosynthesis inhibitor yucasin ((Nishimura et al., 2013) 10, 25, or 75 µM, in 0.01% DMSO, 0.1% Tween-20), auxin transport inhibitor 1-*N*-naphthylphthalamic acid (NPA (Duchefa Haarlem); 25 µM, in 0.1% EtOH, 0.1% Tween-20), auxin perception inhibitor α-(phenylethyl-2-one)-IAA (PEO-IAA (Hayashi et al., 2008); 50 µM, 0.01% DMSO, 0.1% Tween-20), and BR biosynthesis inhibitor brassinazole (BRZ (Sigma-Aldrich); 10, 50, or 100 µM, in 0.02% DMSO, 0.1% Tween-20) were sprayed on the plants 24 hr prior to, and at the start of, the light treatment. For GA biosynthesis inhibition, pots were not watered for 2 days and then treated with 20 ml of paclobutrazol (PAC (Duchefa Haarlem); 100 or 500 µM, in 0.5% EtOH), or mock, 72 hr prior to the start of the light treatment. Mock treatments consisted of the appropriate concentration of solvent.

For seedling experiments, sterile solutions of PEO-IAA (final concentration 10, 50, or 100 µM, 0.1% DMSO), NPA (final concentration 25, 50, or 100 µM, 0.1% DMSO), BRZ (final concentration 0.5 or 5 µM, 0.1% DMSO), or PAC (final concentration 10, 50, or 100 µM, 0.1% EtOH) were added to sterile agar medium. Mock plates contained the appropriate concentration of solvent. Seedlings were transferred to these plates 24 hr prior to the start of the light treatment.

2.7 | Growth measurements

Geranium petiole elongation was measured over a given time with a digital caliper (t_x–t_0). Leaf laminas were harvested after 48 hours of light treatment and scanned (600dpi). The area was measured using ImageJ. In seedling experiments, cotyledon petioles were measured using the digital caliper at the end of the light treatment.

2.8 | Statistical analysis

Differences between measurements were analyzed in Microsoft Excel with a Student's *t* test, preceded by an f test to check equal variances. Data in Figure 5 and Supporting Information Figure S5 are analyzed in R with a 1-way ANOVA and a post hoc Tukey test, preceded by Levene's test, to check equal variances. Data was LN-transformed if needed.

3 | RESULTS

To get a first impression of auxin, BR, and GA regulation in response to FR enrichment in *G. pyrenaicum* and *G. robertianum*, we reanalyzed the RNA sequencing dataset previously published by Gommers et al. (2017). We applied gene ontology (GO) clustering to find (over) representation of biological processes among the up- and downregulated transcripts in *Geranium* petioles after 2 or 11.5 hr exposure to

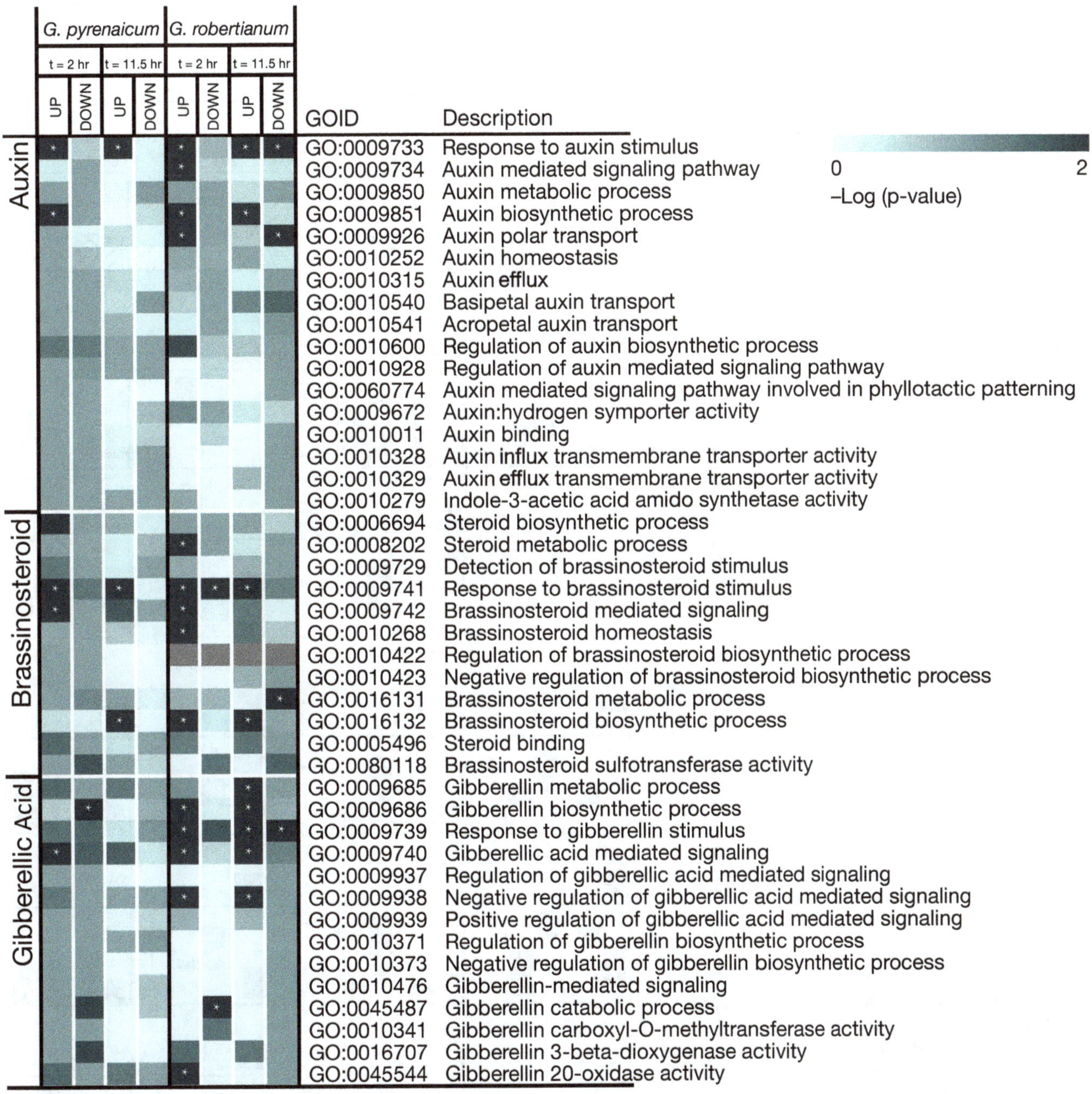

FIGURE 1 *Geranium pyrenaicum* and *G. robertianum* transcriptionally induce phytohormone-related processes in FR-enriched light. Heat map presenting (over-) representation of Gene Ontology terms associated with auxin-, brassinosteroid-, and gibberellin-related processes among up- and downregulated OMCL groups after 2 hr or 11.5 hr of FR-enriched (R:FR 0.2) light in *G. pyrenaicum* and *G. robertianum*. Terms are clustered according to the different hormones. Data from the RNA sequencing study published in Gommers et al., 2017. Colors represent -Log of the *p*-value, and significantly overrepresented terms (-Log(p) > 2) are marked with asterisks

FR-enriched light. A complete list of significantly enriched GO terms upon FR light enrichment in both species and both time points is presented in Gommers et al., 2017;. The heat map in Figure 1 shows the subset for auxin-, BR-, and GA-associated gene ontology clusters. Several GO terms are significantly overrepresented among especially FR-induced transcripts in both species.

Several FR light-regulated genes are selected from these GO clusters and show clear differences in expression between the

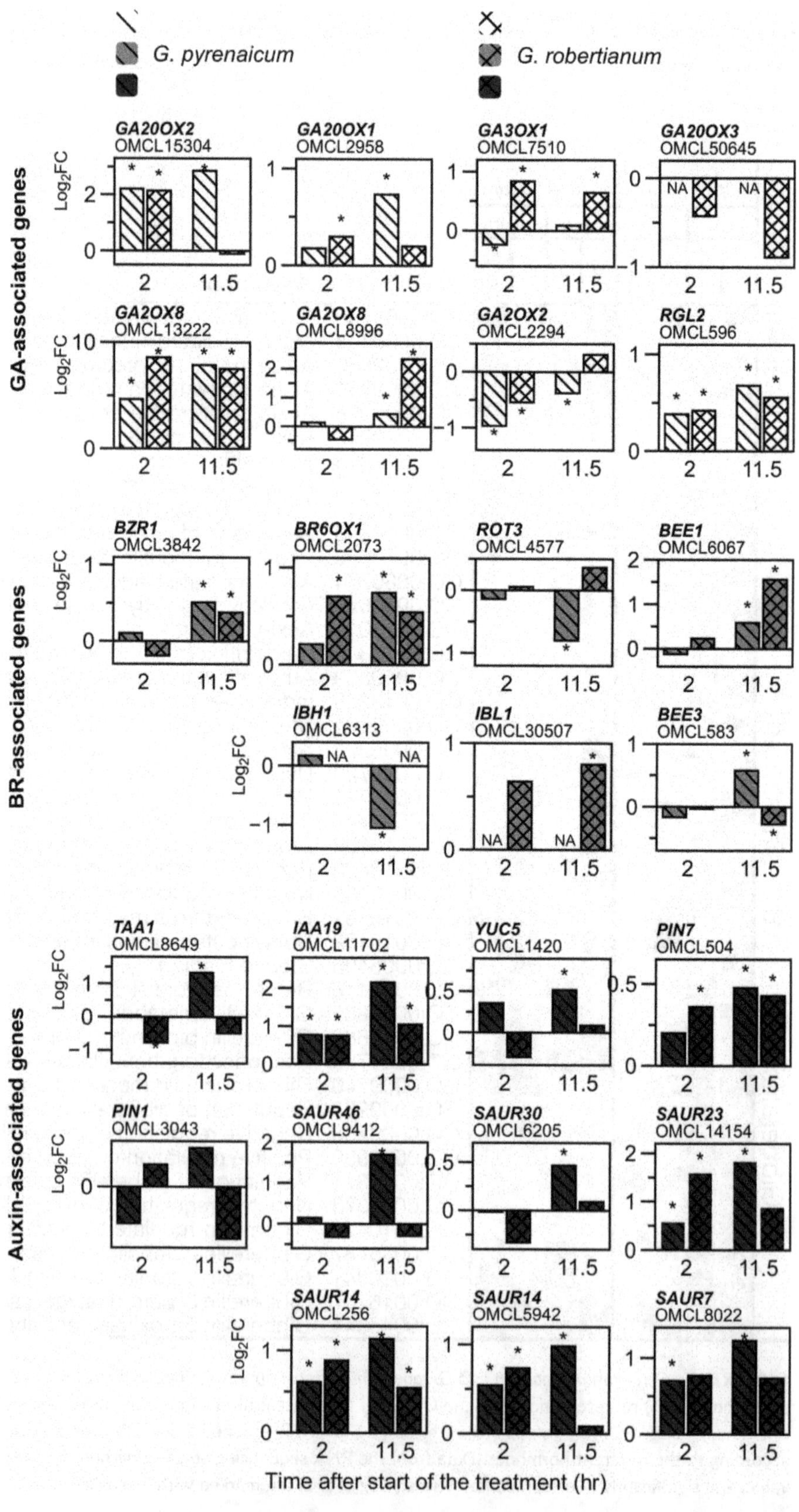

FIGURE 2 Expression patterns of gibberellin-, brassinosteroid-, and auxin-associated OMCL groups in *Geranium pyrenaicum* and *G. robertianum* after 2 or 11.5 hr of FR-enriched light. Log2-transformed fold changes (WL + FR/WL) of OMCL families subtracted from the GO categories presented in Figure 1. Data from the RNA sequencing study published in Gommers et al., 2017. Asterisks mark significant induction/suppression ($p < 0.01$)

two species, as shown by more closely looking into the RNA-Seq data (Figure 2) and by additional RT-qPCR experiments (Figure 3). FR enrichment especially enhances transcription of GA synthesis gene orthologues (*GA20ox2*, *GA20ox1*, *GA3ox1*) in both species but to a lesser extent, specifically at the later time point, in *G. robertianum*, whereas this species did show stronger induction of GA catabolism gene orthologues (*GA2ox8*, *GA2ox2*) than *G. pyrenaicum*. Interestingly, the orthologue of BR synthesis gene *ROTUNDIFOLIA 3* (*ROT3*) is strongly suppressed in *G. pyrenaicum*, while *BR6OX1*, just like BR signal transducers *BZR1*, *BEE1*, and *BEE3* are marginally induced in the RNA sequencing dataset (Figure 2). This induction was not reproduced in the separate RT-qPCR experiment in both species (Figure 3). Orthologues of negative regulators of BR signaling (*IBH1* and *IBH-LIKE1* (*IBL*)) are suppressed in *G. pyrenaicum* but induced in *G. robertianum*. Transcripts for genes involved in auxin synthesis (*TAA1*, *YUC5*), transport (*PIN7*, *PIN1*), and signaling (*AUX/IAA* and several *SAUR* orthologues), are clearly more strongly induced by supplemental FR light in *G. pyrenaicum* than in *G. robertianum*, as shown by RNA-Seq and RT-qPCR data (Figures 2 and 3).

This transcriptome analysis suggested that strong differences exist between the two species in regulation of hormone synthesis and signaling in response to FR enrichment. Next, we measured free auxin, BR, and GA levels in *G. pyrenaicum* and *G. robertianum* petioles and laminas after 2 or 11.5 hr of supplemental FR exposure. The data for the most active forms of all three hormone groups (indole-3-acetic acid, IAA; brassinolide, BL; and gibberellin A_1, GA_1) are presented in Figure 4. After 2 hr of FR light enrichment, no major differences were detectable in both species and tissues. After 11.5 hr, however, IAA and GA levels were strongly increased, while BL decreased, in *G. pyrenaicum* petioles. Interestingly, the hormone levels in *G. pyrenaicum* laminas were unaffected by the light treatment. In great contrast, in *G. robertianum*, IAA, BL, and GA_1 levels were relatively low in petioles and generally unaffected by light treatment. More detailed analysis of several biosynthetic precursors and metabolites of all three hormone groups follow a similar pattern (Supporting Information Figures S1, S2, S3). Interestingly, multiple BL biosynthetic precursors showed great differences in absolute quantity between the species (Supporting Information Figure S2): C_{27} BRs originating from the cholesterol-dependent biosynthesis pathway (28-*nor*castasterone and 28-*nor*teasterone, Joo et al., 2012) were more abundant in *G. pyrenaicum*, whereas C_{28} and C_{29}-BRs formed by the campesterol- and sitosterol-dependent pathway (typhasterol and 28-homocastasterone), respectively, were found to be more abundant in *G. robertianum*. As seen in Supporting Information Figure S3, several gibberellins were not detectable at all in both

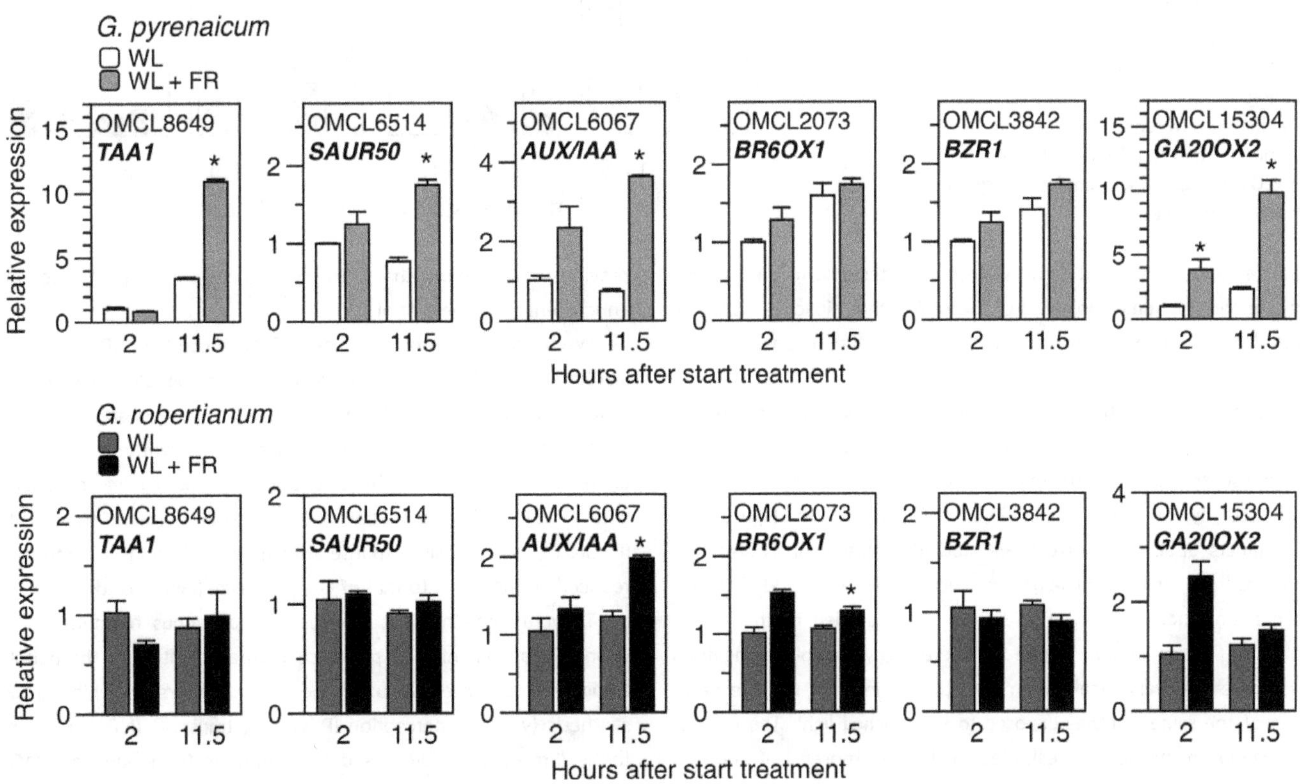

FIGURE 3 FR light enrichment affects hormone synthesis- and signaling-associated gene expression in *Geranium pyrenaicum*, but less so in *G. robertianum*. Expression of *Geranium* orthologues of *TAA1*, an *AUX/IAA*, *SAUR50*, *BR6OX1*, *BZR1*, and *GA20OX2* in *G. pyrenaicum* (upper graphs) and *G. robertianum* (lower graphs) petioles upon 2 or 11.5 hr of FR light enrichment (WL + FR). Data are relative to the expression of a reference gene (orthologue of *AT3G57890*) and the WL control at 2 hr and represent means ± SE, $n = 3$ biological replicates. Asterisks indicate a significant difference between WL and WL + FR at the same time point, Student's t test, $p < 0.05$

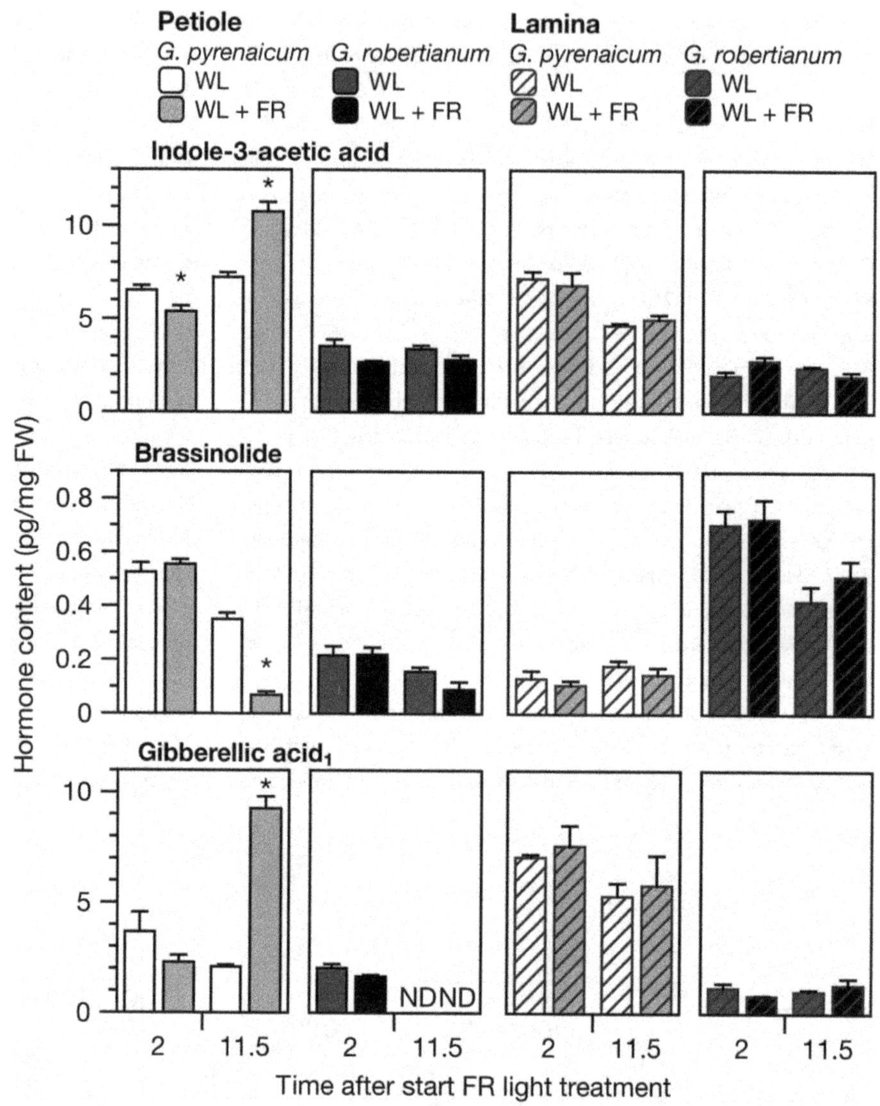

FIGURE 4 FR-enriched light affects hormone levels in the *Geranium pyrenaicum* petiole, but not in the lamina, nor *G. robertianum*. Free indole-3-acetic acid (IAA), brassinolide (BL), and gibberellin A_1 (GA_1) levels (pg/mg fresh weight) in *G. pyrenaicum* and *G. robertianum* petioles (plain bars) and lamina (striped bars) exposed to 2 or 11.5 hr of control white light (WL; R:FR 1.8) or WL supplemented with far-red light (WL + FR; R:FR 0.2). Data represent means ± *SEM*, $n = 3$ biological replicates. Asterisks represent significant differences in WL + FR compared to WL (Student's *t* test, $p < 0.05$). ND marks samples where no hormone was detected

species (GA_4, GA_7), and some were only detectable in one of the treatments or time points. For example, the bioactive GA_5 in *G. robertianum* was solely detectable in the tissues exposed to FR light enrichment.

The FR-induced hormonal changes seem to be restricted to the petiole, which is also the organ that changes its growth in response to this light treatment, given that lamina size was not affected by FR enrichment (Supporting Information Figure S4) (Gommers et al., 2017). Next, we studied if this organ is also the required site of FR light exposure. Local application of FR light to the petiole-induced elongation similar to FR applied to the whole plant in *G. pyrenaicum* (Figure 5). Interestingly, petiole elongation response was very weak when only the lamina, but not the petiole, of the same leaf was exposed to FR-enriched light. The petiole elongation response was restricted to the FR-exposed leaf and did not induce a systemic response in a younger leaf that did not receive FR enrichment (Supporting Information Figure S5). Petiole elongation in *G. robertianum* was completely unresponsive to FR enrichment of the petiole or lamina separately and showed a marginal response to whole-plant FR enrichment, consistent with previous observation that this species hardly responds to FR enrichment (Gommers et al., 2017).

We hypothesized that the lack of FR-induced IAA, BR and/or GA synthesis could be responsible for the lack of petiole elongation in *G. robertianum* to FR enrichment. Therefore, we treated *G. robertianum* petioles with different concentrations of synthetic hormones under control (white-) light conditions. *G. robertianum* responded with a clear petiole elongation response to all hormone treatments (Figure 6), but its auxin response at high concentrations was strongly reduced as compared to that of *G. pyrenaicum* (Figure 6a,d).

To verify whether the changes in endogenous hormone levels upon FR enrichment in *G. pyrenaicum* were essential for the petiole elongation response, we used chemical inhibitors to inhibit auxin synthesis (yucasin), perception (PEO-IAA), transport (NPA), BR synthesis (brassinazole, BRZ), and GA synthesis (paclobutrazol, PAC). Application of these inhibitors in different concentrations to the *G. pyrenaicum* leaf (petiole and lamina; auxin inhibitors and BRZ) or potting soil (PAC) prior to and during the experiment had no effect on the petiole growth over 24 hr, both in control and FR-enriched light (Supporting Information Figure S6).

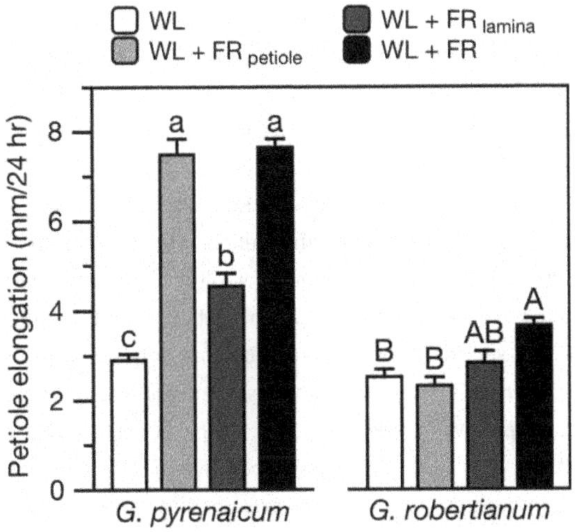

FIGURE 5 FR light illumination of the *Geranium* petiole, and not the lamina, induces elongation. Petiole elongation (mm 24 hr $^{-1}$) of *G. pyrenaicum* and *G. robertianum* plants grown in control white light (WL), WL supplemented with FR (WL + FR), or WL supplemented with a local FR treatment to only the petiole (WL + $FR_{petiole}$) or the lamina (WL + FR_{lamina}) for 24 hr. The treated leaf is measured. Data represent means ± *SEM*, $n = 8$ biological replicates. Different letters represent significant differences (1-way ANOVA, $p < 0.05$)

To overcome the possibility that perhaps these chemicals would not be absorbed through the trichome-rich epidermis of the *Geranium* leaf, we studied these chemicals in seedlings rather than mature plants. Unlike *Arabidopsis*, seedlings of both *G. pyrenaicum* and *G. robertianum* had nonresponsive hypocotyls in FR-enriched light. Nevertheless, *G. pyrenaicum*, but not *G. robertianum*, showed a strong elongation response in the petioles of the cotyledons, and we used this response to assay shade and hormone responses in the *G. pyrenaicum* seedlings (Supporting Information Figure S7).

Inhibition of auxin perception by PEO-IAA and transport by NPA, inhibited the growth of *G. pyrenaicum* cotyledon petioles, especially in seedlings grown under FR-enriched light (Figure 7a). This indicates that intact perception and distribution of this hormone is necessary for the SAS.

Even stronger growth inhibitions were visible in seedlings treated with PAC. Although elongation growth in general was inhibited, this effect was particularly strong for the supplemental FR light-triggered growth responses, already by the lowest concentration. More severe treatments made the light effect disappear completely (Figure 7c).

Unlike IAA and GA, BR seems to play a less important role in the elongation responses to FR enrichment as BRZ hardly affected elongation growth in the seedling assays, irrespective of the light treatment (Figure 7b).

4 | DISCUSSION

The plant hormones auxin, BR, and GA are well-established regulators of shade avoidance responses (de Wit, Costa Galvão, et al., 2016). In the current study, we aimed to look for differences in hormone synthesis and signaling upon plant neighbor detection through FR light enrichment in two species with opposite growth strategies during shade. *G. pyrenaicum* is a shade-intolerant species, often found in grasslands, and responds to simulated shade with strong petiole elongation. The closely related *G. robertianum* is a so-called shade-tolerant species that lacks the shade avoidance response. Reanalysis of a previously published RNA-Seq dataset (Gommers et al., 2017), and additional RT-qPCR analysis, confirms that supplemental FR light affects auxin-, BR-, and GA-associated gene expression in both these species, similar to what was shown for *Arabidopsis* (de Wit, Spoel, et al., 2013; de Wit, Keuskamp, et al., 2016; Kohnen et

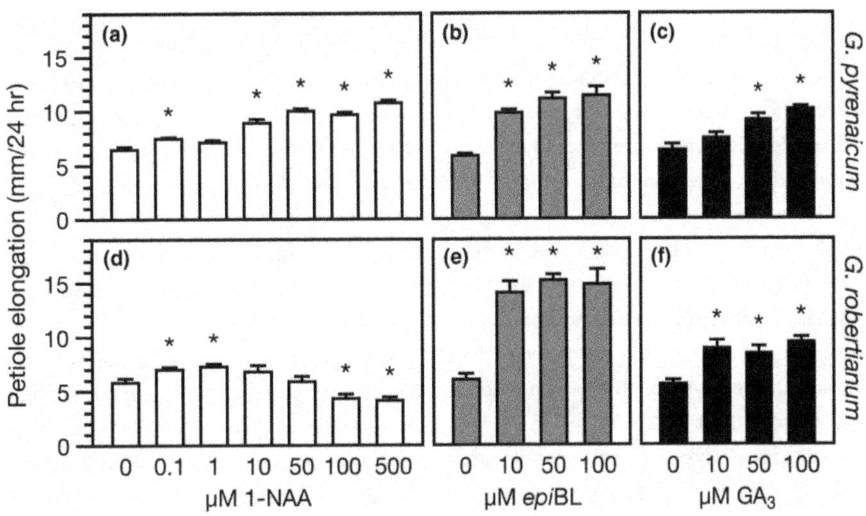

FIGURE 6 Auxin, brassinosteroid, and gibberellin induce growth in *Geranium pyrenaicum* and *G. robertianum* petioles. *G. pyrenaicum* (a–c) and *G. robertianum* (d–f) petiole elongation (mm 24 hr $^{-1}$) after treatment with different concentrations of 1-naphthaleneacetic acid (1-NAA; 0.1, 1, 10, 50, 100, 500 μM; a & d), 24-*epi*brassinolide (*epi*BL; 10, 50, 100 μM; b,e), or gibberellic acid (GA_3; 10, 50, 100 μM; c,f), which were sprayed on the petiole and lamina (total 250 μL/leaf). Data represent means ± *SEM*, $n = 10$ (1-NAA) or $n = 5$ (*epi*BL and GA_3) biological replicates. Asterisks represent significant differences with the mock control (0 μM) (Student's *t* test, $p < 0.05$)

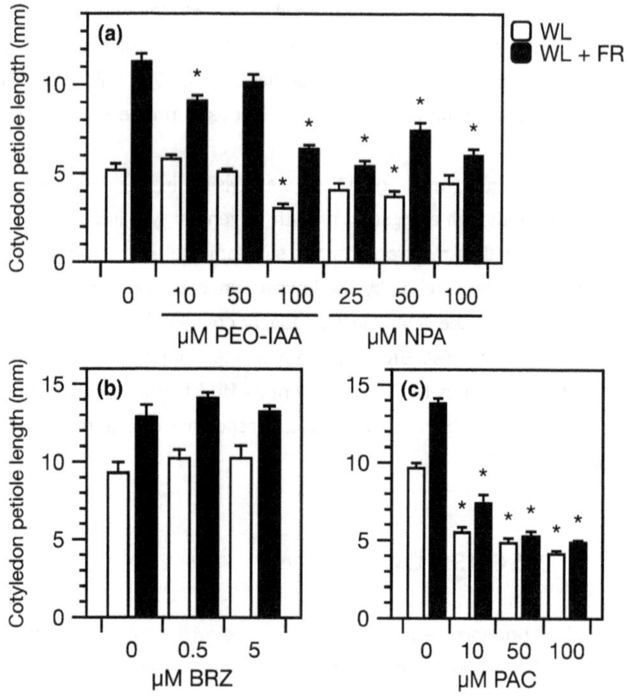

FIGURE 7 Inhibition of auxin or gibberellins, but not brassinosteroids, represses supplemental FR light-induced elongation in *Geranium pyrenaicum*. Cotyledon petiole length of *G. pyrenaicum* seedlings after five (a) or four (b–c) days in control white light (WL; R:FR 1.8) or WL supplemented with far-red light (WL + FR; R:FR 0.2), combined with different concentrations of hormone inhibitors: (a) α-(phenylethyl-2-one)-IAA (PEO-IAA; inhibitor of auxin perception), 1-*N*-naphthylphthalamic acid (NPA; inhibitor of auxin transport), (b) brassinazole (BRZ; inhibitor of brassinosteroid synthesis), or (c) paclobutrazol (PAC; inhibitor of gibberellin synthesis). Chemicals were added to a medium, to which plants were transferred 24 hr prior to the start of the light treatments. Data represent means ±SEM, $n = 6–10$. Asterisks mark significant differences compared to mock in the same light treatment (Student's *t* test, $p < 0.05$)

al., 2016). The strong induction of auxin and GA_1 synthesis genes in *G. pyrenaicum* are consistent with increased levels of these hormones in FR-enriched light, while in *G. robertianum* both the gene expression patterns and auxin and GA_1 levels are hardly affected by the light treatment. Remarkably, another biologically active gibberellin, GA_5, was undetectable in control white light, but abundant in FR light-treated petioles of *G. robertianum*. In this species, this induction clearly does not lead to enhanced petiole elongation. Consistent with this observation, an enrichment of GA_1 in the shade did also not positively correlate with hypocotyl growth in a non-shade-avoiding *Arabidopsis* mutant *shadow-1* (Li et al., 2016).

Although hormone-related gene expression patterns already change soon (2 hr) after the start of the shade treatment, hormone abundance changes later. In *Arabidopsis* seedlings, auxin and CS levels rapidly change after FR light exposure, but GA levels increase only after prolonged treatments (Bou-Torrent et al., 2014; Tao et al., 2008). In addition, genes involved in hormone-signaling pathways are rapidly upregulated upon exposure to FR-enriched light (Kohnen et al., 2016). Our data might indicate that in *Geranium*, early FR light-induced petiole growth would be associated with enhanced sensitivity to the already present pool of hormones (via enhanced expression of signaling genes) and that changing hormone levels would promote petiole growth in prolonged FR treatments.

Interestingly, unlike *Arabidopsis*, FR-induced IAA accumulation in *G. pyrenaicum* is restricted to the petiole, and not the lamina (de Wit et al., 2015). Consistent with our differential auxin accumulation data showing accumulation in the petiole, also FR detection in the petiole specifically is required for the elongation response in *G. pyrenaicum* (Figure 5). In *Arabidopsis*, the elongation response is induced solely by FR light perceived in the petiole as well (Pantazopoulou et al., 2017), while lamina-perceived FR light induces local auxin synthesis in the lamina, probably followed by transport toward the petiole and ultimately leading to upward petiole movement (hyponasty) (Michaud et al., 2017; Pantazopoulou et al., 2017). *G. pyrenaicum* does not induce hyponasty in response to FR enrichment (data not shown), possibly because of the putative absence of FR-triggered auxin synthesis in the lamina in this species.

The important role for auxin in supplemental FR-induced petiole elongation is supported by (a) the lack of IAA synthesis and IAA signaling gene expression as well as growth response in FR-treated *G. robertianum* and (b) the inhibition of a petiole elongation response to FR enrichment in *G. pyrenaicum* seedlings treated with auxin transport- and perception inhibitors.

Even though auxin is considered to be the key player mediating supplemental FR-induced shade avoidance (de Wit, Lorrain, & Fankhauser, 2013), our data show that GA levels also strongly increase in *G. pyrenaicum* petioles when exposed to FR enrichment. Similar to auxin, GA levels also increase specifically at the site of perception and response, the petiole (Figure 5, Supporting Information Figure S4). Albeit to a lesser extent, GA_4 levels were found to be induced in whole *Arabidopsis* seedlings exposed to far-red light (Bou-Torrent et al., 2014), consistent with increased expression of *GA20ox* genes (Hisamatsu, King, Helliwell, & Koshioka, 2005). Upon binding of GA to its receptor GID1, DELLA proteins are degraded in low R:FR light conditions, and this releases the DELLA-mediated inhibition of PIFs and thus shade avoidance (de Lucas et al., 2008; Djakovic-Petrovic et al., 2007; Murase, Hirano, Sun, & Hakoshima, 2008). Consistent with this growth-promoting role for GA, petiole elongation of both *Geranium* species is enhanced by exogenous GA_3 application (Figure 6) and the supplemental FR-induced elongation of cotyledon petioles in *G. pyrenaicum* is prevented upon chemical inhibition of endogenous GA synthesis (Figure 7). Importantly, the lack of additional GA_1 accumulation in *G. robertianum* under FR-enriched light corresponds with its lack of a petiole elongation response to this light treatment.

The observation that *G. robertianum* and *G. pyrenaicum* can both respond to exogenous IAA and GA treatments and indicates that the differences in shade avoidance between these are more likely occurring at the level of local hormone synthesis than at the level of hormone response.

Unlike IAA and GA, bioactive BRs brassinolide (BL) and castasterone (CS) were reduced upon FR light enrichment in *G. pyrenaicum* petioles. This result seems contradictory, as BRs are growth-promoting substances (Figure 6) and earlier studies have shown that functional BR synthesis is essential for the full growth responses to shade cues (Keuskamp et al., 2011; Kozuka et al., 2010). Interestingly, other studies to measure CS levels in low R:FR-exposed plants found a similar reduction in *Arabidopsis* (Bou-Torrent et al., 2014) and unchanged levels in sunflower seedlings (Kurepin, Joo, Kim, Pharis, & Back, 2012). Furthermore, dark-grown pea seedlings that elongate rapidly had reduced CS levels compared to light-grown controls (Symons et al., 2002). Perhaps, petiole elongation responses depend more strongly on BR signaling than on their biosynthesis. Light- and brassinosteroid signaling are known to converge at the level of the PIF4 transcription factor complex. Low R:FR light-stabilized PIF4 interacts with the BR-activated transcription factor BZR1 and, among others, enhances auxin signaling, GA biosynthesis, and DELLA suppression (Oh et al., 2012, 2014; Shahnejat-Bushehri et al., 2016).

To conclude, our analysis of transcription patterns, hormone synthesis, and growth in FR-enriched light environments show that local light perception and synthesis of auxin and GA, but not BR, account for the SAS in the shade-intolerant *G. pyrenaicum*. Likewise, the lack of a FR light effect on hormone levels in shade-tolerant *G. robertianum* seems responsible for its lack of the SAS (Figure 8). Future studies could be focused at identifying if the previously identified differential regulation of the atypical bHLH protein KIDARI upon FR light enrichment (Gommers et al., 2017) is functionally associated with the differential hormone patterns described here between these two species.

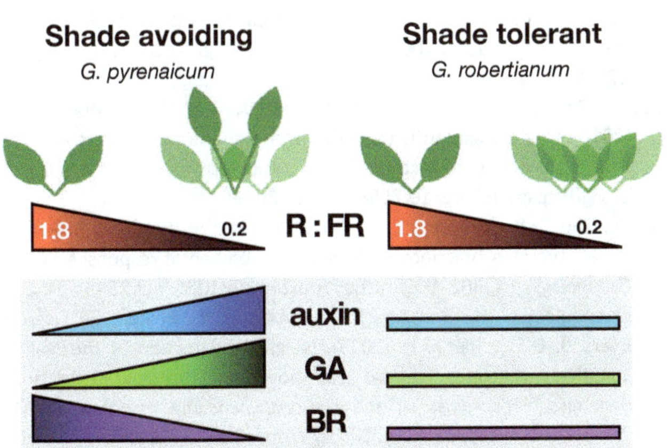

FIGURE 8 Graphic summary. When plants grow in dense communities, the R:FR light ratio decreases through R absorption and FR reflection, which is perceived in the petioles and causes a local increase in auxin and gibberellin (GA), but a reduction in brassinosteroid (BR) in shade-avoiding plant species *Geranium pyrenaicum* (left). These changing hormone levels regulate the elongation of petioles, which facilitates shade escape. When shade-tolerant species *G. robertianum* (right) is exposed to FR-enriched light, this does not affect hormone synthesis or signaling, which results in no elongation response in its petioles

ACKNOWLEDGMENTS

Funding was provided by the Netherlands Organisation for Scientific Research (NWO) to CMMG (EPS talent grant nr. 022.001.018) and RP (VIDI grant nr. 864.12.003), and by the Ministry of Education, Youth and Sports of the Czech Republic (grant no. LO 1204) to DT and AP.

AUTHOR CONTRIBUTIONS

CG and RP conceived and designed the project. CG, SB, DT, AP, JB, and VA performed the experiments and analyzed the data. CG and RP wrote the manuscript. All authors discussed the data and gave substantial comments on the manuscript.

ORCID

Charlotte M. M. Gommers http://orcid.org/0000-0002-0817-5134
Sara Buti http://orcid.org/0000-0003-4879-2265
Danuše Tarkowská http://orcid.org/0000-0003-1478-1904
Aleš Pěnčík http://orcid.org/0000-0002-1314-2249
Jason P. Banda http://orcid.org/0000-0002-5641-2675
Vincent Arricastres http://orcid.org/0000-0002-1905-9542
Ronald Pierik http://orcid.org/0000-0002-5320-6817

REFERENCES

Ballaré, C. L., & Pierik, R. (2017). The shade-avoidance syndrome: Multiple signals and ecological consequences. *Plant, Cell and Environment*, *40*, 2530–2543. https://doi.org/10.1111/pce.12914

Bernardo-García, S., de Lucas, M., Martínez, C., Espinosa-Ruiz, A., Davière, J. M., & Prat, S. (2014). BR-dependent phosphorylation modulates PIF4 transcriptional activity and shapes diurnal hypocotyl growth. *Genes & Development*, *28*, 1681–1694. https://doi.org/10.1101/gad.243675.114

Bou-Torrent, J., Galstyan, A., Gallemí, M., Cifuentes-Esquivel, N., Molina-Contreras, M. J., Salla-Martret, M., … Martínez-García, J. F. (2014). Plant proximity perception dynamically modulates hormone levels and sensitivity in Arabidopsis. *Journal of Experimental Botany*, *65*, 2937–2947. https://doi.org/10.1093/jxb/eru083

Casal, J. J. (2013). Photoreceptor signaling networks in plant responses to shade. *Annual Review of Plant Biology*, *64*, 403–427. https://doi.org/10.1146/annurev-arplant-050312-120221

Cifuentes-Esquivel, N., Bou-Torrent, J., Galstyan, A., Gallemí, M., Sessa, G., Salla Martret, M., … Martínez-García, J. F. (2013). The bHLH proteins BEE and BIM positively modulate the shade avoidance syndrome in Arabidopsis seedlings. *The Plant Journal*, *75*, 989–1002. https://doi.org/10.1111/tpj.12264

de Lucas, M., Davière, J.-M., Rodríguez-Falcón, M., Pontin, M., Iglesias-Pedraz, J. M., Lorrain, S., … Prat, S. (2008). A molecular framework for light and gibberellin control of cell elongation. *Nature*, *451*, 480–484. https://doi.org/10.1038/nature06520

de Wit, M., Costa Galvão, V., & Fankhauser, C. (2016). Light-mediated hormonal regulation of plant growth and development. *Annual Review of Plant Biology*, *67*, 513–537. https://doi.org/10.1146/annurev-arplant-043015-112252

de Wit, M., Keuskamp, D. H., Bongers, F. J., Hornitschek, P., Gommers, C. M. M., Reinen, E., … Pierik, R. (2016). Integration of phytochrome and cryptochrome signals determines plant growth during competition for light. *Current Biology, 26*, 3320–3326. https://doi.org/10.1016/j.cub.2016.10.031

de Wit, M., Ljung, K., & Fankhauser, C. (2015). Contrasting growth responses in lamina and petiole during neighbor detection depend on differential auxin responsiveness rather than different auxin levels. *New Phytologist, 208*, 198–209.

de Wit, M., Lorrain, S., & Fankhauser, C. (2013). Auxin-mediated plant architectural changes in response to shade and high temperature. *Physiologia Plantarum, 105*, 13–24. https://doi.org/10.1111/ppl.12099

de Wit, M., Spoel, S. H., Sanchez-Perez, G. F., Gommers, C. M. M., Pieterse, C. M. J., Voesenek, L. A. C. J., & Pierik, R. (2013). Perception of low red:Far-red ratio compromises both salicylic acid- and jasmonic acid-dependent pathogen defences in Arabidopsis. *The Plant Journal, 75*, 90–103. https://doi.org/10.1111/tpj.12203

Djakovic-Petrovic, T., de Wit, M., Voesenek, L. A. C. J., & Pierik, R. (2007). DELLA protein function in growth responses to canopy signals. *The Plant Journal, 51*, 117–126. https://doi.org/10.1111/j.1365-313X.2007.03122.x

Feng, S., Martinez, C., Gusmaroli, G., Wang, Y., Zhou, J., Wang, F., … Deng, X. W. (2008). Coordinated regulation of arabidopsis thaliana development by light and gibberellins. *Nature, 451*, 475–479. https://doi.org/10.1038/nature06448

García-Martinez, J. L., & Gil, J. (2001). Light regulation of gibberellin biosynthesis and mode of action. *Journal of Plant Growth Regulation, 20*, 354–368. https://doi.org/10.1007/s003440010033

Gommers, C. M. M., Keuskamp, D. H., Buti, S., van Veen, H., Koevoets, I. T., Reinen, E., … Pierik, R. (2017). Molecular profiles of contrasting shade response strategies in wild plants: Differential control of immunity and shoot elongation. *Plant Cell, 29*, 331–344. https://doi.org/http:/Qdx. doi. org/10. 1105/tpc. 16.00790

Gommers, C. M. M., Visser, E. J. W., St Onge, K. R., Voesenek, L. A. C. J., & Pierik, R. (2013). Shade tolerance: When growing tall is not an option. *Trends in Plant Science, 18*, 65–71. https://doi.org/10.1016/j.tplants.2012.09.008

Goyal, A., Karayekov, E., Galvão, V. C., Ren, H., Casal, J. J., & Fankhauser, C. (2016). Shade promotes phototropism through phytochrome B-controlled auxin production. *Current Biology, 26*(24), 1–8. https://doi.org/10.1016/j.cub.2016.10.001

Hayashi, K.-i., Tan, X., Zheng, N., Hatate, T., Kimura, Y., Kepinski, S., & Nozaki, H. (2008). Small-molecule agonists and antagonists of F-box protein-substrate interactions in auxin perception and signaling. *Proceedings of the National Academy of Sciences, 105*, 5632–5637. https://doi.org/10.1073/pnas.0711146105

Hisamatsu, T., King, R. W., Helliwell, C. A., & Koshioka, M. (2005). The involvement of gibberellin 20-oxidase genes in phytochrome-regulated petiole elongation of Arabidopsis. *Plant Physiology, 138*, 1106–1116. https://doi.org/10.1104/pp.104.059055

Hornitschek, P., Kohnen, M. V., Lorrain, S., Rougemont, J., Ljung, K., López-Vidriero, I., … Fankhauser, C. (2012). Phytochrome interacting factors 4 and 5 control seedling growth in changing light conditions by directly controlling auxin signaling. *The Plant Journal, 71*, 699–711. https://doi.org/10.1111/j.1365-313X.2012.05033.x

Hornitschek, P., Lorrain, S., Zoete, V., Michielin, O., & Fankhauser, C. (2009). Inhibition of the shade avoidance response by formation of non-DNA binding bHLH heterodimers. *EMBO Journal, 28*, 3893–3902. https://doi.org/10.1038/emboj.2009.306

Joo, S. H., Kim, T. W., Son, S. H., Lee, W. S., Yokota, T., & Kim, S. K. (2012). Biosynthesis of a cholesterol-derived brassinosteroid, 28-norcastasterone, in Arabidopsis thaliana. *Journal of Experimental Botany, 63*, 1823–1833. https://doi.org/10.1093/jxb/err354

Keuskamp, D. H., Pollmann, S., Voesenek, L. A. C. J., Peeters, A. J. M., & Pierik, R. (2010). Auxin transport through PIN-FORMED 3 (PIN3) controls shade avoidance and fitness during competition. *Proceedings of the National Academy of Sciences, 107*, 22740–22744. https://doi.org/10.1073/pnas.1013457108/-/DCSupplemental.www.pnas.org/cgi/doi/10.1073/pnas.1013457108.

Keuskamp, D. H., Sasidharan, R., Vos, I., Peeters, A. J. M., Voesenek, L. A. C. J., & Pierik, R. (2011). Blue-light-mediated shade avoidance requires combined auxin and brassinosteroid action in Arabidopsis seedlings. *The Plant Journal, 67*, 208–217. https://doi.org/10.1111/j.1365-313X.2011.04597.x

Kohnen, M. V., Schmid-Siegert, E., Trevisan, M., Petrolati, L. A., Sénéchal, F., Müller-Moulé, P., … Fankhauser, C. (2016). Neighbor detection induces organ-specific transcriptomes, revealing patterns underlying hypocotyl-specific growth. *Plant Cell, 28*, 2889–2904. https://doi.org/10.1105/tpc.16.00463

Kozuka, T., Kobayashi, J., Horiguchi, G., Demura, T., Sakakibara, H., Tsukaya, H., & Nagatani, A. (2010). Involvement of auxin and brassinosteroid in the regulation of petiole elongation under the shade. *Plant Physiology, 153*, 1608–1618. https://doi.org/10.1104/pp.110.156802

Kurepin, L. V., Joo, S. H., Kim, S. K., Pharis, R. P., & Back, T. G. (2012). Interaction of brassinosteroids with light quality and plant hormones in regulating shoot growth of young sunflower and arabidopsis seedlings. *Journal of Plant Growth Regulation, 31*, 156–164. https://doi.org/10.1007/s00344-011-9227-7

Li, W., Katin-Grazzini, L., Krishnan, S., Thammina, C., El-Tanbouly, R., Yer, H., … Li, Y. (2016). A novel two-step method for screening shade tolerant mutant plants via dwarfism. *Frontiers in Plant Science, 7*, 1–12. https://doi.org/10.3389/fpls.2016.01495

Michaud, O., Fiorucci, A.-S., Xenarios, I., & Fankhauser, C. (2017). Local auxin production underlies a spatially restricted neighbor-detection response in Arabidopsis. *Proceedings of the National Academy of Sciences, 114*, 7444–7449. https://doi.org/10.1073/pnas.1702276114

Murase, K., Hirano, Y., Sun, T., & Hakoshima, T. (2008). Gibberellin-induced DELLA recognition by the gibberellin receptor GID1. *Nature, 456*, 459–463. https://doi.org/10.1038/nature07519

Nishimura, T., Hayashi, K., Suzuki, H., Gyohda, A., Takaoka, C., Sakaguchi, Y., … Koshiba, T. (2013). Yucasin is a potent inhibitor of YUCCA, a key enzyme in auxin biosynthesis. *The Plant Journal, 77*, 352–366. https://doi.org/10.1111/tpj.12399

Nozue, K., Tat, A. V., Kumar Devisetty, U., Robinson, M., Mumbach, M. R., Ichihashi, Y., … Maloof, J. N. (2015). Shade avoidance components and pathways in adult plants revealed by phenotypic profiling. *PLoS Genetics, 11*, e1004953. https://doi.org/10.1371/journal.pgen.1004953

Oh, E., Zhu, J.-Y., Bai, M.-Y., Arenhart, R. A., Sun, Y., & Wang, Z.-Y. (2014). Cell elongation is regulated through a central circuit of interacting transcription factors in the Arabidopsis hypocotyl. *Elife, 3*, 1–19. https://doi.org/10.7554/eLife. 03031

Oh, E., Zhu, J.-Y., & Wang, Z.-Y. (2012). Interaction between BZR1 and PIF4 integrates brassinosteroid and environmental responses. *Nature Cell Biology, 14*, 802–809. https://doi.org/10.1038/ncb2545

Pantazopoulou, C. K., Bongers, F. J., Küpers, J. J., Reinen, E., Das, D., Evers, J. B., … Pierik, R. (2017). Neighbor detection at the leaf tip adaptively regulates upward leaf movement through spatial auxin dynamics. *Proceedings of the National Academy of Sciences, 114*, 7450–7455. https://doi.org/10.1073/pnas.1702275114

Pěnčík, A., Rolčík, J., Novák, O., Magnus, V., Barták, P., Buchtík, R., … Strnad, M. (2009). Isolation of novel indole-3-acetic acid conjugates by immunoaffinity extraction. *Talanta, 80*, 651–655. https://doi.org/10.1016/j.talanta.2009.07.043

Reed, J. W., Foster, K. R., Morgan, P. W., & Chory, J. (1996). Phytochrome B affects responsiveness to giberellins in Arabidopsis. *Plant Physiology, 112*, 337–342.

Rittenberg, D., & Foster, G. L. (1940). A new procedure for quantitative analysis by isotope dilution, with application to the determination of amino acids and fatty acids. *Journal of Biological Chemistry, 133*, 737–744.

Roig-Villanova, I., Bou-Torrent, J., Galstyan, A., Carretero-Paulet, L., Portolés, S., Rodríguez-Concepción, M., & Martínez-García, J. F. (2007). Interaction of shade avoidance and auxin responses: A role for two novel atypical bHLH proteins. *EMBO Journal*, *26*, 4756–4767. https://doi.org/10.1038/sj.emboj.7601890

Shahnejat-Bushehri, S., Tarkowska, D., Sakuraba, Y., & Balazadeh, S. (2016). Arabidopsis NAC transcription factor JUB1 regulates GA/BR metabolism and signalling. *Nature Plants*, *2*, 16013. https://doi.org/10.1038/nplants.2016.13

Symons, G. M., Schultz, L., Kerckhoffs, L. H. J., Davies, N. W., Gregory, D., & Reid, J. B. (2002). Uncoupling brassinosteroid levels and de-etiolation in pea. *Physiologia Plantarum*, *115*, 311–319. https://doi.org/ppl1150219 [pii]

Tao, Y., Ferrer, J.-L., Ljung, K., Pojer, F., Hong, F., Long, J. A., … Chory, J. (2008). Rapid synthesis of auxin via a new tryptophan-dependent pathway is required for shade avoidance in plants. *Cell*, *133*, 164–176. https://doi.org/10.1016/j.cell.2008.01.049

Tarkowská, D., Novák, O., Oklestkova, J., & Strnad, M. (2016). The determination of 22 natural brassinosteroids in a minute sample of plant tissue by UHPLC–ESI–MS/MS. *Analytical and Bioanalytical Chemistry*, *408*, 6799–6812. https://doi.org/10.1007/s00216-016-9807-2

Urbanová, T., Tarkowská, D., Novák, O., Hedden, P., & Strnad, M. (2013). Analysis of gibberellins as free acids by ultra performance liquid chromatography-tandem mass spectrometry. *Talanta*, *112*, 85–94. https://doi.org/10.1016/j.talanta.2013.03.068

Young, M. D., Wakefield, M. J., Smyth, G. K., & Oshlack, A. (2010). Gene ontology analysis for RNA-seq: Accounting for selection bias. *Genome Biology*, *11*, R14. https://doi.org/10.1186/gb-2010-11-2-r14

12

A sorghum NAC gene is associated with variation in biomass properties and yield potential

Jingnu Xia[1] | Yunjun Zhao[2] | Payne Burks[1] | Markus Pauly[2] | Patrick J. Brown[1]

[1]Department of Crop Sciences, University of Illinois at Urbana Champaign, Urbana, Illinois

[2]Department of Plant and Microbial Biology, University of California, Berkeley, Berkeley, California

Correspondence
Jingnu Xia and Patrick Brown, Department of Crop Sciences, University of Illinois at Urbana Champaign, Urbana, IL 61801.
Emails: jxia5@illinois.edu and pjbrown@ucdavis.edu

Present addresses
Jingnu Xia, Department of Biochemistry, University of Oxford, Oxford, UK.
Yunjun Zhao, Brookhaven National Lab, Upton, New York
Payne Burks, Chromatin Inc., Lubbock, Texas.
Markus Pauly, Heinrich-Heine University, Duesseldorf, Germany.
Patrick J. Brown, University of California, Davis, Davis, California.

Funding information
This research was supported by the Office of Science (BER), U.S. Department of Energy, Grant no DE-SC0012400. This funding body played no role in the design of the study, in the collection, analysis, and interpretation of data, or in the writing of the manuscript.

Abstract

Sorghum bicolor is a C_4 grass widely cultivated for grain, forage, sugar, and biomass. The sorghum *Dry Stalk (D)* locus controls a qualitative difference between juicy green *(dd)* and dry white *(D-)* stalks and midribs, and co-localizes with a quantitative trait locus for sugar yield. Here, we apply fine-mapping and genome-wide association study (GWAS) to identify a candidate gene underlying *D*, and use nearly isogenic lines (NILs) to characterize the transcriptional, compositional, and agronomic effects of variation at the *D* locus. The *D* locus was fine-mapped to a 36 kb interval containing four genes. One of these genes is a NAC transcription factor that contains a stop codon in the NAC domain in the recessive *(dd)* parent. Allelic variation at *D* affects grain yield, sugar yield, and biomass composition in NILs. Green midrib *(dd)* NILs show reductions in lignin in stalk tissue and produce higher sugar and grain yields under well-watered field conditions. Increased yield potential in *dd* NILs is associated with increased stalk mass and moisture, higher biomass digestibility, and an extended period of grain filling. Transcriptome profiling of midrib tissue at the 4–6 leaf stages, when NILs first become phenotypically distinct, reveals that *dd* NILs have increased expression of a miniature zinc finger (MIF) gene. MIF genes dimerize with and suppress zinc finger homeodomain (ZF-HD) transcription factors, and a ZF-HD gene is associated with midrib color variation in a GWAS analysis across 1,694 diverse sorghum inbreds. A premature stop codon in a NAC gene is the most likely candidate polymorphism underlying the sorghum *D* locus. More detailed understanding of the sorghum *D* locus could help improve agronomic potential in cereals.

KEYWORDS
biomass composition, biomass moisture, grain-filling, secondary cell wall

1 | INTRODUCTION

Sorghum bicolor, a versatile and resilient C_4 grass, is an important staple cereal in semiarid areas of Africa and Asia (Paterson et al., 2009). In addition to its use as a cereal, sorghum is widely grown for production of forage, ethanol, heat, and electricity (Rooney, Blumenthal, Bean, & Mullet, 2007). Reflecting this diversity of end uses, sorghum varieties vary greatly in plant height, flowering time, grain yield and harvest index, and sorghum vegetative biomass varies greatly in sugar content, composition, digestibility, and mechanical strength (Mullet et al., 2014; Stefaniak et al., 2012). Despite the identification of several major loci affecting sorghum plant height (Multani et al.,

2003; Yamaguchi et al., 2016) and flowering time (Murphy et al., 2011; Murphy et al., 2014; Yang, n.d), no major loci affecting vegetative biomass properties in sorghum have previously been identified.

Vegetative biomass properties are largely determined by the architecture of the vascular system, xylem, and phloem, which provide mechanical support, allow water and nutrient acquisition, and mediate transport of photoassimilates and signaling molecules from source to sink tissues (Bihmldlne, Hunter, Johns, Koch, & Braun, 2013). Grass leaves and inflorescences are major source and sink organs, and the genetic control of their architecture has been intensely studied (Kellogg, 2007; Pautler, Tanaka, Hirano, & Jackson, 2013; Tanaka, Pautler, Jackson, & Hirano, 2013). However, the architecture of the vascular system also plays a crucial role in determining cereal grain yields. During grain filling, continued acquisition of water, sugars, and nitrogen is accompanied by re-mobilization of transient storage reserves, and stalk strength must be maintained throughout senescence to protect the drying grain from lodging (Peiffer et al., 2013). The genetic control of vascular system architecture has been studied primarily in model systems (Handakumbura & Hazen, 2012), where genetic effects on agronomic performance may not be easily detected.

The large family of NAC transcription factors (TFs), named after founding members NAM, ATAF, and CUC (Fang, You, Xie, Xie, & Xiong, 2008), plays a critical role in plant vascular development (Ko, Jeon, Kim, Kim, & Han, 2014; Nakano, Yamaguchi, Endo, Rejab, & Ohtani, 2015). A subset known as the secondary wall NACs (SWNs) act as master regulators of secondary cell wall development in vascular tissues (Handakumbura & Hazen, 2012; Zhong, Demura, & Ye, 2006). Conservation of SWN function in moss hydroids and stereids, which, respectively, conduct water and lend mechanical support to the moss gametophyte, provides evidence that these structures are homologous to sporophytic vascular systems in higher plants (Xu et al., 2014). SWNs activate a subset of MYB transcription factors that contain secondary wall NAC binding elements (SNBEs) in their promoters, inducing secondary cell wall thickening and deposition of cellulose, hemicellulose and lignin (Zhong, Lee, & Ye, 2010). NAC TFs also act downstream of MYB TFs. During sieve element maturation, for example, the *Altered Phloem Development (APL)* gene both drives the expression of NAC045 and NAC086 and is itself regulated by NAC020 (Bonke, Thitamadee, Mahonen, Hauser, & Helariutta, 2003; Furuta et al., 2014). The extreme complexity of the secondary cell wall gene regulatory network in Arabidopsis is driven by functional redundancy, feed-back and feed-forward loops, and combinatorial control enabling functional fine-tuning (Taylor-Teeples et al., 2014).

The *Dry Stalk (D)* locus in sorghum conditions a difference between dry, pithy white stems and midribs *(D–)*, and juicy green stems and midribs *(dd)* (Smith & Frederiksen, 2000). A previous genome-wide association study (GWAS) in sweet sorghum mapped large-effect QTL for sugar yield, juice volume, and stalk moisture to the *D* locus (Burks, Kaiser, Hawkins, & Brown, 2015). In this study, we fine-map the *D* locus to a 36 kb region, identify a premature stop codon in a NAC transcription factor as a candidate polymorphism, and show that nearly isogenic lines (NILs) segregating for the premature stop codon differ significantly in sugar yield, grain yield, and biomass composition.

2 | METHODS

2.1 | Plant material and controlled growth conditions

A dominant white midrib allele *(D)* was introduced into a genetic male sterile *(ms3/ms3)* version of Tx623, which carries a recessive green midrib allele *(d)*, through four generations of backcrossing. Homozygous *D/D-Ms3/Ms3* and *d/d-Ms3/Ms3* seed stocks were derived from a single BC_4 plant by selfing. Three replications of greenhouse plantings were performed on 1/7/16, 2/16/16, and 3/16/16, planted 2 seeds per cell in 6 by 8 flats and thinning to 1 plant per cell at 7 days after planting. Midrib color changes were recorded using a MiScope (Zarbeco, USA) at V4, V6, and V8, and midrib tissue for RNAseq was sampled at V4 and V6 into liquid N.

2.2 | Field experiments

For sugar yield measurements, paired rows of NILs were planted in six replicates at a single location in Urbana, IL in summer 2014, and phenotyped as previously described (Burks et al., 2015). For grain and stalk dry matter and moisture measurements, paired 2-row plots of NILs were planted in four replicates at each of two locations (Urbana and Savoy, IL), and samples were pooled from 2-6 individual representative plants per row for each time point. Immature grain weight was estimated by clipping individual panicle branches from the rachis, stalk weight was measured after stripping off leaf blades, and moisture was estimated by weighing samples before and after oven-drying at 45°C for 5 days. Midrib color in the GWAS panel was phenotyped as a binary trait (green vs white, Supporting information Table S7) in the youngest leaf at ~45 days after planting. All field experiments were machine planted in rows 10′ long with 30″ row spacing. Seeds were treated with Apron fungicide (Syngenta, USA) and Concep II seed safener (Syngenta, USA) before planting, and weeds were controlled through use of a pre-emergent herbicide (Bicep). All raw agronomic data (Supporting information Tables S5 and S8) and raw compositional data (Supporting information Table S9) were analyzed by ANOVA in R using a linear model with location, block nested within location, and NIL genotype as fixed effects. We present *p*-values for genotype effects, and boxplots show residuals from a model including location and block effects.

2.3 | Fine-mapping and GWAS

Two MITE-based indel markers (Supporting information Table S10) were used to screen 1,132 BC_3 and BC_2F_2 individuals. Subsequent genotyping-by-sequencing was performed in both putative recombinants and a GWAS panel of 1,624 diverse sorghum accessions using the two-enzyme GBS protocol with PstI-HF and HinP1I enzymes (Poland, Brown, Sorrells, & Jannink, 2012), followed by alignment to v3 of the reference genome (www.phytozome.org) using bowtie2 (Langmead & Salzberg, 2012) and SNP calling using the TASSEL5 GBSv2 pipeline (Glaubitz et al., 2014). Imputation of biparental recombinants was performed using FSFHap (Swarts et al., 2014) in TASSEL, and imputation of the GWAS panel was performed using Beagle4 (Browning & Browning, 2007). Midrib color GWAS was performed using the mixed linear model implemented in GAPIT, using model selection to choose the optimal number of principal components, which was zero. A minor allele frequency cutoff of 5% resulted in 50,899 SNPs for testing, and only associations significant at a false discovery rate of less than 5% ($q < 0.05$) are reported. GBS data for this project have been deposited at the Illinois Data Bank (https://databank.illinois.edu/).

2.4 | Lignocellulosic compositional analysis

Lyophilized tissue samples were ground to a fine powder using three 5 mm metal balls in 2 ml plastic tubes (Retsch Ball mill, 2 times at 25 Hz for 2.5 min) and washed sequentially with 70% ethanol, 1:1 (v:v) methanol:chloroform, and acetone. The alcohol-insoluble cell wall material was further destarched with alpha amylase (Sigma) and Pullulanase M2 (Megazyme) in 0.1 M Citrate buffer pH 5.0. The destarched material was aliquoted for different compositional assays. 1 mg of destarched cell wall was hydrolyzed in 2 M Tri-fluoroacetic acid (TFA) heated to 121°C for 90 min followed by a stream of dried air using nitrogen. Dried samples were re-suspended in water, centrifuged, and the supernatant was collected for monosaccharides analysis and the pellet was dried for crystalline cellulose measurements. TFA-soluble samples were analyzed using High Performance Anion Exchange liquid Chromatography with Pulse Amperometric Detection (HPAEC-PAD) according to de Souza, Hull, Gille and Pauly (2014). Neutral sugars were separated via a CarboPac PA20 column, while a CarboPac PA200 was used to separate uronic acids. Three distinct programs were used to resolve the sugars of interest. Samples were run at a flow rate of 0.4 ml/min and gradients consisted of (a) 2 mM NaOH for 20 min followed by a 5 min 100 mM flush and subsequent 5 min at 2 mM (neutral sugar separation 1; excludes xylose and mannose); (b) 18 mM NaOH for 15 min followed by a 5 min 100 mM flush and subsequent 7 min at 18 mM (neutral sugar separation 2; excludes rhamnose and arabinose); (c) 0.1 M NaOH with a gradient of 50–200 mM sodium acetate from 0 to 10 min followed by a 2 min 200 mM sodium acetate flush returning to 50 mM for 2.9 min (uronic acid separation). Crystalline cellulose was measured using the Updegraf method and the released glucose was measured using the anthrone assay (Updegraff, 1969; Laurentin & Edwards, 2003). Lignin content and lignin composition were measured using the ultra-violet acetyl bromide lignin method and a thioacidolysis procedure, respectively, according to Foster, Martin and Pauly (2010). Saccharification yield of cell wall materials was determined after enzymatic treatment. In brief, 1 mg of destached cell wall was incubated with 0.5 μl Accellerase 1,500 enzyme mix (Gencor) in 1 ml of 50 mM citrate buffer (pH 4.5), shaking at 250 rpm at 50°C for 20 hr. Solubilized glucose and xylose were detected on a Bio-Rad HPX-87H, 300 × 7.8 mm column in an Shimadzu UFLC chromatography system. The elution profile encompassed 0.01 N sulfuric acid in 15 min at 0.6 ml/min and column temperature is 50°C.

2.5 | Phylogenetic tree construction and sequence alignment

SbNAC074a homologs from *Arabidopsis thaliana*, *Oryza sativa subsp. Japonica*, *Sorghum bicolor*, *Zea mays*, *Setaria italica*, and *Glycine max* were extracted from UNIPROT (http://www.uniprot.org). The full length sequence of SbNAC074a from the *DD* NIL, which doesn't have the null mutation, was BLASTED in UNIPROT (http://www.uniprot.org) and homologs from *Arabidopsis thaliana (sequence version 1), Oryza sativa subsp. Japonica (sequence version 2 for Os04 g43560, others were sequence version 1), Sorghum bicolor (sequence version 1), Zea mays (sequence version 1), Setaria italica (sequence version 1), and Glycine max (sequence version 1)* were identified from the BLAST result. Alignment was performed using MUSCLE in MEGA7 with default settings, and CLUSTALW2_Phylogeny was used with distance correction off, gaps excluded, and the UPGMA clustering method to create the tree. Visualization was performed using EVOLVIEW (http://www.evolgenius.info/evolview/#login).

2.6 | Gene expression analysis

Total RNA was extracted from greenhouse-grown midrib tissue of the youngest fully expanded leaf of *DD* and *dd* NILs at the four-leaf and six-leaf stages using Spectrum™ Plant Total RNA Kit (Sigma-Aldrich, USA), following the manufacturer's protocol. All RNA samples were digested with DNase I (New England Biolabs, USA), and rtPCR was performed using M-MuLV reverse transcriptase (New England Biolabs, USA). RNAseq libraries were prepared and sequenced at Roy J. Carver Biotechnology Center at the University of Illinois using single-end, 100 bp reads on a HiSeq2500 instrument. Reads were processed using HISAT2 and StringTie, and the stattest function in the R package "Ballgown" was used to test for differential gene expression between *DD* and *dd* NILs using 3 biological replicates at each developmental stage, controlling the false discovery rate at 0.05. The R package "WGCNA" was used to construct co-expression modules using the arguments power = 8, minModuleSize = 20, and mergeCutHeight = 0.05. RNAseq reads for this project have been deposited at the Illinois Data Bank (https://databank.illinois.edu/).

3 | RESULTS

3.1 | Creation and characterization of nearly isogenic lines for the *D* locus

A white midrib allele *(D)* was introduced into the green midrib Tx623 background *(dd)* through four generations of backcrossing, and homozygous *DD* and *dd* seed stocks were derived from a single BC_4 plant by selfing. NILs appear identical until the sixth-leaf stage, when a narrow band of pithy white tissue first appears in *DD* midribs, becoming wider and more distinct in successive leaves (Figure 1a). At anthesis, *DD* stalks are visibly drier and pithier than *dd* stalks (Figure 1b). All sampling times and tissues presented in this study are summarized in Supporting information Table S1.

3.2 | GWAS and fine-mapping

Midrib color was scored as a binary trait (green/white) in the flag leaf at anthesis in a large panel (n = 1624) of sorghum inbreds, revealing a single very strong association at ~ 51 Mb on chromosome 6 (Figure 2a; Supporting information Table S2). Polymorphic markers flanking this region were used to screen 1132 BC_3 and BC_2F_2 individuals. Putative recombinant individuals were subjected to genotyping-by-sequencing (GBS), confirming 17 recombinants and defining a ~36 kb interval that co-segregates perfectly with the phenotype (Figure 2b, Supporting information Table S3). This interval contains four predicted genes, of which two are expressed in midrib tissue at either the fourth- or sixth-leaf stage (Supporting information Table S4): a NAC transcription factor (Sobic.006G147400) and a threonine aldolase (Sobic.006G147600). The two most significant hits in the GWAS analysis are the two closest flanking SNPs to the NAC gene. Moreover, the NAC gene is the only gene in the interval with a different annotated exon-intron structure compared to its closest homologs in other cereals, which is relevant because the sorghum reference genome is derived from Tx623, a recessive *dd* mutant. Reverse-transcriptase PCR and cDNA sequencing confirms that the annotated first intron in Sobic.006G147400 does not exist, and that a T/C SNP in this region produces a stop codon in the *d* allele from Tx623 but not the contrasting D allele (Figure 2c). The premature stop codon in Tx623 eliminates most of the NAC domain.

3.3 | Phylogeny and identification of homologs

The full-length NAC protein, derived from an allele of Sobic.006G147400 lacking the premature stop codon, was used to search for homologous proteins in four grasses (*Oryza sativa subsp. japonica, Sorghum bicolor, Zea mays, Setaria italica)* and two dicots (*Arabidopsis thaliana* and *Glycine max*), which were aligned using MUSCLE in MEGA7 and used to construct a neighbor-joining tree using UPGMA clustering (Figure 3). Syntenic regions on *Oryza* chromosome 4 and *Setaria* chromosome 7 each contain single orthologous copies of our sorghum NAC candidate, and co-syntenic regions on *Zea* chromosomes 2 and 10 each contain intact homeologues derived from segmental duplication. These proteins are part of the NAC1 sub-clade (Peng et al., 2015), and contain five conserved motifs in the N-terminal NAC domains whereas the C-termini are highly variable (Supporting information Figure S1). The closest *Arabidopsis* homolog is NAC074 (At4G28530), and we hereafter refer to the three co-orthologous grass clades as NAC074a, NAC074b, and NAC074c, and to Sobic.006G147400 as *SbNAC074a*.

3.4 | *dd* NILs show increases in stalk dry matter, soluble sugar yield, and grain yield

To confirm the previously-reported association between the *D* locus and sugar yield, we grew NILs in six replications of paired rows, extracted total sugar from 1 meter (m) of row 30 days after anthesis

FIGURE 1 Phenotypic differences between *DD* and *dd* nearly-isogenic lines (NILs). (a) Midribs of the youngest fully expanded leaf in *dd* and *DD* NILs at V4, V6, and V8 stages at 2, 3, and 4 weeks after planting, respectively. (b) Cross-sections through the second stalk internode above ground level at anthesis (10 weeks after planting (WAP)), with 0.5 cm scale bar

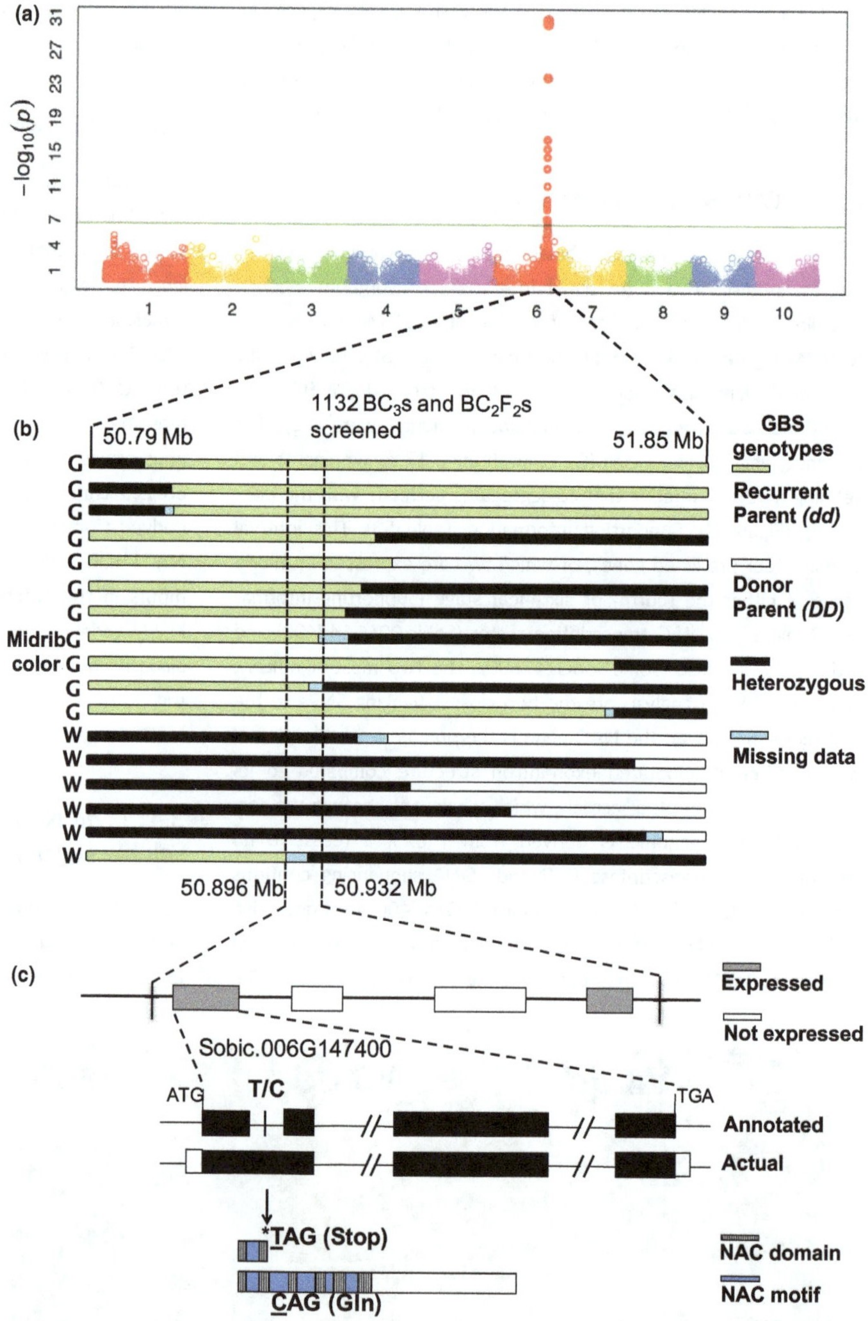

FIGURE 2 Genome-wide association study (GWAS) and fine-mapping of the sorghum *D* locus. (a) Midrib color GWAS in 1,624 sorghum inbreds, with a major peak at ~51 Mb on chr6. (b) Screening 1,132 individuals with flanking markers yielded 17 recombinants, which were GBSed to define a 36 kb interval. (c) Two of four genes in the 36 kb interval are expressed, one of which is a NAC transcription factor (Sobic.006G147400) with a stop codon in the first exon disrupting the NAC domain. This gene was annotated incorrectly in the sorghum reference genome (Tx623), which has the stop codon and a recessive *d* allele

(see Methods), and found that *dd* NILs yield nearly twice the sugar of *DD* NILs (Figure 4a, $p < 0.001$) despite being identical in plant height and flowering time (Supporting information Table S5). Sugar yield per meter of row (g/m; Figure 4a) is a function of juice volume (ml/m; Figure 4b) and the concentration of simple sugars in the juice (degrees Brix; Figure 4c), and we observe that increased yield in *dd* NILs is driven by increased juice volume, which in turn is associated with higher vegetative weight (kg/m; Figure 4d). Given the dramatic effect of the *D* locus on sugar yield, we next conducted a time series experiment to quantify its effects on the rate and extent of grain filling. Paired, two-row plots of NILs were grown in four replicates at each of two locations, and dry matter and moisture of stalk and grain were monitored at 2-week intervals beginning at anthesis. Overall, this grain filling period is characterized by movement of dry matter from the stalk to the grain, and by decreases in grain moisture relative to stalk moisture (Figure 5). Stalk dry matter is significantly higher in *dd* NILs at all stages, most notably at 4 weeks after anthesis when it is >58% higher ($p = 0.004$; Figure 5a). *dd* NILs also have higher grain moisture at 4 weeks after anthesis ($p = 0.042$; Figure 5d), and higher grain yield at 6 weeks after anthesis ($p = 0.025$; Figure 5b).

3.5 | Compositional differences drive increased biomass digestibility of *dd* NILs

Nearly-isogenic lines stalk tissue was subjected to detailed lignocellulosic compositional analysis at two growth stages: developing internodes from a 1 cm plug of tissue immediately below the shoot apical meristem at 6 weeks after planting (WAP), and the third internode below the inflorescence at 9 WAP, which coincided with the boot stage shortly before flowering. While tissues show relatively few compositional differences between NILs at 6 WAP, by 9 WAP *dd* NILs show reduced lignin (p = 0.007; Figure 6a) and increased glucose release following enzymatic saccharification (p = 0.049; Figure 6b), while their decrease in crystalline cellulose is not significant at $p < 0.05$ (p = 0.055; Figure 6c). *dd* NILs also show a lower ratio of syringyl to guaiacyl lignin monomers at 9 WAP (p = 0.029), and differences in acid-hydrolyzed lignocellulosic monosaccharides at both stages (Supporting information Figure S2).

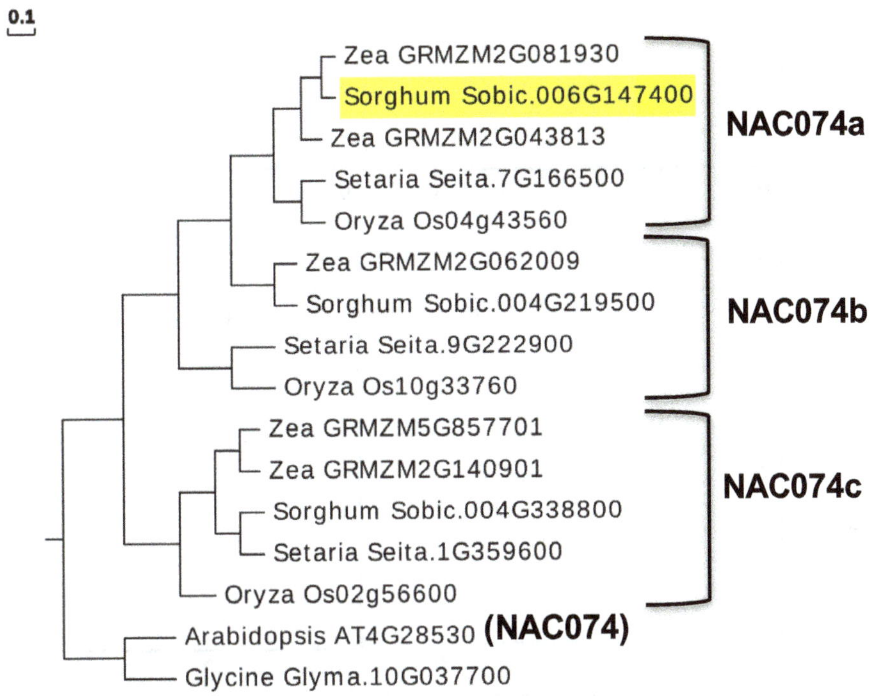

FIGURE 3 UPGMA tree of NAC074 genes in grasses. *SbNAC074a* is highlighted

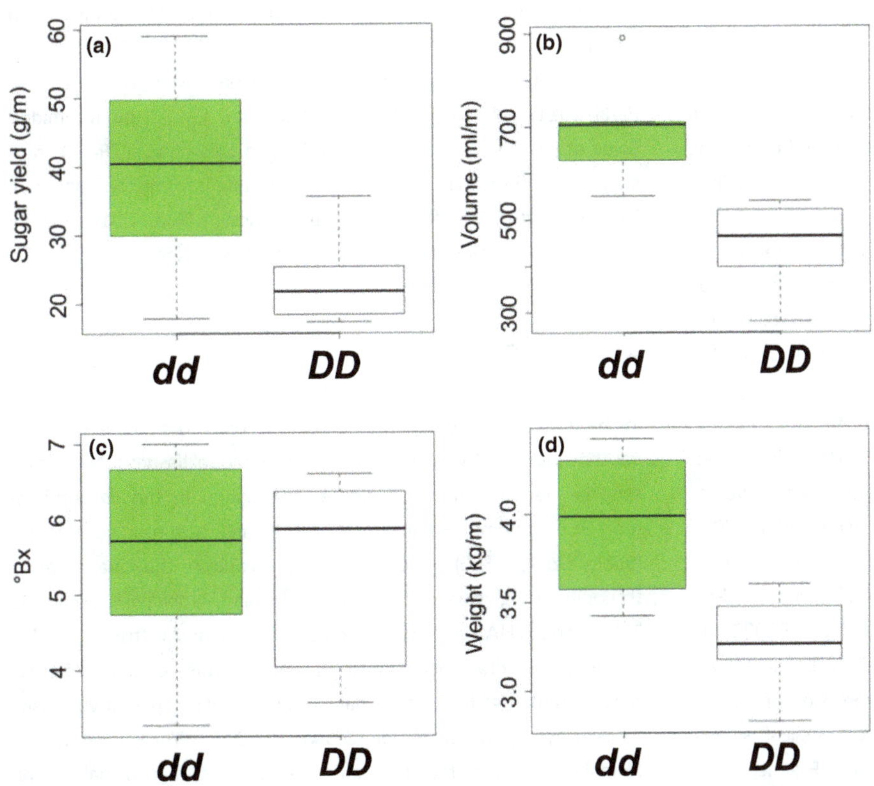

FIGURE 4 Sugar accumulation in *DD* and *dd* nearly-isogenic lines (NILs). Total stalk tissue from 1 m of row was sampled at the hard dough stage, 4 weeks after anthesis. (a) Sugar yield (g/m); (b) Juice volume (ml/m); (c) Brix (g/ml); (d) Vegetative Wet Weight (kg/m). Sugar yield is calculated by multiplying juice volume and brix

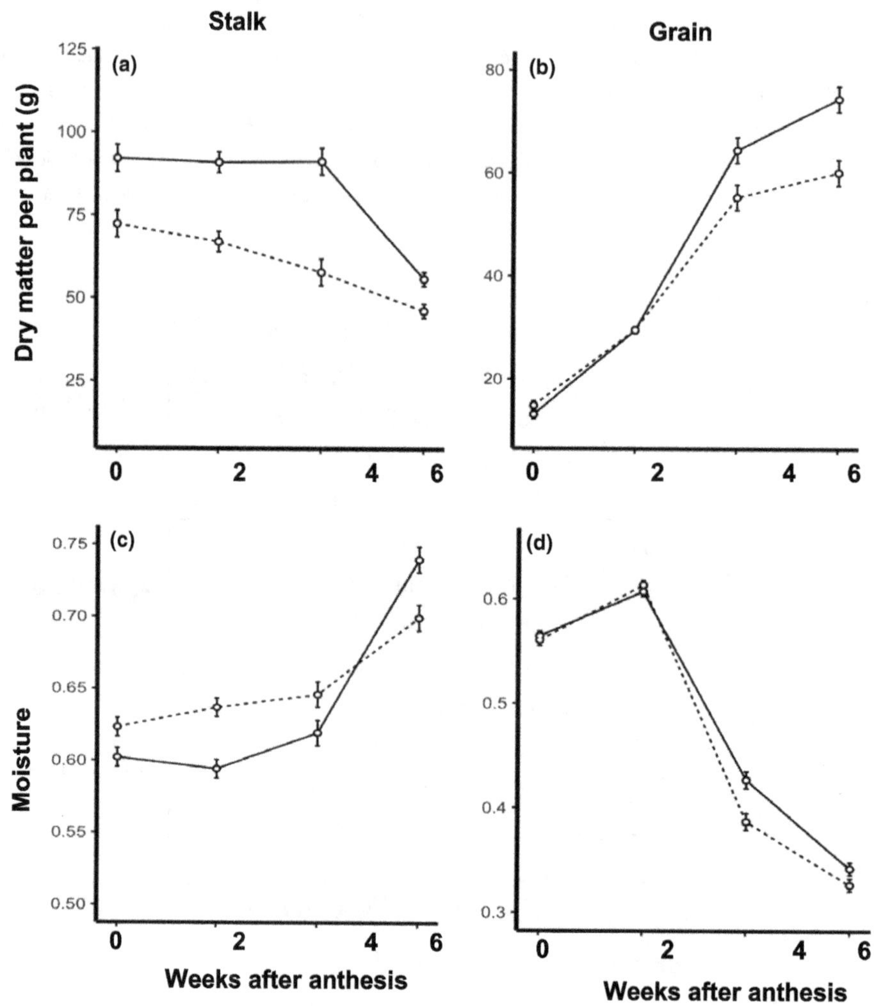

FIGURE 5 Grain filling of *DD* nearly-isogenic lines (NILs) (dashed lines) and *dd* NILs (solid lines) under field conditions. Changes in dry matter (a, b) and moisture (c, d) were monitored at 2-week intervals in stalk (a, c) and grain (b, d) from 0–6 weeks after anthesis

3.6 | A miniature zinc finger (MIF) gene is upregulated in *dd* NILs

Transcriptome profiling was performed on mRNA from NIL midrib tissue at the four-leaf and six–leaf stages, just before and during the first appearance of phenotypic differences between the NILs (Figure 1a). Three independent biological replicates were obtained from pooled tissue from three separate greenhouse plantings, for a total of 12 samples across the two NILs and two stages. Co-expression analysis using WGCNA's step by step network construction and module detection (Langfelder & Horvath, 2008) shows that *SbNAC074a* shares a regulatory module with 263 other annotated genes, including a group of five NAC transcription factors (Figure 7; Supporting information Table S6) that includes *SbNAC074c* but not *SbNAC074b* as well as two NACs associated with xylem development in Arabidopsis, *Xylem NAC Domain 1 (XND1)* and *NAC075* (Endo et al., 2015; Zhao, Avci, Grant, Haigler, & Beers, 2007). However, using the HiSAT-Stringtie-Ballgown pipeline (Pertea, Kim, Pertea, Leek, & Salzberg, 2016), only a single gene (Sobic.008G020700) is differentially expressed between DD and *dd* NILs ($q < 0.01$), showing no expression in *DD* NILs and mean expression of 21 and 35 FPKM in *dd* NIL midribs at the four-leaf and six-leaf stages, respectively. Sobic.008G020700 is annotated as a MIF gene, which are seed plant-specific, truncated versions of ZF-HD transcription factors (Hu & Ma, 2006) that dimerize with ZF-HDs and suppress their transcriptional activation activity (Hong, Kim, Kim, Yang, & Park, 2011). Intriguingly, the only significant GWAS hit for midrib color other than our candidate NAC gene falls near a ZF-HD transcription factor on chromosome 1 (Supporting information Table S2). However, this ZF-HD transcription factor (Sobic.001G112500) shows no expression in any of the *DD* or *dd* samples.

4 | DISCUSSION

In this study, we fine-map the sorghum *D* locus to a four-gene interval that includes two genes expressed in midrib tissue: a NAC transcription factor with a premature stop codon in the *dd* sorghum reference genome (Sobic.006G147400), and a threonine aldolase (Sobic.006G147600). Threonine aldolase controls the catabolism of threonine to glycine and acetaldehyde, and *Arabidopsis* homologs THA1 and THA2 are both expressed in vascular tissue. *tha1-2* mutants show dramatic increases in seed threonine content, whereas *tha2-1* mutants have a lethal albino seedling phenotype(Joshi, Laubengayer, Schauer, Fernie, & Jander, 2006). While we have not formally excluded the threonine aldolase as a candidate gene

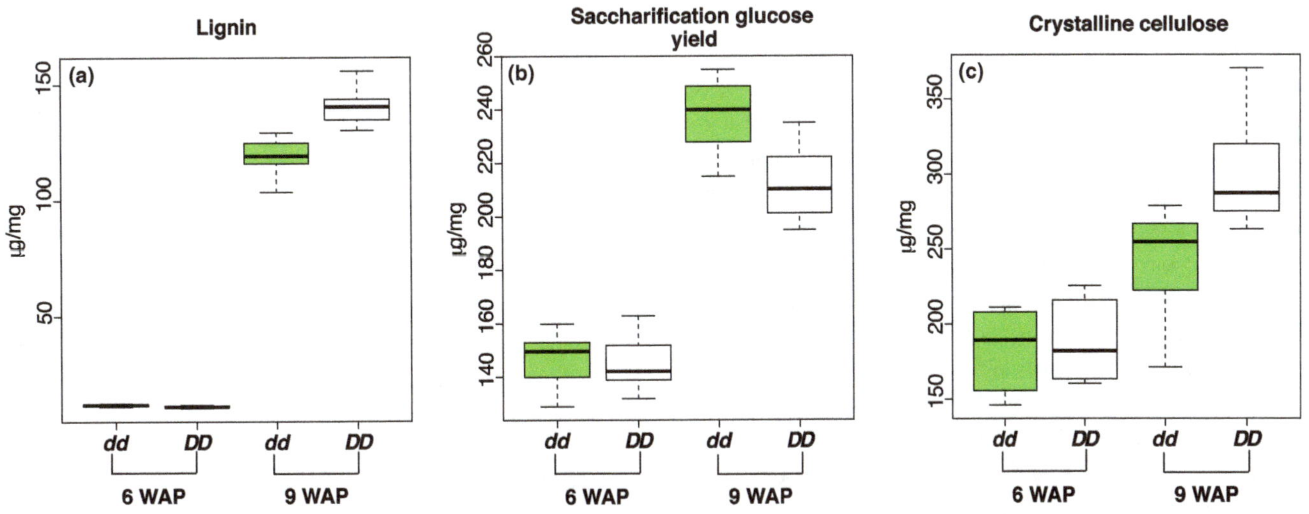

FIGURE 6 Compositional analysis of developing internodes in *DD* and *dd* nearly-isogenic lines (NILs). (a) Lignin; (b) Saccharification glucose yield; (c) Crystalline cellulose. Tissues sampled included developing internodes at 6 weeks after planting (6 weeks after planting (WAP)) and the third internode below the inflorescence at 9 WAP

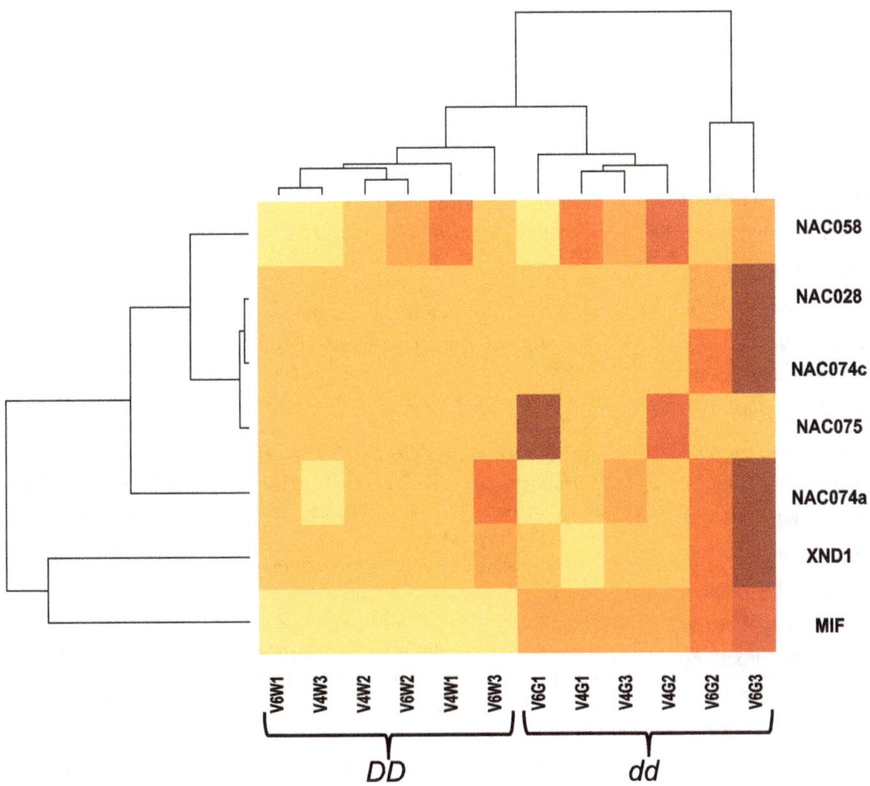

FIGURE 7 A Co-expression of SbNAC074a, the differentially-expressed miniature zinc finger (MIF) gene, and 5 other NAC transcription factors in V4 and V6 midribs of *DD* and *dd* nearly-isogenic lines (NILs). Light and dark colors indicate low and high gene expression, respectively. Dendrograms reflect Euclidean distance and determine row and column order

underlying the *D* locus in sorghum, our data from *dd* and *DD* NILs is more consistent with perturbation of NAC gene function. First, we note that the two most significant SNPs in our GWAS analysis are the two closest flanking markers to this NAC gene. Second, the most obvious mutant phenotype in *dd* NILs is the persistence of a green midrib throughout development. In *Arabidopsis*, a suite of NAC genes including ORE1, ANAC046, ANAC087, and ANAC100 positively regulate leaf de-greening and senescence by binding directly to the promoters of chlorophyll catabolic genes (Oda-Yamamizo et al., 2016, p. 46; Qiu et al., 2015). Therefore, we propose that loss of NAC gene function in *dd* NILs represses chlorophyll catabolism, causing the persistence of a green midrib. We have named this NAC gene *SbNAC074a*, after its closest homolog in Arabidopsis.

Previous study of genes orthologous to and co-expressed with *SbNAC074a* suggest a role in xylem development. *Arabidopsis* NAC074 (At4G28530) is upregulated in xylem relative to phloem-cambium and non-vascular tissues (Zhao, 2005), and is one of many NAC genes upregulated during leaf senescence (Podzimska-Sroka, O'Shea, Gregersen, & Skriver, 2015). The rice ortholog of *SbNAC074a* (Os04 g43560) is upregulated in panicle and root under drought stress in drought-tolerant, but not drought-susceptible, NIL backgrounds (Nuruzzaman et al., 2012). Here, we show that *SbNAC074a* is co-expressed with other NAC transcription factors involved in xylem development, including homologs of Arabidopsis *XND1* and *NAC075*. *xnd1* mutants display a mild dwarfing phenotype associated with a reduction in tracheary element length, whereas overexpression of XND1 results in reduced formation of xylem vessels, expansion of the phloem, and increased starch storage in amyloplasts (Zhao et al., 2007).The authors suggest that XND1 may promote vessel elongation by repressing their terminal differentiation. Overexpression of NAC075 results in ectopic formation of xylem vessel elements (Endo et al., 2015) and rescues the pendent stem phenotype of *nst1-nst3* double mutants, which results from complete loss of secondary cell wall deposition in xylem fibers.

Lignocellulosic compositional data are consistent with a role for *SbNAC074a* as a positive regulator of xylem development. *dd* NILs internodes have significantly reduced lignin. Lignin content has been positively correlated with xylem development and inversely correlated with lignocellulosic saccharification yields (Chen & Dixon, 2007), consistent with the observed increase in glucose yield following enzymatic saccharification in *dd* NILs. The increased digestibility of *dd* NILs comes with no obvious agronomic penalty in the dwarf grain sorghum background of Tx623, though we observe that green midrib accessions have higher stalk lodging than white midrib accessions in a biomass sorghum panel (Supporting information Figure S3). Strikingly, yields of soluble sugar, grain, and vegetative biomass are all significantly increased in *dd* NILs under well-watered field conditions. These results are not easily explained and should be validated.

A premature stop codon in the NAC domain of SbNAC074a likely underlies allelic variation at the sorghum *Dry Stalk (D)* locus. NILs at the *D* locus show agronomic, compositional, and transcriptional differences. While *dd* NILs showed superior agronomic performance at our Illinois field sites with ample rainfall, the relative performance of *DD* and *dd* genotypes under terminal drought stress might differ. Better understanding of developmental perturbations mediated by the sorghum D locus could lead to enhancement of yield potential and climatic adaptation in cereals.

ACKNOWLEDGMENTS

The authors acknowledge the assistance of Dr. Alvaro Hernandez and Dr. Chris Wright at the W.M. Keck Center with DNA sequencing, and the assistance of Tim Mies and Derek Lewis with establishment and maintenance of field plots.

COMPETING INTERESTS

The authors declare that they have no competing interests.

AUTHORS' CONTRIBUTIONS

JX, PB, and PJB analyzed and interpreted agronomic data. YZ and MP analyzed and interpreted compositional data. JX and PJB analyzed and interpreted genomic data and wrote much of the manuscript. All authors read and approved the final manuscript.

AVAILABILITY OF DATA AND MATERIALS

Sequence data generated during the current study are available at the Illinois Data Bank (https://databank.illinois.edu/). Seed stocks used during the current study are available from the corresponding author on reasonable request.

REFERENCES

Bihmidine, S., Hunter, C. T., Johns, C. E., Koch, K. E., & Braun, D. M. (2013). Regulation of assimilate import into sink organs: Update on molecular drivers of sink strength. *Frontiers in Plant Science*, *4*, 177. https://doi.org/10.3389/fpls.2013.00177

Bonke, M., Thitamadee, S., Mahonen, A. P., Hauser, M. T., & Helariutta, Y. (2003). APL regulates vascular tissue identity in Arabidopsis. *Nature*, *426*, 181–186. https://doi.org/10.1038/nature02100

Browning, S. R., & Browning, B. L. (2007). Rapid and accurate haplotype phasing and missing-data inference for whole-genome association studies by use of localized haplotype clustering. *American Journal of Human Genetics*, *81*, 1084–1097. https://doi.org/10.1086/521987

Burks, P. S., Kaiser, C. M., Hawkins, E. M., & Brown, P. J. (2015). Genomewide association for sugar yield in sweet sorghum. *Crop Science*, *55*, 2138–2148. https://doi.org/10.2135/cropsci2015.01.0057

Chen, F., & Dixon, R. A. (2007). Lignin modification improves fermentable sugar yields for biofuel production. *Nature Biotechnology*, *25*, 759–761. https://doi.org/10.1038/nbt1316

de Souza, A., Hull, P. A., Gille, S., & Pauly, M. (2014). Identification and functional characterization of the distinct plant pectin esterases PAE8 and PAE9 and their deletion mutants. *Planta*, *240*, 1123–1138 https://doi.org/10.1007/s00425-014-2139-6

Endo, H., Yamaguchi, M., Tamura, T., Nakano, Y., Nishikubo, N., Yoneda, A., ... Demura, T. (2015). Multiple classes of transcription factors regulate the expression of VASCULAR-RELATED NAC-DOMAIN7, a master switch of xylem vessel differentiation. *Plant and Cell Physiology*, *56*, 242–254. https://doi.org/10.1093/pcp/pcu134

Fang, Y., You, J., Xie, K., Xie, W., & Xiong, L. (2008). Systematic sequence analysis and identification of tissue-specific or stress-responsive genes of NAC transcription factor family in rice. *Molecular Genetics and Genomics*, *280*, 547–563. https://doi.org/10.1007/s00438-008-0386-6

Foster, C. E., Martin, T. M., & Pauly, M. (2010). Comprehensive compositional analysis of plant cell walls (Lignocellulosic biomass) Part I: Lignin. *Journal of Visualized Experiments: JoVE*, *37*, 1745. *Advance online publication*. https://doi.org/10.3791/1745

Furuta, K. M., Yadav, S. R., Lehesranta, S., Belevich, I., Miyashima, S., Heo, J., ... Helariutta, Y. (2014). Plant development. Arabidopsis NAC45/86 direct sieve element morphogenesis culminating in enucleation. *Science*, *345*, 933–937. https://doi.org/10.1126/science.1253736

Glaubitz, J. C., Casstevens, T. M., Lu, F., Harriman, J., Elshire, R. J., Sun, Q., & Buckler, E. S. (2014). TASSEL-GBS: A high capacity genotyping by sequencing analysis pipeline. *PLoS ONE*, *9*, e90346. https://doi.org/10.1371/journal.pone.0090346

Handakumbura, P. P., & Hazen, S. P. (2012). Transcriptional regulation of grass secondary cell wall biosynthesis: Playing catch-up with arabidopsis thaliana. *Frontiers in Plant Science*, *3*, 74. https://doi.org/10.3389/fpls.2012.00074

Hong, S.-Y., Kim, O.-K., Kim, S.-G., Yang, M.-S., & Park, C.-M. (2011). Nuclear import and DNA binding of the ZHD5 transcription factor is modulated by a competitive peptide inhibitor in Arabidopsis. *Journal of Biological Chemistry*, *286*, 1659–1668. https://doi.org/10.1074/jbc.M110.167692

Hu, W., & Ma, H. (2006). Characterization of a novel putative zinc finger gene MIF1: involvement in multiple hormonal regulation of Arabidopsis development. *The Plant Journal*, *45*, 399–422. https://doi.org/10.1111/j.1365-313X.2005.02626.x

Joshi, V., Laubengayer, K. M., Schauer, N., Fernie, A. R., & Jander, G. (2006). Two Arabidopsis threonine aldolases are nonredundant and compete with threonine deaminase for a common substrate pool. *Plant Cell*, *18*, 3564–3575. https://doi.org/10.1105/tpc.106.044958

Kellogg, E. A. (2007). Floral displays: Genetic control of grass inflorescences. *Current Opinion in Plant Biology*, *10*, 26–31. https://doi.org/10.1016/j.pbi.2006.11.009

Ko, J.-H., Jeon, H.-W., Kim, W.-C., Kim, J.-Y., & Han, K.-H. (2014). The MYB46/MYB83-mediated transcriptional regulatory programme is a gatekeeper of secondary wall biosynthesis. *Annals of Botany*, *114*, 1099–1107. https://doi.org/10.1093/aob/mcu126

Langfelder, P., & Horvath, S. (2008). WGCNA: An R package for weighted correlation network analysis. *BMC Bioinformatics*, *9*, 559. https://doi.org/10.1186/1471-2105-9-559

Langmead, B., & Salzberg, S. L. (2012). Fast gapped-read alignment with Bowtie 2. *Nature Methods*, *9*, 357–359. https://doi.org/10.1038/nmeth.1923

Laurentin, A., & Edwards, C. A. (2003). A microtiter modification of the anthrone-sulfuric acid colorimetric assay for glucose-based carbohydrates. *Analytical Biochemistry*, *315*, 143–145. https://doi.org/10.1016/S0003-2697(02)00704-2

Mullet, J., Morishige, D., McCormick, R., Truong, S., Hilley, J., McKinley, B., ... Rooney, W. (2014). Energy Sorghum–a genetic model for the design of C4 grass bioenergy crops. *Journal of Experimental Botany*, *65*, 3479–3489. https://doi.org/10.1093/jxb/eru229

Multani, D. S., Briggs, S. P., Chamberlin, M. A., Blakeslee, J. J., Murphy, A. S., & Johal, G. S. (2003). Loss of an MDR transporter in compact stalks of maize br2 and Sorghum dw3 Mutants. *Science*, *302*, 81–84. https://doi.org/10.1126/science.1086072

Murphy, R. L., Morishige, D. T., Brady, J. A., Rooney, W. L., Yang, S., Klein, P. E., & Mullet, J. E. (2014). Ghd7 (Ma6) represses sorghum flowering in long days: Ghd7 alleles enhance biomass accumulation and grain production. *Plant Genome*, *7*, https://doi.org/10.3835/plantgenome2013.11.0040

Murphy, R. L., Klein, R. R., Morishige, D. T., Brady, J. A., Rooney, W. L., Miller, F. R., ... Mullet, J. E. (2011). Coincident light and clock regulation of pseudoresponse regulator protein 37 (PRR37) controls photoperiodic flowering in sorghum. *Proceedings of the National Academy of Sciences*, *108*, 16469–16474. https://doi.org/10.1073/pnas.1106212108

Nakano, Y., Yamaguchi, M., Endo, H., Rejab, N. A., & Ohtani, M. (2015). NAC-MYB-based transcriptional regulation of secondary cell wall biosynthesis in land plants. *Frontiers in Plant Science*, *6*, 288. https://doi.org/10.3389/fpls.2015.00288

Nuruzzaman, M., Sharoni, A. M., Satoh, K., Moumeni, A., Venuprasad, R., Serraj, R., ... Kikuchi, S. (2012). Comprehensive gene expression analysis of the NAC gene family under normal growth conditions, hormone treatment, and drought stress conditions in rice using near-isogenic lines (NILs) generated from crossing Aday Selection (drought tolerant) and IR64. *Molecular Genetics and Genomics*, *287*, 389–410. https://doi.org/10.1007/s00438-012-0686-8

Oda-Yamamizo, C., Mitsuda, N., Sakamoto, S., Ogawa, D., Ohme-Takagi, M., & Ohmiya, A. (2016). Corrigendum: The NAC transcription factor ANAC046 is a positive regulator of chlorophyll degradation and senescence in Arabidopsis leaves. *Scientific Reports*, *6*, 35125. https://doi.org/10.1038/srep35125

Paterson, A. H., Bowers, J. E., Bruggmann, R., Dubchak, I., Grimwood, J., Gundlach, H., ... Rokhsar, D. S. (2009). The Sorghum bicolor genome and the diversification of grasses. *Nature*, *457*, 551–556. https://doi.org/10.1038/nature07723

Pautler, M., Tanaka, W., Hirano, H.-Y., & Jackson, D. (2013). Grass meristems I: Shoot apical meristem maintenance, axillary meristem determinacy and the floral transition. *Plant and Cell Physiology*, *54*, 302–312. https://doi.org/10.1093/pcp/pct025

Peiffer, J. A., Flint-Garcia, S. A., De Leon, N., McMullen, M. D., Kaeppler, S. M., & Buckler, E. S. (2013). The genetic architecture of maize stalk strength. *PLoS ONE*, *8*(6), e67066. https://doi.org/10.1371/journal.pone.0067066

Peng, X., Zhao, Y., Li, X., Wu, M., Chai, W., Sheng, L., ... Cheng, B. (2015). Genomewide identification, classification and analysis of NAC type gene family in maize. *Journal of Genetics*, *94*, 377–390. https://doi.org/10.1007/s12041-015-0526-9

Pertea, M., Kim, D., Pertea, G. M., Leek, J. T., & Salzberg, S. L. (2016). Transcript-level expression analysis of RNA-seq experiments with HISAT, StringTie and Ballgown. *Nature Protocols*, *11*, 1650–1667. https://doi.org/10.1038/nprot.2016.095

Podzimska-Sroka, D., O'Shea, C., Gregersen, P. L., & Skriver, K. (2015). NAC transcription factors in senescence: From molecular structure to function in crops. *Plants*, *4*, 412–448. https://doi.org/10.3390/plants4030412

Poland, J. A., Brown, P. J., Sorrells, M. E., & Jannink, J.-L. (2012). Development of high-density genetic maps for barley and wheat using a novel two-enzyme genotyping-by-sequencing approach. *PLoS ONE*, *7*, e32253. https://doi.org/10.1371/journal.pone.0032253

Qiu, K., Li, Z., Yang, Z., Chen, J., Wu, S., Zhu, X., ... Zhou, X. (2015). EIN3 and ORE1 accelerate degreening during ethylene-mediated leaf senescence by directly activating chlorophyll catabolic genes in Arabidopsis. *PLoS Genetics*, *11*, e1005399. https://doi.org/10.1371/journal.pgen.1005399

Rooney, W. L., Blumenthal, J., Bean, B., & Mullet, J. E. (2007). Designing sorghum as a dedicated bioenergy feedstock. *Biofuels Bioproducts and Biorefining*, *1*, 147–157. https://doi.org/10.1002/bbb.15

Smith, C. W., & Frederiksen, R. A. (2000). *Sorghum: Origin, history, technology, and production*. New York, NY: Wiley.

Stefaniak, T. R., Dahlberg, J. A., Bean, B. W., Dighe, N., Wolfrum, E. J., & Rooney, W. L. (2012). Variation in biomass composition components among forage, biomass, sorghum-sudangrass, and sweet sorghum types. *Crop Science*, *52*, 1949. https://doi.org/10.2135/cropsci2011.10.0534

Swarts, K., Li, H., Romero Navarro, J. A., An, D., Romay, M. C., Hearne, S., ... Bradbury, P. J. (2014). Novel methods to optimize genotypic imputation for low-coverage, next-generation sequence data in crop plants. *Plant Genome*, *7*, https://doi.org/10.3835/plantgenome2014.05.0023

Tanaka, W., Pautler, M., Jackson, D., & Hirano, H.-Y. (2013). Grass meristems II: Inflorescence architecture, flower development and meristem fate. *Plant and Cell Physiology*, *54*, 313–324. https://doi.org/10.1093/pcp/pct016

Taylor-Teeples, M., Lin, L., de Lucas, M., Turco, G., Toal, T. W., Gaudinier, A., ... Brady, S. M. (2014). An Arabidopsis gene regulatory network for secondary cell wall synthesis. *Nature*, *517*, 571–575. https://doi.org/10.1038/nature14099

Updegraff, D. M. (1969). Semimicro determination of cellulose in biological materials. *Analytical Biochemistry*, *32*, 420–424. https://doi.org/10.1016/S0003-2697(69)80009-6

Xu, B., Ohtani, M., Yamaguchi, M., Toyooka, K., Wakazaki, M., Sato, M., … Demura, T. (2014). Contribution of NAC transcription factors to plant adaptation to land. *Science*, *343*, 1505–1508. https://doi.org/10.1126/science.1248417

Yamaguchi, M., Fujimoto, H., Hirano, K., Araki-Nakamura, S., Ohmae-Shinohara, K., Fujii, A., … Sazuka, T. (2016). Sorghum Dw1, an agronomically important gene for lodging resistance, encodes a novel protein involved in cell proliferation. *Scientific Reports*, *6*, 28366. https://doi.org/10.1038/srep28366

Yang, S. (n.d). CONSTANS is a photoperiod regulated activator of flowering in sorghum. *BMC Plant Biology*, *14*, 148. https://doi.org/doi:10.1186/1471-2229-14-148

Zhao, C. (2005). The Xylem and Phloem transcriptomes from secondary tissues of the Arabidopsis Root-Hypocotyl. *Plant Physiology*, *138*, 803–818. https://doi.org/10.1104/pp.105.060202

Zhao, C., Avci, U., Grant, E. H., Haigler, C. H., & Beers, E. P. (2007). XND1, a member of the NAC domain family in Arabidopsis thaliana, negatively regulates lignocellulose synthesis and programmed cell death in xylem: XND1 blocks tracheary element differentiation. *The Plant Journal*, *53*, 425–436. https://doi.org/10.1111/j.1365-313X.2007.03350.x

Zhong, R., Demura, T., & Ye, Z.-H. (2006). SND1, a NAC domain transcription factor, is a key regulator of secondary wall synthesis in fibers of Arabidopsis. *Plant Cell*, *18*, 3158–3170. https://doi.org/10.1105/tpc.106.047399

Zhong, R., Lee, C., & Ye, Z.-H. (2010). Global analysis of direct targets of secondary wall NAC master switches in Arabidopsis. *Molecular Plant*, *3*, 1087–1103. https://doi.org/10.1093/mp/ssq062

13

AcFT promotes kiwifruit in vitro flowering when overexpressed and Arabidopsis flowering when expressed in the vasculature under its own promoter

Sarah M. A. Moss[1,2] | Tianchi Wang[1] | Charlotte Voogd[1] | Lara A. Brian[1] | Rongmei Wu[1] | Roger P. Hellens[1] | Andrew C. Allan[1,2] | Joanna Putterill[2] | Erika Varkonyi-Gasic[1]

[1]The New Zealand Institute for Plant & Food Research Limited (Plant & Food Research) Mt Albert, Auckland Mail Centre, Auckland, New Zealand

[2]School of Biological Sciences, University of Auckland, Auckland, New Zealand

Correspondence
Erika Varkonyi-Gasic, The New Zealand Institute for Plant & Food Research Limited (Plant & Food Research) Mt Albert, Private Bag 92169, Auckland Mail Centre, Auckland 1142, New Zealand.
Email: erika.varkonyi-gasic@plantandfood.co.nz

Present addresses
Sarah M. A. Moss, The New Zealand Institute for Plant & Food Research Limited (Plant & Food Research) Palmerston North, Palmerston North, New Zealand.

Roger P. Hellens, Centre for Tropical Crops and Biocommodities, Queensland University of Technology, Brisbane, Queensland, Australia.

Funding information
This work was funded by the New Zealand Ministry of Business, Innovation and Employment grants C10X0816 and C11X1602

Abstract

Kiwifruit (*Actinidia chinensis*) has three *FLOWERING LOCUS T* (*FT*) genes, *AcFT*, *AcFT1*, and *AcFT2*, with differential expression and potentially divergent roles. *AcFT* was previously shown to be expressed in source leaves and induced in dormant buds by winter chilling. Here, we show that *AcFT* promotes flowering in *A. chinensis*, despite a short sequence insertion not present in other *FT*-like genes. A 3.5-kb *AcFT* promoter region contained all the regulatory elements required to mediate vascular expression in transgenic *Arabidopsis thaliana* (Arabidopsis). The promoter activation was initially confined to the veins in the distal end of the leaf, before extending to the veins in the base of the leaf, and was detected in inductive and noninductive photoperiods. The 3-kb and 2.7-kb promoter regions of *AcFT1* and *AcFT2*, respectively, demonstrated different activation patterns in Arabidopsis, corresponding to differential expression in kiwifruit. Expression of *AcFT* cDNA from the *AcFT* promoter was capable to induce early flowering in transgenic Arabidopsis in noninductive photoperiods. Further, expression of *AcFT* cDNA fused to the green fluorescent protein was detected in the vasculature and was also capable to advance flowering in noninductive photoperiods. Taken together, these studies implicate *AcFT* in regulation of kiwifruit flowering time and as a candidate for kiwifruit florigen.

KEYWORDS
Actinidia chinensis, Florigen, *FLOWERING LOCUS T*, *FT*, Kiwifruit, promoter

1 | INTRODUCTION

The transition to flowering is a highly regulated process in which developmental and environmental signals are integrated to ensure reproduction. *FLOWERING LOCUS T* (*FT*) (Kardailsky et al., 1999; Kobayashi, Kaya, Goto, Iwabuchi, & Araki, 1999) is the key floral integrator gene in *Arabidopsis thaliana* (Arabidopsis), encoding a major mobile flowering hormone 'florigen' (Chailakhyan, 1968; Zeevaart, 2008). *FT* is transcribed in the leaf vasculature, and the FT protein is subsequently transported in the phloem to the shoot apex, where it unloads and moves cell to cell to interact with the bZIP transcription factor FD and initiate flowering (reviewed in Abe et al., 2005; Andres & Coupland, 2012). The conservation of FT-like protein role as florigen has been well established, but it has become clear that many FT-like proteins perform other roles, including regulation of vegetative growth, storage organ differentiation, and fruit set (reviewed in Pin & Nilsson, 2012; Wickland & Hanzawa, 2015).

Previous studies have demonstrated that transcriptional regulation of *FT* was essential to ensure flowering time control in response to environmental conditions (Adrian et al., 2010; Takada & Goto, 2003) and more recent studies have begun to unravel the detailed mechanisms controlling FT movement and action (reviewed in Putterill & Varkonyi-Gasic, 2016). In Arabidopsis and rice, the CONSTANS (CO)-FT pathway is critical for photoperiodic regulation of *FT* transcription (Hayama, Yokoi, Tamaki, Yano, & Shimamoto, 2003; Samach et al., 2000; Suarez-Lopez, 2001), but this pathway is not conserved in legumes (Putterill et al., 2013; Weller & Ortega, 2015) and little is known about activators of *FT* transcription in other species. Repressors such as Arabidopsis MADS-box proteins FLOWERING LOCUS C (FLC) and SHORT VEGETATIVE PHASE (SVP) bind to the CArG box motifs in the first intron and the promoter region of *FT* (Lee et al., 2007; Li et al., 2008; Searle et al., 2006) to repress flowering, but this role may not be universal. Further layers of complexity in regulation of FT include chromatin-mediated changes (Adrian et al., 2010; He, 2015; Kotake, Takada, Nakahigashi, Ohto, & Goto, 2003) and posttranslational regulation (Kim et al., 2016). It is unclear whether these mechanisms are conserved across species, but it has been established that FT-like proteins interact with conserved protein partners including FD homologs (Abe et al., 2005; Pnueli et al., 2001; Wigge et al., 2005) and other transcription factors (Ho & Weigel, 2014; Mimida et al., 2011; Niwa et al., 2013).

The long life span, large size, coexistence of vegetative, and floral fate on the same plant, combined with the lack of natural mutants and lengthy transformation, have resulted in a much slower progress in studies of *FT* homologs in woody perennial plants, where many questions about gene diversification and florigen function remain unresolved (Putterill & Varkonyi-Gasic, 2016). However, the importance of FT in regulation of flowering time was demonstrated by expression of endogenous or heterologous *FT* genes, which dramatically reduced the juvenile period and provided a means to accelerate breeding in trees (Böhlenius et al., 2006; Endo et al., 2005; Hsu, Liu, Luthe, & Yuceer, 2006; Klocko et al., 2016; Kotoda et al., 2010; Song, Walworth, Zhao, Jiang, & Hancock, 2013; Srinivasan, Dardick, Callahan, & Scorza, 2012; Wenzel, Flachowsky, & Hanke, 2013), but these approaches have often proven to be unreliable (Zhang et al., 2010). Most of the studies utilized strong overexpression, which generally induced very early flowering and failed to distinguish between *FT* paralogs. However, in some cases overexpression of endogenous *FT* genes did not result in precocity, despite their ability to promote flowering in model annuals. Instead, they affected dormancy and leaf senescence in apple (Freiman et al., 2015) or shoot vigor in kiwifruit (Varkonyi-Gasic et al., 2013), suggesting roles besides flowering time control. Moderate overexpression comparable to normal peak expression revealed functional divergence among *FT* paralogs to determine reproductive onset and vegetative growth in poplar (Hsu et al., 2011). However, for a large number of *FT*-like genes from woody perennial species, the function, regulation, and modes of action remain largely unknown.

In this study, we are focusing on *FT*-mediated regulation of growth and flowering in kiwifruit. Kiwifruit are deciduous woody perennial vine species of the genus *Actinidia*, recently domesticated as a horticultural crop, with increasing importance but very short history of breeding and limited genetic resources (Datson & Ferguson, 2011). *Actinidia* are characterized by a juvenile unproductive period and a temperate flowering phenology. The growth cycle is spread between two seasons, interrupted by winter dormancy. In the first season, meristems with a potential to differentiate flowers are initiated in the axils of leaf primordia, within the lateral buds of developing shoots (Varkonyi-Gasic et al., 2011; Walton, Fowke, Weis, & Mcleay, 1997; Walton, Podivinsky, & Wu, 2001). In natural conditions, autumn short days and cooler temperatures induce growth cessation and these lateral buds become dormant (Lionakis & Schwabe, 1984). Accumulation of chilling during the winter dormancy period is essential to resume growth and flowering in spring (Brundell, 1976). Inflorescences develop in the lower leaf axils of shoots emerging after the chilling period, from the preestablished meristems. Photoperiod has no clear role in regulation of flowering in spring (Snelgar, Clearwater, & Walton, 2007), but insufficient chilling results in sporadic budbreak, low flower numbers, and low fruit yield.

Recently, we identified and characterized three *FT* genes from a commercial kiwifruit species *Actinidia chinensis*, designated *AcFT*, *AcFT1*, and *AcFT2* (Varkonyi-Gasic et al., 2013; Voogd, Brian, Wang, Allan, & Varkonyi-Gasic, 2017). They all promoted flowering when expressed in Arabidopsis, but were differentially expressed in kiwifruit, suggesting that they evolved to perform distinct roles. *AcFT* expression domains included leaves and dormant buds, implicating it in regulation of budbreak and flowering. In contrast, *AcFT1* and *AcFT2* were fully excluded from the dormant bud. *AcFT1* was detected in the tip and small apical leaves of actively growing shoots, and *AcFT2* was detected in large basal leaves. Strong ectopic expression of both *AcFT1* and *AcFT2* promoted in vitro flowering in *A. chinensis*, indicating that AcFT1 and AcFT2 may function as flowering activators in kiwifruit, although their expression patterns could not fully explain this role. Conversely, ectopic overexpression of the third *Actinidia FT* gene, *AcFT*, did not cause another species of kiwifruit, *A. eriantha*, to flower more rapidly. Therefore, it remained unclear whether *AcFT* was a kiwifruit flowering gene.

Here, we ectopically overexpressed *AcFT* in *A. chinensis* to establish whether its strong ectopic expression promoted flowering. Next, we studied the expression, function, and localization of *AcFT* using reporter gene fusions and gene expression constructs driven by the *AcFT* promoter. These studies were performed in Arabidopsis to circumvent the inefficient transformation, large size and long juvenility of kiwifruit and to take advantage of absence of endogenous *FT* in noninductive photoperiods. Constructs driven by the *AcFT1* and *AcFT2* promoters were also included in the study and the *SUCROSE TRANSPORTER 2* (*SUC2*) promoter (Stadler, Lauterbach, & Sauer, 2005), which is specific to the phloem companion cells of the major veins and commonly used to determine whether genes encode florigen-like activity, was included in control experiments.

2 | MATERIALS AND METHODS

2.1 | Amplification of genes and vector construction

The full-length *AcFT* coding sequence was amplified from kiwifruit *A. chinensis* var. *chinensis* 'Hort16A' cDNA, cloned into pUC19 vector and verified by sequence analysis as previously described (Varkonyi-Gasic et al., 2013). After addition of attB sites by PCR, it was recombined using Gateway into pDONR221, verified by sequence analysis and recombined into pHEX2 (Hellens et al., 2005), placing the cDNA between the CaMV *35S* promoter and the *ocs 3′* transcriptional terminator. The *AcFT** sequence with added attL sites was synthesized by GenScript (http://www.genscript.com) and recombined into pHEX2 as above. The resulting plant transformation vectors were transformed into *Agrobacterium tumefaciens* strain EHA105 by electroporation.

The 3.5-kb sequence upstream of the *AcFT* translation start site (GenBank accession KJ439053) was obtained using inverse PCR (Ochman, Gerber, & Hartl, 1988) from kiwifruit 'Hort16A' genomic DNA digested with HindIII and circularized using T4 DNA ligase (Invitrogen, Life Science Technologies, Carlsbad, CA). For the cloning of *AcFT1* and *AcFT2* promoter sequences, multiple primers were designed based on the draft *Actinidia* genome (Huang et al., 2013) and used for amplification from the kiwifruit 'Hort16A' genomic DNA template. The 3.0- and 2.7-kb sequence upstream of the *AcFT1* and *AcFT2* translation start sites, respectively (GenBank accession numbers KX611592 and KX611593) were chosen for subsequent study as the longest promoter fragments. All promoters included the 5′UTR sequences of unknown size. They were cloned into pGEM-T (Promega, Madison, WI), verified by sequence analysis, reamplified to include appropriate restriction sites and subsequently cloned to generate fusion constructs.

The 1.8-kb fragment of the Arabidopsis *FT* promoter was amplified from Col-0 genomic DNA and the Arabidopsis *FT* coding region was amplified from Col-0 seedling cDNA. The vascular-specific *SUC2* promoter, *AcFT1* and *AcFT2* cDNA (Voogd et al., 2017) were previously described. The promoter-less *GUS* and *pro35S:GUS* vector controls were previously constructed from pHEX2 (Hellens et al., 2005). pHEX2S:GFP derivative of pHEX2, designed previously to include the *35S* promoter-driven sequence encoding GFP(S65T) (Heim, Cubitt, & Tsien, 1995), was used to generate GFP fusion constructs and pGreenII 0800-LUC vector (Hellens et al., 2005) was used to generate LUC fusions. Oligonucleotide primer sequences used for amplification and as linkers are presented in Table S1 and generation of constructs used in this study in Table S2. All the resulting plant transformation vectors were transformed into *Agrobacterium tumefaciens* strain GV3101 by electroporation.

2.2 | Plant material and transformation

Agrobacterium-mediated transformation of *A. chinensis* 'Hort16A' was as previously described (Voogd, Wang, & Varkonyi-Gasic, 2015; Wang, Atkinson, & Janssen, 2007), using *A. tumefaciens* strain EHA105. *Agrobacterium*-mediated transformation of Arabidopsis Col-0 (Clough & Bent, 1998; Martinez-Trujillo, Limones-Briones, Cabrera-Ponce, & Herrera-Estrella, 2004) was performed using *A. tumefaciens* strain GV3101.

2.3 | Plant growth conditions and sampling

Seeds of transgenic Arabidopsis plants were selected on half-strength Murashige and Skoog (½ MS) medium supplemented with kanamycin and placed in a growth room under a long-day (LD, 21°C, 16/8-hr light/dark) or short-day (*SD*, 21°C, 8/16 hr light/dark) regime. Subsequently, plants were grown in soil using a standard potting mix in LD and SD regime. Because *proAcFT* is most highly active during early stages of development, expression analysis in kanamycin-resistant lines was performed 7 days after germination (DAG) for transgenic Arabidopsis, using 10 T2 kanamycin-resistant Arabidopsis seedlings of each line. All sampling was performed at the end of the light cycle to minimize variation.

2.4 | RNA extraction and expression studies

Total RNA from Arabidopsis was isolated using the Trizol reagent (Invitrogen) and from kiwifruit using the Spectrum Plant Total RNA Kit (Sigma-Aldrich, St. Louis, MO, USA). Reverse transcription (RT) was performed using a 1 μg aliquot of RNA treated with DNase I (Invitrogen), an oligo(dT) primer and the SuperScript III reverse transcriptase (Invitrogen). Quantifications using real-time PCR were performed with the FastStart DNA Master SYBR Green I mix (Roche Diagnostics, Mannheim, Germany) using the LightCycler 1.5 instrument and the LightCycler Software version 4 (Roche Diagnostics). Amplification was carried out using a 10^{-3} dilution of the cDNA template, with an initial denaturing step at 95°C for 5 min, then 40–50 cycles of 95°C for 5 s, 60°C for 5 s, and 72°C for 10 s. A non-template control was included in each run. Oligonucleotide primers (Table S1) were designed to produce amplification products of 100–150 nucleotides. The specificity of primer pairs was confirmed by melting curve analysis of PCR products and agarose gel electrophoresis followed by sequence analysis. The expression was normalized to Arabidopsis *ACT2* (At3 g18780) and kiwifruit *A. chinensis* var. *deliciosa ACTIN* (GenBank accession FG403300).

2.5 | Histochemical localization of GUS activity

Detection of GUS activity was performed according to Jefferson (1987) with some modifications. Briefly, whole seedlings or excised tissue were soaked in the staining solution (50 mM $NaPO_4$ pH7.0, 0.05% X-Gluc (Gold Biotechnology, St. Louis, MO), 0.2% dimethylformamide, 10% methanol and 0.1% Tween 20) for 24 hr at 37°C, rinsed twice with water followed by 50% ethanol and soaked in ethanol to remove chlorophyll, stored in fresh ethanol and photographed. Whole seedlings and plant organs or hand sections of tissue were used for photographing.

To analyze transgenic Arabidopsis plants carrying the GUS fusion constructs, a minimum of 10 independent transgenic lines for each GUS fusion construct were screened to determine if GUS activity could be detected, but allowing plants to grow and produce seed. For the detailed screen, progeny of at least three independent lines were used for staining at six time points during development in LD conditions, to represent the cotyledon stage (7 DAG), rosette leaf development (4-5 rosette leaves, ~12 DAG), floral transition (9-11 rosette leaves, ~26 DAG), early and late flower development (inflorescence stem 10–20 cm long), and late maturity (multiple inflorescences, silique development). For the first four time points, whole seedlings on ½ MS medium were used. For the last two time points, staining was performed using aerial parts of plants grown in soil. Progeny of the same lines were also grown in SD regime and stained at corresponding time points.

2.6 | Promoter transactivation assays

Nicotiana benthamiana plants were grown in LD conditions, and transient leaf assays carried out as described in Hellens et al. (2005). Assays were performed using three biological replicates and four technical replicates. Sampling of infiltrated leaves was performed 3 days post infiltration. The quantification was performed using the Dual-Glo Luciferase Assay System (Promega) and the firefly luciferase (LUC) and *Renilla* luciferase (REN) luminescence were measured using the Orion Microplate Luminometer and Simplicity software (Berthold Detection Systems). The promoter activity was quantified as the LUC/REN chemiluminescence ratio.

2.7 | Imaging

Photography of whole seedlings or plant organs was performed using a Nikon D80 DSLR with 60 mm Micro-Nikkor lens. GFP activity was examined using the Leica MZ FLIII stereo fluorescence microscope with a GFP2 and GFP3 filter (Leica Microsystems, Wetzlar, Germany) equipped with the Infinity2 camera (Lumenera, Ottawa, ON, Canada).

3 | RESULTS

3.1 | Ectopic expression of *AcFT* in *A. chinensis* promotes in vitro flowering

The *AcFT* coding sequence contains an unusual region, not identified in any other *FT* gene, giving rise to an insertion of 6 amino acids close to the N-terminus (Figure S1). Transformation of *A. chinensis* was therefore initiated using two constructs, the full-length *AcFT* coding sequence and a modified *AcFT* coding sequence designed to exclude the unusual region, both driven by the cauliflower mosaic virus (CAMV) *35S* promoter (Figure 1a). Proliferation of flower-like initials which never differentiated further and died within 2 months of the transformation experiment was observed with the *AcFT* construct (*pro35S:AcFT*; Figure 1b). The modified *AcFT* construct (*pro35S:AcFT**) gave rise to in vitro flowers (Figure 1c) and very early flowering plantlets (Figure 1d). In contrast, development of leaf-like structures (Figure 1e) followed by normal vegetative growth (Figure 1f) were seen in control (*pro35S:GUS*) experiments. To confirm floral identity of *pro35S:AcFT* flower-like initials, amplification of flower-specific MADS-box transcripts (Varkonyi-Gasic et al., 2011) and *SEPALLATA4* (*SEP4*) abundant during fruit development (Richardson et al., 2011) was performed. The presence or absence of kiwifruit *SEP1*, *APETALA3* (*AP3*), *PISTILLATA* (*PI*) and *AGAMOUS* (*AG*) transcripts revealed floral and vegetative fates in appropriate tissues (Figure 1g). Therefore, both constructs strongly promoted flowering in vitro and the slightly reduced activity after removal of the unusual small insertion in *AcFT** resulted in distinguishable flower organ development, but neither construct gave rise to viable plants.

3.2 | The 3.5-kb sequence upstream of the *AcFT* translation start site was sufficient for specific vascular expression in Arabidopsis

To study the regulation of *AcFT* expression, a 3.5-kb sequence upstream of the *AcFT* translation start site was cloned from *A. chinensis* genomic DNA and designated *proAcFT*. To evaluate the activity of this promoter fragment, a transcriptional fusion with the reporter gene *uidA* (*GUS*) was prepared. A translational fusion including the *AcFT* first exon, first intron and the first eight codons of the second exon was also evaluated to examine whether regulatory regions exist in the first intron. Constructs containing the constitutive CaMV *35S* and the vascular-specific *SUC2* promoters fused to the *GUS* gene were used as positive controls and the promoter-less *GUS* construct was used as a negative control. In addition, to compare the regulation of *AcFT* expression with the other two kiwifruit *FT* genes, a 3-kb and 2.7-kb sequence upstream of the *AcFT1* and *AcFT2* translation start sites, respectively were isolated from *A. chinensis* genomic DNA, designated *proAcFT1* and *proAcFT2*, and fused with the *GUS* gene (Figure 2a).

GUS expression driven by the *proAcFT* fragment (*proAcFT:GUS*) was developmentally regulated and restricted to the vascular tissue of transgenic Arabidopsis. Initially, strong expression was detected at the distal end of the cotyledons (Figure 2b), followed by expression throughout the vasculature of cotyledons and veins in the distal half of developing leaves (Figure 2c,d). At later stages of development, the signal was detected in the veins in the distal half of sepals (Figure 2e) and weak staining was observed in the phloem of the vasculature at the base of rosette leaves. In some lines, GUS staining was detected in siliques (Figure 2f). The promoter activity

FIGURE 1 Transgenic kiwifruit flowering in vitro. (a) Schematic diagram of the constructs of the full-length *AcFT* coding sequence (*pro35S:AcFT*) and a modified *AcFT* coding sequence designed to exclude the unusual region (*pro35S:AcFT**). (b, c) Proliferation of in vitro flowers within 2 months from initiation of transformation with both constructs. (d) Flower and leaf development in *pro35S:AcFT** plantlets. (e) Leaf initials in the control transformation. (f) Vegetative control plantlets. (g) RT–PCR of *AcFT* and kiwifruit *FRUITFULL-like* (*FUL-like*), *SEPALLATA1* (*SEP1*), *SEPALLATA4* (*SEP4*), *APETALA3* (*AP3*), *PISTILLATA* (*PI*), *AGAMOUS* (*AG*), and *ACTIN* (*ACT*). Flower and leaf tissue are indicated by arrows and arrowheads, respectively. Bars, 1 mm (b, e), 2 mm (c, d, f)

in the vasculature during transgenic Arabidopsis development is summarized in a schematic (Figure 2g). GUS signal was not detected in the shoot apical meristem, leaf primordia, hypocotyls or roots. A similar pattern was obtained with the construct that included the first intron (Figure S2). The control CaMV *35S* provided the expected constitutive activity and the *SUC2* promoter demonstrated strong vascular expression throughout development (Figures 2h, S2).

In contrast to *proAcFT*, *GUS* expression driven by the *proAcFT1* fragment was observed in the shoot tip of transgenic Arabidopsis plants (Figure 2i), while *GUS* expression driven by the *proAcFT2* fragment was detected in the vascular tissue of the cotyledon petiole and the hypocotyl (Figure 2j), followed by expression in the petioles and midribs at the basal end of the leaf (Figures 2k, S3). Therefore, different expression patterns were demonstrated for each kiwifruit *FT* gene, similar to their differential expression in kiwifruit (Varkonyi-Gasic et al., 2013; Voogd et al., 2017), but only the *GUS* expression driven by the *proAcFT* fragment was following the basipetal leaf development pattern, closely resembling the pattern of the Arabidopsis *FT* promoter (Adrian et al., 2010).

3.3 | The *AcFT* promoter is active under the noninductive photoperiod in transgenic Arabidopsis

In Arabidopsis, *FT* is expressed at a much higher level in long-day (LD) than in short-day (SD) conditions due to direct activation by CO

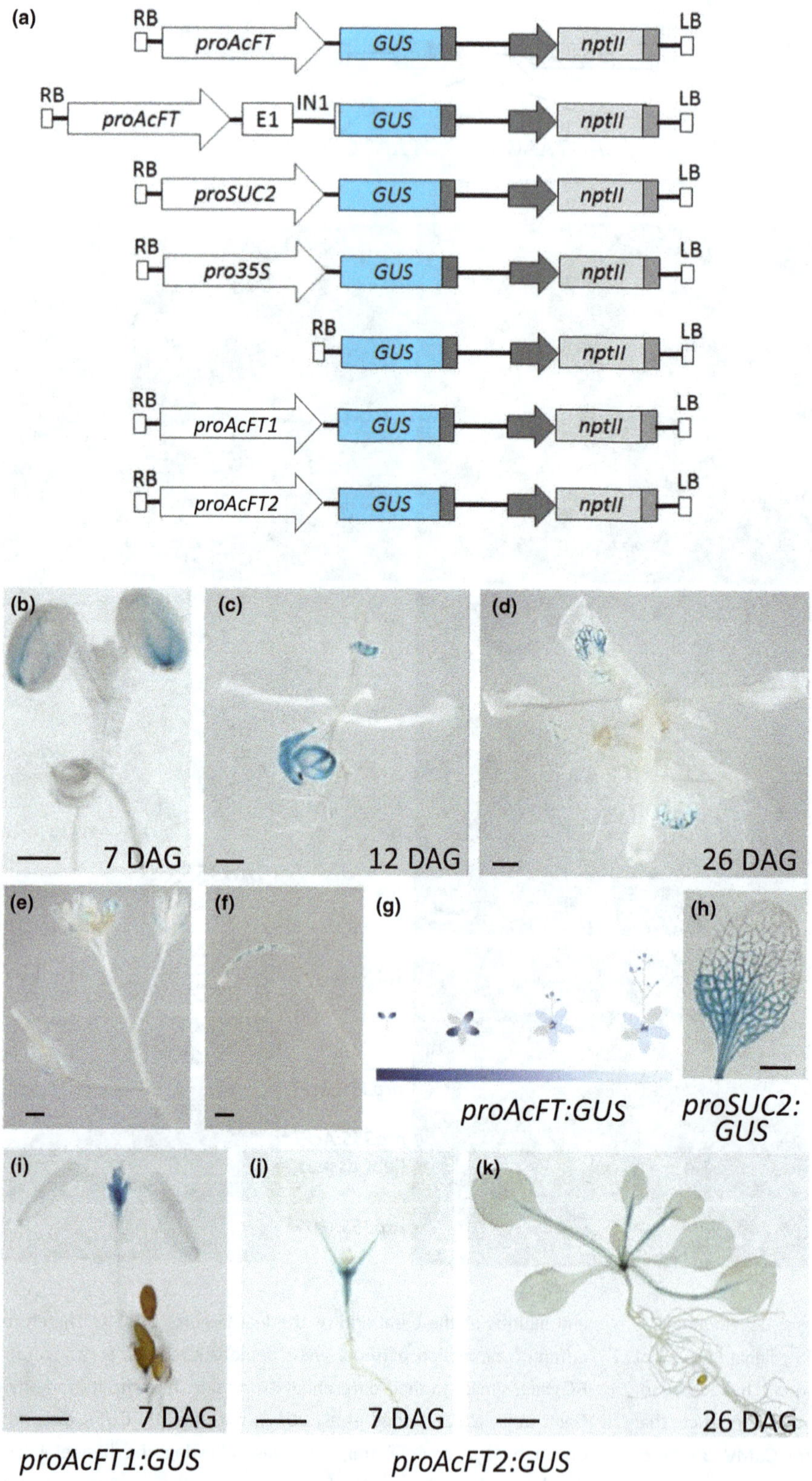

FIGURE 2 Temporal and spatial regulation of the *AcFT* promoter (*proAcFT*) in transgenic Arabidopsis in long-day conditions (LD). (a) Promoter fusion constructs in the binary vector: 3.5-kb sequence upstream of the *AcFT* translation start sites (*proAcFT*), a fragment containing *proAcFT*, *AcFT* first exon (E1), first intron (IN1) and the first eight codons of the second exon, *SUCROSE TRANSPORTER 2* (*SUC2*) promoter (*proSUC2*), CaMV *35S* promoter (*pro35S*), 3-kb sequence upstream of the *AcFT1* translation start sites (*proAcFT1*), and 2.7-kb sequence upstream of the *AcFT2* translation start sites (*proAcFT2*) fused with *uidA* (*GUS*) reporter gene. (b–f) Histochemical localization of GUS activity in transgenic *proAcFT:GUS* Arabidopsis grown in long-day conditions (LD) during the cotyledon stage (b), rosette leaf development (c), and floral transition (d), in flowers (e), and in siliques (f). (g) Schematic summary of the position and intensity of GUS staining. (h) Histochemical localization of GUS activity in transgenic *proSUC2:GUS* Arabidopsis leaf. (i–k) Histochemical localization of GUS activity in transgenic *proAcFT2:GUS* Arabidopsis during the cotyledon stage (i, j) and at floral transition (k). Bars, 1 mm; DAG, days after germination

(Samach et al., 2000). In contrast, the expression of *GUS* transcript and intensity of GUS staining of the *proAcFT:GUS* transgenic plants grown in different photoperiods were similar (Figure 3a,b), suggesting that it is not regulated by the Arabidopsis photoperiodic "switch," CO. To further investigate, transactivation studies of a firefly luciferase (*LUC*) gene under the control of the Arabidopsis *FT* or *AcFT* promoters were performed in *Nicotiana benthamiana*. A 1.8-kb fragment upstream of the Arabidopsis *FT* translation start site was isolated and a transcriptional fusion with *LUC* was generated. A transcriptional fusion of the 3.5-kb *AcFT* promoter fragment with *LUC* was also prepared. Both constructs contained the *Renilla* luciferase gene (*REN*) driven by the *35S* promoter fragment to normalize the luminescence recording

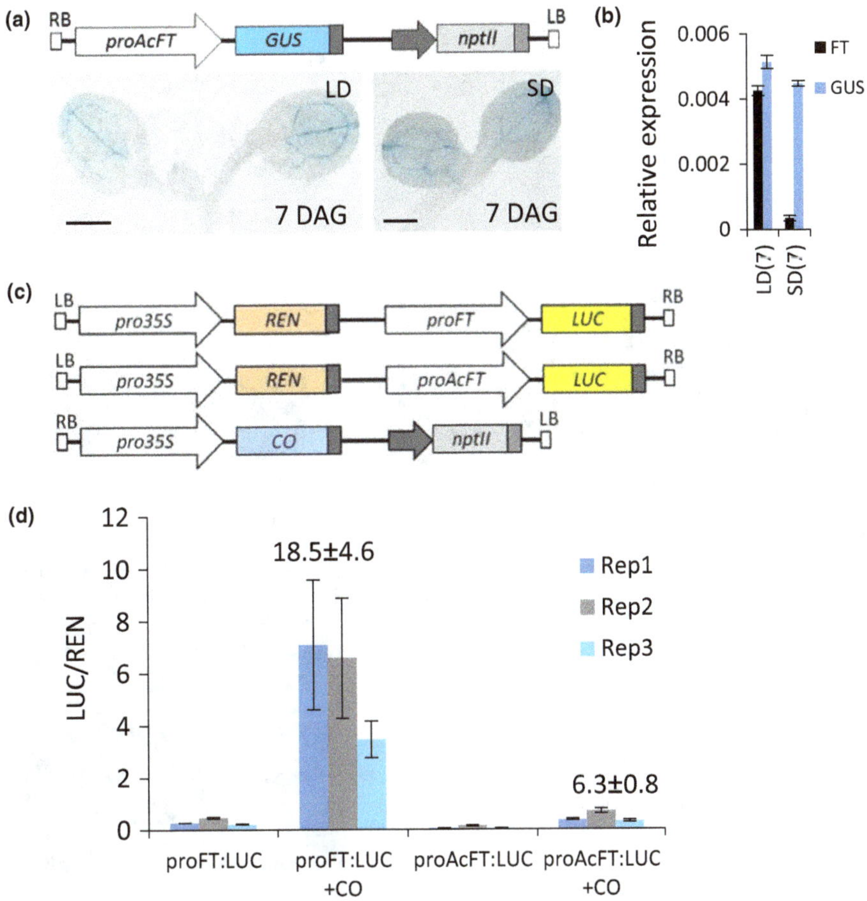

FIGURE 3 The *AcFT* promoter is not strongly regulated by photoperiod in transgenic Arabidopsis. (a) Histochemical localization of GUS activity in cotyledons of transgenic *proAcFT:GUS* Arabidopsis grown in inductive LD and noninductive SD. Bars, 1 mm. (b) Relative expression ± *SE* of Arabidopsis *FT* and *GUS* in transgenic *proAcFT:GUS* Arabidopsis grown in LD and SD for 7 days. The sampling was performed at the end of the light phase. (c) Schematic diagram of constructs used in *Agrobacterium*-mediated transient promoter transactivation assays performed in *Nicotiana benthamiana*. Fusion constructs of promoters of Arabidopsis *FT* (*proFT*) and kiwifruit *AcFT* (*proAcFT*) with firefly luciferase (*LUC*) gene were coinfiltrated with Arabidopsis *CONSTANS* (*CO*) driven by the *35S* promoter. *Renilla* luciferase (*REN*) driven by *35S* promoter is used to normalize the assays. (d) *CO*-mediated induction of Arabidopsis *FT* and *AcFT* promoters measured as LUC/REN ratio. The average values of fold change for all three biological replicates are indicated above. Error bars represent standard deviations for four technical replicates

(Figure 3c). The *Agrobacterium* carrying appropriate promoter constructs was infiltrated into *N. benthamiana* leaves on plants growing in LD photoperiods, to determine basal luminescence and coinfiltrated with *Agrobacterium* carrying a construct of Arabidopsis *CO* cDNA under the control of the *35S* promoter (*pro35S:CO*; Figure 3c) to determine the effect of CO on *LUC* expression. Coinfiltration with *pro35S:CO* resulted in a significant increase in luminescence with the Arabidopsis *FT* promoter, but only a small increase with *proAcFT* (Figure 3d). We conclude that the *proAcFT* fragment is largely independent of photoperiodic regulation in transgenic Arabidopsis and is active in conditions when endogenous *FT* is not expressed.

3.4 | *AcFT* can promote flowering in Arabidopsis when expressed under its own promoter

The activity in SD provided the opportunity to evaluate if *FT* expression driven by the *proAcFT* fragment was sufficient to supplement for the lack of endogenous *FT* in noninductive conditions. A transcriptional fusion with Arabidopsis *FT* cDNA under control of *proAcFT* (*proAcFT:FT*; Figure 4a) was introduced into Arabidopsis Col-0. Three out of seven lines flowered early in SD conditions, after the plants produced between 10 and 16 leaves, in contrast to >30 in control lines (Figure 4b), demonstrating that expression of *FT* from *proAcFT* could induce flowering under noninductive conditions, although the flowers often aborted and some aerial rosette growth was observed (Figure 4c).

Next, to evaluate if the *AcFT* coding sequence under control of its own promoter (*proAcFT:AcFT*; Figure 4a) may also be sufficient to promote flowering, ten *proAcFT:AcFT* lines in Col-0 were generated. To evaluate if they can promote flowering in absence of endogenous *FT* under noninductive conditions, kanamycin-resistant progeny expressing *AcFT* (Figure S4) was monitored. Progeny of three lines showed early bolting in SD (Figure 4b), followed by growth of numerous aerial rosettes (Figure 4d). We conclude that *AcFT* under its own promoter can activate flowering in Arabidopsis,

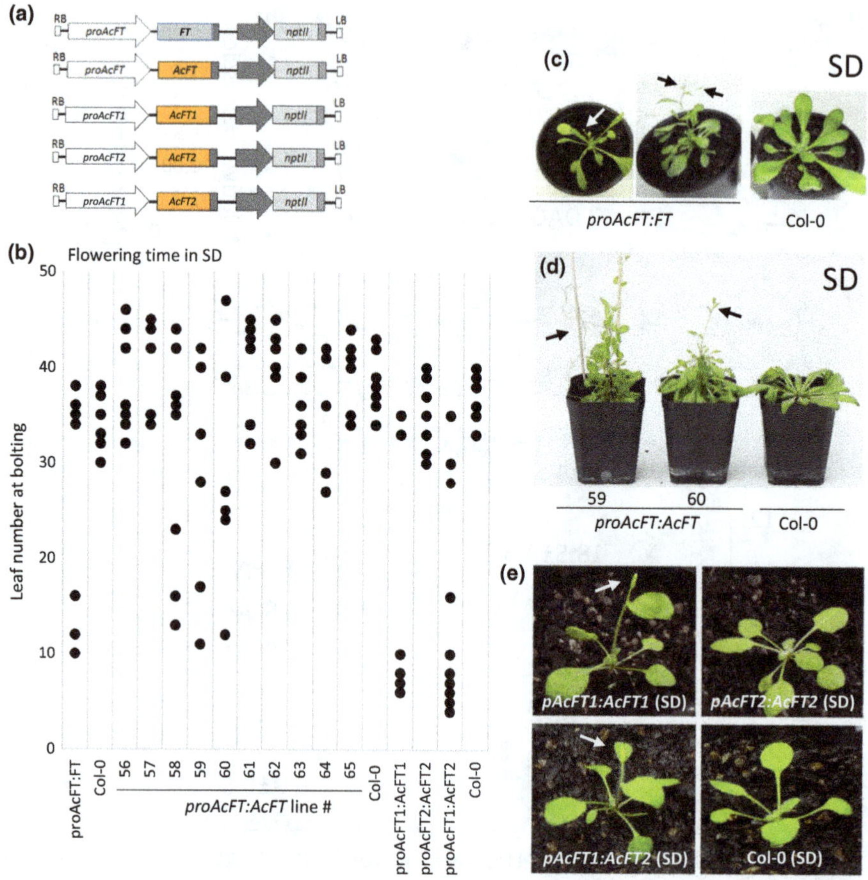

FIGURE 4 Expression of an *FT* gene from the *AcFT* and *AcFT1* promoters is sufficient for flowering in noninductive conditions in transgenic Arabidopsis. (a) Fusion constructs of kiwifruit (*Ac*) *FT* promoters with Arabidopsis *FT* and kiwifruit (*Ac*) *FT* cDNAs. (b) The range of flowering time in Arabidopsis grown in SD conditions. Flowering time was recorded as the total leaf number when the primary inflorescence stems were 0.2 cm long. Each dot represents the flowering time of at least one line; multiple lines with the same flowering time are presented as single dots. (c) Early bolting and growth of aerial rosettes in transgenic *proAcFT:FT* Arabidopsis in SD. (d) Early bolting and growth of aerial rosettes in transgenic *proAcFT:AcFT* Arabidopsis in SD. (e) Early flowering in *proAcFT1:AcFT1* and *proAcFT1:AcFT2* Arabidopsis in SD. Flowers and siliques are indicated by arrows

but may not be sufficient to maintain flowering in noninductive conditions.

We also evaluated if *AcFT1* and *AcFT2* under the control of their own promoters affect flowering, using *proAcFT1:AcFT1* and *proAcFT2:AcFT2* constructs transformed into Arabidopsis (Figure 4a, S4). Very early flowering in SD conditions was recorded for some *proAcFT1:AcFT1*, without reversion to vegetative growth, but not for *proAcFT2:AcFT2* plants (Figure 4b,e). Expression of *AcFT2* under control of *proAcFT1* (*proAcFT1:AcFT2*) was also sufficient for very early flowering in SD conditions (Figure 4b,e). Taken together, the data show that expression of an *FT* cDNA from *proAcFT* in leaf veins and *proAcFT1* in the shoot tip may be sufficient for promotion of flowering, and in case of *proAcFT1* maintenance of floral fate, while the *AcFT2* promoter fragment used in this study did not show such effect.

3.5 | Vascular expression of *AcFT:GFP* promotes flowering in Arabidopsis

To study the AcFT protein localization, *AcFT* cDNA fused to the green fluorescent protein (GFP) coding sequence (*AcFT:GFP*) was placed under control of *AcFT* and *SUC2* promoters (Figure 5a) and transformed into Arabidopsis. The expected vascular GFP activity was observed in cotyledons of *proSUC2:AcFT:GFP* transgenic lines and plants flowered significantly earlier in SD (Figure 5b,c), confirming functionality and potential mobility of the fusion AcFT:GFP protein in Arabidopsis. Much weaker green florescence was detected in the veins in cotyledons of *proAcFT:AcFT:GFP* transgenic lines (Figure 5b), but earlier flowering compared to controls was recorded (Figure 5c), confirming flower-promoting activity of *AcFT:GFP* expressed from *proAcFT*.

4 | DISCUSSION

4.1 | *AcFT* promotes floral fate in kiwifruit *A. chinensis*

In this study, we investigated the ability of *AcFT* to promote flowering in kiwifruit and Arabidopsis and characterized its expression using fusion constructs in Arabidopsis. Previous studies implicated *AcFT* in regulation of floral transition, but this was not confirmed by our initial transgenic studies in kiwifruit *A. eriantha*. In addition, *AcFT*

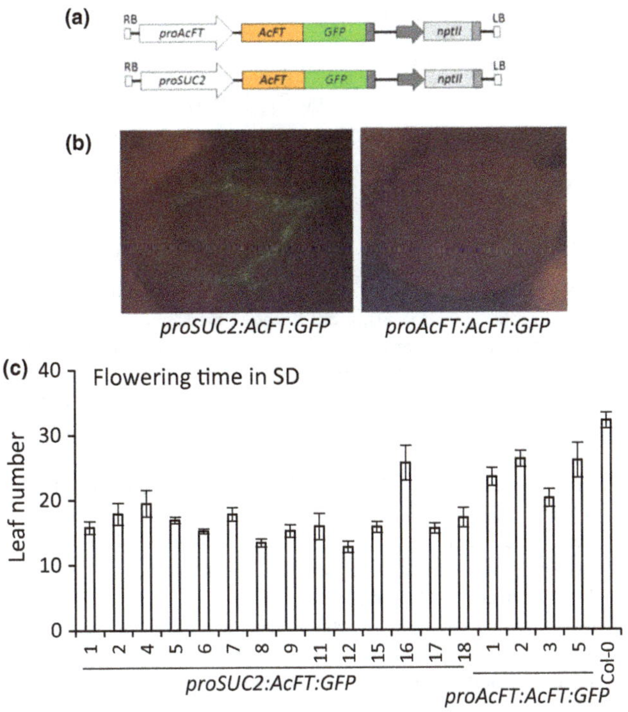

FIGURE 5 Expression of *AcFT:GFP* in the phloem promotes flowering in noninductive conditions in transgenic Arabidopsis. (a) Fusion constructs with *GFP* reporter gene used in this study. (b) GFP (green fluorescence) detected in cotyledons of transgenic Arabidopsis. (c) Early flowering in transgenic Arabidopsis. The minimum of eight T2 kanamycin-resistant lines were monitored for flowering time, which was recorded as the total leaf number when the primary inflorescence stems were 0.2 cm long and presented as means ± *SE*. ANOVA *p*-value of ≤ 0.01 was used to determine statistical significance for presented lines

was also expressed in developing flowers and fruit, suggesting additional roles in regulation of floral fate and fruit development (Varkonyi-Gasic et al., 2013).

Here we overexpressed *AcFT* in *A. chinensis* and obtained very early flowering in vitro, indicating that *AcFT*, at least when ectopically overexpressed, is able to promote flowering in kiwifruit. The unusual sequence not present in any other *FT* gene did not have a detrimental effect on the AcFT function. Rather, it enhanced its capacity to induce the flowering fate and terminate growth in tissue culture conditions, in agreement with the impaired growth and shoot tip abortion previously observed in transgenic *A. eriantha*. Regeneration resulting in in vitro floral initials and no viable plant development might have been the reason why only a small number of *A. eriantha* lines expressing low transgene levels have been recovered, and expression in kiwifruit flowers combined with in vitro flowering upon overexpression indicate that *AcFT* might also maintain the floral fate when expressed in flowers.

4.2 | The *AcFT* promoter activity in Arabidopsis is restricted to the vascular tissue in a manner similar to the Arabidopsis *FT* promoter

The 3.5-kb sequence upstream of the *AcFT* translation start site was sufficient to drive *GUS* expression in the vasculature of the distal half of cotyledons and the rosette leaves in Arabidopsis, before extending in the basipetal direction as the leaf ages. This is replicating the pattern of Arabidopsis *FT* promoter (Adrian et al., 2010; Takada & Goto, 2003) and is consistent with the basipetal pattern of dicot leaf development. The expression in Arabidopsis rosette leaves is comparable to expression profiles described in kiwifruit, where *AcFT* accumulated in basal leaves but was absent from terminal bud, distal leaves and root tissue (Varkonyi Gasic et al., 2013). On the other hand, differential expression recorded using *GUS* fused to a 3-kb and 2.7-kb sequence upstream of the *AcFT1* and *AcFT2* translation start sites, respectively, was consistent with *AcFT1* expression in kiwifruit terminal buds and *AcFT2* restriction to mature leaves (Voogd et al., 2017), suggesting that all promoter fragments in this study contained the regulatory elements required for specific expression patterns, although additional regulatory regions may exist outside these fragments.

Expression of *FT* in minor veins in the distal end of cotyledons and leaves was sufficient to promote flowering in Arabidopsis (Corbesier et al., 2007; Yoo, Hong, Jung, & Ahn, 2013). In minor veins, sucrose is loaded into the phloem, providing the osmotic potential difference required for bulk flow and long-distance transport of phloem sap compounds (Sauer, 2007). The activity of *proAcFT* in minor veins during early development of Arabidopsis implies that the promoter might be regulated in similar manner in kiwifruit and that AcFT protein may be loaded into the vascular stream to perform a role outside its domain of expression. This is unlikely for the shoot tip-associated *AcFT1* and less clear for *AcFT2*, which is expressed in kiwifruit mature leaves but potentially confined to major veins. Therefore, the role of *AcFT2* at native expression levels and domains needs to be studied further. Similarly, expression of *AcFT* in kiwifruit woody stem and dormant bud during winter chilling, which precedes spring bud break and flowering, cannot be addressed in Arabidopsis.

4.3 | The *AcFT* promoter is active in Arabidopsis regardless of the photoperiod

Very little is known about the regulation of transcription of *FT* gene homologs in species other than Arabidopsis and rice. It has become obvious that while some commonalities exist, the regulatory mechanisms differ greatly between species (reviewed in Andres & Coupland, 2012). In spite of a remarkable similarity in expression domains detected with Arabidopsis *FT* and *AcFT* promoters in transgenic Arabidopsis, their activity in SD photoperiod was very different. In contrast to the Arabidopsis *FT* promoter with its day–length response (Adrian et al., 2010; Takada & Goto, 2003), *proAcFT* had similar activity under inductive and noninductive day–length conditions in Arabidopsis, providing the opportunity for functional studies in absence of endogenous FT.

The photoperiod-dependent *FT* expression in Arabidopsis results from interaction with CO at the *cis*-acting element in the *FT* proximal promoter (Tiwari et al., 2010) and elements in the distal promoter (Adrian et al., 2010). The CO binding site (Tiwari et al., 2010) has not been identified in *proAcFT* and the small increase in LUC activity detected in transactivation experiments could have resulted from the presence of multiple CCAAT box motifs, previously shown to bind the CO-HAP3-HAP5 complex to upregulate *FT* expression in Arabidopsis (Wenkel et al., 2006). In Arabidopsis, *FT* expression is repressed by MADS-box proteins FLC and SVP as a result of binding to the CArG box motifs in the first intron and the promoter region (Lee et al., 2007; Li et al., 2008; Searle et al., 2006). Consistent with its small size (Varkonyi-Gasic et al., 2013) and lack of CArG box motifs, the first intron did not affect expression of *AcFT*. Therefore, at this stage, it is unclear if *AcFT* interacts with kiwifruit SVP or FLC-like MADS-box proteins previously implicated in regulation of bud dormancy (Varkonyi-Gasic et al., 2014; Wu et al., 2012, 2017).

4.4 | AcFT is functional when expressed under its own promoter

Expression of *AcFT* cDNA under control of the 3.5-kb *proAcFT* fragment was able to advance flowering in transgenic Arabidopsis Col-0 in noninductive SD in a proportion of plants, suggesting that *AcFT* encoded an activator of flowering at native expression levels. The very early flowering confirmed that cotyledons were the main source of *AcFT* sufficient to induce bolting and flowering (Yoo et al., 2013). At later stages of development, *proAcFT* was insufficient to maintain flowering in noninductive conditions, leading to flower abortion and development of aerial rosettes with both Arabidopsis *FT* and *AcFT*, potentially reflecting the kiwifruit vegetative flushing that follows the first spring floral flush (Grant & Ryugo, 1982). Additionally, the fusion with GFP was capable of advancing flowering time, but the weak fluorescence prevented monitoring of the potential translocation of the AcFT:GFP fusion protein. Therefore, it cannot be distinguished if AcFT:GFP acts as a mobile signal, or affects other flower-promoting mechanisms by acting locally.

In conclusion, *AcFT* promotes floral fate when overexpressed in kiwifruit, the combination of expression levels and protein activity of AcFT has the capacity to drive reproductive onset, while the conserved pattern in minor veins is consistent with vascular export and activity outside the domain of gene expression. Taken together, the data suggest that AcFT performs a role in orchestrating kiwifruit growth and flowering and is a candidate for kiwifruit florigen.

ACKNOWLEDGMENTS

We thank Sakuntala Karunairetnam, Daisy Wang, and Andrew P. Gleave for the cloning support, Lindy Guo for statistical analysis, Monica Dragulescu for assistance with transgenic plants, Tim Holmes for photography, and Anne Gunson and Zac Hanley for critically reading the manuscript. The Arabidopsis *SUC2* promoter was kindly provided by Ruth Stadler, the pHEX2S:GFP by Revel Drummond, and *pro35S:CO* by George Coupland's laboratory.

CONFLICT OF INTEREST

The authors declare that the research was conducted in the absence of any commercial or financial relationships that could be construed as a potential conflict of interest.

AUTHOR CONTRIBUTIONS

SM, JP, RH, and EV-G conceived the research and designed the experiments. SM, TW, CV, LB, and RW performed the experiments. SM and EV-G analyzed the data and wrote the manuscript. JP and AA revised the manuscript.

REFERENCES

Abe, M., Kobayashi, Y., Yamamoto, S., Daimon, Y., Yamaguchi, A., Ikeda, Y., … Araki, T. (2005). FD, a bZIP protein mediating signals from the floral pathway integrator FT at the shoot apex. *Science*, *309*, 1052–1056. https://doi.org/10.1126/science.1115983

Adrian, J., Farrona, S., Reimer, J. J., Albani, M. C., Coupland, G., & Turck, F. (2010). cis-regulatory elements and chromatin state coordinately control temporal and spatial expression of *FLOWERING LOCUS T* in *Arabidopsis*. *Plant Cell*, *22*, 1425–1440. https://doi.org/10.1105/tpc.110.074682

Andres, F., & Coupland, G. (2012). The genetic basis of flowering responses to seasonal cues. *Nature Reviews Genetics*, *13*, 627–639. https://doi.org/10.1038/nrg3291

Böhlenius, H., Huang, T., Charbonnel-Campaa, L., Brunner, A. M., Jansson, S., Strauss, S. H., & Nilsson, O. (2006). CO/FT regulatory module controls timing of flowering and seasonal growth cessation in trees. *Science*, *312*, 1040–1043. https://doi.org/10.1126/science.1126038

Brundell, D. J. (1976). The effect of chilling on the termination of rest and flower bud development of the chinese gooseberry. *Scientia horticulturae*, *4*, 175–182. https://doi.org/10.1016/S0304-4238(76)80010-6

Chailakhyan, M. K. (1968). Internal factors of plant flowering. *Annual Review of Plant Physiology*, *19*, 1–37. https://doi.org/10.1146/annurev.pp.19.060168.000245

Clough, S. J., & Bent, A. F. (1998). Floral dip: A simplified method for *Agrobacterium*-mediated transformation of *Arabidopsis thaliana*. *Plant Journal*, *16*, 735–743. https://doi.org/10.1046/j.1365-313x.1998.00343.x

Corbesier, L., Vincent, C., Jang, S., Fornara, F., Fan, Q., Searle, I., … Coupland, G. (2007). FT protein movement contributes to long-distance signaling in floral induction of *Arabidopsis*. *Science*, *316*, 1030–1033. https://doi.org/10.1126/science.1141752

Datson, P. M., & Ferguson, A. R. (2011). *Actinidia*. In C. Kole (Ed.), *Wild Crop Relatives: Genomic and Breeding Resources* (pp. 1–20). Berlin Heidelberg, NY: Springer.

Endo, T., Shimada, T., Fujii, H., Kobayashi, Y., Araki, T., & Omura, M. (2005). Ectopic expression of an *FT* homolog from citrus confers an early flowering phenotype on trifoliate orange (*Poncirus trifoliata* L. Raf.). *Transgenic research*, *14*, 703–712. https://doi.org/10.1007/s11248-005-6632-3

Freiman, A., Golobovitch, S., Yablovitz, Z., Belausov, E., Dahan, Y., Peer, R., … Flaishman, M. A. (2015). Expression of flowering locus T2 transgene from *Pyrus communis* L. delays dormancy and leaf senescence in *Malus × domestica* Borkh, and causes early flowering in tobacco. *Plant Science*, *241*, 164–176. https://doi.org/10.1016/j.plantsci.2015.09.012

Grant, J. A., & Ryugo, K. (1982). Influence of developing shoots on flowering potential of dormant buds of *Actinidia chinensis*. *HortScience*, *17*, 977–978.

Hayama, R., Yokoi, S., Tamaki, S., Yano, M., & Shimamoto, K. (2003). Adaptation of photoperiodic control pathways produces short-day flowering in rice. *Nature*, *422*, 719–722. https://doi.org/10.1038/nature01549

He, Y. (2015). Enabling photoperiodic control of flowering by timely chromatin silencing of the florigen gene. *Nucleus*, *6*, 179–182. https://doi.org/10.1080/19491034.2015.1038000

Heim, R., Cubitt, A. B., & Tsien, R. Y. (1995). Improved green fluorescence. *Nature*, *373*, 663–664. https://doi.org/10.1038/373663b0

Hellens, R., Allan, A., Friel, E., Bolitho, K., Grafton, K., Templeton, M., ... Laing, W. (2005). Transient expression vectors for functional genomics, quantification of promoter activity and RNA silencing in plants. *Plant Methods*, *1*, 13. https://doi.org/10.1186/1746-4811-1-13

Ho, W. W. H., & Weigel, D. (2014). Structural features determining flower-promoting activity of Arabidopsis FLOWERING LOCUS T. *Plant Cell*, *26*, 552–564. https://doi.org/10.1105/tpc.113.115220

Hsu, C.-Y., Adams, J. P., Kim, H., No, K., Ma, C., Strauss, S. H., ... Yuceer, C. (2011). *FLOWERING LOCUS T* duplication coordinates reproductive and vegetative growth in perennial poplar. *Proceedings of the National Academy of Sciences*, *108*, 10756–10761. https://doi.org/10.1073/pnas.1104713108

Hsu, C. Y., Liu, Y., Luthe, D. S., & Yuceer, C. (2006). Poplar *FT2* shortens the juvenile phase and promotes seasonal flowering. *Plant Cell*, *18*, 1846–1861. https://doi.org/10.1105/tpc.106.041038

Huang, S., Ding, J., Deng, D., Tang, W., Sun, H., Liu, D., ... Liu, Y. (2013). Draft genome of the kiwifruit *Actinidia chinensis*. *Nature Communications*, *4*, https://doi.org/10.1038/ncomms3640

Jefferson, R. (1987). Assaying chimeric genes in plants: The GUS gene fusion system. *Plant Molecular Biology Reporter*, *5*, 387–405. https://doi.org/10.1007/bf02667740

Kardailsky, I., Shukla, V., Ahn, J. H., Dagenais, N., Christensen, S. K., Nguyen, J. T., ... Weigel, D. (1999). Activation tagging of the floral inducer *FT*. *Science*, *286*, 1962–1965. https://doi.org/10.1126/science.286.5446.1962

Kim, S.-J., Hong, S. M., Yoo, S. J., Moon, S., Jung, H. S., & Ahn, J. H. (2016). Post-translational regulation of FLOWERING LOCUS T protein in Arabidopsis. *Molecular Plant*, *9*, 308–311. https://doi.org/10.1016/j.molp.2015.11.001

Klocko, A. L., Ma, C., Robertson, S., Esfandiari, E., Nilsson, O., & Strauss, S. H. (2016). *FT* overexpression induces precocious flowering and normal reproductive development in *Eucalyptus*. *Plant Biotechnology Journal*, *14*, 808–819. https://doi.org/10.1111/pbi.12431

Kobayashi, Y., Kaya, H., Goto, K., Iwabuchi, M., & Araki, T. (1999). A pair of related genes with antagonistic roles in mediating flowering signals. *Science*, *286*, 1960–1962. https://doi.org/10.1126/science.286.5446.1960

Kotake, T., Takada, S., Nakahigashi, K., Ohto, M., & Goto, K. (2003). *Arabidopsis TERMINAL FLOWER 2* gene encodes a heterochromatin protein 1 homolog & represses both *FLOWERING LOCUS T* to regulate flowering time and several floral homeotic genes. *Plant and Cell Physiology*, *44*, 555–564. https://doi.org/10.1093/pcp/pcg091

Kotoda, N., Hayashi, H., Suzuki, M., Igarashi, M., Hatsuyama, Y., Kidou, S.-I., ... Abe, K. (2010). Molecular characterization of FLOWERING LOCUS T-like genes of apple (Malus × domestica Borkh.). *Plant and Cell Physiology*, *51*, 561–575. https://doi.org/10.1093/pcp/pcq021

Lee, J. H., Yoo, S. J., Park, S. H., Hwang, I., Lee, J. S., & Ahn, J. H. (2007). Role of SVP in the control of flowering time by ambient temperature in Arabidopsis. *Genes & Development*, *21*, 397–402. https://doi.org/10.1101/gad.1518407

Li, D., Liu, C., Shen, L., Wu, Y., Chen, H., Robertson, M., ... Yu, H. (2008). A repressor complex governs the integration of flowering signals in Arabidopsis. *Developmental Cell*, *15*, 110–120. https://doi.org/10.1016/j.devcel.2008.05.002

Lionakis, S. M., & Schwabe, W. (1984). Bud dormancy in the kiwifruit, *Actinidia chinensis* planch. *Annals of Botany*, *54*, 467–484. https://doi.org/10.1093/oxfordjournals.aob.a086818

Martinez-Trujillo, M., Limones-Briones, V., Cabrera-Ponce, J. L., & Herrera-Estrella, L. (2004). Improving transformation efficiency of *Arabidopsis thaliana* by modifying the floral dip method. *Plant Molecular Biology Reporter*, *22*, 63–70. https://doi.org/10.1007/BF02773350

Mimida, N., Kidou, S.-I., Iwanami, H., Moriya, S., Abe, K., Voogd, C., ... Kotoda, N. (2011). Apple FLOWERING LOCUS T proteins interact with transcription factors implicated in cell growth and organ development. *Tree Physiology*, *31*, 555–566. https://doi.org/10.1093/treephys/tpr028

Niwa, M., Daimon, Y., Kurotani, K.-I., Higo, A., Pruneda-Paz, J. L., Breton, G., ... Araki, T. (2013). BRANCHED1 interacts with FLOWERING LOCUS T to repress the floral transition of the axillary meristems in *Arabidopsis*. *Plant Cell*, *25*, 1228–1242. https://doi.org/10.1105/tpc.112.109090

Ochman, H., Gerber, A. S., & Hartl, D. L. (1988). Genetic applications of an inverse polymerase chain reaction. *Genetics*, *120*, 621–623.

Pin, P. A., & Nilsson, O. (2012). The multifaceted roles of FLOWERING LOCUS T in plant development. *Plant, Cell & Environment*, *35*, 1742–1755. https://doi.org/10.1111/j.1365-3040.2012.02558.x

Pnueli, L., Gutfinger, T., Hareven, D., Ben-Naim, O., Ron, N., Adir, N., & Lifschitz, E. (2001). Tomato SP-interacting proteins define a conserved signaling system that regulates shoot architecture and flowering. *Plant Cell*, *13*, 2687–2702. https://doi.org/Doi 10.2307/3871528

Putterill, J., & Varkonyi-Gasic, E. (2016). FT and florigen long-distance flowering control in plants. *Current Opinion in Plant Biology*, *33*, 77–82. https://doi.org/10.1016/j.pbi.2016.06.008

Putterill, J., Zhang, L., Yeoh, C., Balcerowicz, M., Jaudal, M., & Varkonyi Gasic, E. (2013). *FT* genes and regulation of flowering in the legume *Medicago truncatula Funct. Plant Biol*, *40*, 1199–1207.

Richardson, A., Boldingh, H., Mcatee, P., Gunaseelan, K., Luo, Z., Atkinson, R., ... Schaffer, R. (2011). Fruit development of the diploid kiwifruit, *Actinidia chinensis* 'Hort16A'. *BMC Plant Biology*, *11*, 182. https://doi.org/10.1186/1471-2229-11-182

Samach, A., Onouchi, H., Gold, S. E., Ditta, G. S., Schwarz-Sommer, Z., Yanofsky, M. F., & Coupland, G. (2000). Distinct roles of *CONSTANS* target genes in reproductive development of Arabidopsis. *Science*, *288*, 1613–1616. https://doi.org/10.1126/science.288.5471.1613

Sauer, N. (2007). Molecular physiology of higher plant sucrose transporters. *FEBS Letters*, *581*, 2309–2317. https://doi.org/10.1016/j.febslet.2007.03.048

Searle, I., He, Y., Turck, F., Vincent, C., Fornara, F., Kröber, S., ... Coupland, G. (2006). The transcription factor FLC confers a flowering response to vernalization by repressing meristem competence and systemic signaling in Arabidopsis. *Genes & Development*, *20*, 898–912. https://doi.org/10.1101/gad.373506

Snelgar, W. P., Clearwater, M. J., & Walton, E. F. (2007). Flowering of kiwifruit (*Actinidia deliciosa*) is reduced by long photoperiods. *New Zealand Journal of Crop and Horticultural Science*, *35*, 33–38. https://doi.org/10.1080/01140670709510165

Song, G.-Q., Walworth, A., Zhao, D., Jiang, N., & Hancock, J. F. (2013). The *Vaccinium corymbosum FLOWERING LOCUS T*-like gene (*VcFT*): A flowering activator reverses photoperiodic and chilling requirements in blueberry. *Plant Cell Reports*, *32*, 1759–1769. https://doi.org/10.1007/s00299-013-1489-z

Srinivasan, C., Dardick, C., Callahan, A., & Scorza, R. (2012). Plum (*Prunus domestica*) trees transformed with poplar FT1 result in altered architecture, dormancy requirement, and continuous flowering. *PLoS ONE*, *7*, e40715. https://doi.org/10.1371/journal.pone.0040715

Stadler, R., Lauterbach, C., & Sauer, N. (2005). Cell-to-cell movement of green fluorescent protein reveals post-phloem transport in the outer integument and identifies symplastic domains in *Arabidopsis* seeds and embryos. *Plant Physiology*, *139*, 701–712. https://doi.org/10.1104/pp.105.065607

Suarez-Lopez, P. (2001). *CONSTANS* mediates between the circadian clock and the control of flowering in *Arabidopsis*. *Nature*, *410*, 1116–1120. https://doi.org/10.1038/35074138

Takada, S., & Goto, K. (2003). TERMINAL FLOWER2, an arabidopsis homolog of HETEROCHROMATIN PROTEIN1, counteracts the activation of FLOWERING LOCUS T by CONSTANS in the vascular tissues of leaves to regulate flowering time. *Plant Cell, 15*, 2856–2865. https://doi.org/10.1105/tpc.016345

Tiwari, S. B., Shen, Y., Chang, H.-C., Hou, Y., Harris, A., Ma, S. F., … Ratcliffe, O. J. (2010). The flowering time regulator CONSTANS is recruited to the *FLOWERING LOCUS T* promoter via a unique cis-element. *New Phytologist, 187*, 57–66. https://doi.org/10.1111/j.1469-8137.2010.03251.x

Varkonyi-Gasic, E., Moss, S. M. A., Voogd, C., Wang, T., Putterill, J., & Hellens, R. P. (2013). Homologs of *FT, CEN* and *FD* respond to developmental and environmental signals affecting growth and flowering in the perennial vine kiwifruit. *New Phytologist, 198*, 732–746. https://doi.org/10.1111/nph.12162

Varkonyi-Gasic, E., Moss, S., Voogd, C., Wu, R., Lough, R., Wang, Y.-Y., & Hellens, R. (2011). Identification and characterization of flowering genes in kiwifruit: Sequence conservation and role in kiwifruit flower development. *BMC Plant Biology, 11*, 72. https://doi.org/10.1186/1471-2229-11-72

Varkonyi-Gasic, E., Wang, T., Karunairetnam, S., Nain, B., Wu, R., & Hellens, R. P. (2014). Analysis of kiwifruit MADS box genes with potential roles in bud dormancy and flower development. *Acta Horticulturae, 1048*, 107–112. https://doi.org/10.17660/ActaHortic.2014.1048.12

Voogd, C., Brian, L. A., Wang, T., Allan, A. C., & Varkonyi-Gasic, E. (2017). Three *FT* and multiple *CEN* and *BFT* genes regulate maturity, flowering, and vegetative phenology in kiwifruit. *Journal of Experimental Botany, 68*, 1539–1553. https://doi.org/doi:10.1093/jxb/erx044

Voogd, C., Wang, T., & Varkonyi-Gasic, E. (2015). Functional and expression analyses of kiwifruit *SOC1*-like genes suggest that they may not have a role in the transition to flowering but may affect the duration of dormancy. *Journal of Experimental Botany, 66*, 4699–4710. https://doi.org/10.1093/jxb/erv234

Walton, E., Fowke, P., Weis, K., & Mcleay, P. (1997). Shoot axillary bud morphogenesis in kiwifruit (*Actinidia deliciosa*). *Annals of Botany, 80*, 13–21. https://doi.org/10.1006/anbo.1996.0381

Walton, E. F., Podivinsky, E., & Wu, R. M. (2001). Bimodal patterns of floral gene expression over the two seasons that kiwifruit flowers develop. *Physiol Plantarum, 111*, 396–404. https://doi.org/DOI 10.1034/j.1399-3054.2001.1110318.x

Wang, T., Atkinson, R., & Janssen, B. (2007). The choice of *Agrobacterium* strain for transformation of kiwifruit. *ISHS Acta Horticulturae, 753*, 227–232. https://doi.org/10.17660/ActaHortic.2007.753.26

Weller, J. L., & Ortega, R. (2015). Genetic control of flowering time in legumes. *Frontiers in Plant Science, 6*, https://doi.org/10.3389/fpls.2015.00207

Wenkel, S., Turck, F., Singer, K., Gissot, L., Le Gourrierec, J., Samach, A., & Coupland, G. (2006). CONSTANS and the CCAAT box binding complex share a functionally important domain and interact to regulate flowering of *Arabidopsis*. *Plant Cell, 18*, 2971–2984. https://doi.org/10.1105/tpc.106.043299

Wenzel, S., Flachowsky, H., & Hanke, M.-V. (2013). The Fast-track breeding approach can be improved by heat-induced expression of the *FLOWERING LOCUS T* genes from poplar (*Populus trichocarpa*) in apple (*Malus × domestica* Borkh.). *Plant Cell, Tissue and Organ Culture, 115*, 127–137. https://doi.org/10.1007/s11240-013-0346-7

Wickland, D. P., & Hanzawa, Y. (2015). The *FLOWERING LOCUS T/TERMINAL FLOWER* gene family: Functional evolution and molecular mechanisms. *Molecular Plant, 8*, 983–997. https://doi.org/10.1016/j.molp.2015.01.007

Wigge, P. A., Kim, M. C., Jaeger, K. E., Busch, W., Schmid, M., Lohmann, J. U., & Weigel, D. (2005). Integration of spatial and temporal information during floral induction in *Arabidopsis*. *Science, 309*, 1056–1059. https://doi.org/10.1126/science.1114358

Wu, R., Tomes, S., Karunairetnam, S., Tustin, S. D., Hellens, R. P., Allan, A. C., … Varkonyi-Gasic, E. (2017). *SVP*-like MADS box genes control dormancy and budbreak in apple. *Frontiers in Plant Science, 8*, https://doi.org/10.3389/fpls.2017.00477

Wu, R.-M., Walton, E. F., Richardson, A. C., Wood, M., Hellens, R. P., & Varkonyi-Gasic, E. (2012). Conservation and divergence of four kiwifruit *SVP*-like MADS-box genes suggest distinct roles in kiwifruit bud dormancy and flowering. *Journal of Experimental Botany, 63*, 797–807. https://doi.org/10.1093/jxb/err304

Yoo, S. J., Hong, S. M., Jung, H. S., & Ahn, J. H. (2013). The cotyledons produce sufficient FT protein to induce flowering: Evidence from cotyledon micrografting in arabidopsis. *Plant and Cell Physiology, 54*, 119–128. https://doi.org/10.1093/pcp/pcs158

Zeevaart, J. A. D. (2008). Leaf-produced floral signals. *Current Opinion in Plant Biology, 11*, 541–547. https://doi.org/10.1016/j.pbi.2008.06.009

Zhang, H., Harry, D. E., Ma, C., Yuceer, C., Hsu, C.-Y., Vikram, V., … Strauss, S. H. (2010). Precocious flowering in trees: The *FLOWERING LOCUS T* gene as a research and breeding tool in *Populus*. *Journal of Experimental Botany, 61*, 2549–2560. https://doi.org/10.1093/jxb/erq092

14

Salt and oxidative stresses uniquely regulate tomato cytokinin levels and transcriptomic response

Erika A. Keshishian[1] | H. Tucker Hallmark[1] | Thiruvarangan Ramaraj[2] | Lenka Plačková[3] | Anitha Sundararajan[2] | Faye Schilkey[2] | Ondřej Novák[3] | Aaron M. Rashotte[1]

[1]Department of Biological Sciences, Auburn University, Auburn, Alabama

[2]National Center for Genome Resources, Santa Fe, New Mexico

[3]Laboratory of Growth Regulators, Centre of the Region Haná for Biotechnological and Agricultural Research, Faculty of Science of Palacký University & Institute of Experimental Botany of the Czech Academy of Sciences, Olomouc, Czech Republic

Correspondence
Aaron M. Rashotte, Department of Biological Sciences, Auburn University, 101Rouse Life Sciences, Auburn AL.
Email: rashotte@auburn.edu

Funding information
Alabama Agricultural Experiment Station HATCH, Grant/Award Number: 370225-31006-2055; AgR SEED, Grant/Award Number: 370228-310006-2055; Ministry of Education, Youth and Sports of the Czech Republic, Grant/Award Number: LO1204

Abstract

Cytokinins are well-known to be involved in processes responsible for plant growth and development. More recently, these hormones have begun to be associated with stress responses as well. However, it is unclear how changes in cytokinin biosynthesis, signaling, or transport relate to stress effects. This study examines in parallel how two different stresses, salt, and oxidative stress, affect changes in both cytokinin levels and whole plant transcriptome response. *Solanum lycopersicum* seedlings were given a short-term (6 hr) exposure to either salt (150 mM NaCl) or oxidative (20 mM H_2O_2) stress and then examined to determine both changes in cytokinin levels and transcriptome. LC-MS/MS was used to determine the levels of 22 different types of cytokinins in tomato plants including precursors, active, transported, and conjugated forms. When examining cytokinin levels we found that salt treatment caused an increase in both active and inactive cytokinin levels and oxidative stress caused a decrease in these levels. RNA-sequencing analyses of these same stress-treated tissues revealed 6,643 significantly differentially expressed genes (DEGs). Although many DEGs are similar between the two stresses, approximately one-third of the DEGs in each treatment were unique to that stress. Several cytokinin-related genes were among the DEGs. Examination of photosystem II efficiency revealed that cytokinins affect physiological response to stress in tomato, further validating the changes in cytokinin levels seen *in planta*.

KEYWORDS
abiotic stress, cytokinin, oxidative stress, RNA-sequencing, salt, *Solanum lycopersicum*, tomato, transcriptome

1 | INTRODUCTION

Abiotic stress results in billions of dollars of crop losses every year, including in the important crop plant tomato (Acosta-Motos et al., 2017; Pineda et al., 2012; Qin, Shinozaki, & Yamaguchi-Shinozaki, 2011). Tomato growth and production is highly dependent on healthy functioning and photosynthesizing plants that are resistant to or tolerant of abiotic environmental stresses. Two primary stresses affecting tomato are salt stress and the broad stress of oxidative damage (Acosta-Motos et al., 2017; Pineda et al., 2012). These stresses are often thought of as overlapping in their effects, yet are also thought to be distinct in mode of action. In tomato, salt stress has been examined by a number of researchers, while oxidative stress is virtually unstudied (Acosta-Motos et al., 2017).

Salt stress-related crop losses are a problem growing in magnitude as salt levels in farm lands are increasing due to abundant irrigation that continues to add residual salt. In tomatoes, salt stress is well known to reduce plant size, fresh weight, and photosynthesis (as recently reviewed in Acosta-Motos et al., 2017). As such attempts to understand how plants respond to salt stress to generate resistant or tolerant lines is of great importance. Salt stress is known to affect plants in two basic ways: either through osmotic complications or through ion toxicity (Acosta-Motos et al., 2017; Ismail & Horie, 2017). However, what remains unclear are mechanistic actions that occur as a plant responds to salt stress from either or both of these effects, particularly in regard to the genes involved in salt tolerance and resistance. Although there has been a great deal of progress made in understanding some of the basic components, signaling pathways, and initial genes involved in plant salt stress response, there is still much that is unknown (Acosta-Motos et al., 2017; Ismail & Horie, 2017; Singh, Singh, Prasad, & Singh, 2017).

Oxidative stress produced during abiotic stresses is highly detrimental to growth at all stages of development and can greatly reduce crop yields in plants (as reviewed in Pandey et al., 2011; Qin et al., 2011; Duque et al., 2013). In tomato, oxidative stress is a major abiotic factor (Pandey et al., 2011), which causes damage at both tissue and cellular levels that can rupture membranes, breakdown photosynthetic machinery, and cause cell death (Zwack & Rashotte, 2015). Plants utilize the redox potential of molecular oxygen (O_2) and water (H_2O) to drive both reductive (photosynthesis) and oxidative (respiration) energy production. During each process, partially reduced/oxidized O_2/H_2O can form. Such redox intermediates readily oxidize other molecules and are known as reactive oxygen species (ROS) (Apel & Hirt, 2004; Gill & Tuteja, 2010). ROS or oxidative stress damage can occur during various abiotic stresses, such as extreme temperatures that alter membrane fluidity disrupting electron transport chain reactions (Mittler, Finka, & Goloubinoff, 2012) and high light levels that overwhelm photosystem II and its light-harvesting complexes with excitation energy (Gill & Tuteja, 2010). ROS can additionally be produced directly through O_2 interactions with heavy metals, UV radiation and pollutants such as ozone and sulfur dioxide (Gill & Tuteja, 2010; Mittler, 2002). As many stresses stimulate a response that includes the production of ROS, it is often involved in general stress responses (Apel & Hirt, 2004; Mittler, Vanderauwera, Gollery, & Van Breusegem, 2004; Saxena, Srikanth, & Chen, 2016). Yet for some crop species such as tomato, only a handful of genes have been linked to oxidative stress and require additional study.

An underexplored avenue for regulation of both salt and oxidative stress responses is the plant hormone cytokinin. Cytokinins are essential to plant hormones involved in numerous plant growth and developmental processes of great agronomic importance (Mok & Mok, 2001; Davies, 2004; To & Kieber, 2008; Muller, 2011). Notably, while both salt and oxidative stresses are detrimental to plant health and linked to abiotic stress senescence, cytokinins are connected to delaying plant senescence. Cytokinins have also been strongly linked to plant abiotic stress responses in studies examining cytokinin biosynthesis, metabolism, and signaling, where changes in cytokinin levels, responsive genes, or receptor mutants have dramatically altered growth under stress conditions (as seen or reviewed in Peleg & Blumwald, 2011; Wilkinson, Kudoyarova, Veselov, Arkhipova, & Davies, 2012; Nishiyama et al., 2011; Ramireddy, Chang, & Schmülling, 2014; Zwack & Rashotte, 2015; Veselov, Kudoyarova, Kudryakova, & Kusnetsov, 2017). This has also been shown in tomato for salt stress, albeit to a lesser degree (Ghanem et al., 2008, 2011; Pandey et al., 2011; Pineda et al., 2012; Žižková et al., 2015). Oxidative stress remains largely unstudied in tomato (Bose, Rodrigo-Moreno, & Shabala, 2014; Mittova, Tal, Volokita, & Guy, 2003). As such, little is known in tomato of the mechanisms behind or pertinent genes involved in cytokinin-based regulation of salt or oxidative stress.

This study examines connections between salt stress, oxidative stress, and cytokinins. Distinct changes in cytokinin levels after salt and oxidative stress along with distinct transcriptome changes were found. Specific alterations were identified in cytokinin-related genes under each stress condition. These distinctive changes could be further linked to the efficiency of photosystem II (F_v/F_m) under salt and oxidative stress conditions. Furthermore, cytokinins have effects on F_v/F_m under these stress conditions, possibly connecting the changes in cytokinin levels *in planta* to the observed physiological stress responses.

2 | MATERIALS AND METHODS

2.1 | Tomato seedling growth and stress exposure

Solanum lycopersicum cv. MicroTom seeds were planted in damp sunshine #8 soil mix, grown in 16 hr, 26°C light and 8 hr, 22°C dark periods with light supplemented at 150 μE m^{-2} s^{-1}. For stress treatments seedlings, 10 days after germination were excised from soil with roots intact and soil was carefully removed. Initial examinations of tomato pTCS::VENUS lines, a generous gift from Dr. Yuval Eshed, were made after seedlings were exposed to one of four treatments for 24 hr under gentle shaking conditions at room temperature: MES + 5 μM *trans*-zeatin, MES + 150 mM NaCl, MES + 20 mM H_2O_2, or MES buffer (3 mM MES buffer, pH 5.7) as a control. Roots from three plants in three biological treated replicates were viewed using a Nikon Eclipse 80i microscope epifluorescence microscope with a UV source and a yellow fluorescent protein (YFP) filter. A representative photo of a root from each treatment was taken with a Qimaging Fast 1394 digital camera and cropped using Adobe Photoshop CS3 without altering the original photo integrity.

Further examination of plants for LC-MS/MS and RNA-sequencing have grown as above then exposed to one of three treatments for 6 hr under gentle shaking conditions at room temperature: MES + 150 mM NaCl, MES + 20 mM H_2O_2, or MES buffer (3 mM MES buffer, pH 5.7) as a control. After treatment, plants were carefully patted dry to determine fresh weight, then flash frozen in liquid nitrogen and ground by mortar and pestle into a fine powder. Samples were then split to allow RNA extraction/sequencing and as well as measurement of cytokinin levels from the same samples as described below.

2.2 | RNA extraction, library preparation, and illumina GAIIX sequencing

Three independent biological replicates were used to isolate total RNA using the Qiagen RNeasy Plant Mini-kit according to the manufacturer's instructions. Total RNA then used for messenger RNA isolation with polyA selection and subsequent library construction with the TruSeq RNA sample preparation protocol from Illumina (San Diego, CA). Three biological replicates were sequenced and analyzed for each of the three treatment combinations. Single-end sequencing was performed on the nine samples by the Illumina GAIIX platform, generating 186,460,653 1×54 bp reads. Raw sequence data is available for download at NCBI Sequence Read Archive under the BioProject ID: PRJNA476376, SRA accession: SRP150651.

2.3 | Illumina mRNA sequence data analysis

High-quality sequence data generated for each of the nine samples were aligned to Solanum lycopersicum genome downloaded from NCBI as performed in Gupta et al., 2013 (ftp://ftp.ncbi.nlm.nih.gov/genomes/Solanum_lycopersicum/). The associated annotation file, GFF format, was used to obtain genic information for downstream analysis. BAM alignments were generated using GSNAP (Genomic Short-read Nucleotide Alignment Program) (version released on 2013_05_09) (Wu & Watanabe, 2005) with the following parameters; indel penalty = 2, maximum mismatches = 0.06, terminal threshold = 1,000, novel splicing = 1, local splice distance = 10,000, distant splice penalty = 1,000 and everything else set to default. Read counts were generated using NCGR's in-house pipeline, ALPHEUS (Miller et al., 2008) as previously performed in Gupta et al., 2013;. Gene expression for each of the nine samples was computed as a measure of the total number of reads uniquely aligning to the reference genome, binned by genic coordinates (information acquired from the annotation GFF3 file). Differential gene expression analysis was performed using the R (R core Team 2013) (http://www.R-project.org/) Bioconductor package DESEQ (Anders & Huber, 2012). Raw read counts obtained were normalized to account for differences in sequencing depth and composition using methods implemented within DESEQ package. Differential expression of pairwise comparisons (combinations of the different conditions) was assessed using the negative binomial test with the Benjamani–Hochberg false discovery rate (FDR) adjustment applied for multiple testing corrections. For this study, an FDR of 0.05 was applied and any candidate that had a *p*-adjusted value of ≤ 0.05 was considered to be significantly regulated. Full lists of gene expression are given in Supporting Information Table S2.

Generation of a tomato cytokinin-regulated list was generated by combined lists of all RNA-sequenced DEGs after cytokinin treatment previously performed in tomato (Gupta et al., 2013; Shi et al., 2013) and removing duplicates as shown in Supporting Information Table S5. This list of nearly 1,100 DEGs was directly compared to the lists of stress DEGs found in this study.

Gene ontology GO analysis for overrepresentation was examined using the Overrepresentation Test at the PantherDB.org webpage using default settings. GO complete categories (Molecular Function, Biological Process, and Cellular Component) were examined using a False Discovery Rate (FDR < 0.05) for each set of DEGs for salt and oxidative stress separated using distinct induced and repressed lists (Supporting Information Table S4). Additional examination of the combined cytokinin-regulated list was performed in the same manner (Supporting Information Table S5).

2.4 | Confirmation of RNA-sequencing by qPCR

Five genes identified as DEG under stress conditions vs buffer control were selected to validate the RNA-sequencing results using quantitative real-time PCR analysis (qPCR) following a modified protocol from Shi et al., 2013;. Total RNA for each treatment was extracted as described above then used to generate cDNA for qPCR by reverse transcription using Quanta qScript cDNA supermix. qPCR was performed using PerfeCTa SYBR Green Supermix (QuantaBio, Beverly, MA) in 20 μL reactions on an Eppendorf realplex2 (Hamburg, Germany). The qPCR conditions were as follows: 15 s 95°C, 20 s 58°C, 30 s 68°C (40 cycles), followed by melt curve analysis. All qPCR reactions were performed using two biological replicates and three technical replicates. For these replicates plants were grown and treated under identical conditions as for transcriptome analysis. Fold change was calculated using the delta CT method with TIP41 as a control gene. Analysis for change was analyzed using a one-tailed *t*-Test and $p < 0.05$ as a cut off for significance. Primer sequences for the genes which were verified through qPCR are presented in Supporting Information Table S3.

2.5 | Measurements of cytokinin level

Extraction and quantification of cytokinins were performed as described previously (Svačinová et al., 2012) using the LC-MS/MS system consisting of an ACQUITY UPLC System and Xevo TQ-S triple quadrupole mass spectrometer (Waters). Results are presented as the average of five biological replicates ± standard deviation in pmol/g FW for samples treated and described above. Statistical examinations were made between buffer control treated and stress treatments using an ANOVA analysis with $p < 0.05$.

2.6 | Cytokinin-abiotic stress senescence assay

For salt stress priming and recovery, 50-60-days-old plants (Micro-Tom) were used to make leaf disks (1 mm diameter) and then floated in 3 mM MES buffer in cell culture plates. Full treatments were made either in buffer as noted, + salt (150 mM NaCl) or + cytokinin (5 μM BA) for the entire experiment. Pretreatment included treatment at the initiation of the experiment followed by an addition of another treatment or buffer at 48 hr indicated as a post-treatment. Pre and post-treatments were performed using the same levels of salt and cytokinin as in full treatments. Oxidative stress priming and recovery were conducted in a similar manner as for salt stress, except 15-days-old plants (M82) were used and 20 mM H_2O_2 was used as an oxidative stress treatment. Leaf disks were grown under standard conditions. F_v/F_m measurements were performed in a manner similar to as in Zwack et al., 2016 with 9–18 leaf disks were used for all experimental treatments examined and conducted in triplicate biological replicates with analysis using Student's *t*-test. Additional experiments performed under each stress were performed in both M82 and MicroTom cultivars with similar findings as presented in the results section (data not shown).

3 | RESULTS

3.1 | Examination of cytokinin regulation by salt and oxidative stress reveals distinct responses

Cytokinins have been previously connected to various stress responses in plants. To further examine these, several approaches were taken. A simple examination of the cytokinin-responsive reporter line pTCS::VENUS that is well-known to indirectly indicate cytokinin levels in tomato roots (Bar et al., 2016; Zürcher et al., 2013) was examined under salt and oxidative stress treatments. We found as previously shown that cytokinin (5 μM *trans*-zeatin) strongly induces YFP (Yellow Florescence Protein) fluorescence over a buffer only control treatment in young seedlings after 24 hr (Figure 1). There was also a smaller qualitative change in YFP fluorescence in response to stress versus buffer control: fluorescence appeared to be induced with salt (150 mM NaCl) and repressed with oxidative stress (20 mM H_2O_2) (Figure 1). Although these stress-induced changes are not as dramatic as the change seen with cytokinin treatment, they do suggest that cytokinin levels might be affected by stress.

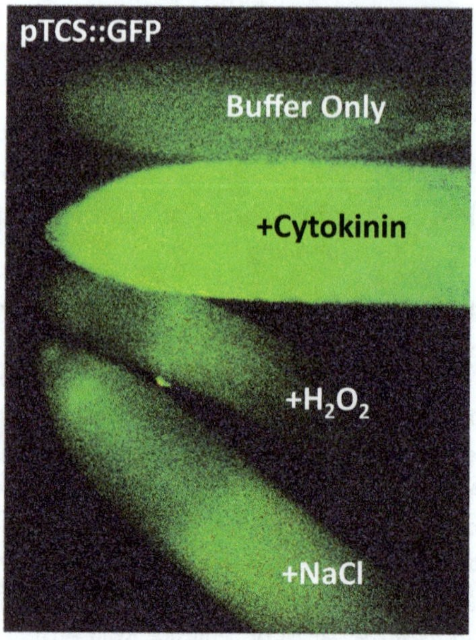

FIGURE 1 Stress interactions with a cytokinin-responsive YFP reporter line (TCS::VENUS). Visualization of the stable cytokinin-responsive YFP reporter line (pTCS::VENUS) in tomato roots. pTCS::VENUS was treated for 24 hr in MES buffer, Cytokinin (5 μM *trans*-zeatin), Oxidative (20 mM H_2O_2) or Salt (150 mM NaCl) stress. Representative fluorescing root tips are shown. Results show a strong increase in YFP expression after cytokinin treatment as expected. Oxidative stress shows a reduction, while Salt stress shows a moderate increase in expression compared to Buffer only, indicating stress interaction with cytokinin

To directly examine cytokinin level changes in response to these stress treatments, LC-MS/MS measurements of salt and oxidative stress-treated seedlings (10 days) were performed. Soil grown seedlings were gently washed and placed in MES buffer with light shaking, then exposed to either salt (150 mM NaCl), oxidative stress (20 mM H_2O_2), or the buffer control for 6 hr. Twenty-two different cytokinin forms were detected in this study including precursors, active, transported, and conjugated forms of isopentenyladenine (iP), *trans*-zeatin (*tZ*), dihydrozeatin (DHZ), and *cis*-zeatin (*cZ*). Cytokinin levels in buffer-treated control plants show general similarities to those previously found in tomato (Žižková et al., 2015) and other eudicot plants with some variation (Sakakibara, 2006). The most abundant components are the N-glucoside forms (iP7G, tZ7G, and DHZ7G), with the active bases iP, *tZ*, and *cZ* making up a small part of the total measured cytokinins (Figure 2). Salt treated plants showed a significant increase in 11 individual cytokinins, including the important active form *tZ*, as well as the total amount of cytokinin bases, ribosides, and nucleotide forms (Figure 2). Although there is more than a 25% increase in overall cytokinin levels compared to control, this was not significant, likely due to the lack of a significant increase in total levels of cytokinin conjugates. Oxidative stress-treated plants showed the opposite result with significantly decreased levels of seven different cytokinin forms, including *tZ*. There is also an opposing overall 20% decrease in cytokinin levels, although not significant, again likely due to nonsignificant changes in the high levels of conjugate cytokinin forms. However, a comparison of overall cytokinin levels in plants between salt and oxidative stress treatment is significantly different ($p < 0.01$) indicating a clear difference in cytokinin levels between these two stresses. This result is similarly true when comparing of all general classes of cytokinins to each other (bases, ribosides, nucleotides, O-glucosides, and N-glucosides) at $p < 0.05$ level (Supporting Information Table S1). Interestingly, the only significantly regulated cytokinin under both stresses was *cZ*, which has previously been suggested to play a role in environmental stress response (Schäfer et al., 2015).

3.2 | Transcriptomic effects of salt and oxidative stresses

As part of the split experimental design, plants from the same stress treatments described above were examined using RNA-sequencing.

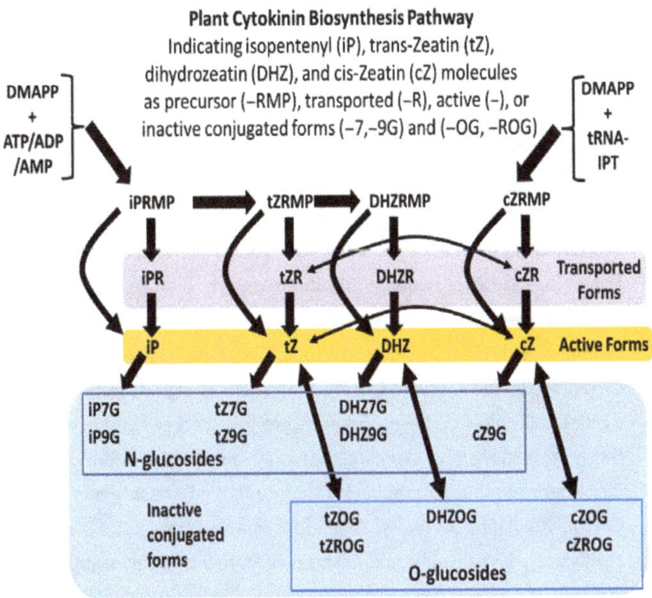

MT - Control	iP	tZ	DHZ	cZ
RMP	0.69±0.21	0.32±0.08	ND	0.98±0.17
R	1.27±0.42	0.27±0.08	0.033±0.008	0.31±0.07
-	0.50±0.08	0.16±0.02	ND	0.09±0.01
7G	26.82±5.27	15.69±3.38	4.03±1.13	
9G	0.15±0.04	0.42±0.10	1.36±0.48	0.12±0.04
OG		0.13±0.03	0.137±0.031	0.21±0.02
ROG		0.32±0.09	ND	0.04±0.01

MT + NaCl	iP	tZ	DHZ	cZ
RMP	1.18±0.28	0.42±0.13	ND	2.46±0.64
R	2.70±0.72	0.39±0.12	0.050±0.009	0.87±0.19
-	0.58±0.11	0.30±0.09	ND	0.41±0.06
7G	35.07±6.83	18.53±4.76	2.60±0.76	
9G	0.22±0.03	0.52±0.11	0.74±0.23	0.13±0.02
OG		0.20±0.02	0.103±0.023	0.30±0.06
ROG		0.28±0.08	ND	0.08±0.02

MT + H_2O_2	iP	tZ	DHZ	cZ
RMP	0.52±0.14	0.22±0.07	ND	1.44±0.37
R	0.65±0.19	0.11±0.03	0.046±0.010	0.44±0.12
-	0.49±0.13	0.12±0.02	ND	0.13±0.01
7G	23.47±4.85	12.54±3.37	1.46±0.42	
9G	0.14±0.03	0.33±0.07	0.34±0.07	0.10±0.02
OG		0.12±0.02	0.076±0.019	0.20±0.04
ROG		0.18±0.06	ND	0.04±0.01

FIGURE 2 Cytokinin levels are altered by abiotic stress. (Left) Biosynthetic pathway for cytokinins in plants. Abbreviations in text. (Right) Specific amounts of cytokinin compounds measured by LC/MS in pmol/g FW from 10 days tomato seedlings without (Control), Salt (150 mM NaCl) or Oxidative Stress (20 mM H_2O_2) treatment for 6 hr (n = 5). Pastel colors indicate cytokinin form shown as in the biosynthetic pathway. Green/Red colors indicate significant ($p < 0.05$) increase/decrease versus untreated control. Results show general increases in various cytokinin levels in response to salt stress, in contrast to decreasing levels seen for oxidative stresses

RNA was extracted from these seedlings to produce cDNA libraries from which single-end Illumina GAIIX RNA-sequencing was performed. This generated a total of 186,460,653 1 × 54 bp reads from all samples, which were aligned to the *S. lycopersicum* reference genome. Further analysis resulted on average in 14.3 million uniquely aligning reads per sample from which gene expression was quantified, using the total number of reads per sample that uniquely aligned to the reference binned by gene (Table 1). Genes used for differential expression (DE) analysis were restricted to those found to be significantly regulated based on a $p_{adj} < 0.05$ (Supporting Information Table S2) as compared between stress-treated and untreated control.

qPCR was performed to confirm the results of RNA-sequencing on five DE genes affected by salt and oxidative stress treatments (Table 2). This comparison yielded similar expression directionality and level of regulation for genes examined indicating that the $\log_2$ fold change values obtained from RNA-sequencing were accurate. In addition to confirming the overall validity of the RNA-sequencing data, these qPCR data suggest that a number of hormone-related genes are affected under stress conditions, including ACC synthase (ethylene), IAA-Amido Synthetase (auxin), and His-containing phosphotransmitter Hpt4-like (cytokinins). Universal Stress Protein A and a sodium-proton antiporter were also altered by stress.

TABLE 1 Total reads from RNA-sequencing runs

Sample	Total RNA reads	Uniquely aligned reads
MT-Control rep1	20,105,395	12,102,357
MT-Control rep2	8,608,669	5,648,071
MT-Control rep3	29,611,517	17,630,048
MT + H_2O_2 rep1	18,838,675	14,791,987
MT + H_2O_2 rep2	22,658,651	17,233,910
MT + H_2O_2 rep3	15,577,188	12,031,139
MT + NaCl rep1	18,086,167	11,616,237
MT + NaCl rep2	21,937,106	15,517,603
MT + NaCl rep3	31,037,285	22,555,867

Note. Below is the total number of RNA-sequencing reads from each biological replicate of as well as the number of those reads that were then uniquely aligned to the Heinz tomato reference genome. Samples are of the *Solanum lycopersicum* cv. Microtom (MT) either buffer-treated (Control), or stress treated (+20 mM H_2O_2), (+150 mM NaCl) for 6 hr.

Overall transcriptomic changes under each stress condition were compared using the lists of significantly differentially expressed genes

(DEGs) in a Venn diagram, 3,950 DEGs were identified under oxidative stress and 4,617 DEG were found under salt stress. A similar number of DEGs has been found to be regulated by other stresses in other systems (Nishiyama et al., 2011). While abiotic stresses are sometimes thought of as triggering similar gene expression changes in plants, here we see strong differences between the overall set of stress-regulated DEGs where 71.0% of all DEGs show unique stress regulation and only a 29% overlap. A similar trend is found when the stress induced and repressed gene lists are separately compared to each other: 67.6% of induced and 75.6% of repressed DEGs show unique stress regulation (Figure 3a). In contrast, almost no induced genes from one stress are repressed by the other list: only 38 oxidative stress induced and salt repressed, and 74 salt-induced and oxidative stress repressed. Additionally, a Principal component analysis (PCA) and variance decomposition (both as implemented in SAS JMP Genomics 5.1) of the overall, full transcriptome dataset ($n = 9$) showed distinct differences between each set of treatment samples and the buffer control (Figure 3b). Approximately 35% of the variation appears to be from treatment, with at least 20% of the variation appears as a difference between salt and oxidative stress-treated samples (Figure 3b). Together this suggests a distinct transcript pattern of gene regulation from salt versus oxidative stress.

To additionally compare patterns of gene regulation between salt and oxidative stress, GO (Gene Ontology) enrichment analysis was performed on the lists of DEGs for each stress. As each gene list contained a large number of DEGs, lists were further divided into induced and repressed lists for each stress. Importantly, in each of these lists, response to abiotic stress was found as significantly enriched 2.0-2.5-fold (Table 3). Several other stress or stimulus-response GO terms were also found as enriched (Table 3). Distinct GO terms were found as significantly enriched for each stress (Supporting Information Table S4). One such example is for ethylene-related GO categories (biosynthetic, metabolic, and response to ethylene) that were significantly enriched for oxidative yet not for salt stress. There are also a number of GO terms that are enriched for both stresses only when comparing the induced lists (calcium ion/calmodulin binding, glucosyltransferase activity, phosphatase activity) and similarity only when comparing the repressed lists (chlorophyll binding, structural constituent of cytoskeleton). Based on changes that were seen from examination of direct and indirect measurements of cytokinins in Figures 1 and 2, it would be predicted that some cytokinin-related category would be found as enriched from this GO analysis, although that was not the case. We believe that this is due in part to poor annotation of GO terms in descriptions of cytokinin-related genes in tomato.

To more thoroughly examine the regulation of cytokinins under stress conditions, a manual examination of DEG lists was made, using search terms such as "cytokinin" and "zeatin" as well as searching by Solyc# based on previous publications of cytokinin-related gene lists and phylogenetic comparisons to between tomato and Arabidopsis (Capua & Eshed, 2017; Sun et al., 2017). From that examination 35 cytokinin-related DEGs could be identified as significantly regulated by stress (Table 4). These 35 genes are involved in cytokinin biosynthesis, transport, and signaling, yet only three were found with GO identifiers relating to cytokinin. Twenty-five of these genes show a unique regulation between salt and oxidative stresses: 10 oxidative stress only, 15 salt only, and 10 both oxidative and salt regulated.

3.3 | Examination of cytokinin treatment on photosystem II efficiency under oxidative and salt stress

Based on the findings that cytokinin levels can be affected under specific stress conditions as seen from both cytokinin measurements and transcriptomic results (Figures 1-3, Table 3), we further examined the effects of cytokinin on abiotic stress response. The classic cytokinin senescence bioassay, normally used to examine developmental senescence, was modified to examine abiotic stress responses of tomato leaf disks (Mok & Mok, 2001; Zwack, Robinson, Risley, & Rashotte, 2013). Previous work, again verified here, shows that cytokinin treatment of leaf disks reduces leaf senescence as measured using a chlorophyll fluorimeter to determine the efficiency of photosystem II or F_v/F_m (Zwack et al., 2016; Figures 4 and 5). Oxidative stress (20 mM H_2O_2) reduces F_v/F_m by 10%–30% over simple buffer treatment. However, application of exogenous cytokinin (5 μM BA) either before (pre – initial stress treatment 0 hr) or after (post 48 hr) this oxidative stress treatment significantly increases F_v/F_m levels over stress treatment alone (Figure 4). This is consistent with findings here that cytokinin levels are reduced in oxidative stress-treated plants (Figures 1 and 2), such that the

TABLE 2 qPCR Confirmation of RNA-sequencing transcriptomic results

Solyc #	Gene	Description	MT vs. MT+NaCl	MT vs. MT+H2O2
Solyc01g107400	IAA-AS	Indole-3-Acetic Acid-Amido Synthetase	3.29 ± 0.05	2.04 ± 0.02
Solyc08g081550	ACS	1-Aminocyclopropane-1-Carboxylate Synthase-like	1.96 ± 0.10	1.21 ± 0.02
Solyc06g008820	NHX1	Sodium Hydrogen Exchanger-like	1.23 ± 0.01	1.30 ± 0.01
Solyc09g011670	USPA	Universal Stress Protein A	0.82 ± 0.01	1.28 ± 0.00
Solyc08g066350	HPT4-like	Histidine-Containing Phosphotransfer 4-like	0.72 ± 0.00	1.18 ± 0.04

Note. Five genes found to be DE for at least one stress comparison to buffer-treated samples by RNA-sequencing were selected for further verification of expression change by qPCR. qPCR was performed using two biological replicates and three technical replicates treated in the same manner as for RNA-sequencing experiments and normalized to TIP41 gene expression as a control. All qPCR expression changes follow RNA-sequencing expression changes in a significant manner ($p < 0.05$, T-test) as indicated by font color (green-induced, red-repressed, gray-unchanged).

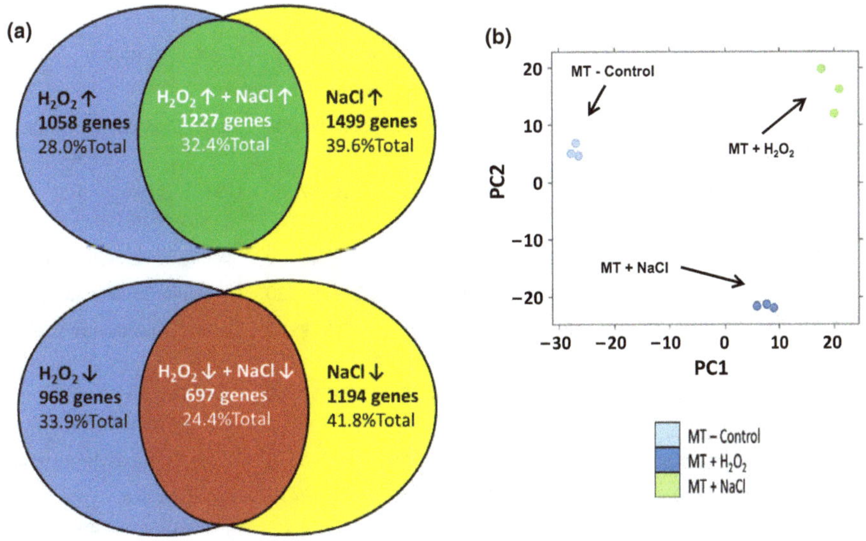

FIGURE 3 Transcriptome analysis reveals distinct patterns abiotic stress gene regulation. (a) Venn diagrams indicating the numbers of tomato genes significantly regulated by H_2O_2 and NaCl stress treatments: Top – induced genes, Bottom – repressed genes. While 32.4% of the induced and 24.4% of the repressed genes overlap between these stresses, the majority 67.6% of the induced and 75.6% of the repressed genes show unique stress regulation. (b) PCA analysis of each transcript replicate indicates a strong and distinct clustering of due to treatment (PC1) as well as a major component of transcript differences due to type of stress (PC2)

TABLE 3 Gene ontology enrichment analysis of salt and oxidative stress DE genes

	Fold enrichment	FDR		Fold enrichment	FDR
Salt stress induced			*Oxidative stress induced*		
Response to abiotic stimulus	2.07	5.29E-03	Oxidation-reduction process	1.68	1.94E-09
Response to stress	1.53	2.80E-04	regulation of response to stimulus	2.25	3.03E-02
Regulation of response to stress	3.27	1.81E-02	Response to stimulus	1.77	2.88E-17
Regulation of response to stimulus	2.99	2.14E-05	Response to abiotic stimulus	2.37	4.26E-04
Response to stimulus	1.64	1.80E-14	Response to oxygen-containing compound	2.45	1.38E-03
Cellular response to stimulus	1.43	1.25E-04	Cellular response to oxidative stress	3.03	4.25E-02
			Cellular response to stimulus	1.35	8.88E-03
Salt stress repressed			Response to oxidative stress	2.15	1.41E-02
Response to abiotic stimulus	2.25	8.95E-03			
Response to oxidative stress	2.2	4.60E-02	*Oxidative stress repressed*		
Response to stimulus	1.38	1.02E-03	Response to abiotic stimulus	2.42	7.66E-03

Note. Gene ontology GO analysis for overrepresentation was examined using an Overrepresentation Test at the PantherDB.org webpage using default settings. Stress and oxidative categories with a False Discovery Rate (FDR < 0.05) for each stress separated by induced and repressed DEGs are shown.

exogenous application to raise that level could potentially restore plants to a normal functioning level.

Salt stress (150 mM) treatments show a similar reduction in F_v/F_m to 30%–40% of buffer-treated leaf disks (Figure 5). A cytokinin pretreatment is able to significantly reduce the drop in F_v/F_m levels after a salt treatment, although salt still has an effect compared to buffer or cytokinin treatment alone (Figure 5). While a cytokinin application postsalt stress treatment was unable to increase F_v/F_m levels. As these results show cytokinin levels were increased after salt stress treatments (Figure 2), it seems that further increasing cytokinin levels after stress has begun has no remediation effect and a cytokinin application has only a minor effect once the salt stress is given.

4 | DISCUSSION

Tomato production and yield is highly dependent on healthy photosynthesizing and fully functioning plants that can resist or tolerate abiotic stress. A greater understanding of the genes that regulate salt and oxidative stress resistance and tolerance is highly valuable toward the improvement of healthy tomato plants and fruits grown under field conditions. Whereas many studies of tomato have focused on older reproductively active plants or on fruits, here, we examined stress response in young seedlings that are sensitive to both salt and oxidative effects. Although fruit production is the ultimate goal of tomato cultivation, abiotic stress in young tomato

TABLE 4 Cytokinin-related genes are altered by salt and oxidative stress treatment

Gene ID	Name/Description	FC	Function	Gene ID	Name/Description	FC	Function
Salt induced				*Oxidative stress induced*			
Solyc01g105360	UGT85A1-like glucosyltransferase	22.94	Biosynthesis	Solyc04g016230	Zeatin O-xylosyltransferase	18.30	Biosynthesis
Solyc10g079320	Zeatin O-glucosyltransferase	15.50	Biosynthesis	Solyc12g057080	UGT85A1-like glucosyltransferase	17.38	Biosynthesis
Solyc04g016230	Zeatin O-xylosyltransferase	13.74	Biosynthesis	Solyc07g006800	Zeatin O-glucosyltransferase	5.11	Biosynthesis
Solyc08g061930	Cytokinin Oxidase	6.75	Biosynthesis	Solyc04g080820	Cytokinin Oxidase (SlCKX4)	4.24	Biosynthesis
Solyc03g078490	UGT85A1-like glucosyltransferase	6.16	Biosynthesis	Solyc10g079930	Zeatin O-glucosyltransferase	3.82	Biosynthesis
Solyc07g006800	Zeatin O-glucosyltransferase	5.96	Biosynthesis	Solyc12g057060	UGT85A1-like glucosyltransferase	3.77	Biosynthesis
Solyc10g079930	Zeatin O-glucosyltransferase	5.80	Biosynthesis	Solyc11g066670	Zeatin O-glucosyltransferase	3.47	Biosynthesis
Solyc03g120320	F-box kelch-repeat KMD-like	5.11	Signaling	Solyc03g078490	UGT85A1-like glucosyltransferase	2.09	Biosynthesis
Solyc02g071220	Type-A Response regulator ARR8-like	3.45	Signaling	Solyc10g079600	Type-A Response regulator ARR9-like	2.00	Signaling
Solyc04g081290	LOG1-like	3.24	Biosynthesis	Solyc01g088550	PUP11-like Transporter	1.92	Transport
Solyc10g079600	Type-A Response regulator ARR9-like	2.72	Signaling	Solyc08g062820	LOG8-Like	1.66	Biosynthesis
Solyc02g090400	Type-B Response regulator ARR13-like	2.64	Signaling	Solyc05g054390	Type-B Response regulator ARR1-like	1.49	Signaling
Solyc11g066670	Zeatin O-glucosyltransferase	2.55	Biosynthesis				
Solyc07g005660	PUP5-like Cytokinin Transporter	2.42	Transport				
Solyc04g074870	PUP3-like Cytokinin Transporter	2.30	Transport				
Solyc08g062820	LOG8-Like	1.94	Biosynthesis				
Solyc05g054390	Type-B Response regulator ARR1-like	1.90	Signaling				
Solyc12g057060	UGT85A1-like glucosyltransferase	1.64	Biosynthesis				
Solyc05g015610	Cytokinin Receptor HK3	1.55	Signaling				
Salt repressed				*Oxidative stress reduced*			
Solyc08g066350	Histophosphotransfer Protein Hpt4-like	Neg Inf	Signaling	Solyc10g079700	Type-A Response regulator ARR9-like	−4.41	Signaling
Solyc06g048600	Type-A Response regulator ARR17-like	−5.89	Signaling	Solyc12g087870	PUP3-like Cytokinin Transporter	−2.51	Transport
Solyc01g098400	Histophosphotransfer Protein Hpt1-like	−2.89	Signaling	Solyc05g006420	Type-A Response regulator ARR5-like	−2.51	Signaling
Solyc02g079330	ENT3-like Cytokinin Transporter	−2.60	Transport	Solyc08g081960	Cytokinin Response Factor SlCRF2	−2.42	Signaling
Solyc04g016190	Zeatin O-glucosyltransferase	−2.24	Biosynthesis	Solyc04g008110	Cytokinin Receptor HK4	−1.77	Signaling
				Solyc07g047770	Cytokinin Receptor HK2	−1.69	Signaling
				Solyc04g016190	Zeatin O-glucosyltransferase	−1.68	Biosynthesis

Note. Genes with cytokinin-related functions (biosynthesis (biosynthesis/metabolism), signaling, or transport) show regulation after stress treatment. Results are presented in average Fold Change (FC) compared to a buffer-treated control of transcriptome (RNAseq) analyses of 10-days-old Salt (150 mM NaCl) or Oxidative (20 mM H_2O_2) stress treatment for 6 hours, n = 3 biological replicates, similar to plants with cytokinin levels measured.

seedlings can lead to highly reduced plant growth, fruit production, and death (reviewed in Acosta-Motos et al., 2017). As cytokinins are well known to be involved in growth and development with links to delaying senescence, this study examined connection between cytokinins and abiotic stress in tomato seedlings.

The first indication we identified of a connection between cytokinins and both salt and oxidative stress was seen in the cytokinin-responsive transgenic reporter line in tomato: pTCS::VENUS (Capua & Eshed, 2017). Here we found changes in YFP fluorescence in response to each stress in the roots of this stable tomato

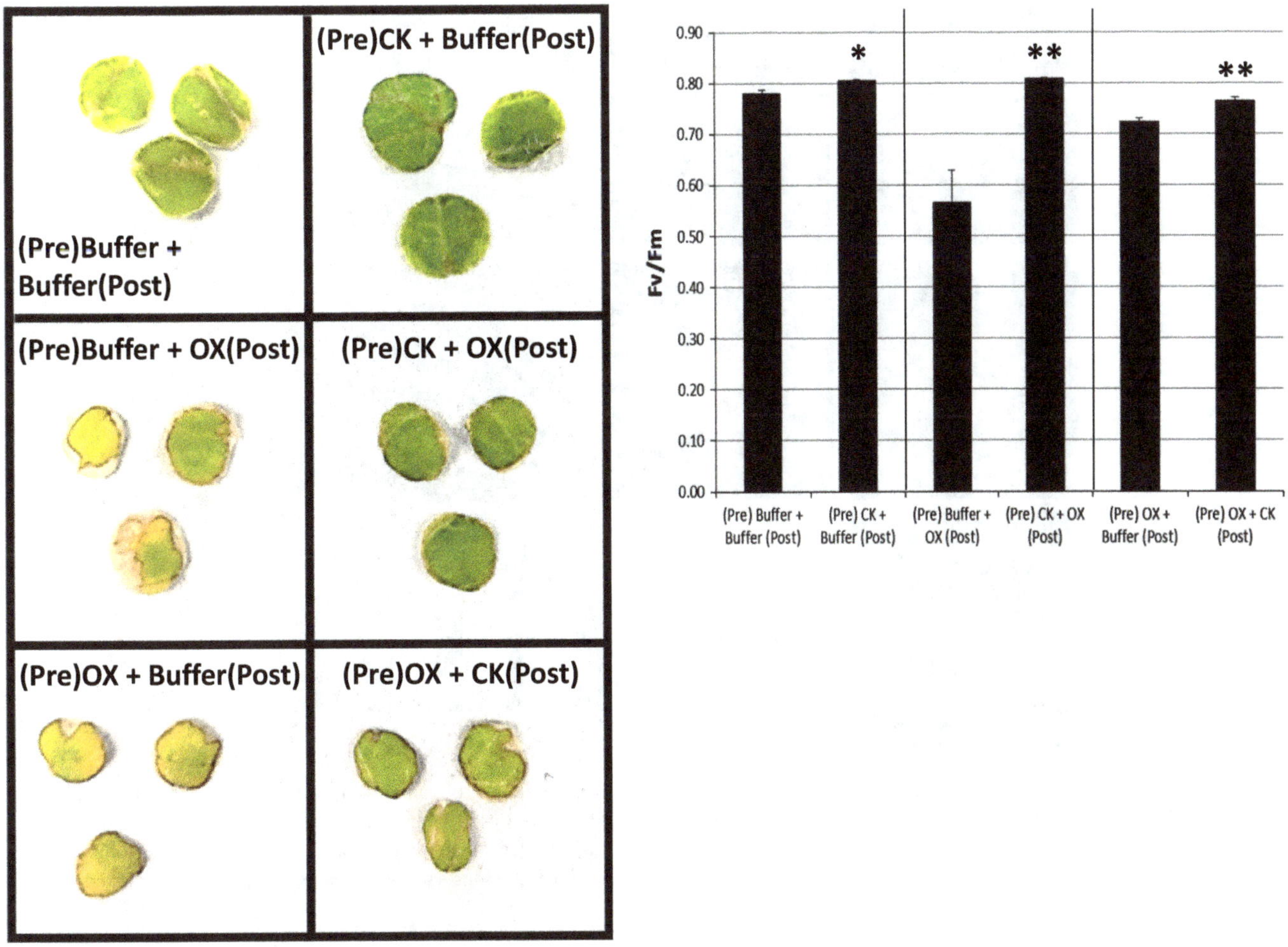

FIGURE 4 Cytokinin and oxidative stress treatment interactions. Visualization and Ave ± SE of F_v/F_m of Tomato leaf disks treated with cytokinin (CK, 5 μM *trans*-zeatin) or oxidative stress (OX, 20 mM H_2O_2). Initial treatment (Pre) was followed, after 48 hr by an additional treatment (Post). * indicates significance at $p < 0.05$, ** $p < 0.01$. Results show CK improves F_v/F_m, while oxidative stress decreases it. Both pre- and post-treatments with CK significantly lessen oxidative stress reduction in F_v/F_m levels

pTCS::VENUS line, specifically designed to show cytokinin responsiveness (Figure 1). Salt (NaCl) treatment resulted in some increase in YFP fluorescence over buffer control treatment. Although this induction is not nearly as strong as with cytokinins, it does indicate there is an interaction between salt stress and cytokinin response. In contrast, a H_2O_2 oxidative stress treatment results in reduced levels of cytokinin-responsive YFP fluorescence, suggesting that cytokinin activity is negatively regulated by oxidative stress treatment. While these findings indicate indirect cytokinin connections to each stress through activation/repression of this type-B response regulator binding site linked to YFP, it provided a basis to conduct direct and quantifiable measurements of cytokinin levels under these same stress conditions.

Direct measurements of cytokinin levels using LC-MS/MS of stress-treated tomato seedlings found similar evidence of distinct regulation for each stress (Figure 2) as seen with the pTCS::VENUS reporter line over a longer exposure. Seedlings exposed to salt stress (150 mM NaCl for 6 hr) showed a significant increase in several different cytokinin forms over buffer-treated samples, including precursors, transported, active and conjugated forms (Figure 2). Importantly the highly active form *tZ* is significantly increased to nearly 200% the level of untreated plants and levels of *cZ* are even more highly increased. Although *cZ* is known to be a less active cytokinin form it has been previously connected to salt and other stress responses (Prerostova et al., 2017; Schäfer et al., 2015). Together this shows that cytokinin levels are increased in young tomato seedlings in response to salt stress. Similar increases in cytokinin levels have been previously found in connection to salt stress (Joshi et al., 2018; Prerostova et al., 2017; Šimura et al., 2018; Veselov et al., 2017; Žižková et al., 2015). However, there are also a number of studies where decreased levels of cytokinins are seen after salt stress. Interestingly many of those studies were of developmentally older plants or specific tissues, such as fruits exposed to stress treatments of greater length, sometimes weeks in duration (Ghanem et al., 2008, 2011; Nishiyama et al., 2011). This may suggest developmental differences in response to salt or possibly a more pronounced effect of general oxidative stress effects after a lengthy salt exposure. Additionally, it has been found that plants with reduced cytokinin levels, due to decreased cytokinin synthesis or increased degradation, can have increased tolerance to abiotic stresses including salt stress

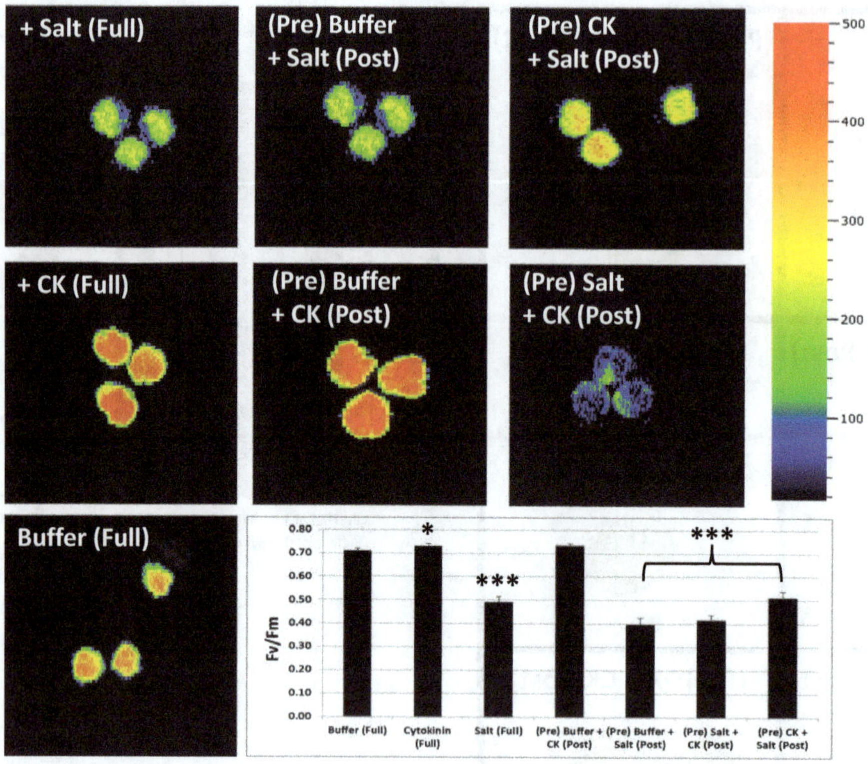

FIGURE 5 Cytokinin and salt treatment interactions. Visualization and Ave ± SE of F_v/F_m of tomato leaf disks treated with cytokinin (5 μM BA) or salt (150 mM NaCl) on a relative color scale. Total treatment was for 96 hr: (Pre) pretreatments for the first 48 hr, followed by additional 48 hr treatment (Post). * indicates significance at $p < 0.05$, *** $p < 0.001$. Results show CK treatment improves F_v/F_m, while salt decreases it. Buffer pretreatments are similar to full treatment times. CK pretreatment lessens salt reduction in F_v/F_m, but post-treatment cannot rescue salt treatment

(Ghanem et al., 2008, 2011; Mackova et al., 2013; Nishiyama et al., 2011; Veselov et al., 2017; Žižková et al., 2015). Taken together the results of our examinations of salt stress and cytokinins indicate a clear interaction, which may vary depending on the experimental parameters being addressed.

Oxidative stress treatment (20 mM H_2O_2 for 6 hr) resulted in an overall reduction in the levels several different cytokinin forms (active, transported, and conjugated) in contrast to the increases seen under salt stress (Figure 2). The active *tZ* form was found to be significantly reduced 25% after oxidative stress compared to buffer treated and a full 2.5-fold lower when compared to levels after salt stress treatment (Figure 2). It is interesting to note that the one cytokinin form found to be at significantly increased levels was *cZ*, which was also induced by salt, although to a much higher level. It is possible that *cZ* functions in a general stress role as previously noted (Schäfer et al., 2015). As this is the first study specifically examining effects of oxidative stress treatment on cytokinin levels it is difficult to find comparable measurement data, however, a general reduction in cytokinin levels is often found in response to different stresses, as seen in Nishiyama et al., 2011. Our finding of reduced cytokinin levels after oxidative stress treatment is in line with this and we believe consistent with its role as a general stressor.

Total levels of cytokinin in plants showed an increase of about 25% after salt treatment and a reduction of about 20% after oxidative stress compared to buffer treatment (Supporting Information Table S1). Although each of these levels is not significantly different from buffer treatment, a comparison of total cytokinin levels between salt treated and oxidative stress-treated plant is significantly different ($p = 0.011$, *T*-test). This supports a general finding of distinct changes in cytokinin levels for salt versus oxidative stress. While several cytokinin compounds that were measured are not affected in this snapshot of cytokinin levels after stress treatment, the prominent active cytokinin, *tZ* follows a differentially regulated pattern: significantly induced by salt and significantly reduced by oxidative stress. It is additionally important to note that stress responses to both salt and oxidative stress are known to have different dynamics in particular in relation to cytokinin levels, as reviewed in Zwack & Rashotte, 2015. Examinations of cytokinin levels and transcriptomic alterations in this study were performed at 6 hours after stress exposure to ensure initial changes in response had occurred. The generally similar findings of the pTCS reporter line at 24 hr and F_v/F_m over longer periods needed to see alterations due to stress overall support the early findings, although additional work is needed to fully understand changes in cytokinin dynamics during stress response.

Transcriptomic findings similarly indicate distinct profiles of gene regulation in salt and oxidative stress-treated seedlings. This can most easily be seen in Figure 3 from the Venn diagrams, where about 70% of DEGs have a unique salt or oxidative stress regulation pattern. A principal components analysis (PCA) of all transcriptome data also revealed three distinct groupings of results, with at least 30% of the differences explained by stress-specific treatment (Figure 3b).

It is difficult to compare the transcriptome findings here with previous results in tomato as there have been relatively few genomewide transcriptomic studies in tomato and fewer under abiotic stress conditions. This is the first RNA-sequencing based experiment that has been conducted in tomato on salt and oxidative stress. Previous large-scale transcript experiments performed in tomato were often focused on older plants, aspects related to fruit, and most were conducted using early tomato microarrays, which allowed for only partial (~1/3) genome analysis (Ouyang et al., 2007; Sun et al., 2010). As such, expression of nearly 2/3 of tomato genes remained previously unexamined. Findings here do appear to generally parallel those for other plant species, with large numbers of genes being found to be stress regulated. Core sets of stress-related genes that are affected under the salt and oxidative stress conditions are examined here, as evidenced from GO enrichment analysis (Supporting Information Table S4). One set of genes that we specifically looked for to be stress regulated were cytokinin-related genes, although they were not identified as overrepresented from GO enrichment analysis. We believe this is largely due to incomplete annotation of the tomato genome as well as incomplete identification of cytokinin-related genes in tomato. Despite these barriers, a manual search revealed 35 cytokinin-related genes involved in biosynthesis/metabolism, transport, and signaling, which showed various levels of regulation under stress conditions (Table 4). Interestingly the pattern of gene regulation mirrored that found in the full DEG list, with about 70% showing distinct, nonoverlapping regulation between stresses. In all, 113 cytokinin-related genes were identified, with 35 (31%) being found as DEG under stress conditions. A similar examination of Arabidopsis seedlings exposed to short-term salt stress revealed 25 of 92 cytokinin-related genes were affected (27%), compared here to 21% for salt alone in this study (Nishiyama et al., 2011). This suggests 20%–30% of cytokinin-related genes appear to be stress regulated.

To examine this problem from another direction, we looked at how many cytokinin-regulated genes in tomato could be connected to stress response. To do this we generated a combined list of DEGs after treatment by cytokinin in tomato, from the only two cytokinin-based transcriptome experiments previously performed in tomato (Gupta et al., 2013; Shi et al., 2013) as shown in Supporting Information Table S5. From this list of nearly 1,100 DEGs just over a third overlap with the stress DEGs found in this study. Additionally, this combined tomato cytokinin-regulated list was also found to be significantly over enriched for many different stress-related terms, such as "response to abiotic stress" from a GO analysis (Supporting Information Table S5). While this is in many ways is an imperfect comparison based on a number of experimental sampling differences, it does suggest again the connection between cytokinins and abiotic stress regulation. Interesting most (70%) of the overlap between these lists is unique to salt or oxidative stress-treated lists. Further supporting the finding that cytokinin-stress interactions appear to occur in a stress-specific manner.

We additionally examined if these cytokinin salt/oxidative stress-specific patterns could be tested with a physiological parameter. Both salt and oxidative stresses are well known to be detrimental to plant growth and general photosynthetic processes, as previously detailed. One simple, nondestructive examination of photosystem II efficiency is the measurement of chlorophyll fluorescence or F_v/F_m using a fluorimeter. The classic cytokinin senescence bioassay that was modified to study abiotic stress responses was used in this study (Letham, 1971; Zwack et al., 2013, 2016). Tomato leaf disks were floated in buffer in the presence or the absence of cytokinin, either before (pre) or after (post) salt (NaCl), or oxidative (H_2O_2) stress treatment. Found here, and as previously shown, photosystem II efficiency (F_v/F_m) is correlated with leaf senescence in response to the addition of cytokinin, which delays it, and stress, which promotes senescence (Figures 4 and 5; Zwack et al., 2016). Figure 4 shows this for oxidative stress and cytokinin interactions (visual images of leaf disks used for F_v/F_m measurements are shown). Figure 5 shows this specifically for salt and cytokinin interactions (false-colored images of leaf disks generated by the fluorimeter used for F_v/F_m measurements are shown). Under salt stress, there is a rapid decrease in F_v/F_m levels or an increase in senescence (Figure 5). Importantly a cytokinin pretreatment lessens the effect of salt-induced reduction of F_v/F_m levels, indicating there is an interaction between cytokinins and salt stress, potentially involved in tolerance (Figure 5). However, the addition of cytokinin postsalt treatment shows no positive effect to lessen senescence. Oxidative stress causes a similar increase in senescence that can be lessened by the addition of a cytokinin pretreatment (Figure 4). Interestingly, after oxidative stress treatment has begun a cytokinin post-treatment is able to reduce senescence, which does not occur with salt treatment (Figure 4).

Overall we present the first examination of both cytokinin levels and changes in whole genome transcript levels to salt and oxidative stress in tomato. While we find some overlap between these two stresses, there are many distinct, stress-specific effects. Levels of several different cytokinin forms, including the active form *tZ* are increased after salt treatment and reduced after oxidative stress treatment. Quite distinct patterns of DEG regulation were similarly found after the same salt and oxidative stress treatments. Interestingly, a number of cytokinin-related genes were also found to be regulated in salt versus oxidative stress-dependent manners. Additional examination of cytokinin treatments during leaf abiotic stress senescence assays measuring photosystem II efficiency further supports the finding that cytokinins are connected to stress responses in a distinct manner for salt and oxidative stresses.

ACKNOWLEDGMENTS

We would like to thank members of the Rashotte lab for critical reading of the manuscript. This work was supported by the following. The Alabama Agricultural Experiment Station HATCH and AgR SEED grants 370225-31006-2055 and 370228-310006-2055 for funding to E.A.K, H.T.H and A.M.R. E.A.K were supported by an Auburn University Cellular and Molecular Biosciences Research

Fellowship. This work was also funded by the Ministry of Education, Youth and Sports of the Czech Republic (National Program for Sustainability I, grant no. LO1204).

AUTHOR CONTRIBUTIONS

E.A.K. and A.M.R. designed research; E.A.K., H.T.H., T.R., L.P, A.S., F.S., O.N., and A.M.R performed research and analyzed data; E.A.K. and A.M.R. wrote the paper.

REFERENCES

Acosta-Motos, J. R., Ortuño, M. F., Bernal-Vicente, A., Diaz-Vivancos, P., Sanchez-Blanco, M. J., & Hernandez, J. A. (2017). Plant responses to salt stress: Adaptive mechanisms. *Agronomy*, *23*, 18. https://doi.org/10.3390/agronomy7010018

Anders, S., & Huber, W. (2012). *Differential Expression of RNA-Seq Data at the Gene Level–the DESeq Package*. Heidelberg, Germany: European Molecular Biology Laboratory (EMBL).

Apel, K., & Hirt, H. (2004). Reactive oxygen species: Metabolism, oxidative stress, and signal transduction. *Annual Review of Plant Biology*, *55*, 373–399. https://doi.org/10.1146/annurev.arplant.55.031903.141701

Bar, M., Israeli, A., Levy, M., Gera, H. B., Jiménez-Gómez, J., Kouril, S., … Ori, N. (2016). CLAUSA is a MYB transcription factor that promotes leaf differentiation by attenuating cytokinin signaling. *The Plant Cell*, *28*, 1602–1615. https://doi.org/10.1105/tpc.16.00211

Bose, J., Rodrigo-Moreno, A., & Shabala, S. (2014). ROS homeostasis in halophytes in the context of salinity stress tolerance. *Journal of Experimental Botany*, *65*, 1241–1257. https://doi.org/10.1093/jxb/ert430

Capua, Y., & Eshed, Y. (2017). Coordination of auxin-triggered leaf initiation by tomato LEAFLESS. *PNAS*, *114*, 3246–3251. https://doi.org/10.1073/pnas.1617146114

Davies, P. J. (2004). *Plant Hormones: Biosynthesis, Signal Transduction, Action!*. Dordrecht, The Netherlands: Kluwer Academic Publishers.

Duque, A. S., deAlmeida, A. M., da Silva, A. B., da Silva, J. M., Farinha, A. P., Santos, D., … deSousa Araújo, S. (2013). Abiotic stress responses in plants: Unraveling the complexity of genes and networks to survive. In K. Vahdati (Ed.), *Abiotic Stress - Plant Responses and Applications in Agriculture*. London, UK: InTech. ISBN: 978-953-51-1024-8.

Ghanem, M. E., Albacete, A., Martínez-Andújar, C., Acosta, M., Romero-Aranda, R., Dodd, I. C., … Pérez-Alfocea, F. (2008). Hormonal changes during salinity-induced leaf senescence in tomato (Solanum lycopersicum L.). *Journal of Experimental Botany*, *59*, 3039–3050. https://doi.org/10.1093/jxb/ern153

Ghanem, M. E., Albacete, A., Smigocki, A. C., Frébort, I., Pospíšilová, H., Martínez-Andújar, C., … Pérez-Alfocea, F. (2011). Root-synthesized cytokinins improve shoot growth and fruit yield in salinized tomato (Solanum lycopersicum L.) plants. *Journal of Experimental Botany*, *62*, 125–140. https://doi.org/10.1093/jxb/erq266

Gill, S. S., & Tuteja, N. (2010). Reactive oxygen species and antioxidant machinery in abiotic stress tolerance in crop plants. *Plant Physiology and Biochemistry*, *48*, 909–930. https://doi.org/10.1016/j.plaphy.2010.08.016

Gupta, S., Shi, X., Lindquist, I. E., Devitt, N. P., Mudge, J., & Rashotte, A. M. (2013). Transcriptome profiling of cytokinin and auxin regulation in tomato root. *Journal of Experimental Botany*, *64*, 695–704. https://doi.org/10.1093/jxb/ers365

Ismail, A. M., & Horie, T. (2017). Genomics, physiology, and molecular breeding approaches for improving salt tolerance. *Annual Review of Plant Biology*, *68*, 1. https://doi.org/10.1146/annurev-arplant-042916-040936

Joshi, R., Sahoo, K. K., Tripathi, A. K., Kumar, R., Gupta, B. K., Pareek, A., & Singla-Pareek, S. L. (2018). Knockdown of an inflorescence meristem-specific cytokinin oxidase-OsCKX2 in rice reduces yield penalty under salinity stress condition. *Plant, Cell and Environment*, *41*, 936–946. https://doi.org/10.1111/pce.12947

Letham, D. S. (1971). Regulators of cell division in plant tissues XII. A cytokinin bioassay using excised radish cotyledons. *Physiologia Plantarum*, *25*, 391–396. https://doi.org/10.1111/j.1399-3054.1971.tb01462.x

Mackova, H., Hronkova, M., Dobra, J., Tureckova, V., Novak, O., Lubovska, Z., … Vankova, R. (2013). Enhanced drought and heat stress tolerance of tobacco plants with ectopically enhanced cytokinin oxidase/dehydrogenase gene expression. *Journal of Experimental Botany*, *64*, 2805–2815. https://doi.org/10.1093/jxb/ert131

Miller, N. A., Kingsmore, S. F., Farmer, A., Langley, R. J., Mudge, J., Crow, J. A., … May, G. D. (2008). Management of high-throughput DNA sequencing projects: Alpheus. *Journal of Computer Science and Systems Biology*, *1*, 132.

Mittler, R. (2002). Oxidative stress, antioxidants and stress tolerance. *Trends in Plant Science*, *7*, 405–410. https://doi.org/10.1016/S1360-1385(02)02312-9

Mittler, R., Finka, A., & Goloubinoff, P. (2012). How do plants feel the heat? *Trends in Biochemical Sciences*, *37*, 118–125. https://doi.org/10.1016/j.tibs.2011.11.007

Mittler, R., Vanderauwera, S., Gollery, M., & Van Breusegem, F. (2004). Reactive oxygen gene network of plants. *Trends in Plant Science*, *9*, 490–498. https://doi.org/10.1016/j.tplants.2004.08.009

Mittova, V., Tal, M., Volokita, M., & Guy, M. (2003). Up-regulation of the leaf mitochondrial and peroxisomal antioxidative systems in response to salt-induced oxidative stress in the wild salt-tolerant tomato species Lycopersicon pennellii. *Plant, Cell and Environment*, *26*, 845–856. https://doi.org/10.1046/j.1365-3040.2003.01016.x

Mok, D. W., & Mok, M. C. (2001). Cytokinin metabolism and action. *Annual Review of Plant Physiology and Plant Molecular Biology*, *89*, 89–118. https://doi.org/10.1146/annurev.arplant.52.1.89

Muller, B. (2011). Generic signal-specific responses: Cytokinin and context-dependent cellular responses. *Journal of Experimental Botany*, *62*, 3273–3288.

Nishiyama, R., Watanabe, Y., Fujita, Y., Le, D. T., Kojima, M., Werner, T., … Tran, L.-S. P. (2011). Analysis of cytokinin mutants and regulation of cytokinin metabolic genes reveals important regulatory roles of cytokinins in drought, salt and abscisic acid responses, and abscisic acid biosynthesis. *The Plant Cell*, *23*, 2169–2183. https://doi.org/10.1105/tpc.111.087395

Ouyang, B., Yang, T., Li, H., Zhang, L., Zhang, Y., Zhang, J., … Ye, Z. (2007). Identification of early salt stress response genes in tomato root by suppression subtractive hybridization and microarray analysis. *Journal of Experimental Botany*, *58*, 507–520. https://doi.org/10.1093/jxb/erl258

Pandey, S. K., Nookaraju, A., Upadhyaya, C. P., Gururani, M. A., Venkatesh, J., Kim, D.-H., & Park, S. W. (2011). An update on biotechnological approaches for improving abiotic stress tolerance in tomato. *Crop Science*, *51*, 2303–2324. https://doi.org/10.2135/cropsci2010.10.0579

Peleg, Z., & Blumwald, E. (2011). Hormone balance and abiotic stress tolerance in crop plants. *Current Opinion in Plant Biology*, *14*, 290–295. https://doi.org/10.1016/j.pbi.2011.02.001

Pineda, B., Garcia-Abellan, J. O., Anton, T., Perez, F., Moyano, E., Sogo, B. G., … Atares, A. (2012). Tomato: Genomic approaches for salt and drought stress tolerance. In N. Tuteja, S. S. Gill, A. F. Tiburcio & R. Tuteja (Eds.), *Improving Crop Resistance to Abiotic Stress*. Singapore: Wiley-VCH.

Prerostova, S., Dobrev, P. I., Gaudinova, A., Hosek, P., Soudek, P., Knirsch, V., & Vankova, R. (2017). Hormonal dynamics during salt stress responses of salt-sensitive Arabidopsis thaliana and salt-tolerant *Thellungiella salsuginea*. *Plant Science*, *264*, 188–198. https://doi.org/10.1016/j.plantsci.2017.07.020

Qin, F., Shinozaki, K., & Yamaguchi-Shinozaki, K. (2011). Achievements and challenges in understanding plant abiotic stress responses and tolerance. *Plant and Cell Physiology*, *52*, 1569–1582. https://doi.org/10.1093/pcp/pcr106

R Core Team (2013). *R: A Language and Environment for Statistical, Computing*. Vienna, Austria: R Foundation for Statistical Computing.

Ramireddy, E., Chang, L., & Schmülling, T. (2014). Cytokinin as a mediator for regulating root system architecture in response to environmental cues. *Plant Signaling and Behavior*, *9*, 5021–5032. https://doi.org/10.4161/psb.27771

Sakakibara, H. (2006). Cytokinins: Activity, biosynthesis, and translocation. *Annual Review of Plant Biology*, *57*, 431–449. https://doi.org/10.1146/annurev.arplant.57.032905.105231

Saxena, I., Srikanth, S., & Chen, Z. (2016). Cross talk between H_2O_2 and interacting signal molecules under plant stress response. *Frontiers in Plant Science*, *7*, 570. https://doi.org/10.3389/fpls.2016.00570

Schäfer, M., Brütting, C., Meza-Canales, I. D., Großkinsky, D. K., Vankova, R., Baldwin, I. T., & Meldau, S. (2015). The role of cis-zeatin-type cytokinins in plant growth regulation and mediating responses to environmental interactions. *Journal of Experimental Botany*, *66*, 4873–4884. https://doi.org/10.1093/jxb/erv214

Shi, X., Gupta, S., Lindquist, I. E., Cameron, C. T., Mudge, J., & Rashotte, A. M. (2013). Transcriptome analysis of cytokinin response in tomato leaves. *PLoS ONE*, *8*, e55090. https://doi.org/10.1371/journal.pone.0055090

Šimura, J., Antoniadi, I., Široká, J., Tarkowska, D., Strnad, M., Ljung, K., & Novak, O. (2018). Plant hormonomics: Multiple phytohormone profiling by targeted metabolomics. *Plant Physiology*, https://doi.org/10.1104/pp.18.00293https://doi.org/10.1104/pp.18.00293

Singh, M., Singh, A., Prasad, S. M., & Singh, R. K. (2017). Regulation of plants metabolism in response to salt stress: An omics approach. *Acta Physiologiae Plantarum*, *39*, 48. https://doi.org/10.1007/s11738-016-2345-x

Sun, Y., Ji, K., Liang, B., Du, Y., Jiang, L., Wang, J., … Wang, H. (2017). Suppressing ABA uridine diphosphate glucosyltransferase (SlUGT75C1) alters fruit ripening and the stress response in tomato. *The Plant Journal*, *91*, 574–589. https://doi.org/10.1111/tpj.13588

Sun, W., Xu, X., Zhu, H., Liu, A., Liu, L., Li, J., & Hua, X. (2010). Comparative transcriptomic profiling of a salt-tolerant wild tomato species and a salt-sensitive tomato cultivar. *Plant and Cell Physiology*, *51*, 997–1006. https://doi.org/10.1093/pcp/pcq056

Svačinová, J., Novák, O., Lenka Plačková, L., Lenobel, R., Holík, J., Strnad, M., & Doležal, K. (2012). A new approach for cytokinin isolation from Arabidopsis tissues using miniaturized purification: Pipette tip solid-phase extraction. *Plant Methods*, *8*, 17.

To, J. P. C., & Kieber, J. J. (2008). Cytokinin signaling: Two-components and more. *Trends in Plant Science*, *13*, 85–92. https://doi.org/10.1016/j.tplants.2007.11.005

Veselov, D. S., Kudoyarova, G. R., Kudryakova, N. V., & Kusnetsov, V. V. (2017). Role of cytokinins in stress resistance of plants. *Russian Journal of Plant Physiology*, *64*, 15–27. https://doi.org/10.1134/S1021443717010162

Wilkinson, S., Kudoyarova, G. R., Veselov, D. S., Arkhipova, T. N., & Davies, W. J. (2012). Plant hormone interactions: Innovative targets for crop breeding and management. *Journal of Experimental Botany*, *63*, 3499–3509. https://doi.org/10.1093/jxb/ers148

Wu, T. D., & Watanabe, C. K. (2005). GMAP: A genomic mapping and alignment program for mRNA and EST sequences. *Bioinformatics*, *21* (9), 1859–1875. https://doi.org/10.1093/bioinformatics/bti310

Žižková, E., Dobrev, P. I., Muhovski, Y., Hošek, P., Hoyerová, K., Haisel, D., … Hichri, I. (2015). Tomato (Solanum lycopersicum L.) SlIPT3 and SlIPT4 isopentenyltransferases mediate salt stress response in tomato. *BMC Plant Biology*, *15*, 85.

Zürcher, E., Tavor-Deslex, D., Lituiev, D., Enkerli, K., Tarr, P. T., & Müller, B. (2013). A robust and sensitive synthetic sensor to monitor the transcriptional output of the cytokinin signaling network in planta. *Plant Physiology*, *161*, 1066–1075. https://doi.org/10.1104/pp.112.211763

Zwack, P. J., De Clercq, I., Howton, T. C., Hallmark, H. T., Hurny, A., Keshishian, E. A., … Rashotte, A. M. (2016). Cytokinin response factor 6 represses cytokinin-associated genes during oxidative stress. *Plant Physiology*, *172*, 1249–1258.

Zwack, P. J., & Rashotte, A. M. (2015). Interactions between cytokinin signalling and abiotic stress responses. *Journal of Experimental Botany*, *66*, 4863–4871. https://doi.org/10.1093/jxb/erv172

Zwack, P. J., Robinson, B. R., Risley, M. G., & Rashotte, A. M. (2013). Cytokinin response factor 6 negatively regulates leaf senescence and is induced in response to cytokinin and numerous abiotic stresses. *Plant and Cell Physiology*, *54*, 971–981. https://doi.org/10.1093/pcp/pct049

Transcriptome assembly and annotation of johnsongrass (*Sorghum halepense*) rhizomes identify candidate rhizome-specific genes

Nathan Ryder[1] | Kevin M. Dorn[2] | Mark Huitsing[1] | Micah Adams[1] | Jeff Ploegstra[1] | Lee DeHaan[3] | Steve Larson[4] | Nathan L. Tintle[1]

[1]Departments of Biology and Statistics, Dordt College, Sioux Center, Iowa

[2]Department of Plant Pathology, Kansas State University, Manhattan, Kansas

[3]The Land Institute, Salina, Kansas

[4]USDA-ARS, Forage and Range Research Laboratory, Utah State University, Logan, Utah

Correspondence
Nathan Ryder, Department of Statistics, Dordt College, Sioux Center, IA.
Email: nathan.ryder@dordt.edu

Abstract

Rhizomes facilitate the wintering and vegetative propagation of many perennial grasses. *Sorghum halepense* (johnsongrass) is an aggressive perennial grass that relies on a robust rhizome system to persist through winters and reproduce asexually from its rootstock nodes. This study aimed to sequence and assemble expressed transcripts within the johnsongrass rhizome. A de novo transcriptome assembly was generated from a single johnsongrass rhizome meristem tissue sample. A total of 141,176 probable protein-coding sequences from the assembly were identified and assigned gene ontology terms using Blast2GO. Estimated expression analysis and BLAST results were used to reduce the assembly to 64,447 high-confidence sequences. The johnsongrass assembly was compared to *Sorghum bicolor*, a related nonrhizomatous species, along with an assembly of similar rhizome tissue from the perennial grain crop *Thinopyrum intermedium*. The presence/absence analysis yielded a set of 98 expressed johnsongrass contigs that are likely associated with rhizome development.

KEYWORDS
next-generation sequencing, perennial plants, rhizomes, RNA-Seq, *Sorghum halepense*, transcriptomics/transcriptome analysis

1 | INTRODUCTION

Rhizomes are the horizontally aligned subterranean stems which allow a perennial plant to grow back from dormancy after a period of harsh seasonal conditions. With rhizomes at sufficient depth, an herbaceous plant may reemerge each year without having to make the extensive investment into root growth required by germination. The growing season for a rhizomatous plant is lengthened when its roots reach below the top layer of the soil, reducing the effects of environmental stresses such as temperature. Furthermore, a large persisting root network allows perennial plants to limit soil erosion, reduce runoff, and store more carbon underground as compared to an annual crop (Cox, Glover, Van Tassel, Cox, & DeHaan, 2006).

Annual species such as wheat or corn require frequent disruptions from tillage and can only achieve diminished root depth and length before harvest. Soils planted with annual crops are then susceptible to excessive erosion. A 100-year agricultural experiment revealed that plots with continuously cropped corn retained less than half the amount of topsoil as plots with only perennial grasses (Gantzer, Anderson, Thompson, & Brown, 1990). Annual root systems also exacerbate nutrient runoff. Corn, wheat, and rice have shown nitrogen fertilizer uptake efficiencies ranging from 18 to only 49% (Cassman, Dobermann, & Walters, 2002). In fact, the loss of nitrate N through subsurface drainage may be 30 to 50 times greater in annual than in perennial crops (Randall & Mulla, 2001).

Perennial plants present further ecological advantages via carbon sequestration. Typical annual crops increase soil organic carbon by 0–30 $\frac{g}{m^2}$ in a year, while perennials plants were found to accumulate 32–44 $\frac{g}{m^2}$ (Robertson, Paul, & Harwood, 2000). Perennial crops not only store more carbon, but also have no requirement to expend fossil fuels on tillage. They have a negative global warming potential, in CO_2 equivalents, while annual crops actually increase atmospheric carbon (Robertson et al., 2000).

Sorghum halepense is a perennial grass which has been classified as an invasive weed in 53 countries (Holm, Plucknett, Pancho, & Herberger, 1977). *Sorghum halepense* rhizomes, which are known to regenerate even when cut into pieces, allow the plant to endure winters and extend locally (Howard, 2004). One plant may grow 275 feet of rhizomes in a single growing season. Moreover, johnsongrass is self-fertile with a high rate of seed production, yielding more than 80,000 seeds in the same growing season (Howard, 2004; Johnson, Kendig, Smeda, & Fishel, 1997). This propensity for reproduction, coupled with rapid growth relative to native grasses, can give johnsongrass a competitive edge and lead to monocultures in unsupervised areas. Johnsongrass also exhibits the ability to grow in a wide range of environmental conditions and has resistances to many common herbicides and pathogens (Howard, 2004; Johnson et al., 1997). With such vigorous root systems and high fecundity, this species is an excellent model for understanding one important strategy of the perennial growth habit. Sequences that are expressed specifically in Johnsongrass rhizomes and found to be related to resource management or asexual reproduction could improve prediction of perennial behavior in perennial grasses (Jang et al., 2009).

Attempts at breeding a perennial hybrid from diploid *S. bicolor* and tetraploid *S. halepense* began in earnest before 2003 at the Land Institute in Salina, Kansas (Cox et al., 2006). Developing a hybrid with sufficient yield may require an extensive time-frame, but this can be expedited through the use of genetically informed techniques. Marker-assisted selection could make use of QTLs discovered in johnsongrass that are linked to rhizome activity (Xiong et al., 2007).

Previous studies have aligned johnsongrass RNA transcripts with QTLs likely related to rhizome function in sorghum and rice (Jang et al., 2006, 2009), but lack the sheer quantity of data expected from sequencing technologies today. This study provides a large, novel transcriptome for rhizome tissue from johnsongrass. The sequenced rhizome RNA was assembled, filtered, and annotated with up-to-date tools for nonmodel organisms. An annotated transcriptome for johnsongrass rhizomes then facilitates the discovery of additional rhizome-enriched transcripts. Further analysis with BLAST compared assembled johnsongrass sequences with nonrhizomatous and rhizomatous species and has led to a reduced set of candidate genes for rhizome-related function.

2 | MATERIALS AND METHODS

2.1 | Plant origin and RNA extraction

Johnsongrass plants were harvested at the coordinates 38°46′15.46″ N and 97°34′21.77″ W near Salina, Kansas on July 7, 2015. They were shipped overnight on dry ice to Sioux Center, Iowa, where the rhizome apical meristems were removed, immersed in liquid nitrogen, and placed in a −80°C freezer. RNA was extracted from four samples of rhizome node buds using the RNeasy Plant Mini Kit (Qiagen) precisely following recommended protocols. DNase digestion was then performed with the TURBO DNA-*free* kit (Thermo Fisher Scientific), followed by the RNA cleanup procedure from the RNeasy mini handbook. Each sample received an A_{260}/A_{280} ratio greater than 2 in a NanoDrop 2000 Spectrophotometer (Thermo Scientific). All four samples were then sent to the University of Minnesota Genomics Center for sequencing.

Thinopyrum intermedium rhizome tissue was obtained from a clone of a plant derived from The Land Institute's breeding program (C3-3471). RNA was extracted as described above.

2.2 | Library construction and next-generation sequencing

The four johnsongrass samples were run through a denaturing agarose gel to visualize RNA integrity (Supporting Information Figure S1). Samples 3 and 4, which depicted intact RNA, were pooled and submitted for library prep with the TruSeq Stranded mRNA Library Prep kit (Illumina). The resulting cDNA library was sequenced in a single lane of High Output (2 × 125 bp) on the Illumina HiSeq 2500 platform. The raw reads were trimmed in BBduk (Bushnell, 2014), removing adaptors and reads that were short or low quality.

Thinopyrum intermedium rhizome RNA libraries were prepared using the Illumina TruSeq RNA Library Prep Kit v2 by the University of Minnesota Genomics Center and sequenced across three lanes of an Illumina HiSeq 2000 (2 × 100 bp) run. Libraries were size-selected with a Caliper LabChip XT (PerkinElmer) to have an approximate insert size of 200 bp. The resulting reads were also trimmed with bbduk to produce 5.7 gigabases of high-quality (>Q30) data as 65,413,146 paired-end reads (NCBI Sequence Read Archive SRX3529031).

2.3 | Trinity assembly and TransDecoder identification of protein-coding sequences

The cleaned reads from each organism were de novo assembled in Trinity (version 2.4.0) (Grabherr et al., 2011; Haas et al., 2013). The johnsongrass assembly was run through TransDecoder (version 3.0.1), a companion to Trinity which uses open reading frames within assembly transcripts to find likely coding regions (Haas & Papanicolaou, 2017).

2.4 | Annotation with Blast2GO PRO

The predicted coding sequences from TransDecoder were annotated in Blast2GO PRO, an all-in-one tool that befits large-scale functional annotation with Gene Ontology (GO), especially for an original assembly of a nonmodel organism. The suite performs the BLAST algorithm to align FASTA-formatted sequence inputs with homologs in specified databases. Annotations are mapped to query sequences by their association with BLAST hits (Conesa et al., 2005).

The BLASTX-fast search was used, aligning the coding sequences against the NCBI nonredundant protein database (NR). Query sequences with BLAST hits were mapped to GO terms and assigned functional annotations where available. See Supporting Information Data S2 for Blast2GO results on the TransDecoder predicted coding sequences.

2.5 | Estimated transcript abundance with kallisto

To produce approximate abundances of the assembly transcripts, the quantification software kallisto (version 0.44.0) was used via the Trinity toolkit. Kallisto makes an index and performs pseudoalignment to quickly quantify the expression of a set of transcripts, while maintaining similar or greater accuracy than existing methods (Bray, Pimentel, Melsted, & Pachter, 2016). See Supporting Information Data S1 for the top 20 most expressed *S. halepense* sequences from kallisto with descriptions and annotations.

2.6 | Assembly filtering

The johnsongrass assembly was further reduced by filtration of estimated expression levels and top BLAST hit organisms. All lowly expressed transcripts, having a FPKM (fragments per kilobase transcript per million) less than 1, were removed from the assembly as possible background transcription (Davidson et al., 2012). This threshold may have also excluded truly expressed sequences, but was supported by the distribution of estimated abundances (Supporting Information Figure S2). All transcripts with a top BLAST hit that was not a plant were also removed, being potential contaminants. The final, reduced assembly then contained sequences that were very likely to be truly expressed and very unlikely to be from contaminant reads.

2.7 | Comparative transcriptomics with *S. bicolor* and *Th. intermedium* rhizomes

Two sequenced genomes closely related to *S. halepense* were selected: *S. bicolor* (a close annual relative and possible ancestor of *S. halepense* (Jang et al., 2009)) and *Th. intermedium* (a recently sequenced rhizomatous perennial grass). The JGI Plant Flagship genome of nonrhizomatous *S. bicolor* (version 3.1.1) was accessed from the NCBI RefSeq database (Accession ID: ABXC03000000). The protein model reference sequences were formatted into a protein BLAST database. Predicted transcripts for the rhizomatous species *Th. intermedium* were obtained from a prepublication annotated assembly (JGI annotation version 2.1) (Dorn et al., unpublished—pending release onto Phytozome) and were formatted into a second protein BLAST database for comparison.

Using the *S. halepense* proposed protein-coding sequences as queries, two BLASTX searches were performed against the custom protein databases of *Th. intermedium* and *S. bicolor*. The query sequences that received one or more BLAST hit (e-value < 10^{-20}) were cross-listed between the two searches, and removed. The *S. halepense* sequences that aligned in rhizomatous *Th. intermedium* but had no BLAST hits in nonrhizomatous *S. bicolor* were selected as candidates for rhizome-related function.

A rhizome-specific assembly for *Th. intermedium* was also generated in this study (NCBISequence Read Archive SRX3529031) and used to further reduce candidates. The assembly was aligned to the *Th. intermedium* protein BLAST database with a BLASTX search, identifying the sequences from rhizomes in the full *Th. intermedium* gene model set. The *Th. intermedium* predicted peptides declared present in rhizomes were then intersected with the *S. halepense* queries that had a *Th. intermedium* BLAST hit but not a hit from *S. bicolor*. The sequences that resulted are likely related to rhizome function, as they were expressed in *Th. intermedium* and *S. halepense* rhizomes, but do not appear in the genome of a closely related annual grain, *S. bicolor*.

3 | RESULTS

3.1 | Assembly

Johnsongrass rhizome RNA was extracted, sequenced, and trimmed to yield 283,167,847 paired reads in 72.1 gigabases with an average Phred score > 30 (NCBI Sequence Read Archive SRX3427471). The reads were assembled de novo in Trinity to produce 196,266 contigs. Using TransDecoder (version 3.0.1), 141,176 of the assembled contigs were identified as possible protein-coding sequences. The proposed coding sequences have an average length of 825 base pairs (bp) and range from 297 to 15,174 bp. See Supporting Information Figure S3 for an illustration of the distribution of contig lengths for the filtered assembly. An N50 value of 1,071 bp indicates that the contigs of this length or longer make up 50% of the total assembly length in base pairs. Table 1 lists the statistics for the sequencing/assembly process as a whole.

3.2 | Estimated abundance and assembly filtering

Quantification with kallisto identified 75,153 lowly expressed transcripts (FPKM < 1), which were removed from the assembly. After removing a further 1,576 sequences which had BLAST top hit results from nonplant organisms, 64,447 remained in a reduced assembly which had high likelihood to contain relevant functional information on the genetics of rhizome development in *S. halepense* (Supporting Information Data S3).

3.3 | Annotation

A BLASTX-fast search against the NCBI nonredundant protein database (NR) aligned 57,985 (90%) of the johnsongrass filtered assembly sequences with at least one hit (e-value < 10^{-20}). GO terms associated with BLAST hits were then mapped to 49,368 (76%) of the query sequences. Lastly, functional annotations were found for 46,792 (73%) of the sequences (Table 2).

Sorghum bicolor was the most prominently aligned species by far, receiving 51,544 of the top BLASTX hits (89%). The next highest amount of hits for a plant species was 3,953 (7%), from *Zea mays* (maize). A collection of fungi, bacteria, and animals accounted for 1,576 (2%) of the top alignments, but many of these may be from symbiotes or pathogens dwelling within the roots. See Supporting Information Figure S4 for the distribution of BLASTX top hits among the twenty most-hit plant species.

A total of 322,100 gene ontology terms were mapped to the filtered johnsongrass contigs. These GO terms were split into three categories by their associations: 104,705 were for biological processes, 125,645 were for molecular functions, and 91,750 were for cellular components. The most-mapped term was cellular "membrane" and was associated with 17,252 sequences. The results of gene ontology mapping and most common terms are listed in Table 2. Supporting Information Figure S5 depicts the distribution of GO level 1 terms over the three categories.

TABLE 1 Quantitative overview of sequenced reads before and after cleaning, assembled contigs, and predicted coding sequences from TransDecoder

Sequencing and assembly statistics	
Total number of raw reads	286,173,708 × 2
Total number of cleaned reads	283,167,847 × 2
Total number of assembled contigs	196,226
Total number of assembled genes	93,708
Total number of protein coding sequences from TransDecoder	141,176
Mean length of coding sequences (bp)	825
Median length of coding sequences (bp)	570
N50 value of coding sequences (bp)	1,071
Range of coding sequence length (bp)	297–15,174
GC content	53.37%

TABLE 2 Results from the Blast2GO annotation suite

Blast2GO statistics	
Sequences with BLASTX hits	57,985 (90%)
Mapped sequences	49,368 (76%)
Annotated sequences	46,792 (73%)
Gene ontology (GO)	
GO terms mapped	322,100
Biological processes (BP)	104,705
Molecular functions (MF)	125,645
Cellular components (CC)	91,750
10 most-mapped GO terms	
Membrane (CC)	17,252
Integral component of membrane (CC)	16,388
ATP binding (MF)	7,651
Plastid (CC)	6,831
Nucleotide binding (MF)	6,816
Nucleus (CC)	6,463
Mitochondrion (CC)	6,201
Metal ion binding (MF)	6,157
Transferase activity (MF)	5,432
Hydrolase activity (MF)	4,705

3.4 | Comparative transcriptomics

Of the 141,176 johnsongrass coding sequences, only 991 of the johnsongrass contigs had a BLASTX hit from rhizomatous *Th. intermedium* and no hit from annual *S. bicolor*. Of these selected contigs, 259 had the same BLASTX hit from *Th. intermedium* as a sequence from the *Th. intermedium* rhizome assembly. After removing sequences which may be lowly expressed or from nonplant organisms, 98 transcripts remained, which corresponded with 58 unique, putative peptide sequences from the *Th. intermedium* reference genome (Supporting Information Data S4 and S6, respectively). Figure 1 depicts the BLASTX comparisons of the johnsongrass contigs. Table 3 lists the specific results of these searches, and Supporting Information Data S5 lists the Blast2GO annotation results for the 98 candidate sequences grouped by function.

4 | DISCUSSION/CONCLUSION

This study was meant to be an exploration of the rhizome transcriptome in a nonmodel organism. The lack of sample replication and of

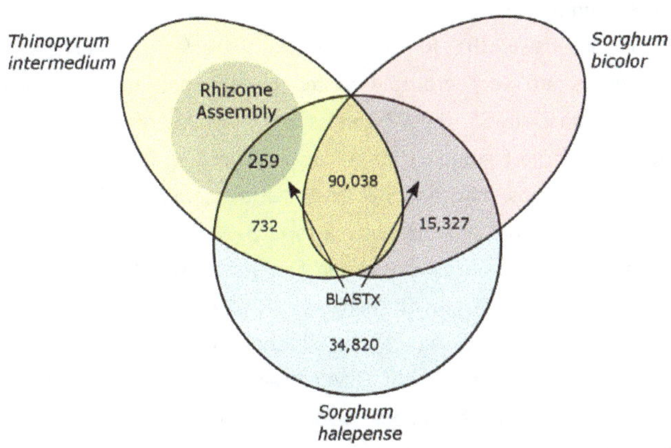

FIGURE 1 *Sorghum halepense* transcriptome comparison with *S. bicolor* and *Thinopyrum intermedium*. A Venn diagram depicts the distribution of *S. halepense* coding sequences as they have BLASTX hits against *Th. intermedium*, *S. bicolor*, both, or neither. An assembly of rhizome RNA from *Th. intermedium* was cross-listed with the *S. halepense* sequences that had BLASTX hits from *Th. intermedium* but not *S. bicolor*. All quantities refer to unique *S. halepense* contigs

TABLE 3 Overview of BLASTX comparisons of johnsongrass contigs with *Sorghum bicolor*, *Thinopyrum intermedium*, and a rhizome assembly from *Th. intermedium*

BLASTX comparisons results	
Total number of *S. halepense* query sequences	141,176
Sequences with > 1 hit from *S. bicolor*	105,365
Sequences with > 1 hit from *Th. intermedium*	91,029
Sequences with > 1 hit from both organisms	90,038
Sequences with no hits	34,820
Sequences with > 1 hit from only *S. bicolor*	15,327
Sequences with > 1 hit from only *Th. intermedium*	732
Sequences with > 1 hit from only *Th. intermedium* and in the rhizome assembly	259
Sequences from above 259 with an FPKM value > 1	98

a reference genome or transcriptome have, however, led to some limitations in analysis.

Modern sequencing and assembly techniques were used, but as no reference exists for alignment methods, a de novo assembler, Trinity, was chosen. RNA-sequencing and de novo assemblers tend produce to very large amounts of assembled transcripts, many of which are lowly expressed gene isoforms. To make the transcript set more likely to be biologically relevant, all lowly expressed sequences were removed, even though some of these still may have been relevant. Furthermore, as only a single tissue sample was sequenced, the calculated expression values have unverified accuracy.

Using presence/absence analysis via BLASTX searches against a *Th. intermedium* rhizome assembly and other genomes, 98 sufficiently expressed *S. halepense* sequences were selected as potentially related to rhizome behavior. This comparison method is dependent on BLASTX search parameters and a stringent e-value $< 10^{-20}$, so some sequences without or with lower quality matches may also have been relevant.

In any case, the functions of the 98 selected *S. halepense* sequences are very probably related to rhizomes. In Supporting Information Data S5, the 98 contigs are in groups of similar functions. The largest groupings of contigs are based on their relation to retrotransposons, mitochondrial DNA, cell membrane, and an F-box protein. These are not obviously associated with rhizome function, although F-box proteins are known to regulate signal transduction and will affect plant growth (Peng et al., 2012). The transcript with the largest expression value by far is in the ZIM (zinc finger inflorescence meristem) family motif. ZIM domain proteins are part of the signaling pathway of jasmonic acid, a hormone involved in plant growth, development, and responses to environmental stress (Liu et al., 2017; Major et al., 2017). The second most expressed transcript in the set is annotated as a member of the Armadillo (ARM)-repeat superfamily. ARM proteins have shown differential expression under abiotic stress and in developmental conditions (Sharma et al., 2014). That these two sequences are highly expressed in harvested rhizomes is not surprising.

To gain more insight into the annotations of the rhizome-specific contigs, GO enrichment analysis was performed. With the integrated FatiGO package within Blast2GO (Al-Shahrour, Diaz-Uriarte, & Dopazo, 2004), a Fisher's exact test was used to compare the GO terms mapped to the 98 rhizome-specific contigs against those mapped to the johnsongrass filtered assembly (no lowly expressed sequences and no probable contaminants). The two sets are from the same tissue sample, meaning over- or underexpression should be due to the selection of rhizome-specific sequences. A total of 12 terms were found to be overexpressed and 6 to be under-expressed. The overexpressed terms included karyogamy and a variety of DNA processes and activities, which may be linked to cell division and coincide with rhizome behavior. The rest of the terms were for processes and activities of vitamin B6, which may be linked to development and nitrogen metabolism (Colinas et al., 2016). The results of the Fisher's exact test are listed in Supporting Information Data S7.

The 58 peptide sequences from *Th. intermedium* that corresponded with the rhizome-related *S. halepense* contigs were run through the CD-Search Tool (Marchler-Bauer & Bryant, 2004). A total of 300 conserved domains were aligned with 40 of the sequences. The alignment results, as well as the *S. halepense* contig IDs associated with each protein sequence are listed in Supporting Information Data S8. Broadly, the alignments have a theme of DNA modification, including some transcription factors, transposases, zinc fingers, and variety of other DNA/RNA binding domains. Transcription factors such as WRKY4 (transcript TRINITY_DN40459_c3_g1_i5_g.83411_m.83411) are notable in this study, as a main anticipated difference between *S. halepense* and recently rhizomatous *S. bicolor* would be in expression levels. In Hu et al., 2011; there were 24 transcription factor genes upregulated in rhizome tips, some of which also encoded WRKY domains. Another upregulated gene from the same study encoded kinase family proteins, just as several transcripts in the rhizome-specific johnsongrass set do. The large range of functional categories for the rhizome-related johnsongrass transcripts complies with the suggestion of Jang et al., 2006; —that members within gene families may specialize in an area such as the rhizome meristems. In fact, the putative lipid transfer protein used as an example in the previous study may also be present in the 98 rhizome-specific contigs as TRINITY_DN33084_c0_g1_i2_g.31287_m.31287. Seven genes encoding such lipid transfer proteins were found to be upregulated in rhizome tips (Hu et al., 2011).

Also in Jang et al., 2006, developmental and environmental responses specific to rhizomes were cited as reasoning for numerous highly expressed genes affecting hormones, signaling proteins, abiotic stimuli, and development. Highly expressed contigs in the 98 rhizome-specific set such as those of the ZIM family, ARM repeats, or F-box annotations fit this description well.

The identification of genes essential to perenniality remains an important, and yet, elusive biological problem, especially in efforts to perennialize grain crops and enhance the sustainability of food production. Recent efforts to gather expression data on rhizomatous tissue are providing important lists of candidate genes to accelerate the process of uncovering the biological pathways related to perenniality. This study provides an important step in the process, as well as providing a high-confidence set of contigs (64,447 predicted contigs) for the, as yet, unsequenced *S. halepense*.

The 98 candidate sequences will now be the focus of further bioinformatics analysis (e.g., protein functional prediction, enhanced annotation, and BLASTing). In particular, promising candidates may become the target of genetic modification (e.g., knockout; overexpression) experiments to elucidate genes critically part of grain crop perenniality.

Quantitative trait loci associated with rhizome function may be used to improve breeding practices for perennial traits in a Sorghum hybrid. This study has produced a comprehensive dataset which facilitates further research with *S halepense* and other perennial grains.

DATA ACCESSION

The johnsongrass dataset has been registered with NCBI as a BioProject (PRJNA417857). The raw read files are available in the Sequence Read Archive (SRP125786), and this Transcriptome Shotgun Assembly project has been deposited at DDBJ/EMBL/GenBank under the accession GGDZ00000000. The version described in this article is the first version, GGDZ01000000.

The *Th. intermedium* dataset has been registered with NCBI as a BioProject (PRJNA428355). The raw read files are available in the Sequence Read Archive (SRP127982), and this Transcriptome Shotgun Assembly project has been deposited at DDBJ/EMBL/GenBank under the accession GGEN00000000. The version described in this article is the first version, GGEN01000000.

ACKNOWLEDGMENTS

None.

AUTHOR CONTRIBUTIONS

NLT, KMD, and JP conceived and supervised this study. NR and MH extracted *S. halepense* RNA from the rhizomes harvested by LDH and processed by JP. KMD extracted from the *Th. intermedium* rhizomes and assembled both rhizome assemblies after sequencing. SL, MA, NLT, and NR performed the annotation and analysis. NR wrote the manuscript with contributions from all other authors.

REFERENCES

Al-Shahrour, F., Diaz-Uriarte, R., & Dopazo, J. (2004). FatiGO: A web tool for finding significant associations of Gene Ontology terms with groups of genes. *Bioinformatics*, *20*(4), 578–580. https://doi.org/10.1093/bioinformatics/btg455

Bray, N. L., Pimentel, H., Melsted, P., & Pachter, L. (2016). Near-optimal probabilistic RNA-seq quantification. *Nature Biotechnology*, *34*(5), 525–527. https://doi.org/10.1038/nbt.3519

Bushnell, B. (2014). BBMap. Retrieved from https://sourceforge.net/projects/bbmap/

Cassman, K. G., Dobermann, A., & Walters, D. T. (2002). Agroecosystems, nitrogen-use efficiency, and nitrogen management. *Ambio*, *31*(2), 132–140Retrieved from http://www.jstor.org/stable/4315226%0Ahttp://www.jstor.org/stable/. https://doi.org/10.1579/0044-7447-31.2.132

Colinas, M., Eisenhut, M., Tohge, T., Pesquera, M., Fernie, A. R., Weber, A. P. M., & Fitzpatrick, T. B. (2016). Balancing of B6 vitamers is essential for plant development and metabolism in arabidopsis. *The Plant Cell*, *28*(2), 439–453. https://doi.org/10.1105/tpc.15.01033

Conesa, A., Gotz, S., Garcia-Gomez, J. M., Terol, J., Talon, M., & Robles, M. (2005). Blast2GO: A universal tool for annotation, visualization and analysis in functional genomics research. *Bioinformatics*, *21*(18), 3674–3676. https://doi.org/10.1093/bioinformatics/bti610

Cox, T. S., Glover, J. D., Van Tassel, D. L., Cox, C. M., & DeHaan, L. R. (2006). Prospects for developing perennial grain crops. *BioScience*, *56* (8), 649–659. https://doi.org/10.1641/0006-3568(2006) 56[649:pfdpgc]2.0.co;2

Davidson, R. M., Gowda, M., Moghe, G., Lin, H., Vaillancourt, B., Shiu, S.-H., ... Robin Buell, C. (2012). Comparative transcriptomics of three Poaceae species reveals patterns of gene expression evolution. *The Plant Journal*, *71*(3), 492–502. https://doi.org/10.1111/j.1365-313X.2012.05005.x

Gantzer, C. J., Anderson, S. H., Thompson, A. L., & Brown, J. R. (1990). Estimating soil erosion after 100 years of cropping on Sanborn Field. *Journal of Soil and Water Conservation*, *45*(6), 641–644Retrieved from http://www.jswconline.org/content/45/6/641.abstract.

Grabherr, M. G., Haas, B. J., Yassour, M., Levin, J. Z., Thompson, D. A., Amit, I., ... Regev, A. (2011). Full-length transcriptome assembly from RNA-Seq data without a reference genome. *Nature Biotechnology*, *29* (7), 644–652. https://doi.org/10.1038/nbt.1883

Haas, B. J., & Papanicolaou, A. (2017). TransDecoder. Retrieved from http://transdecoder.github.io

Haas, B. J., Papanicolaou, A., Yassour, M., Grabherr, M., Blood, P. D., Bowden, J., ... Regev, A. (2013). De novo transcript sequence reconstruction from RNA-seq using the Trinity platform for reference generation and analysis. *Nature Protocols*, *8*(8), 1494–1512. https://doi.org/10.1038/nprot.2013.084

Holm, L. G., Plucknett, D. L., Pancho, J. V., & Herberger, J. P. (1977). *The world's worst weeds: Distribution and biology. East-West Center.* Honolulu, HI: University Press of Hawaii.

Howard, J. L. (2004). *Sorghum halepense. Fire Effects Information System [Online].* U.S. Department of Agriculture, Rocky Mountain Research Station, Fire Sciences Laboratory. Retrieved from https://www.fs.fed.us/database/feis/plants/graminoid/sorhal/all.html

Hu, F., Wang, D., Zhao, X., Zhang, T., Sun, H., Zhu, L., ... Li, Z. (2011). Identification of rhizome-specific genes by genome-wide differential expression Analysis in Oryza longistaminata. *BMC Plant Biology*, *11*(1), 18. https://doi.org/10.1186/1471-2229-11-18

Jang, C. S., Kamps, T. L., Skinner, D. N., Schulze, S. R., Vencill, W. K., & Paterson, A. H. (2006). Functional classification, genomic organization, putatively cis-acting regulatory elements, and relationship to quantitative trait loci, of sorghum genes with rhizome-enriched expression. *Plant Physiology*, *142*(3), 1148–1159. https://doi.org/10.1104/pp.106.082891

Jang, C. S., Kamps, T. L., Tang, H., Bowers, J. E., Lemke, C., & Paterson, A. H. (2009). Evolutionary fate of rhizome-specific genes in a non-rhizomatous Sorghum genotype. *Heredity*, *102*(3), 266–273. https://doi.org/10.1038/hdy.2008.119

Johnson, B., Kendig, A., Smeda, R., & Fishel, F. (1997). Johnsongrass Control. In Weed Identification and Herbicide Injury Guide for Corn and Soybean (p. 128). University of Missouri Extension. Retrieved from http://extension.missouri.edu/p/G4872

Liu, F., Sun, T., Wang, L., Su, W., Gao, S., Su, Y., ... Que, Y. (2017). Plant jasmonate ZIM domain genes: Shedding light on structure and expression patterns of JAZ gene family in sugarcane. *BMC Genomics*, *18*(1), 771. https://doi.org/10.1186/s12864-017-4142-3

Major, I. T., Yoshida, Y., Campos, M. L., Kapali, G., Xin, X.-F., Sugimoto, K., ... Howe, G. A. (2017). Regulation of growth-defense balance by the JASMONATE ZIM-DOMAIN (JAZ)-MYC transcriptional module. *New Phytologist*, *215*(4), 1533–1547. https://doi.org/10.1111/nph.14638

Marchler-Bauer, A., & Bryant, S. H. (2004). CD-Search: Protein domain annotations on the fly. *Nucleic Acids Research*, *32*(Web Server), W327–W331https://doi.org/10.1093/nar/gkh454

Peng, J., Dashi, Y., Liqun, W., Minmin, X., Congying, Y., Yu, W., ... Xuanming, L. (2012). Arabidopsis F-box gene FOA1 involved in ABA signaling. *Science China Life Sciences*, *55*(6), 497–506. https://doi.org/10.1007/s11427-012-4332-9

Randall, G. W., & Mulla, D. J. (2001). Nitrate nitrogen in surface waters as influenced by climatic conditions and agricultural practices. *Journal of Environmental Quality*, *30*, 337–344. https://doi.org/10.2134/jeq2001.302337x

Robertson, G. P., Paul, E. A., & Harwood, R. R. (2000). Greenhouse gases in intensive agriculture: Contributions of greenhouse gases to the radiative forcing of the atmosphere. *Science*, *289*, 1922–1925. https://doi.org/10.1126/science.289.5486.1922

Sharma, M., Singh, A., Shankar, A., Pandey, A., Baranwal, V., Kapoor, S., ... Pandey, G. K. (2014). Comprehensive expression analysis of rice Armadillo gene family during abiotic stress and development. *DNA Research*, *21*(3), 267–283. https://doi.org/10.1093/dnares/dst056

Xiong, Y., Fei, S., Arora, R., Brummer, E. C., Barker, R. E., Jung, G., & Warnke, S. E. (2007). Identification of quantitative trait loci controlling winter hardiness in an annual × perennial ryegrass interspecific hybrid population. *Molecular Breeding*, *19*(2), 125–136. https://doi.org/10.1007/s11032-006-9050-1

16

Effects of δ-aminolevulinic acid dehydratase silencing on the primary and secondary metabolisms of citrus

Nabil Killiny | Faraj Hijaz | Yasser Nehela | Subhas Hajeri | Siddarame Gowda

Department of Plant Pathology, Citrus Research and Education Center, IFAS, University of Florida, Lake Alfred, Florida

Correspondence
Nabil Killiny, Department of Plant Pathology, Citrus Research and Education Center, University of Florida, 700 Experiment Station Rd., Lake Alfred, FL 33850.
Email: nabilkilliny@ufl.edu

Present address
Subhas Hajeri, Citrus Pest Detection Program, Central California Tristeza Eradication Agency, Tulare, California.

Funding information
SCRI-NIFA, USDA, Grant/Award Number: 201500955-04

Abstract

δ-aminolevulinic acid dehydratase (ALAD) is an important enzyme in tetrapyrrole synthesis. ALAD combines two δ-aminolevulinic acid (δ-ALA) molecules to form the pyrrole molecule, porphobilinogen, an important precursor for plant pigments involved in photosynthesis, respiration, and nutrient uptake. In this study, we investigated the effects of silencing of *ALAD* gene on citrus leaf pigments and metabolites. The ALAD enzyme was inhibited using virus-induced gene silencing (VIGS) technology using *citrus tristeza* virus (CTV). δ-ALA accumulated in citrus plants inoculated with the recombinant virus (CTV-tALAD) to silence *ALAD* and resulted in discrete yellow spots (yellow islands) and necrosis in leaves and stems. The levels of chlorophylls, starch, sucrose, *trans*- and *cis*-violaxanthin, and α- and β-cryptoxanthin were reduced in CTV-tALAD plants, whereas zeaxanthin was increased. The increase in zeaxanthin and the decrease in its precursors indicated that the reduction in chlorophylls resulted in light damage. Salicylic acid and jasmonic acid levels, as well as emission of (*E*)-α-bergamotene and (*E*)-β-farnesene, increased in CTV-tALAD plants indicating these plants were under stress. Our results showed that silencing of *ALAD* induces stress in plants and that VIGS using mild CTV strains is a promising technique to study biological function of citrus genes.

KEYWORDS
δ-aminolevulinic acid dehydratase, abscisic acid, citrus, leaf pigments, virus-induced gene silencing

1 | INTRODUCTION

Tetrapyrroles are macrocyclic compounds that contain four pyrrole rings with various structural alternatives and numerous biological functions in prokaryotes and eukaryotes. Higher plants possess four classes of tetrapyrroles that play important roles in many biological processes; chlorophyll is essential for photosynthesis, heme is necessary for respiration, siroheme is necessary for sulfur and nitrogen assimilation, and phytochromobilin is required for light sensing (Tanaka & Tanaka, 2007). Tetrapyrrole biosynthesis is a multistep and a multibranched pathway; the first six steps (from glutamate to uroporphyrinogen III) are required by all classes of tetrapyrroles, whereas the next three steps (from uroporphyrinogen III to protoporphyrin IX) are common to phytochromobilin, heme, and chlorophyll biosynthesis (Tanaka & Tanaka, 2007).

In the first step of tetrapyrrole biosynthesis, glutamyl-tRNA synthetase attaches glutamate to t-RNA to form tRNAGlu complex. In the second step, the carboxyl group of glutamyl-tRNA is converted to a formyl group by glutamyl-tRNA reductase to generate glutamate-1-semialdehyde (GSA). In the third step, GSA is converted into δ-ALA by GSA aminotransferase (GSA-AT). In the fourth step, two δ-ALA molecules are combined together by ALA-dehydratase (ALAD) to form a pyrrole molecule, porphobilinogen (PBG) (Tanaka & Tanaka, 2007). The δ-ALA does not accumulate in developing leaves and is directly converted to PBG by ALAD. However, the addition of levulinic acid, a specific competitive inhibitor of ALAD, can induce the accumulation of δ-ALA in greening tissues (Kumar & Söll, 2000).

Functional genomics using mutant plants have been extensively used to investigate the regulatory mechanism of tetrapyrrole synthesis and the role of various enzymes in this pathway. Most of these studies were performed on *Arabidopsis thaliana* (Espineda, Linford, Devine, & Brusslan, 1999; Frick, Su, Apel, & Armstrong, 2003; Hirono & Rédei, 1963; Ishikawa, Okamoto, Iwasaki, & Asahi, 2001; Kumar & Söll, 2000; Molina et al., 1999; Nagata, Tanaka, Satoh, & Tanaka, 2005; Tanaka & Tanaka, 2005) and *Nicotiana tabacum* (Alawady & Grimm, 2005; Höfgen et al., 1994; Kruse, Mock, & Grimm, 1995; Mock, Keetman, Kruse, Rank, & Grimm, 1998; Mock et al., 1999; Papenbrock, Mock, Tanaka, Kruse, & Grimm, 2000; Papenbrock et al., 2001). Other plants species such as *Zea mays* (Hu, Yalpani, Briggs, & Johal, 1998), *Hordeum vulgare* (Jensen et al., 1996; Rzeznicka et al., 2005), *Pisum sativum* (Weller, Terry, Rameau, Reid, & Kendrick, 1996; Weller, Terry, Reid, & Kendrick, 1997), and *Brassica napus* (Tsang et al., 2003) were also investigated. In general, inhibition of δ-ALA formation in mutant plants reduced chlorophyll and heme formation, but did not result in necrotic lesions (Tanaka & Tanaka, 2007). On the other hand, inhibition of later steps in the tetrapyrrole pathway (uroporphyrinogen III decarboxylase to protoporphyrinogen IX oxidase) increased the level of intermediate molecules and resulted in necrotic lesions upon illumination (Tanaka & Tanaka, 2007).

Although the first three enzymes that lead to δ-ALA formation have been extensively targeted (Kim et al., 2005; Kumar & Söll, 2000; Tsang et al., 2003), ALAD rarely has been studied in plants (Chai et al., 2017). In humans, ALAD is necessary for heme biosynthesis and its activity can be inhibited by environmental toxins such as lead (Scinicariello et al., 2007). Lead deactivates ALAD by displacing the zinc from the active site of the enzyme and consequently leads to the accumulation of δ-ALA, which can act as a γ-aminobutyric acid (GABA) receptor agonist in the nervous system resulting in neuropathogenic effects (Brennan & Cantrill, 1979). ALAD enzyme has been isolated and characterized in several organisms, including human, mouse, yeast, bacteria, and many plant species such as pea, soybean, and spinach (Kaczor, Smith, Sangwan, & O'Brian, 1994).

Virus-induced gene silencing (VIGS) technology is a reverse genetic technique that has been widely used to study biological function of plant genes (Romero, Tikunov, & Bovy, 2011). Since its introduction in 1995, VIGS has been applied in about 30 different species, including *A. thaliana*, *N. tabacum*, and *Solanum lycopersicon* (Becker & Lange, 2010). Despite its successful application in many species, VIGS efficiency depends on the distribution and the movement of the virus in plants (Romero et al., 2011). In consequence, it was recommended to use an additional reporter gene such as phytoene desaturase (PDS) to estimate the level of gene silencing (Senthil-Kumar et al., 2007).

VIGS development requires many steps (Becker & Lange, 2010). In the first step, sequences corresponding to the host gene to be silenced are inserted into the viral genome. After creation of the engineered virus, it is inoculated in the host plant. Plant infection with the engineered virus initiates the synthesis of the viral dsRNA. Dicerlike enzymes cleave the dsRNA into short interfering RNAs. These RNAs are recognizable by RNA-induced silencing complex (RISC) and melted into single-stranded RNAs (ssRNAs) which are afterward used as template to target gene degradation.

Because most viruses have a limited host range, numerous virus vectors have been developed for VIGS in plants, including potato virus X (PVX), tobacco mosaic virus (TMV), tomato golden mosaic virus (TGMV), tomato yellow leaf curl virus (TYLCV), and China virus satellite DNA (Igarashi et al., 2009). Although VIGS has been widely used to investigate gene functions in many herbaceous plant species, VIGS has rarely been used in woody plants. Tobacco rattle virus (TRV) was one of the first viruses devolved for VIGS in woody plants such as *Populus euphratica* and *P.* × *canescens* (Shen et al., 2015). In addition, *Citrus leaf blotch virus* (CLBV) has also been used to target the phytoene desaturase (*PDS*) gene in citrus plants (Agüero et al., 2014). Because some of the reliable VIGS vectors can induce similar symptoms to those caused by the silenced gene (Igarashi et al., 2009), it is necessary to develop more VIGS vectors for additional plant species, especially for woody crops. Development of more VIGS vectors will enhance plant genomic studies (Burch-Smith, Anderson, Martin, & Dinesh-Kumar, 2004).

Citrus tristeza virus (CTV) is a closterovirus which is transmitted by the brown citrus aphid (Bar-Joseph, Marcus, & Lee, 1989), and it infects most commercial citrus varieties (Albiach-Marti et al., 2004). T36 and T30 are the most predominant CTV strains in Florida (Folimonova, Robertson, Garnsey, Gowda, & Dawson, 2009; Harper, Cowell, Halbert, Brlansky, & Dawson, 2015). The CTV isolate T36 induces a decline in sweet orange trees grafted onto sour orange rootstock, but it does not induce any appreciable disease syndromes in other common commercial citrus hosts such as grapefruit, and even sweet orange on other rootstocks (Dawson, Robertson, Albiach-Martí, Bar-Joseph, & Garnsey, 2015). Folimonov, Folimonova, Bar-Joseph, and Dawson (2007) developed a CTV strain T36-based vector for transient expression of foreign genes in citrus trees using a green fluorescent protein (GFP) as a reporter. Hajeri, Killiny, El-Mohtar, Dawson, and Gowda (2014) developed CTV strain T36-based vector and demonstrated its capability in silencing transgene *GFP* in *Nicotiana benthamiana* line 16c and citrus endogenous gene *PDS* in citrus. The use of this mild T36 CTV vector offers many advantages: (a) It produces no symptoms in infected plants; (b) has been able to express foreign genes in citrus for many years (Dawson & Folimonova, 2013); (c) can be used for field application to protect against diseases or to treat infected plants (Hajeri et al., 2014); and (d) more than one foreign gene can be inserted into the CTV genome at the same time (Dawson & Folimonova, 2013; El-Mohtar & Dawson, 2014).

In a recent study, we showed that wild-type CTV-T36 (CTV-wt) does not produce dramatic effects on the phloem sap composition and released volatiles of *Citrus macrophylla* (Killiny, Hijaz, Harper, & Dawson, 2017). In the current study, we investigated the effects of silencing of the *ALAD* gene, using the CTV-based silencing vector, on citrus leaf pigments, phytohormones, and volatile and non-volatile metabolites. We hypothesize that silencing of *ALAD* will not only affect chlorophyll biosynthesis, but can also affect the synthesis of other pigments and metabolites.

2 | EXPERIMENTAL PROCEDURES

2.1 | Plant material

Alemow (*C. macrophylla*) plants were grown under controlled greenhouse conditions (temperature of 22–24°C, 16-:8-hr day–light cycle and 60% humidity) at the Citrus Research and Education Centre, Lake Alfred, FL. Plants were about 1 year old (approximately two feet tall and the stem of pencil thickness).

2.2 | *Citrus tristeza virus*-based vectors

The infectious cDNA clones of CTV isolate T36; GenBank accession no. AY170468) in the binary vector pCAMBIA-1380 (Figure 1a) was used as a base plasmid for engineering all of the constructs used in this study (Hajeri et al., 2014). To clone the truncated *ALAD* gene (tALAD) and generate the CTV-tALAD vector, primers were designed based on *Citrus sinensis ALAD* gene cDNA sequence (Phytozome Citrus Gene orange1.1g013861m.g). The antisense of the truncated fragment (400 nucleotides) of the *ALAD* gene was amplified using total RNA from *C. macrophylla* as a template by SuperScript® III One-Step RT-PCR System with Platinum® Taq DNA Polymerase (Life Technologies Corp.) and primers ALAD-PacI.

(5′-GAGC*TTAATTAA*TAAAAAGTGGATATACAAAATTTGCCGGTG AAAG-3′) and ALAD-StuI (5′-GACG*AGGCCT*GCTTCGTCGGTATTT CACGCGCCTTGTG-3′). The PCR product was digested with PacI and StuI restriction enzymes and cloned into similarly digested CTV-tGFP by replacing tGFP with tALAD fragment.

Procedures for agroinfiltration of CTV constructs into *N. benthamiana* were followed as described previously (Hajeri et al., 2014). After 4–6 weeks postinfiltration, systemic leaves from *N. benthamiana* that tested positive for CTV by ELISA were harvested and used in isolating CTV virions and for inoculating *C. macrophylla* by the bark-flap method as described previously (Hajeri et al., 2014). Five plants (about 1 year old) were inoculated with CTV-tALAD virion, five with CTV-wt, and five plants were left as controls. Inoculated plants were kept in the greenhouse

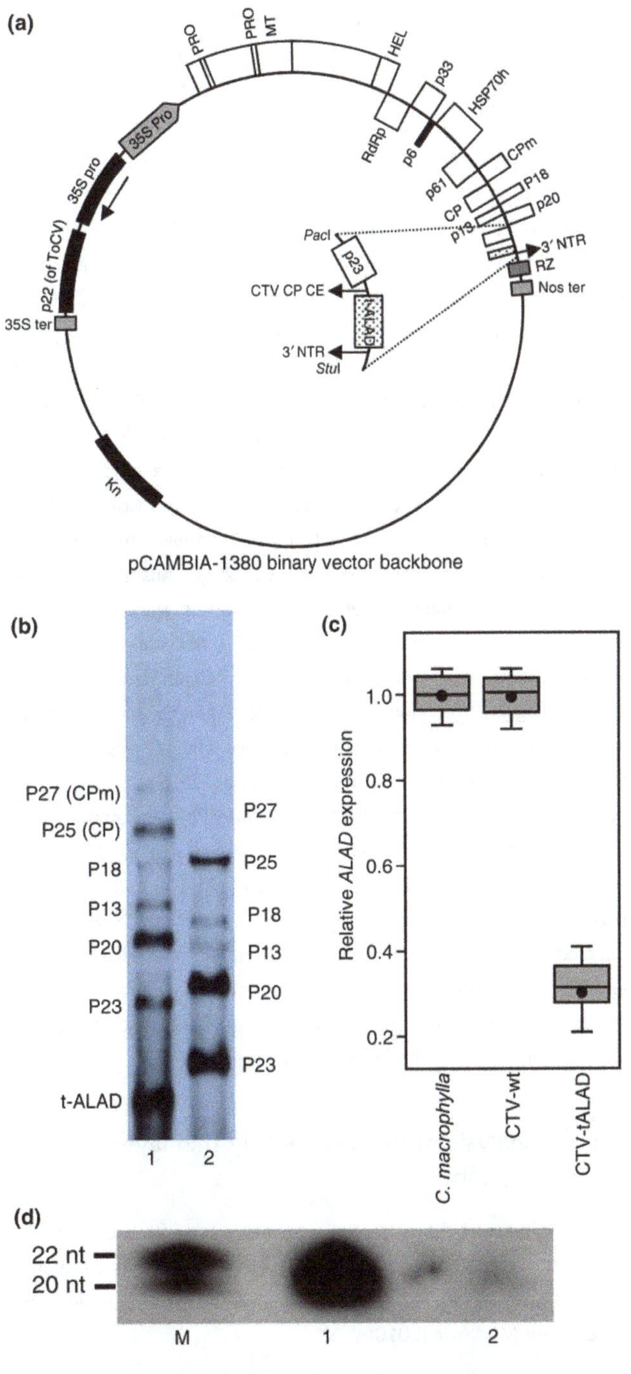

FIGURE 1 Silencing of δ-aminolevulinic acid dehydratase (ALAD) using *Citrus tristeza* virus in a binary vector. (a) Schematic representation of full-length infectious cDNA clones of *C. tristeza* virus (CTV) with its open reading frames (ORF) placed between enhanced 35S promoter of *Cauliflower mosaic* virus at the 5′end, ribozyme (RZ) of Subterranean clover mottle virus satellite RNA and nopaline synthase terminator (Nos ter) at the 3′end in the binary vector pCAMBIA-1380. (b) Northern blot shows the subgenomic RNAs (sgRNAs) of CTV and an extra sgRNA for tALAD, accumulated in CTV-tALAD plants (1) compared to CTV-wt plants (2). The blot was hybridized with digoxigenin labeled minus-sense ribo-probe specific to the 3′nontranslated region of CTV. (c) Gene expression of *ALAD in CTV-tALAD, CTV-wt*, and *Citrus macrophylla.* Elongation factor 1-alpha (*EF1*), F-box/kelch-repeat protein (*F-box*), glyceraldehyde-3-phosphate dehydrogenase GAPC1, cytosolic (*GAPC1*, also known as *GAPDH*), and SAND family protein (*SAND*) were used as endogenous gene (reference gene) to normalize the data of gene expression. Horizontal thick lines indicate the medians, black dots indicate the means, boxes show the interquartile ranges including 25%–75% of the values, and whiskers reflect the highest and the lowest value of data. Experiments in triplicate are presented. (d) Accumulation of ALAD-specific small interfering RNAs (siRNAs) in CTV-tALAD plants (1) compared to CTV-wt (2)

until development of clear genotype (about 6 months). Large RNA northern blot hybridization, small RNA isolation, small RNA northern blot hybridization, reverse transcription quantitative PCR (RT-qPCR) for plant tissue were performed as described previously (Hajeri et al., 2014).

2.3 | High-performance liquid chromatography analysis of citrus leaf pigments

At least three leaves were collected from each plant (five plants were sampled from each treatment, and each sample was analyzed twice). Leaves collected from each plant were placed together in a mortar, mixed with liquid nitrogen, ground into fine powder using a pestle, and citrus leaf pigments were extracted as described previously (Killiny & Nehela, 2017a). In short, about 0.1 g ground material was mixed with 400 μl of acetone 80% and 240 μl of ethyl acetate and vortexed for 30 s. Samples were left in the dark on ice for 10 min and vortexed twice followed by addition of 280 μl of water, and the mixture was centrifuged at 8,500 *g* for 5 min at 4°C. The organic upper layer was removed and dried under a nitrogen stream. The extracted pigments were resuspended in 200 μl ethyl acetate and analyzed immediately by high-performance liquid chromatography (HPLC) as described previously by Wei et al. (2014a,b). The HPLC system consisted of an Agilent 1200 system with photodiode array detector (Agilent Technologies, Santa Clara, CA, USA). Chromatographic separation was performed using a C30 YMC carotenoid column, 250 × 4.6 mm I.D., S-5 μm (YMC America, Allentown, PA, USA). The mobile phase composition and the gradient profile were as described by Mouly, Gaydou, Lapierre, and Corsetti (1999). The column temperature was set at 25°C, mobile phase flow rate was 1 ml/min, and injection volume was 20 μl. Absorbance of the various citrus leaf pigments and peak responses were monitored at several spectral wavelengths (230, 278, 350, 430 and 486 nm). The HPLC output data were analyzed using ChemStation software, B.03.02 (Agilent Technologies, CA, USA). Pigments were identified by comparing experimental retention times and UV–visible spectra with that of published literature (Mouly et al., 1999), as well as with authentic standards. Chlorophyll *a*, chlorophyll *b*, *α*-carotene, *β*-carotene, lutein and zeaxanthin were obtained from Sigma-Aldrich (USA). A set of concentrations (1, 5, 10, 25, 50, and 100 ppm) were prepared from each standard. A 20 μl aliquot from each standard was injected into the HPLC under the same condition described above to establish calibration curves and calculate the concentrations of pigments in citrus leaves. The concentrations of the rest of pigments were estimated using the lutein calibration curve.

2.4 | Analysis of leaf pigment by thin-layer chromatography

A 10 μl aliquot of the extracted pigments was spotted on the silica thin-layer chromatography (TLC) plate, and the pigments were separated using a mixture of n-hexane-acetone (70:30 v/v) as described by Zeb and Murkovic (2010).

2.5 | Gene expression analysis

Total RNA was extracted using TRIzol® reagent (Ambion®, Life Technologies, NY, USA). NanoDrop 2000 spectrophotometer (Thermo Scientific, USA) was used to estimate the quantity and quality of isolated RNA. For synthesizing cDNA, SuperScript first-strand synthesis system (Invitrogen) with random hexamer primers was used as described in the manufacturer's instructions. SYBR® Green PCR master mix (Applied Biosystems) was used to perform quantitative PCR (qPCR) on an ABI 7500 Real-Time PCR System (Applied Biosystems). For each treatment, two technical replicates per biological replicate and five biological replicates per treatment were analyzed in triplicate. Primers for 48 genes were used to measure the gene expression. The $2^{-\Delta\Delta CT}$ method was used to determine the relative expression of genes (Livak & Schmittgen, 2001). Elongation factor 1-alpha (*EF1*), F-box/kelch-repeat protein (*F-box*), glyceraldehyde-3-phosphate dehydrogenase GAPC1, cytosolic (*GAPC1*, also known as *GAPDH*), and SAND family protein (*SAND*) were used as endogenous genes (reference genes) to normalize the data of gene expression (Mafra et al., 2012; Wei et al., 2014a,b).

2.6 | Leaves incubation in levulinic acid

Three young leaves from each plant (five plants from each treatment) were harvested, and each leaf was incubated separately by immersing its petiole in 1.5 ml of 100 mM levulinic in 10% dimethylsulfoxide solution (Beale & Castelfranco, 1974) for 0 and 6 hr at room temperature (20°C).

2.7 | Extraction of *δ*-ALA from citrus leaves

The *δ*-ALA was extracted using 0.1 N HCl as described by Hijaz and Killiny (2016). In short, leaves were ground in liquid nitrogen and 0.2 g of the ground tissues was transferred to 2-ml centrifuge tube. One millilitre of 0.1 N HCl was added, and the sample was vortexed for 2 min. After 10 min of incubation, samples were further vortexed. The procedure was repeated three times, and the sample was centrifuged at 22,000 *g* for 6 min at room temperature.

2.8 | Colorimetric determination of *δ*-ALA

The concentration of *δ*-ALA in the 0.1 N HCl extract was determined colorimetrically as described by Mauzerall and Granick (1956). In short, 0.5 ml of the supernatant was transferred to a new centrifuge tube and was mixed with 0.5 ml of phosphate buffer (equal volume of 0.5 M NaH_2PO_4 and 0.5 Na_2HPO_4; pH: 6.8) and 0.1 ml ethyl acetoacetate. The mixture was placed in a boiling water bath for 10 min. Then, the sample was cooled to room temperature and centrifuged at 14,000 rpm for 6 min at room temperature. After centrifugation, 0.7 ml of the supernatant was mixed with an equal volume of modified Ehrlich's reagent containing 2N perchloric acid. The density of the pink color produced from the reaction of *δ*-ALA-pyrrole with Ehrlich's reagent was measured at 553 nm using a

Shimadzu UV-Vis spectrophotometer (UV-1700 PharmaSpec (Torrance, CA, USA). Each biological sample was analyzed twice (two technical replicates). A set of δ-ALA standards (10.0, 5.0, 2.5, 1.2, 0.6, and 0.0 ppm in water) was reacted with ethyl acetoacetate and Ehrlich's reagent as mentioned above, and results were used to construct the standard curve. The spectra of the standard and the samples were also scanned between 400 and 700 nm.

2.9 | δ-ALA derivatization for GC-MS

The δ-ALA-pyrrole derivative was reacted with N-methyl-(N-trimethylsilyl) trifluoracetamide (MSTFA) as described by Hijaz and Killiny (2016). In brief, a 0.4 ml of the hydrochloric acid extract was reacted with ethyl acetoacetate as mentioned above and δ-ALA-pyrrole derivative was extracted with 3 × 600 µl of ethyl acetate (Tomokuni & Ogata, 1972). Then, the ethyl acetate was evaporated under a nitrogen stream. A 40 µl of N-methyl-(N-trimethylsilyl) trifluoroacetamide (MSTFA) reagent was added to the dried sample and was incubated at 85°C for 35 min. A 400 µl of 20 ppm δ-ALA standard was derivatized in the same way and 1 µl of the derivatized standard or sample was injected into the GC-MS running in the full scan mode as described by Hijaz and Killiny (2016).

2.10 | Colorimetric determination of starch and sucrose

Starch and sucrose were extracted according to Cimò et al. (2013) with slight modifications. In short, about 100 mg of the same ground leaf tissues was mixed with 500 µl distilled water and vortexed for 30 s. The water extract was boiled for 10 min, vortexed for 10 s, and centrifuged at 650 *g* for 5 min. A 300 µl aliquot of the supernatant was mixed with 900 µl of 100% ethanol, vortexed for 10 s, and centrifuged at 18,000 *g* for 10 min. The supernatant was discarded, and the pellet was suspended in 1 ml distilled water and mixed with 50 µl of iodine solution. Starch was determined by monitoring the absorbance at 595 nm as described by Cimò et al. (2013) using rice starch as a standard. The absorbance was measured using a microplate spectrophotometer (Model 680, Bio-Rad Laboratories, CA, USA). Sucrose determination was accomplished at 620 nm as described by van Handel (1968) using PharmaSpec ultraviolet 1700 spectrophotometer (Shimadzu Corporation, Japan). In short, a 100 µl aliquot of the supernatant was mixed with 100 µl 30% KOH and the mixture was boiled for 10 min. When samples cooled to room temperature, 3 ml of anthrone reagent was added and the samples were incubated at 40°C for 10 min for color development.

2.11 | Analysis of carboxylic compounds and phytohormones in leaves by GC-MS

Phytohormones and carboxylic compounds were extracted, from the same ground samples, as described previously (Nehela, Hijaz, Elzaawely, El-Zahaby, & Killiny, 2016), and then derivatized with methyl chloroformate after spiking of each sample with 5 µl aliquot of 200 ppm heptadecanoic acid, which is not found in citrus leaves, as internal standard. Phytohormones and carboxylic acids were analyzed as previously reported (Killiny & Nehela, 2017b; Nehela et al., 2016). Identification of metabolites was further confirmed by comparing their retention time, linear retention indices (LRIs), and mass spectra with authentic standards. Compound peak areas were normalized to the internal standards (heptadecanoic acid). Quantification of metabolites was based on the peak areas obtained from a series of reference standards derivatized and injected under the same conditions as samples. Calibration curves were constructed from the linear regressions obtained by plotting the concentration versus peak area for each standard.

2.12 | Collection of volatile organic compounds from intact citrus leaves

Released leaf volatiles from intact, young intermediate, and mature leaves (hardened) from branches of *C. macrophylla* were collected on a mixed SPME fiber (50/30 µm divinylbenzene/carboxen/polydimethylsiloxane, 1-cm fiber, Supelco) and analyzed using GC-MS as described by Killiny and Jones (2017). Wiley 9th ed., NIST 2011, and Wiley Flavor and Fragrance mass spectral libraries were used for volatile identification as well as by comparing spectra to standards when available. Volatile standard was prepared in n-hexane and was run using the same GC temperature program as samples. Five plants were sampled from each treatment, and three different types of leaves (young, intermediate, and mature) were taken from each plant.

2.13 | Statistical analysis

For HPLC and GC-MS, five biological and two technical replicates per treatment were analyzed. Comparison of pigment concentrations among the three different treatments was performed using the analysis of variance (ANOVA), followed by post hoc pairwise comparisons using a Tukey–Kramer honestly significant difference test (Tukey's HSD). Statistical significance was established as p-value <0.05. The level of other metabolites in CTV-tALAD plants was compared to that of CTV-wt (T36) plants using two-tailed t test, and statistical significance was established as p-value <0.05. Two-way hierarchical cluster analysis (HCA) was also performed with the means of the gene expressions for each treatment. Distance and linkage were measured using the Bray–Curtis similarity measure method (Michie, 1982). 3D-surface plots were performed among the three pigment groups (chlorophylls, carotenes, and xanthophylls) using the total concentration of each group.

3 | RESULTS

3.1 | CTV-tALAD induced gene silencing and produced yellow islands and necrosis in citrus

Northern blot analysis of RNA showed a of the extra subgenomic RNA in CTV-tALAD plants compared to CTV-wt plants (Figure 1b).

In addition, RT-qPCR showed a twofold to threefold downregulation of δ-ALAD mRNA in CTV-tALAD plants compared to CTV-wt (Figure 1c). The northern blot also showed high accumulation of ALAD-specific small interfering RNAs (siRNAs) in CTV-tALAD plants compared to CTV-wt (Figure 1d).

Citrus macrophylla plants inoculated with CTV-tALAD virions displayed yellow islands and necrosis in leaves, thorns and stems, and the apical meristem of citrus plants (Figure 2c–k). No yellow islands or necrosis was observed in CTV-wt plants (Figure 2a). The symptoms start as a few dots in developing young leaves and the number of dots and their intensity grows until they cover most of the leaf surface (Figure 2f–k). The ALAD phenotype was also different from the PDS-silenced plants (Figure 2b). The PDS plants showed a photobleaching phenotype (Figure 2b), whereas the CTV-tALAD plants showed yellow islands and necrosis in leaves, stems, and the apical meristem (Figure 2c–f).

3.2 | Silencing of *ALAD* reduced chlorophyll content and altered the levels of carotenoids

Fifteen pigments were identified in the leaf extract using HPLC-PDA (Supporting Information Figure S1a), whereas only eight of these

FIGURE 2 Phenotype observed in CTV-tALAD plants and CTV-tPDS plants. (a) CTV-wt. (b) Photobleaching in CTV-tPDS plants. (c–e) Yellow islands and necrosis in apical growing and stem of CTV-tALAD plants. (f–k) Development of yellow islands and necrosis in leaves of CTV-tALAD

pigments were identified using TLC (Supporting Information Figure S1b). The CTV-tALAD significantly reduced the levels of chlorophyll *a*, chlorophyll *b*, chlorophyllide *a*, and pheophytin *a* compared to the controls (Figure 3). The reduction in chlorophyll *a*, chlorophyll *b*, and pheophytin in CTV-tALAD plants was observed in the TLC results (Supporting Information Figure S1b).

While α- and β-cryptoxanthin, *cis*- and *trans*-violaxanthin, and neoxanthin were reduced in CTV-tALAD plants, zeaxanthin was increased (Figure 3). The β-carotene and the *cis*-β-carotene were also significantly reduced in CTV-tALAD plants (Figure 3). Lutein, isolutein, and α-carotene were not significantly affected in CTV-tALAD plants (Figure 3). The TLC also showed that violaxanthin and carotenes were reduced in CTV-tALAD plants (Supporting Information Figure S1b).

To investigate the effect of silencing of the *ALAD* gene on the other genes implicated in the biosynthesis pathway of leaf pigments, the transcription levels of 46 genes involved in biosynthetic pathways of chlorophylls (17 genes) and carotenoids (29 genes) were investigated in *C. macrophylla*, CTV-wt, and CTV-tALAD plants (Figure 4). Gene expression data were normalized using four reference genes (*EF1*, *F-box*, *GAPDH*, and *SAND*). These genes showed high stability and were used for transcript normalization in different citrus organs under biotic stress (Mafra et al., 2012; Wei et al., 2014a,b). The normalizing expression levels using the four reference genes were very similar to each other.

The presented heatmap and the hierarchical clustering dendrogram (HCD) between chlorophyll biosynthetic genes in Figure 4a show that all chlorophyll biosynthetic genes were separated into two major clusters. The first cluster (C1; about 0.55 distance) includes only two genes, glutamyl-tRNA reductase 1 (*CitGluTR*) and glutamate-1-semialdehyde 2,1-aminomutase (*CitGSAT*), which were not affected in CTV-tALAD plants compared to the two controls (*C. macrophylla* and *C. macrophylla* with CTV-wt). The second cluster (C2; about 0.60 distance) contains 15 genes, which were downregulated in CTV-tALAD plants compared to the controls. These include δ-aminolevulinic acid dehydratase 1 (*CitALAD*), chlorophyllide *a* oxygenase (*CitCAO*), protochlorophyllide reductase (*CitPOR*), chlorophyllase (*CitChlases*, also known as *CLHs*), chlorophyll(ide) *b* reductase (*CitCBRs*), including chlorophyll(ide) *b* reductase -NON-YELLOW COLORING 1 (*CitNYC1*) and chlorophyll(ide) *b* reductase -NYC1-Like (*CitNOL*), protoporphyrinogen oxidase (*CitPPO*), chlorophyll synthase (*CitChlG*), and divinyl chlorophyllide a 8-vinyl-reductase (*CitDVR*) (Figure 4a). Furthermore, HCD among treatments indicated that while CTV-wt plants were very similar to *C. macrophylla* plants (less than 0.1 distance), CTV-tALAD plants clustered separately (about 1.65 distance) (Figure 4a).

In addition, silencing of the *ALAD* gene altered the expression of several genes implicated in carotenoids biosynthesis pathways. Of the twenty-nine selected genes, ten genes were downregulated, whereas only four genes were upregulated compared to the controls (Figure 4b). Likewise, the HCA revealed that all carotenoid biosynthetic genes were clustered in three main clusters. The first cluster (C1; about 2.45 distance) included 15 unaffected genes, including phytoene synthase (*CitPYSs*), phytoene desaturase (*CitPDSs*), lycopene ε-cyclase (*CitLCYes*), lycopene β-cyclase (*CitLCYb*), and ζ-carotene desaturase (*CitZDSs*) (Figure 4b). The second cluster (C2; about 2.25 distance) included ten downregulated genes implicated in the ABA-biosynthetic pathway, including zeaxanthin epoxidase (*CitZEPs*), short-chain alcohol dehydrogenase (*CitABA2*), violaxanthin de-epoxidase (*CitVDE*), putative 9-*cis*-epoxycarotenoid dioxygenase 3 (*CitNCED*), capsanthin/capsorubin synthase (*CitCCS*), abscisic aldehyde oxidase (*CitAAO3*), and neoxanthin synthase (*CitNSY*) (Figure 4b). The third cluster included four isoforms of carotenoid hydroxylase β-ring (*CitCHYbs*; converts β-carotene to β-cryptoxanthin and then zeaxanthin), which were upregulated in CTV-tALAD plants compared to the controls (Figure 4b). Furthermore, the HCA among treatments in Figure 4b shows that CTV-wt plants were closer to *C. macrophylla* (less than 0.3 distance) than to CTV-tALAD (about 2.4 distance).

As no differences were found in the concentration of studied pigments and their biosynthetic genes between the *C. macrophylla* and *C. macrophylla* infected with CTV-wt, we chose to use CTV-wt as the control in all subsequent analyses. The 3D surface plots were obtained using the whole data matrix to understand the relationships between chlorophyll, xanthophyll, and carotene groups in CTV-wt and CTV-tALAD plant leaves (Figure 5a–f). Overall, the silencing of the *ALAD* gene affected the total net profiles (TNPs) of other pigments, causing a complex relationship among different pigments groups. The effect of xanthophyll and chlorophyll (as two input parameters) on carotene content (as an associated performance metric) is presented in Figure 5a,b. The carotenes TNP of CTV-tALAD was totally different compared to CTV-wt plants. While the carotenes TNP of CTV-wt plants was flatter and had a clear peak in high chlorophylls and high xanthophylls (Figure 5a), the carotenes TNP of CTV-tALAD plants had more plateaus and curvatures without any clear peaks (Figure 5a). Furthermore, the effect of carotene and chlorophyll content on xanthophyll content is presented in Figure 5c,d. The xanthophylls TNP of CTV-tALAD plants was very different compared to CTV-wt, which was flatter with only one small peak under the high-carotene and high-chlorophyll conditions (Figure 5c). The xanthophylls TNP of CTV-tALAD plants had a clear curvature at relatively low levels of carotenes and moderate levels of chlorophylls with a very sharp peak at relatively low-carotene and moderate-chlorophyll conditions (Figure 5d). In addition, the effect of xanthophylls and carotenes on chlorophyll content presented in Figure 5e,f. As observed, the TNP of chlorophyll content in CTV-tALAD plants appeared very different from that of CTV-wt plants, which was flatter with a small plateau (Figure 5e), while the chlorophylls TNP of CTV-tALAD plants had two clear curvatures: the first one at relatively low levels of carotenes and moderate levels of xanthophylls, and the second one at low levels of carotenes and high levels of xanthophylls. In addition, it had a clear peak at high-carotene and high-xanthophyll conditions (Figure 5f).

3.3 | CTV-tALAD plants accumulated δ-ALA in developing leaves

The colorimetric assay showed that δ-ALA accumulated in the developing leaves from CTV-tALAD (6.6 ± 0.5 mg/g FW) plants, whereas

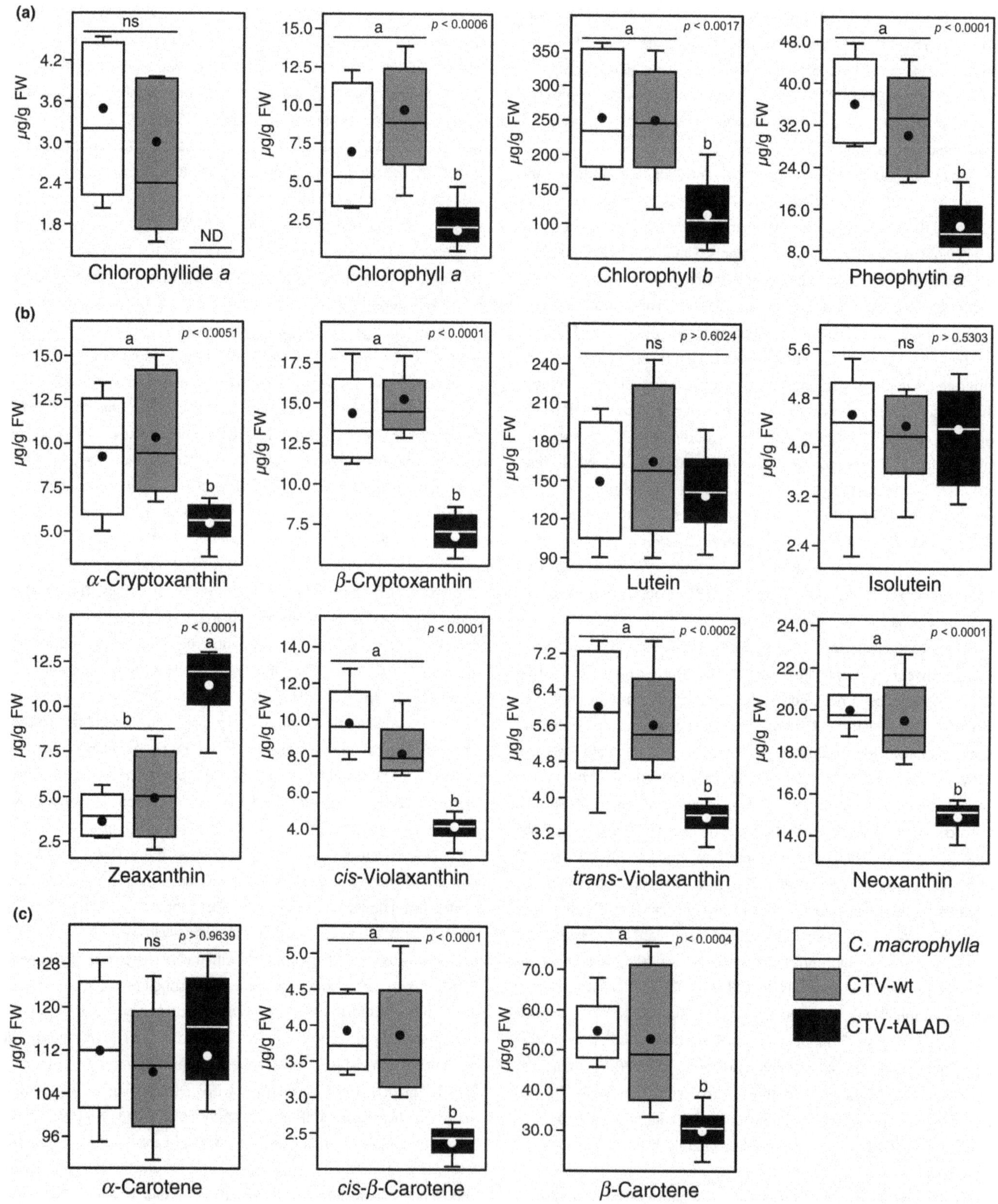

FIGURE 3 Concentrations (μg/g FW) of leaf pigments from control, CTV-wt, and CTV-tALAD *Citrus macrophylla* plants using HPLC (n = 10). Horizontal thick lines indicate the medians, black dots indicate the means, boxes show the interquartile ranges including 25%–75% of the values, whiskers reflect the highest and the lowest value of data. Different letters indicate statistically significant differences among treatments ($p < 0.05$), while "ns" indicates no significant differences among treatments using a Tukey–Kramer honestly significant difference test (Tukey's HSD)

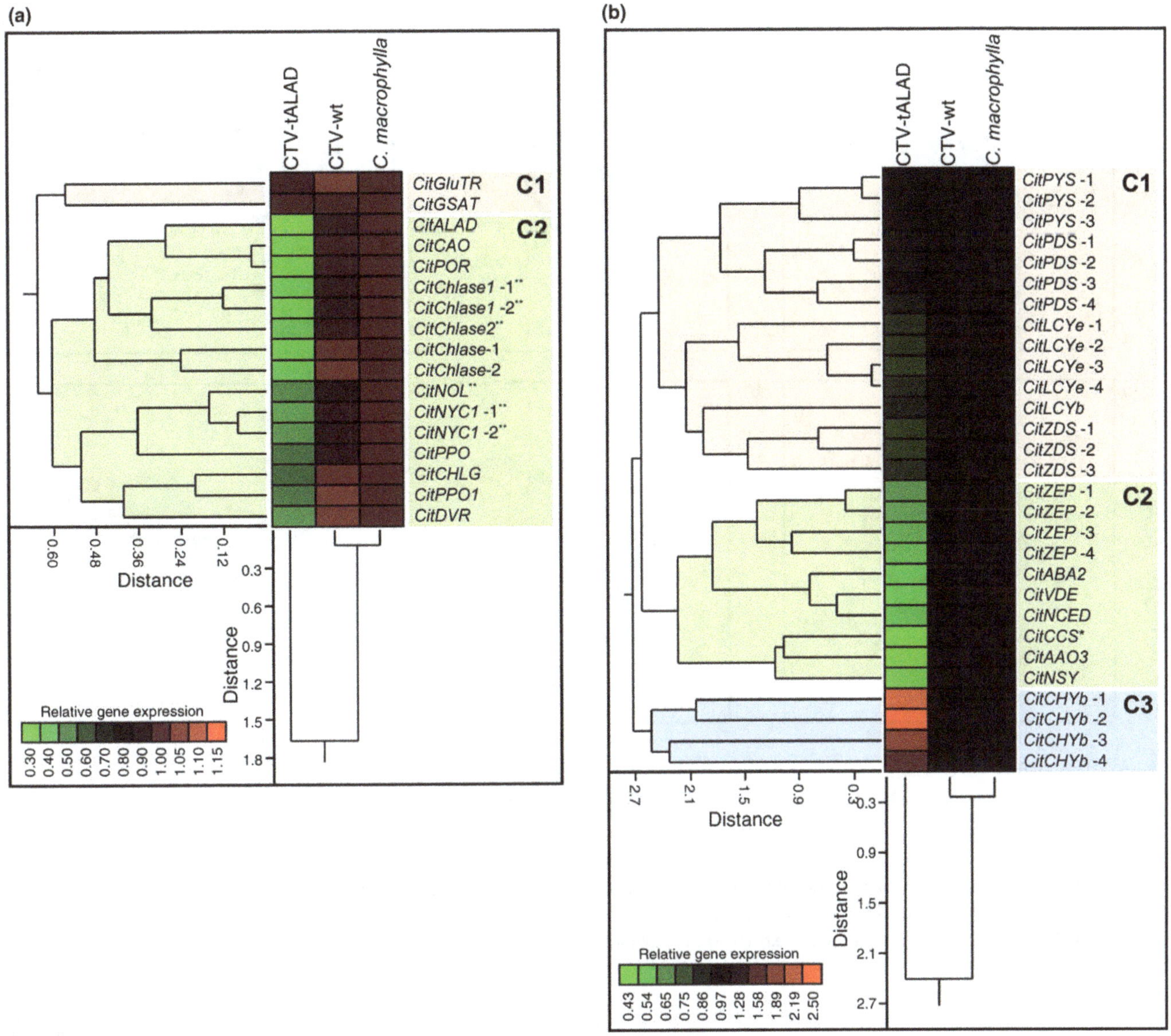

FIGURE 4 Differential biosynthetic gene expression patterns of chlorophylls (a) and carotenoids (b) and detected in leaves of CTV-tALAD, CTV-wt, and *Citrus macrophylla*. Rows represent the genes while columns represent the treatments (n = 30). Lower expressions levels are colored green, and higher expressions are colored red. Treatments and genes were organized using two-way HCA based on similarities in autoscaled values and correlations, respectively. The full lists of expressed genes, accession numbers, and primers are available in our previous study (Killiny & Nehela, 2017a). See text for full names

no δ-ALA was detected in the CTV-wt (Figure 6). No δ-ALA accumulation was detected in mature leaves from CTV-tALAD or CTV-wt plants. In addition, the amount of δ-ALA detected in CTV-tALAD plants after incubation with levulinic acid was higher (138 ± 51 mg/g FW) than that (11 ± 5 mg/g FW) detected in the CTV-wt after 6-hr incubation with levulinic acid (Figure 6). The spectra of the pink color developed in the leaf extracts from CTV-tALAD plants (without incubation with levulinic acid) were similar to those obtained using δ-ALA standard and leaf extracts from leaves incubated with levulinic acid (Supporting Information Figure S2a). The GC-MS results also showed that CTV-tALAD plants accumulate more δ-ALA when incubated with levulinic acid than the CTV-wt (Supporting Information Figure S2b). The mass spectra of the δ-ALA detected in CTV-tALAD plants and in the controls after incubation in levulinic acid were similar to those obtained with authentic standard (Supporting Information Figure S2c).

3.4 | Silencing of *ALAD* decreased the level of photosynthates and affected citrus leaf metabolites and phytohormones

The starch level was significantly reduced (about fivefold) in CTV-tALAD plants compared to the controls (Figure 7a). The sucrose level was also reduced (about threefold) by CTV-tALAD infection (Figure 7b). No effects were observed on the level of sucrose or starch in CTV-wt.

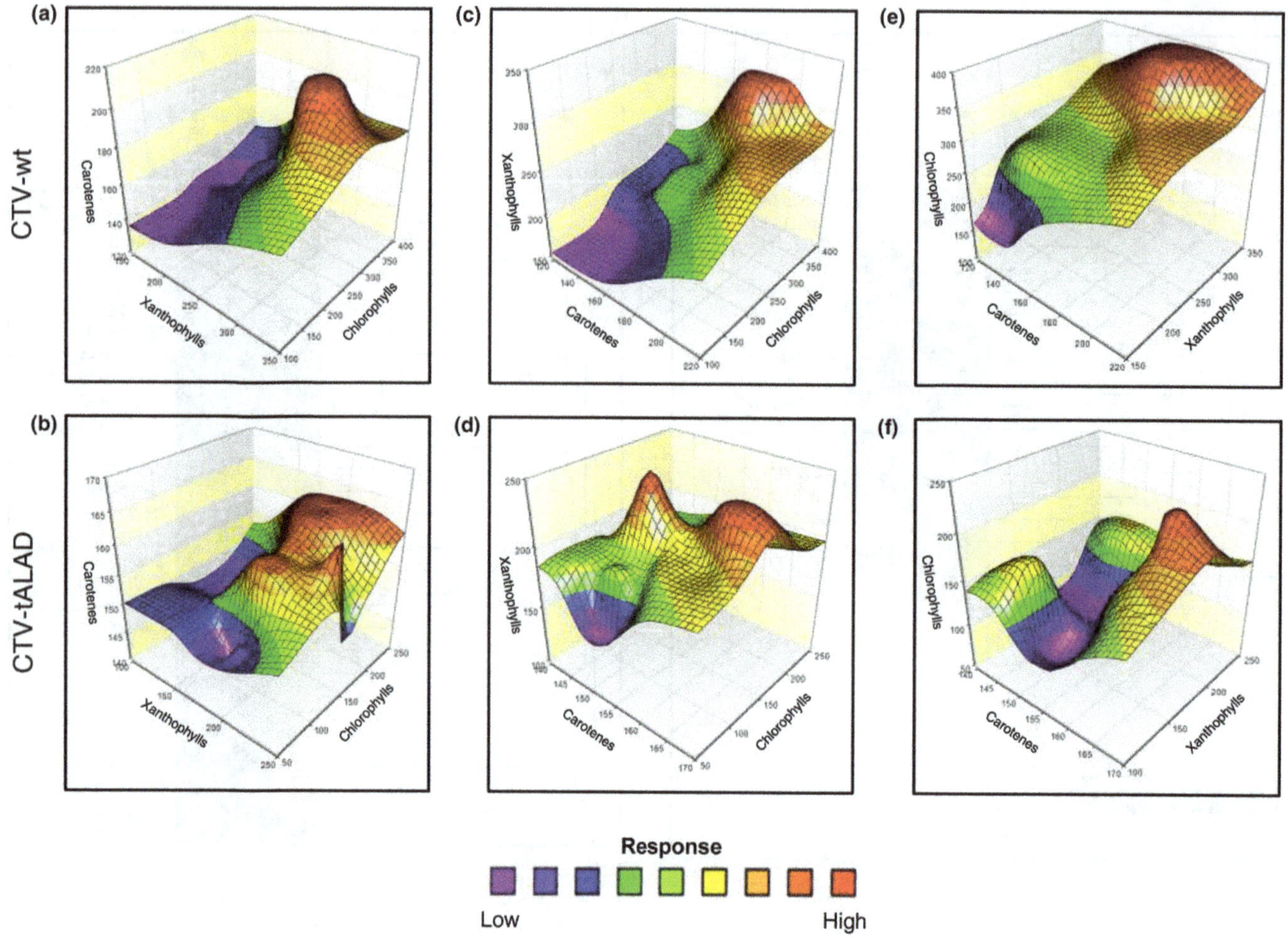

FIGURE 5 Three-dimensional surface plots of leaf pigments groups detected using HPLC in leaves from CTV-wt and CTV-tALAD plants. (a,b) The reciprocal interactions of chlorophylls and xanthophylls on carotene content. (c,d) The reciprocal interactions of carotenes and chlorophylls on xanthophyll content in different treatments. (e,f) The reciprocal interaction of carotenes and xanthophylls on chlorophyll content

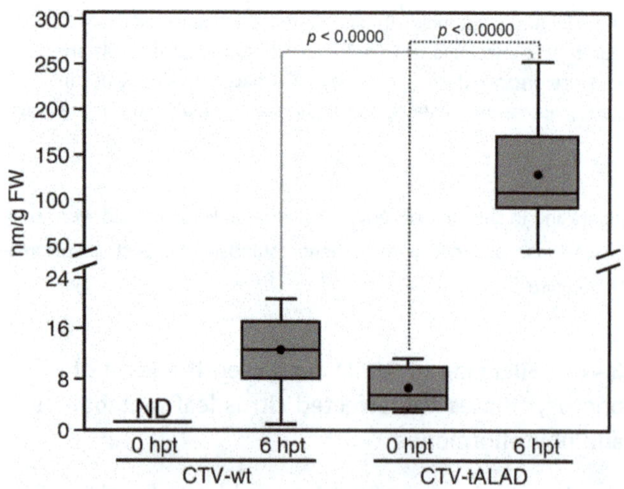

FIGURE 6 Colorimetric determination of δ-ALA. Level of δ-ALA in developing young leaves from CTV-wt and CTV-tALAD plants (n = 5) after 0- and 6-hr incubation with 100 mM levulinic acid. p-value <0.05 indicates statistically significant differences among treatments using two-tailed t test

A few amino acids including L-alanine, *gamma*-aminobutyric acid (GABA), L-threonine, L-glutamic, and L-phenylalanine significantly decreased in leaves of CTV-tALAD plants, whereas L-asparagine increased (Table 1). None of the detected fatty acids was affected in the leaves of CTV-tALAD plants, except linoleic acid, which was significantly increased (Table 1). Fumaric and succinic acid significantly increased in leaves of CTV-tALAD plants (Table 1). In a similar manner, GABA, L-threonine, and L-phenylalanine decreased significantly in the apical meristems of CTV-tALAD plants, whereas L-asparagine increased significantly (Table 1). Linoleic and linolenic acid significantly increased in meristems of CTV-tALAD plants, whereas oleic acid decreased significantly. Fumaric, succinic, and malic also increased significantly in the apical meristems of CTV-tALAD plants (Table 1).

Levels of phytohormones and their precursors were affected in leaves of CT-tALAD plants. BA, *t*CA, IBA, and ABA decreased in leaves from CTV-tALAD plants, whereas SA, JA, *t*JA, IAA, and IPA were not affected (Figure 8a–d). The levels of BA, *t*CA, IBA, and ABA were also reduced in the apical meristems of CTV-tALAD plants, whereas SA, *t*JA, and IPA increased significantly (Figure 8a–d).

The emission of volatile organic compounds was mostly unaffected in juvenile leaves of CTV-tALAD plants. Likewise,

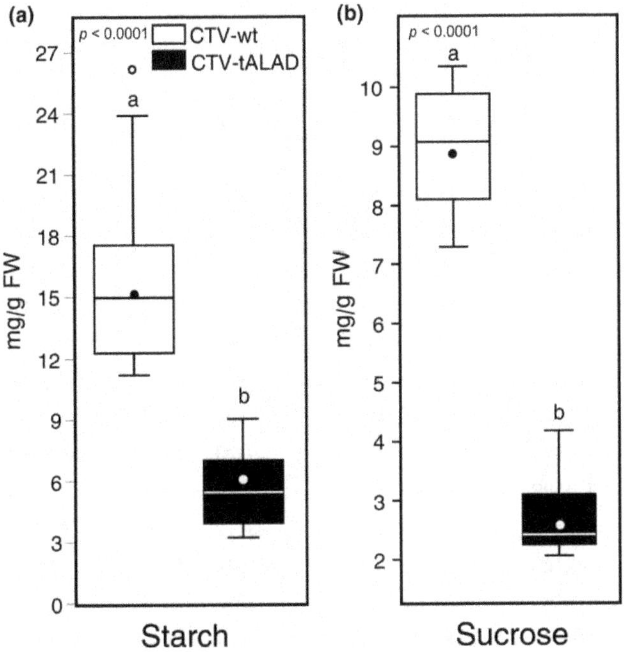

FIGURE 7 Concentrations (mg/g FW) of starch (a) and sucrose (b) in leaves of CTV-wt and CTV-tALA, plants using HPLC (n = 10). Horizontal thick lines indicate the medians, black dots indicate the means, boxes show the interquartile ranges including 25%–75% of the values, and whiskers reflect the highest and the lowest value of data. p-value <0.05 indicates statistically significant differences among treatments using two-tailed t test

intermediate-aged leaves were relatively unaffected, except for (*E*)-*β*-ocimene, (*E*)-*α*-bergamotene, (*E*)-*β*-farnesene, and valencene which increased dramatically (Table 2). Most of the observed differences between the mature leaves of the CTV-wt and the CTV-tALAD plants were small but significant, including increases in *β*-myrcene, (*E*)-*β*-ocimene, allo-ocimene, and (*Z*)-*β*-caryophyllene (Table 2).

4 | DISCUSSION

δ-aminolevulinic acid dehydratase is a key enzyme in plants and is required for the synthesis of all classes of tetrapyrroles, which play important roles in many biological processes, including photosynthesis and respiration. In consequence, we hypothesized that inhibition of this enzyme would not only have an impact on chlorophyll biosynthesis and photosynthesis, but also affect other pigments and could lead to changes in both primary and secondary metabolites. Therefore, we carried out targeted and nontargeted metabolic analysis on CTV-tALAD plants. The level of *δ*-ALA was determined to confirm the successful silencing of the ALAD gene. We evaluated chlorophyll content as accumulation of *δ*-ALA was expected to reduce chlorophyll biosynthesis. In addition, we analyzed carotenoid content because carotenoids are functionally linked to chlorophyll in photosynthetic light harvesting. Reduction in chlorophyll biosynthesis was expected to decrease the rate of photosynthesis; thus, we determined the level of the main photosynthates (sucrose and starch). CTV-tALAD plants showed clear symptoms of necrosis, which indicated they were under stress; therefore, four different groups of phytohormones including the main growth regulators and stress-associated groups (auxins, salicylates, jasmonates, and ABA) were quantified. At last, we performed nontargeted metabolomic analysis on volatile and nonvolatile compounds to get more insights on the metabolic status of CTV-tALAD plants. Our results are discussed in the following paragraphs.

4.1 | Silencing of *ALAD* accumulated *δ*-ALA and reduced chlorophyll and photosynthates levels in citrus leaves

Silencing of *ALAD* led to the accumulation of *δ*-ALA and reduction in chlorophyll and photosynthates in citrus leaves by the CTV-based silencing vector. These results showed that ALAD is a critical enzyme in chlorophyll synthesis as it is required for the synthesis of all classes of tetrapyrroles (Tanaka & Tanaka, 2007). The *δ*-ALA that is formed in developing tissues does not accumulate and is directly converted to PBG by ALAD (Kumar & Söll, 2000). Beale and Castelfranco (1974) showed that the rate of *δ*-ALA accumulation in cucumber cotyledons treated with levulinic acid paralleled the rate of chlorophyll formation, and accumulation of *δ*-ALA ceased when cucumber cotyledons were returned to dark (Beale & Castelfranco, 1974). In agreement with the previous results, our results also showed that the level of *δ*-ALA in young leaves from CTV-tALAD was higher than that in mature leaves, which is in agreement with the high rate of chlorophyll synthesis in young leaves.

Inhibition of *δ*-ALA formation in mutant plants such as tobacco and Arabidopsis reduced chlorophyll and heme biosynthesis, but did not result in necrotic lesions (Tanaka & Tanaka, 2007). In agreement with these observations, our current study showed that silencing of *ALAD* also reduced chlorophyll *a*, chlorophyll *b*, chlorophyllide *a*, and pheophytin *a* synthesis and resulted in yellow islands. However, these changes were accompanied by necrosis in citrus plants. The gene expression results confirmed the reduced levels of pigments and showed that all selected genes in the chlorophyll biosynthesis pathway downstream of the silenced *ALAD* gene were downregulated. Our results indicated that accumulation of *δ*-ALA could be more toxic to the plants than inhibition of its formation.

The reduction in chlorophyll indicated that CTV-tALAD plants have a lower photosynthesis rate than control plants. In fact, our results showed that sucrose and starch levels (photosynthates) were reduced in CTV-tALAD plants. The reduction in chlorophylls reduces the amount of absorbed light and harvested energy (ATP and NADPH); consequently, carbon fixation is reduced and results in a decrease in photosynthates in leaves.

4.2 | Silencing of *ALAD* increased zeaxanthin level in citrus leaves

Interactions between chlorophylls and carotenoids play a major role in photosynthetic light harvesting. Besides absorbing light and harvesting energy, carotenoids, such as xanthophylls and carotenes,

TABLE 1 Concentrations (ng/g FW) of different amino acids, organic acids, and fatty acids compounds detected in CTV-wt and CTV-tALAD leaves and apical meristems using GC-MS (n = 5)

	Mature leaves		t test	Apical meristems		t test
	CTV-wt	CTV-tALA	p-value	CTV-wt	CTV-tALA	p-value
Amino acids						
Glycine	22.4 ± 4.9	21.4 ± 1.7	0.6819	43.2 ± 2.2	43.3 ± 2.3	0.9442
L-Alanine	612.5 ± 83.0	336.8 ± 85.2	0.0008	174.1 ± 70.6	165.6 ± 77.3	0.8605
L-Valine	91.6 ± 7.4	97.7 ± 1.6	0.1417	85.8 ± 1.4	89.7 ± 3.1	0.0441
L-Leucine	41.1 ± 0.9	42.6 ± 0.9	0.0303	37.3 ± 0.4	39.3 ± 0.8	0.0028
GABA	1288.7 ± 171.9	655.0 ± 172.2	0.0004	973.6 ± 57.6	799.0 ± 141.7	0.0486
L-Isoleucine	54.7 ± 7.4	50.1 ± 2.4	0.2524	43.2 ± 0.6	42.7 ± 1.5	0.5286
L-Threonine	142.6 ± 24.5	103.5 ± 12.2	0.0193	84.7 ± 10.2	67.8 ± 11.1	0.0372
L-Proline	6787.8 ± 1362.0	6237.5 ± 1322.9	0.5351	4531.5 ± 972.5	4974.8 ± 473.5	0.3960
L-Asparagine	3657.2 ± 1717.9	4164.1 ± 580.6	0.5599	947.6 ± 173.7	2260.2 ± 941.0	0.0343
L-Aspartic acid	659.8 ± 60.8	526.8 ± 31.7	0.0049	242.1 ± 21.1	317.5 ± 35.5	0.0054
L-Pyroglutamic acid	410.3 ± 85.2	400.7 ± 18.8	0.8166	14.3 ± 3.2	14.3 ± 10.8	0.9913
L-Serine	1588.4 ± 239.5	1692.8 ± 201.7	0.4780	869.8 ± 151.7	717.3 ± 252.5	0.2874
L-Glutamic acid	2115.4 ± 345.7	1417.7 ± 160.3	0.0073	469.3 ± 263.0	462.8 ± 194.5	0.9659
L-Phenylalanine	164.1 ± 20.2	113.6 ± 14.2	0.0024	151.9 ± 14.8	75.3 ± 27.5	0.0014
L-Lysine	216.1 ± 9.9	214.1 ± 9.2	0.7580	208.5 ± 2.0	209.2 ± 2.8	0.6629
L-Tyrosine	157.2 ± 72.4	122.9 ± 0.5	0.3495	120.8 ± 0.7	121.9 ± 3.0	0.4661
L-Tryptophan	49.5 ± 1.1	48.2 ± 0.2	0.0452	57.2 ± 2.2	48.9 ± 2.4	0.0004
Total amino acids	18059.2 ± 3142.7	16245.3 ± 1958.6	0.3112	9054.9 ± 840.9	10449.8 ± 2033.5	0.2121
Fatty acids						
Palmitic acid (C16:0)	1172.6 ± 62.3	1235.6 ± 112.0	0.3121	1009.6 ± 99.0	1254.3 ± 256.6	0.1015
α-Linolenic acid (C18:3)	10653.0 ± 2161.0	8767.2 ± 831.7	0.1264	5342.0 ± 331.6	6535.1 ± 588.8	0.0068
Linoleic acid (C18:2)	1780.6 ± 238.0	2276.9 ± 200.8	0.0077	684.6 ± 68.4	1647.2 ± 208.8	0.0002
Oleic acid (C18:1)	125.4 ± 55.2	88.3 ± 38.1	0.2550	121.6 ± 18.3	52.6 ± 28.7	0.0029
Stearic acid (C18:0)	279.9 ± 68.6	342.3 ± 50.3	0.1430	227.8 ± 17.9	368.0 ± 187.1	0.1694
Total fatty acids	14011.6 ± 2459.0	12710.3 ± 1079.1	0.3239	7385.6 ± 390.5	9857.1 ± 888.0	0.0017
Organic acids						
Fumaric acid	918.8 ± 30.6	981.2 ± 25.8	0.0086	817.7 ± 17.2	913.8 ± 39.7	0.0033
Succinic acid	506.1 ± 77.0	659.4 ± 61.3	0.0089	405.5 ± 10.0	628.0 ± 75.8	0.0025
Malic acid	2817.3 ± 319.0	3117.9 ± 283.7	0.1544	1574.9 ± 168.7	2202.5 ± 228.9	0.0015
Quinic acid	1548.1 ± 101.2	1497.5 ± 292.0	0.7296	1639.6 ± 322.6	1488.9 ± 745.3	0.6939
Citric acid	1846.7 ± 42.7	1851.8 ± 189.8	0.9554	1759.1 ± 52.3	1724.9 ± 60.3	0.3668
Ferulic acid	390.5 ± 189.7	503.5 ± 84.8	0.2732	747.6 ± 225.7	149.6 ± 30.0	0.0038
Total organic acids	8027.4 ± 376.9	8611.4 ± 678.2	0.1413	6944.4 ± 744.2	7107.7 ± 960.1	0.7718

Note. Means were considered significantly different if p-value was less than 0.05.

protect plants from light damage. In fact, absence of carotenoids in plants causes severe photo-oxidation and leads to plant death (Bartley & Scolnik, 1995). Carotenoids reduce photo-oxidation by quenching of chlorophyll triplets to prevent the formation of singlet oxygen that can oxidize lipids, proteins, and other components in plant cells (Bartley & Scolnik, 1995). The xanthophyll cycle plays an important role in photoprotection in plants (Demmig-Adams, 1990). When the sunlight level exceeds the maximum that can be used by chlorophylls, plants can increase their zeaxanthin levels for photoprotection which in turn reduces chlorophyll fluorescence (Demmig-Adams, 1990).

As the levels of chlorophylls were reduced in leaves from CTV-tALAD plants, the amount of light that can be absorbed is expected to decrease. To prevent light damage, the CTV-tALAD plants increased the level of zeaxanthin. These results were in agreement with the increase in the gene expression of the *CitLCHYB* genes, which are implicated in zeaxanthin synthesis. An increase in zeaxanthin and antheraxanthin and a decrease in violaxanthin were reported in *A. thaliana* after treatment with norflurazon, another specific competitive inhibitor of ALAD (Jung, 2004). The increase in the zeaxanthin was accompanied by an increase in

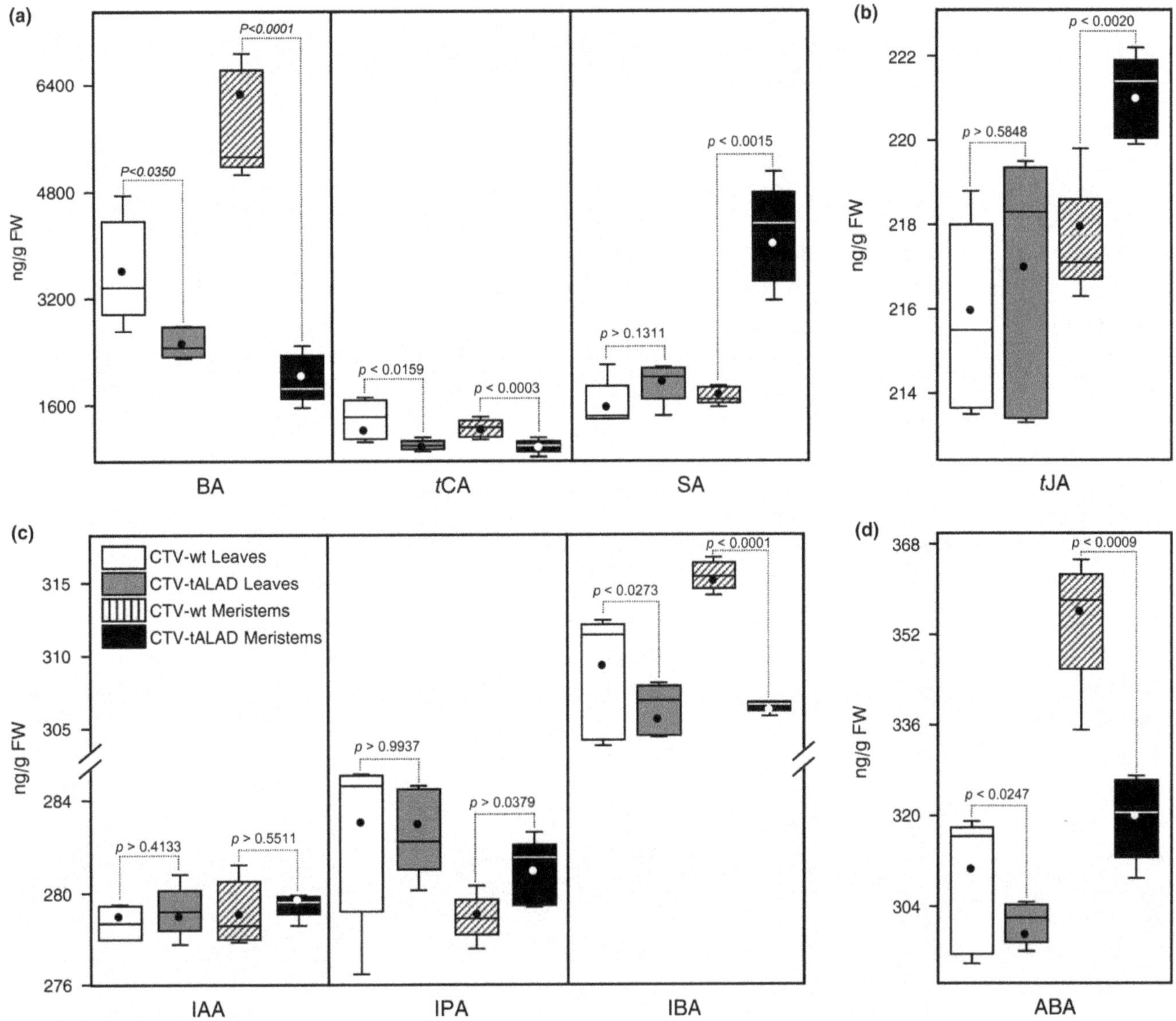

FIGURE 8 Concentrations of different phytohormones and BA, *t*CA in leaves and apical meristems of CTV-wt and CTV-tALA plants a using GC-MS-SIM (n = 10). (a) BA, *t*CA, and SA. (b) *t*JA, (c) auxins. (d) ABA. Horizontal thick lines indicate the medians, black/white dots indicate the means, boxes show the interquartile ranges including 25%–75% of the values, and whiskers reflect the highest and the lowest number of data. p-value <0.05 indicates statistically significant differences among treatments using two-tailed t test. ABA: abscisic acid; BA: benzoic acid; IAA: indole-3-acetic acid; IBA: indole-3-butyric acid; IPA: indole-3-propionic acid; SA: salicylic acid; *t*CA: *trans*-cinnamic acid; *t*JA: *trans*-jasmonic acid

nonphotochemical chlorophyll fluorescence quenching in norflurazon-treated plants (Jung, 2004). In recent times, we also found that the reduction in chlorophyll levels in leaves from *C*Las-infected sweet orange plants was also accompanied by an increase in zeaxanthin (Killiny & Nehela, 2017a). The increase in zeaxanthin indicated that reduction in chlorophyll as a result of silencing of *ALAD* induced light stress in CTV-tALAD plants.

4.3 | The phytohormonal profile indicated that CTV-tALAD plants were under stress

Accumulation of δ-ALA resulted in necrosis in CTV-tALAD, indicating that these plants were under stress. In consequence, we investigated the level of the main phytohormones, which are associated with plant response to biotic and abiotic stresses. Our results showed that several phytohormones were affected by silencing of the *ALAD* gene. SA increased in the apical meristems of CTV-tALAD plants. The *t*JA also increased in the apical meristems of CTV-tALAD plants, and this increase was accompanied by an increase in its precursor, linolenic acid. An increase in *t*JA was observed in citrus plant after *Diaphorina citri* attack and this increase was also accompanied by an increase in linolenic acid and an upregulation of allene oxide synthase (AOS, a JA-biosynthetic enzyme) (Nehela et al., 2018). SA and *t*JA are stress-associated phytohormones in plants (Bari & Jones, 2009; (Nehela et al., 2018)). The SA pathway is associated with plant responses to biotrophic and hemibiotrophic pathogens and plays an

TABLE 2 The relative amounts of volatiles collected from leaves of CTV-tALA plants compared to the CTV-wt (T-36), (n = 5)

		Young leaves		Intermediate leaves		Mature leaves	
Compound	RT	ALAD/T36	t test	ALAD/T36	t test	ALAD/T36	t test
α-Pinene[a]	3.9	0.6	0.4451	1.3	0.5812	0.5	0.0974
β-Myrcene[a]	4.5	0.6	0.4034	1.4	0.4190	0.5	0.0268
β-Phellandrene[a]	5.1	0.4	0.2111	0.3	0.2879	0.3	0.1136
Limonene[a]	5.2	0.6	0.1294	0.7	0.3473	0.8	0.2355
(E)-β-Ocimene[a]	5.3	14.4	0.0535	24.2	0.0002	11.0	0.0213
γ-Terpinene[a]	5.4	0.7	0.4730	1.0	0.9415	0.5	0.1025
α-Terpinolene[a]	5.8	0.5	0.3827	1.2	0.5650	0.3	0.0608
Linalool[a]	6.0	1.5	0.5709	0.7	0.6556	0.6	0.2731
Perillene	6.2	0.5	0.3189	2.3	0.3729	1.2	0.8031
allo-Ocimene	6.4	0.1	0.0831	0.1	0.0439	0.0	0.0105
Limonene oxide	6.6	6.5	0.2807	0.4	0.5352	5.3	0.3174
Citronellal[a]	6.7	0.9	0.8148	0.3	0.2484	0.6	0.4206
Methyl salicylate[a]	7.4	1.1	0.8225	0.4	0.4109	0.5	0.3032
Decanal[a]	7.5	0.7	0.3810	0.3	0.1617	0.7	0.4465
Nerol[a]	7.6	0.4	0.2950	0.6	0.5175	0.3	0.1433
Neral	7.7	4.4	0.2855	2.2	0.2510	4.3	0.1757
Geraniol[a]	7.9	1.3	0.6924	0.5	0.5468	0.7	0.4871
Geranial[a]	8.3	0.8	0.6808	0.4	0.4089	0.7	0.4392
δ-Elemene[a]	9.3	5.5	0.0561	8.8	0.0508	7.1	0.1312
Citronellyl acetate	9.4	0.0	0.2659	0.3	0.3007	2.4	0.4045
α-Cubebene	9.5	10.3	0.1283	3.6	0.0543	10.9	0.1176
Geranyl acetate[a]	9.8	1.4	0.6686	0.6	0.6370	1.8	0.3424
β-Cubebene	9.9	0.3	0.1275	0.2	0.2761	0.2	0.0725
β-Selinene	10.0	1.2	0.8040	0.8	0.8353	0.9	0.7858
(Z)-β-Caryophyllene	10.4	0.1	0.0617	0.1	0.2308	0.1	0.0177
(E)-β-Caryophyllene	10.5	1.0	0.9743	1.0	0.9567	0.7	0.4982
(E)-α-Bergamotene	10.6	207.4	0.1002	104.9	0.0000	82.7	0.0616
(E)-β-Farnesene[a]	10.7	5.5	0.2186	8.7	0.0006	9.9	0.1174
α-Humulene[a]	11.0	1.2	0.8717	0.6	0.5649	3.7	0.4121
α-Selinene	11.3	1.1	0.8653	0.5	0.4726	0.4	0.1386
Valencene[a]	11.5	3.3	0.0386	11.2	0.0018	4.8	0.1414
β-Bisabolene	11.5	1.1	0.8789	1.2	0.7336	1.2	0.7178
Cadinene	11.7	3.4	0.3375	2.2	0.1478	2.1	0.3136
Elemol	12.0	2.4	0.1686	3.6	0.1237	3.3	0.2017
γ-Gurjunene	12.2	0.2	0.2141	4.5	0.1795	0.4	0.1831
Caryophyllene oxide	12.5	2.1	0.4266	2.6	0.0145	0.9	0.7639
Methyl jasmonate[a]	13.1	2.9	0.2930	1.2	0.7969	0.8	0.6841

Notes. The relative amounts were calculated by dividing the peak area of each compound in the CTV-tALAD plants on the peak area of the same compound in the CTV-wt (T-36).

[a]Compounds were confirmed by comparing their retention time and mass spectra with authentic standards. Means were considered significantly different if p-value was less than 0.05.

important role in systemic acquired resistance (SAR) (Bari & Jones, 2009). On the contrary, the *t*JA pathway is associated with plant responses to necrotrophic pathogens and insect herbivory (Bari & Jones, 2009). In addition, it is believed that both SA and *t*JA are implicated in plant responses to abiotic stresses (Ahmad et al., 2016; Horvàth, Szalai, & Janda, 2007). The increase in SA and *t*JA indicated that CTV-tALAD plants were under stress.

In agreement with our results, Chai et al. (2017) reported an increase in the level of ROS and H_2O_2 in *Gossypium hirsutum* (cotton) lesion mimic mutant (*Ghlmm*), which exhibited an accumulation

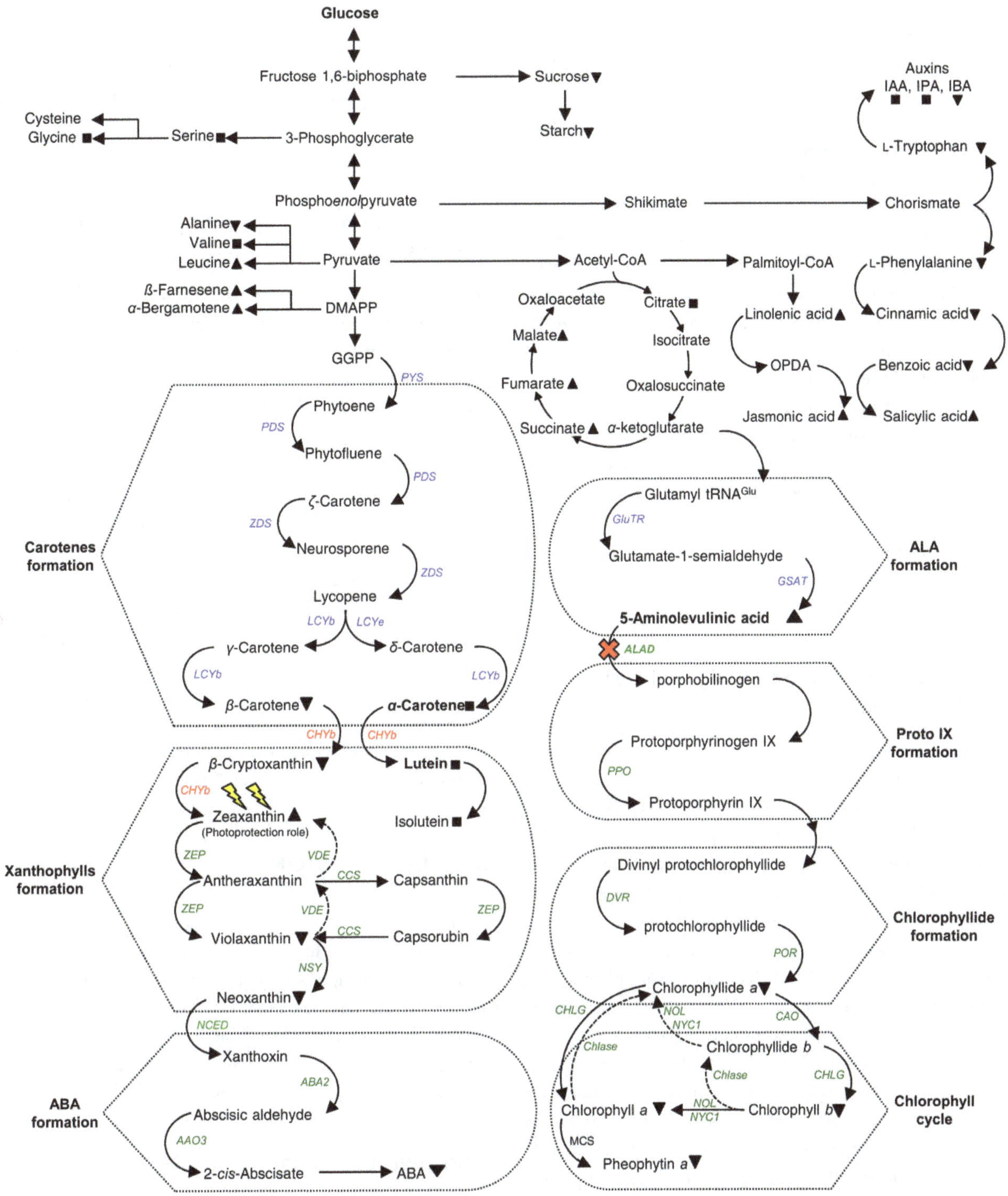

FIGURE 9 Schematic representation of proposed model for the effect of silencing of *ALAD* gene on *Citrus macrophylla* leaf pigments. Silencing of *ALAD* gene (red cross) results in accumulation of δ-ALA and consequently reduces the rate of chlorophylls biosynthesis, photosynthates levels, and induces light stress in CTV-tALAD plants. To prevent light damage, the CTV-tALAD plant increases the rate of zeaxanthin accumulation. The increase in rate of zeaxanthin synthesis decreases the levels of its precursors, upstream pigments, and ABA. Silencing of *ALAD* gene also affected other phytohormones (BA, SA, *t*JA, IBA) and their precursors. In addition, silencing of *ALAD* gene affected the tricarboxylic acid cycle and enhanced the emission of several leaf volatiles. The decrease in upstream pigments reduces ABA synthesis. The up-arrow (▲) indicates increasing, down-arrow (▼) indicates decreasing, and square sign (■) indicates no changes in compound levels. The dotted lines with arrows represent hypothetical mechanisms or uncharacterized elements. Genes in red were upregulated, genes in green were downregulated, whereas those in blue were not affected. ABA: abscisic acid; DMAPP: dimethylallyl diphosphate; GGPP: trans-geranyl-geranyl diphosphate; IAA: indole-3-acetic acid; IBA: indole-3-butaric acid; IPA: indole-3-propionic acid; OPDA: 12-oxo-cis-10,15-phytodienoate

of δ-ALA as a result of low activity of the ALAD enzyme. An increase in methanedicarboxylic aldehyde (MDA) level was also observed in mutant leaves, indicating peroxidation of unsaturated membrane lipids in cells. Silencing of *ALAD* in *G. hirsutum* by the TRV-based VIGS also resulted in the same phenotype observed in mutant plants and showed an accumulation of δ-ALA and H_2O_2 (Chai et al., 2017). The mutant plant also accumulated SA and showed high expression levels of pathogenesis-related genes (PR) and high resistance to *Verticillium dahliae* infection (Chai et al., 2017). In addition, modulation of ALAD activity was found to control programmed cell death (PCD) and PR expression (Chai et al., 2017). Our results, together with the previous results, showed that silencing of *ALAD* induced stress in plants.

The level of ABA decreased in leaves and apical meristems of CTV-tALAD, and this decrease was consistent with the decrease in its precursor (violaxanthin). In agreement with the HPLC results, the gene expression results showed that ten genes implicated in the ABA-biosynthetic pathway were downregulated in CTV-tALAD plants. Previous studies showed that ABA deficiency reduces plant development and elongation (Nitsch et al., 2012). The decrease in ABA indicated that silencing of *ALAD* could affect plant development.

4.4 | Silencing of *ALAD* altered the metabolite profile of citrus leaves

We performed a nontargeted metabolomic analysis on the volatile and nonvolatile profiles of CTV-tALAD plants to gather more information about the effects of silencing of *ALAD* gene in plants. Our results showed that fumaric, succinic, and malic acids were increased in CTV-tALAD plants. The increase in these metabolites could result from the increase in respiration rate, which can be increased to provide more energy during plant responses to stresses (Adi et al., 2012). Citric acid and fumaric acid also increased in CLas-infected citrus plants (Albrecht, Fiehn, & Bowman, 2016). The accumulation of citric acid and fumaric acid was explained as an attempt to enhance nutrient uptake in response CLas-infection (Albrecht et al., 2016).

The level of L-asparagine increased in CTV-tALAD plants compared with CTV-wt plants. L-Asparagine is synthesized in plants by amidation of aspartate using either ammonium or glutamine as ammonium donor (Gaufichon, Reisdorf-Cren, Rothstein, Chardon, & Suzuki, 2010). L-Asparagine is implicated in nitrogen transport, metabolism, recycling and its flow in response to biotic and abiotic stresses (Gaufichon et al., 2010). Light, nitrogen and carbon availability control the three-asparagine synthetase genes in sunflower (*Helianthus annuus*), *HAS1*, *HAS1.1*, and *HAS2* (Herrera-Rodríguez, Maldonado, & Pérez-Vicente, 2004). Light and ample carbon supply induce *HAS2* to supply asparagine to support growth. On the other hand, darkness and a low carbon/nitrogen ratio activate *HAS1* and *HAS1.1* genes to convert excess nitrogen to L-asparagine. Ammonium activates all three genes to enhance its detoxification. The increase in L-asparagine level in CTV-tALAD plants could result from the increase in ammonium, which occurs upon abiotic stresses (Herrera-Rodríguez, Pérez-Vicente, & Maldonado, 2007). It is also possible that the decrease in sucrose and starch decreased the carbon/nitrogen ratio and activated *HAS1* and *HAS1.1* genes to produce more L-asparagine using glutamine as an ammonium donor. The increase in L-asparagine is another indication that accumulation of δ-ALA induces stress in plants.

4.5 | Silencing of *ALAD* enhanced emission of leaf volatiles

Previous studies showed that plant volatiles are triggered by both biotic and abiotic stress (Arimura, Matsui, & Takabayashi, 2009). In agreement with the previous findings, our results showed that the emission of (*E*)-β-ocimene, (*E*)-α-bergamotene and (*E*)-β-farnesene was highly increased in the CTV-tALAD plants compared to the CTV-wt plants. The increased emission of these volatiles indicated that silencing of *ALAD* induced stress in CTV-tALAD. (*E*)-β-farnesene and (*E*)-α-bergamotene emission were increased in *Pinus taeda* upon exposure to high light and temperature (Helmig, Ortega, Guenther, Herrick, & Geron, 2006). In addition, the emission of these two volatiles was also increased in *Z. mays* and *Nicotiana attenuata* upon wounding, herbivory, or application of methyl jasmonate (Gaquerel, Weinhold, & Baldwin, 2009). Terpenes can relieve abiotic stress either by enhancing membrane stability during transient heat stress or by quenching reactive oxygen species (Palmer-Young et al., 2015). Accumulation of ROS upon silencing of *ALAD* (Chai et al., 2017) could be the direct cause of the observed increase in volatile emission (Vickers, Gershenzon, Lerdau, & Loreto, 2009).

5 | CONCLUSION

Silencing of *ALAD* in citrus plants using CTV-tALAD vector resulted in accumulation of the toxic tetrapyrrole intermediate (δ-ALA), decreased chlorophylls and photosynthate levels, altered the levels of carotenoids, phytohormones, volatiles, and many other metabolites and produced yellow islands and necrosis in citrus leaves. These effects are summarized in Figure 9. These changes showed that silencing of *ALAD* gene induces stress in plants.

ACKNOWLEDGMENTS

The authors acknowledge our CREC colleagues for the helpful discussion. We thank Shelley E. Jones for the technical assistance and Laina Lindsey and Floyd Butz for maintaining the trees in the greenhouse. This project is funded by the grant no. 201500955-04 from SCRI-NIFA, USDA.

AUTHOR CONTRIBUTIONS

N.K. and S.G. planned and designed the research. N.K., F.H., Y.N., S.H., and S.G. performed experiments and analyzed data. N.K., F.H., Y.N., S.H., and S.G. wrote the manuscript.

REFERENCES

Adi, M., Jens, P., Brotman, Y., Mikhail, K., Iris, S., Henryk, C., & Rena, G. (2012). Stress Responses to *Tomato Yellow Leaf Curl Virus* (TYLCV) infection of resistant and susceptible tomato plants are different. *Metabolomics*, *S1*(006), 1–13.

Agüero, J., del Carmen Vives, M., Velázquez, K., Pina, J. A., Navarro, L., Moreno, P., & Guerri, J. (2014). Effectiveness of gene silencing induced by viral vectors based on *Citrus leaf blotch virus* is different in *Nicotiana benthamiana* and citrus plants. *Virology*, *460–461*, 154–164. https://doi.org/10.1016/j.virol.2014.04.017

Ahmad, P., Rasool, S., Gul, A., Sheikh, S. A., Akram, N. A., Ashraf, M., … Gucel, S. (2016). Jasmonates: Multifunctional roles in stress tolerance. *Frontiers in Plant Science*, *7*, 813.

Alawady, A. E., & Grimm, B. (2005). Tobacco Mg protoporphyrin IX methyltransferase is involved in inverse activation of Mg porphyrin and protoheme synthesis. *The Plant Journal*, *41*, 282–290.

Albiach-Marti, M. R., Grosser, J. W., Gowda, S., Mawassi, M., Satyanarayana, T., Garnsey, S. M., & Dawson, W. O. (2004). *Citrus tristeza virus* replicates and forms infectious virions in protoplasts of resistant citrus relatives. *Molecular Breeding*, *14*, 117–128. https://doi.org/10.1023/B:MOLB.0000038000.51218.a7

Albrecht, U., Fiehn, O., & Bowman, K. D. (2016). Metabolic variations in different citrus rootstock cultivars associated with different responses to Huanglongbing. *Plant Physiology and Biochemistry*, *107*, 33–44. https://doi.org/10.1016/j.plaphy.2016.05.030

Arimura, G., Matsui, K., & Takabayashi, J. (2009). Chemical and molecular ecology of herbivore-induced plant volatiles: Proximate factors and their ultimate functions. *Plant and Cell Physiology*, *50*, 911–923. https://doi.org/10.1093/pcp/pcp030

Bari, R., & Jones, J. D. G. (2009). Role of plant hormones in plant defence responses. *Plant Molecular Biology*, *69*, 473–488. https://doi.org/10.1007/s11103-008-9435-0

Bar-Joseph, M., Marcus, R., & Lee, R. F. (1989). The continuous challenge of *Citrus tristeza virus* control. *Annual Review of Phytopathology*, *27*, 291–316. https://doi.org/10.1146/annurev.py.27.090189.001451

Bartley, G. E., & Scolnik, P. A. (1995). Plant carotenoids: Pigments for photoprotection, visual attraction, and human health. *The Plant Cell*, *7*, 1027–1038. https://doi.org/10.1105/tpc.7.7.1027

Beale, S. I., & Castelfranco, P. A. (1974). The biosynthesis of δ-aminolevulinic acid in higher plants. *Plant Physiology*, *53*, 291–296. https://doi.org/10.1104/pp.53.2.291

Becker, A., & Lange, M. (2010). VIGS – Genomics goes functional. *Trends in Plant Science*, *15*, 1–4. https://doi.org/10.1016/j.tplants.2009.09.002

Brennan, M. J. W., & Cantrill, R. C. (1979). δ-Aminolaevulinic acid is a potent agonist for GABA autoreceptors. *Nature*, *280*, 514–515. https://doi.org/10.1038/280514a0

Burch-Smith, T. M., Anderson, J. C., Martin, G. B., & Dinesh-Kumar, S. P. (2004). Applications and advantages of virus-induced gene silencing for gene function studies in plants. *The Plant Journal*, *39*, 734–746. https://doi.org/10.1111/j.1365-313X.2004.02158.x

Chai, Q., Shang, X., Wu, S., Zhu, G., Cheng, C., Cai, C., … Guo, W. (2017). 5-Aminolevulinic acid dehydratase gene dosage affects programmed cell death and immunity. *Plant Physiology*, *175*, 511–528. https://doi.org/10.1104/pp.17.00816

Cimò, G., Bianco, R. L., Gonzalez, P., Bandaranayake, W., Etxeberria, E., & Syvertsen, J. P. (2013). Carbohydrate and nutritional responses to stem girdling and drought stress with respect to understanding symptoms of huanglongbing in citrus. *HortScience*, *48*, 920–928.

Dawson, W. O., & Folimonova, S. Y. (2013). Virus-based transient expression vectors for woody crops: A new frontier for vector design and use. *Annual Review of Phytopathology*, *51*, 321–337. https://doi.org/10.1146/annurev-phyto-082712-102329

Dawson, W. O., Robertson, C. J., Albiach-Martí, M. R., Bar-Joseph, M., & Garnsey, S. M. (2015). Mapping sequences involved in induction of decline by *Citrus tristeza virus* T36 on the sour orange rootstock. *Journal of Citrus Pathology*, *2*, 1–6.

Demmig-Adams, B. (1990). Carotenoids and photoprotection in plants: A role for the xanthophyll zeaxanthin. *Biochimica et Biophysica Acta (BBA)-Bioenergetics*, *1020*, 1–24. https://doi.org/10.1016/0005-2728(90)90088-L

El-Mohtar, C., & Dawson, W. O. (2014). Exploring the limits of vector construction based on *Citrus tristeza virus*. *Virology*, *448*, 274–283. https://doi.org/10.1016/j.virol.2013.10.017

Espineda, C. E., Linford, A. S., Devine, D., & Brusslan, J. A. (1999). The *AtCAO* gene, encoding chlorophyll a oxygenase, is required for chlorophyll b synthesis in *Arabidopsis thaliana*. *Proceedings of the National Academy of Sciences of the United States of America*, *96*, 10507–10511. https://doi.org/10.1073/pnas.96.18.10507

Folimonov, A. S., Folimonova, S. Y., Bar-Joseph, M., & Dawson, W. O. (2007). A stable RNA virus-based vector for citrus trees. *Virology*, *368*, 205–216. https://doi.org/10.1016/j.virol.2007.06.038

Folimonova, S. Y., Robertson, C. J., Garnsey, S. M., Gowda, S., & Dawson, W. O. (2009). Examination of the responses of different genotypes of citrus to huanglongbing (citrus greening) under different conditions. *Phytopathology*, *99*, 1346–1354. https://doi.org/10.1094/PHYTO-99-12-1346

Frick, G., Su, Q., Apel, K., & Armstrong, G. A. (2003). An *Arabidopsis porB porC* double mutant lacking light-dependent NADPH:protochlorophyllide oxidoreductases B and C is highly chlorophyll-deficient and developmentally arrested. *The Plant Journal*, *35*, 141–153. https://doi.org/10.1046/j.1365-313X.2003.01798.x

Gaquerel, E., Weinhold, A., & Baldwin, I. T. (2009). Molecular interactions between the specialist herbivore *Manduca sexta* (Lepidoptera, Sphigidae) and its natural host *Nicotiana attenuata*. VIII. An unbiased GCxGC-ToFMS analysis of the plant's elicited volatile emissions. *Plant Physiology*, *149*, 1408–1423. https://doi.org/10.1104/pp.108.130799

Gaufichon, L., Reisdorf-Cren, M., Rothstein, S. J., Chardon, F., & Suzuki, A. (2010). Biological functions of asparagine synthetase in plants. *Plant Science*, *179*, 141–153. https://doi.org/10.1016/j.plantsci.2010.04.010

Hajeri, S., Killiny, N., El-Mohtar, C., Dawson, W. O., & Gowda, S. (2014). *Citrus tristeza virus*-based RNAi in citrus plants induces gene silencing in *Diaphorina citri*, a phloem-sap sucking insect vector of citrus greening disease (Huanglongbing). *Journal of Biotechnology*, *176*, 42–49. https://doi.org/10.1016/j.jbiotec.2014.02.010

Harper, S. J., Cowell, S. J., Halbert, S. E., Brlansky, R. H., & Dawson, W. O. (2015). CTV status in Florida. *Citrus Industry*, 8–14.

Helmig, D., Ortega, J., Guenther, A., Herrick, J. D., & Geron, C. (2006). Sesquiterpene emissions from loblolly pine and their potential contribution to biogenic aerosol formation in the Southeastern US. *Atmospheric Environment*, *40*, 4150–4157. https://doi.org/10.1016/j.atmosenv.2006.02.035

Herrera-Rodríguez, M. B., Maldonado, J. M., & Pérez-Vicente, R. (2004). Light and metabolic regulation of *HAS1*, *HAS1.1* and *HAS2*, three asparagine synthetase genes in *Helianthus annuus*. *Plant Physiology and Biochemistry*, *42*, 511–518. https://doi.org/10.1016/j.plaphy.2004.05.001

Herrera-Rodríguez, M. B., Pérez-Vicente, R., & Maldonado, J. M. (2007). Expression of asparagine synthetase genes in sunflower (*Helianthus annuus*) under various environmental stresses. *Plant Physiology and Biochemistry*, *45*, 33–38. https://doi.org/10.1016/j.plaphy.2006.12.002

Hijaz, F., & Killiny, N. (2016). A gas chromatography–mass spectrometry method for the determination of delta-aminolevulinic acid in plant leaves. *Journal of Chromatography A*, *1447*, 57–63. https://doi.org/10.1016/j.chroma.2016.04.041

Hirono, Y., & Rédei, G. P. (1963). Multiple allelic control of chlorophyll *b* level in *Arabidopsis thaliana*. *Nature*, *197*, 1324–1325. https://doi.org/10.1038/1971324a0

Höfgen, R., Axelsen, K. B., Kannangara, C. G., Schüttke, I., Pohlenz, H. D., Willmitzer, L., … von Wettstein, D. (1994). A visible marker for antisense mRNA expression in plants: Inhibition of chlorophyll synthesis with a glutamate-1-semialdehyde aminotransferase antisense gene. *Proceedings of the National Academy of Sciences of the United States of America*, *91*, 1726–1730. https://doi.org/10.1073/pnas.91.5.1726

Horvàth, E., Szalai, G., & Janda, T. (2007). Induction of abiotic stress tolerance by salicylic acid signaling. *Journal of Plant Growth Regulation*, *26*, 290–300. https://doi.org/10.1007/s00344-007-9017-4

Hu, G., Yalpani, N., Briggs, S. P., & Johal, G. S. (1998). A porphyrin pathway impairment is responsible for the phenotype of a dominant disease lesion mimic mutant of maize. *Plant Cell*, *10*, 1095–1105. https://doi.org/10.1105/tpc.10.7.1095

Igarashi, A., Yamagata, K., Sugai, T., Takahashi, Y., Sugawara, E., Tamura, A., … Yoshikawa, N. (2009). Apple latent spherical virus vectors for reliable and effective virus-induced gene silencing among a broad range of plants including tobacco, tomato, *Arabidopsis thaliana*, cucurbits, and legumes. *Virology*, *386*, 407–416. https://doi.org/10.1016/j.virol.2009.01.039

Ishikawa, A., Okamoto, H., Iwasaki, Y., & Asahi, T. (2001). A deficiency of coproporphyrinogen III oxidase causes lesion formation in Arabidopsis. *The Plant Journal*, *27*, 89–99. https://doi.org/10.1046/j.1365-313x.2001.01058.x

Jensen, P. E., Willows, R. D., Petersen, B. L., Vothknecht, U. C., Stummann, B. M., Kannangara, C. G., … Henningsen, K. W. (1996). Structural genes for Mg-chelatase subunits in barley: *Xantha-f*, *-g* and *-h*. *Molecular and General Genetics*, *250*, 383–394.

Jung, S. (2004). Effect of chlorophyll reduction in *Arabidopsis thaliana* by methyl jasmonate or norflurazon on antioxidant systems. *Plant Physiology and Biochemistry*, *42*, 225–231. https://doi.org/10.1016/j.plaphy.2004.01.001

Kaczor, C. M., Smith, M. W., Sangwan, I., & O'Brian, M. R. (1994). Plant δ-Aminolevulinic acid dehydratase: Expression in soybean root nodules and evidence for a bacterial lineage of the *Alad* gene. *Plant Physiology*, *104*, 1411–1417. https://doi.org/10.1104/pp.104.4.1411

Killiny, N., Hijaz, F., Harper, S. J., & Dawson, W. O. (2017). Effects of *Citrus tristeza closterovirus* infection on phloem sap and released volatile organic compounds in *Citrus macrophylla*. *Physiological and Molecular Plant Pathology*, *98*, 25–36. https://doi.org/10.1016/j.pmpp.2017.03.003

Killiny, N., & Jones, S. E. (2017). Profiling of volatile organic compounds released from individual intact juvenile and mature citrus leaves. *Journal of Plant Physiology*, *208*, 47–51. https://doi.org/10.1016/j.jplph.2016.11.001

Killiny, N., & Nehela, Y. (2017a). Metabolomic response to Huanglongbing: Role of carboxylic compounds in *Citrus sinensis* response to "*Candidatus* Liberibacter asiaticus" and its vector, *Diaphorina citri*. *Molecular Plant-Microbe Interactions*, *30*, 666–678. https://doi.org/10.1094/MPMI-05-17-0106-R

Killiny, N., & Nehela, Y. (2017b). One target, two mechanisms: The impact of "*Candidatus* Liberibacter asiaticus" and its vector, *Diaphorina citri*, on citrus leaf pigments. *Molecular Plant-Microbe Interactions*, *30*, 543–556. https://doi.org/10.1094/MPMI-02-17-0045-R

Kim, Y.-K., Lee, J.-Y., Cho, H. S., Lee, S. S., Ha, H. J., Kim, S., … Pai, H.-S. (2005). Inactivation of organellar glutamyl- and seryl-tRNA synthetases leads to developmental arrest of chloroplasts and mitochondria in higher plants. *Journal of Biological Chemistry*, *280*, 37098–37106. https://doi.org/10.1074/jbc.M504805200

Kruse, E., Mock, H. P., & Grimm, B. (1995). Reduction of coproporphyrinogen oxidase level by antisense RNA synthesis leads to deregulated gene expression of plastid proteins and affects the oxidative defense system. *EMBO Journal*, *14*, 3712–3720.

Kumar, A. M., & Söll, D. (2000). Antisense HEMA1 RNA expression inhibits heme and chlorophyll biosynthesis in Arabidopsis. *Plant Physiology*, *122*, 49–56. https://doi.org/10.1104/pp.122.1.49

Livak, K. J., & Schmittgen, T. D. (2001). Analysis of relative gene expression data using real-time quantitative PCR and the $2^{-\Delta\Delta CT}$ method. *Methods*, *25*, 402–408. https://doi.org/10.1006/meth.2001.1262

Mafra, V., Kubo, K. S., Alves-Ferreira, M., Ribeiro-Alves, M., Stuart, R. M., Boava, L. P., … Machado, M. A. (2012). Reference genes for accurate transcript normalization in citrus genotypes under different experimental conditions. *PLoS One*, *7*, e31263. https://doi.org/10.1371/journal.pone.0031263

Mauzerall, D., & Granick, S. (1956). The occurrence and determination of delta-amino-levulinic acid and porphobilinogen in urine. *Journal of Biological Chemistry*, *219*, 435–446.

Michie, M. G. (1982). Use of the Bray-Curtis similarity measure in cluster analysis of foraminiferal data. *Journal of the International Association for Mathematical Geology*, *14*, 661–667. https://doi.org/10.1007/BF01033886

Mock, H. P., Heller, W., Molina, A., Neubohn, B., Sandermann, H., & Grimm, B. (1999). Expression of uroporphyrinogen decarboxylase or coproporphyrinogen oxidase antisense RNA in tobacco induces pathogen defense responses conferring increased resistance to *tobacco mosaic virus*. *Journal of Biological Chemistry*, *274*, 4231–4238. https://doi.org/10.1074/jbc.274.7.4231

Mock, H.-P., Keetman, U., Kruse, E., Rank, B., & Grimm, B. (1998). Defense responses to tetrapyrrole-induced oxidative stress in transgenic plants with reduced uroporphyrinogen decarboxylase or coproporphyrinogen oxidase activity. *Plant Physiology*, *116*, 107–116. https://doi.org/10.1104/pp.116.1.107

Molina, A., Volrath, S., Guyer, D., Maleck, K., Ryals, J., & Ward, E. (1999). Inhibition of protoporphyrinogen oxidase expression in *Arabidopsis* causes a lesion-mimic phenotype that induces systemic acquired resistance. *The Plant Journal*, *17*, 667–678. https://doi.org/10.1046/j.1365-313X.1999.00420.x

Mouly, P. P., Gaydou, E. M., Lapierre, L., & Corsetti, J. (1999). Differentiation of several geographical origins in single-strength Valencia orange juices using quantitative comparison of carotenoid profiles. *Journal of Agriculture and Food Chemistry*, *47*, 4038–4045. https://doi.org/10.1021/jf990025l

Nagata, N., Tanaka, R., Satoh, S., & Tanaka, A. (2005). Identification of a vinyl reductase gene for chlorophyll synthesis in *Arabidopsis thaliana* and implications for the evolution of prochlorococcus species. *Plant Cell*, *17*, 233–240. https://doi.org/10.1105/tpc.104.027276

Nehela, Y., Hijaz, F., Elzaawely, A. A., El-Zahaby, H. M., & Killiny, N. (2016). Phytohormone profiling of the sweet orange (*Citrus sinensis* (L.) Osbeck) leaves and roots using GC-MS-based method. *Journal of Plant Physiology*, *199*, 12–17. https://doi.org/10.1016/j.jplph.2016.04.005

Nehela, Y., Hijaz, F., Elzaawely, A. A., El-Zahaby, H. M., & Killiny, N. (2018). Citrus phytohormonal response to *Candidatus* Liberibacter asiaticus and its vector *Diaphorina citri*. *Physiological and Molecular Plant Pathology*, *102*, 24–35. https://doi.org/10.1016/j.pmpp.2017.11.004

Nitsch, L., Kohlen, W., Oplaat, C., Charnikhova, T., Cristescu, S., Michieli, P., … Rieu, I. (2012). ABA-deficiency results in reduced plant and fruit size in tomato. *Journal of Plant Physiology*, *169*, 878–883. https://doi.org/10.1016/j.jplph.2012.02.004

Palmer-Young, E. C., Veit, D., Gershenzon, J., Schuman, M. C., Baldwin, I., & Degenhardt, J. (2015). The sesquiterpenes(*e*)-ß-farnesene and (*e*)-α-bergamotene quench ozone but fail to protect the wild tobacco nicotiana attenuata from ozone, UVB, and drought stresses. *PLoS One*, *10*, e0127296. https://doi.org/10.1371/journal.pone.0127296

Papenbrock, J., Mishra, S., Mock, H.-P., Kruse, E., Schmidt, E.-K., Petersmann, A., … Grimm, B. (2001). Impaired expression of the plastidic ferrochelatase by antisense RNA synthesis leads to a necrotic phenotype of transformed tobacco plants. *The Plant Journal*, *28*, 41–50. https://doi.org/10.1046/j.1365-313X.2001.01126.x

Papenbrock, J., Mock, H.-P., Tanaka, R., Kruse, E., & Grimm, B. (2000). Role of magnesium chelatase activity in the early steps of the tetrapyrrole biosynthetic pathway. *Plant Physiology*, *122*, 1161–1169. https://doi.org/10.1104/pp.122.4.1161

Romero, I., Tikunov, Y., & Bovy, A. (2011). Virus-induced gene silencing in detached tomatoes and biochemical effects of phytoene desaturase gene silencing. *Journal of Plant Physiology*, *168*, 1129–1135. https://doi.org/10.1016/j.jplph.2010.12.020

Rzeznicka, K., Walker, C. J., Westergren, T., Kannangara, C. G., von Wettstein, D., Merchant, S., … Hansson, M. (2005). *Xantha-l* encodes a membrane subunit of the aerobic Mg-protoporphyrin IX monomethyl ester cyclase involved in chlorophyll biosynthesis. *Proceedings of the National Academy of Sciences of the United States of America*, *102*, 5886–5891. https://doi.org/10.1073/pnas.0501784102

Scinicariello, F., Murray, H. E., Moffett, D. B., Abadin, H. G., Sexton, M. J., & Fowler, B. A. (2007). Lead and delta-aminolevulinic acid dehydratase polymorphism: Where does it lead? A meta-analysis *Environmental Health Perspectives*, *115*, 35–41.

Senthil-Kumar, M., Hema, R., Anand, A., Kang, L., Udayakumar, M., & Mysore, K. S. (2007). A systematic study to determine the extent of gene silencing in *Nicotiana benthamiana* and other Solanaceae species when heterologous gene sequences are used for virus-induced gene silencing. *New Phytologist*, *176*, 782–791. https://doi.org/10.1111/j.1469-8137.2007.02225.x

Shen, Z., Sun, J., Yao, J., Wang, S., Ding, M., Zhang, H., … Chen, S. (2015). High rates of virus-induced gene silencing by tobacco rattle virus in *Populus*. *Tree Physiology*, *35*, 1016–1029. https://doi.org/10.1093/treephys/tpv064

Tanaka, R., & Tanaka, A. (2005). Effects of chlorophyllide *a* oxygenase overexpression on light acclimation in *Arabidopsis thaliana*. *Photosynthesis Research*, *85*, 327–340. https://doi.org/10.1007/s11120-005-6807-z

Tanaka, R., & Tanaka, A. (2007). Tetrapyrrole biosynthesis in higher plants. *Annual Review of Plant Biology*, *58*, 321–346. https://doi.org/10.1146/annurev.arplant.57.032905.105448

Tomokuni, K., & Ogata, M. (1972). Simple method for determination of urinary -aminolevulinic acid as an index of lead exposure. *Clinical Chemistry*, *18*, 1534–1538.

Tsang, E. W. T., Yang, J., Chang, Q., Nowak, G., Kolenovsky, A., McGregor, D. I., & Keller, W. A. (2003). Chlorophyll reduction in the seed of *Brassica napus* with a glutamate 1-semialdehyde aminotransferase antisense gene. *Plant Molecular Biology*, *51*, 191–201. https://doi.org/10.1023/A:1021102118801

van Handel, E. (1968). Direct microdetermination of sucrose. *Analytical Biochemistry*, *22*, 280–283. https://doi.org/10.1016/0003-2697(68)90317-5

Vickers, C. E., Gershenzon, J., Lerdau, M. T., & Loreto, F. (2009). A unified mechanism of action for volatile isoprenoids in plant abiotic stress. *Nature Chemical Biology*, *5*, 283–291. https://doi.org/10.1038/nchembio.158

Wei, X., Chen, C., Yu, Q., Gady, A., Yu, Y., Liang, G., & Gmitter, F. G. (2014a). Comparison of carotenoid accumulation and biosynthetic gene expression between Valencia and Rohde Red Valencia sweet oranges. *Plant Science*, *227*, 28–36. https://doi.org/10.1016/j.plantsci.2014.06.016

Wei, X., Chen, C., Yu, Q., Gady, A., Yu, Y., Liang, G., & Gmitter, F. G. (2014b). Novel expression patterns of carotenoid pathway-related genes in citrus leaves and maturing fruits. *Tree Genetics and Genomes*, *10*, 439–448. https://doi.org/10.1007/s11295-013-0688-7

Weller, J. L., Terry, M. J., Rameau, C., Reid, J. B., & Kendrick, R. E. (1996). The phytochrome-deficient *pcd1* mutant of pea is unable to convert heme to biliverdin IX[alpha]. *Plant Cell*, *8*, 55–67.

Weller, J. L., Terry, M. J., Reid, J. B., & Kendrick, R. E. (1997). The phytochrome-deficient *pcd2* mutant of pea is unable to convert biliverdin IXalpha to 3(Z)-phytochromobilin. *The Plant Journal*, *11*, 1177–1186. https://doi.org/10.1046/j.1365-313X.1997.11061177.x

Zeb, A., & Murkovic, M. (2010). Thin-layer chromatographic analysis of carotenoids in plant and animal samples. *Journal of Planar Chromatography-Modern TLC*, *23*, 94–103. https://doi.org/10.1556/JPC.23.2010.2.1

Ectopic expression of the *PISTILLATA* homologous MdPI inhibits fruit tissue growth and changes fruit shape in apple

Jia-Long Yao[1]* | Juan Xu[2]* | Sumathi Tomes[1] | Wei Cui[1] | Zhiwei Luo[1] | Cecilia Deng[1] | Hilary S. Ireland[1] | Robert J. Schaffer[1,3] | Andrew P. Gleave[1]

[1]The New Zealand Institute for Plant & Food Research Limited, Auckland, New Zealand

[2]Key Laboratory of Horticultural Plant Biology (Ministry of Education), Huazhong Agricultural University, Wuhan, China

[3]School of Biological Sciences, The University of Auckland, Auckland, New Zealand

Correspondence
Jia-Long Yao, The New Zealand Institute for Plant & Food Research Limited, Auckland, New Zealand.
Email: jia-long.yao@plantandfood.co.nz

Abstract

Fruit shape represents a key trait that consumers use to identify and select preferred cultivars, and although the manipulation of this trait is an opportunity to create novel, differentiated products, the molecular mechanisms regulating fruit shape are poorly understood in tree fruits. In this study, we have shown that ectopic expression of *Malus domestica PISTILLATA (MdPI)*, the apple ortholog of the floral organ identity gene *PISTILLATA (PI)*, regulates apple fruit tissue growth and shape. *MdPI* is a single-copy gene, and its expression is high during flower development but barely detectable soon after pollination. Transgenic apple plants with ectopic expression of *MdPI* produced flowers with white sepals and a conversion of sepals to petals. Interestingly, these plants produced distinctly flattened fruit as a consequence of reduced cell growth at the basipetal position of the fruit. These altered sepal and fruit phenotypes have not been observed in studies using *Arabidopsis*. This study using apple has advanced our understanding of PI functions outside the control of petal and stamen identity and provided molecular genetic information useful for manipulating fruit tissue growth and fruit shape.

KEYWORDS
floral organs, fruit development, MADS-box, *Malus x domestica*, *PISTILLATA*, transgenic plants

1 | INTRODUCTION

In flowering plants, various floral tissues contribute to fruit formation across species resulting in different fruit types (Spjut, 1994). In species with a superior ovary, such as grape and tomato, the berry fruit develops from the ovary tissue alone (Gillaspy, Bendavid, & Gruissem, 1993). In species with an inferior ovary, such as apple and pear, the pome fruit develops from the ovary and hypanthium tissues. The hypanthium surrounding the apple ovary is hypothesized to consist of the fused bases of the sepals, petals and stamens, and contributes tissue to the fruit flesh cortex while the ovary develops into the fruit core (Pratt, 1988).

As fruit is derived from floral organs, the genes regulating floral organ growth may also play important roles in controlling fruit development. Genes regulating floral organ development have been classified into four different classes by function, according to the proposed ABCE model (Krizek & Fletcher, 2005; Weigel & Meyerowitz, 1994). In summary, for *Arabidopsis*, the class A genes *APETALA1* (*AP1*) and *AP2* specify sepal formation; the class A genes and two class B genes [*APETALA3* (*AP3*) and *PISTILLATA* (*PI*)] together specify petal formation; *AP3* and *PI* together with the class C gene *AGAMOUS* (*AG*) are required for stamen formation; and *AG* alone controls carpel formation (Weigel & Meyerowitz, 1994). The class E *SEPALLATA* genes are required for the proper development of all four whorls of floral organs. A mutant plant for all four *SEPALLATA* genes (*sep1*, *sep2*, *sep3*, *sep4*) produces flowers consisting of reiterating whorls of leaf-like organs (Krizek & Fletcher, 2005). All these class A, B, C, and E genes, except for *AP2*, belong to the highly conserved MADS-box gene family. The role of these floral organ genes in regulating fruit development depends on their fruit structures or the floral tissue types contributing to fruit growth (Yao, Tomes, Xu, & Gleave, 2016; Yao et al., 2015).

AP3 and *PI* are single-copy genes in *Arabidopsis*. They are specifically expressed in petals and stamens during their development (Goto & Meyerowitz, 1994; Jack, Brockman, & Meyerowitz, 1992). Loss-of-function mutations of either *PI* or *AP3* produce flowers that contain two whorls of sepals and double numbers of carpels, but without stamens or petals (Goto & Meyerowitz, 1994; Jack et al., 1992). Ectopic expression of *PI* or *AP3* alone in *Arabidopsis* does not significantly change flower development, while ectopic expression of *PI* and *AP3* together converts sepals to petals, and carpels to stamens (Krizek & Meyerowitz, 1996). The functions of PI and AP3 are dependent on the coexistence of the two proteins forming a bifunctional heterodimer that activates genes involved in the control of numerous developmental processes required for organogenesis and represses key regulators of carpel formation (Wuest et al., 2012). In many species, there is a single copy of the *PI* homolog, but several copies of *AP3* as a result of gene duplication events (Lee & Irish, 2011). The single-copy *PI* homolog has maintained its petal- and stamen-specific expression pattern, in apple (Tanaka, Wada, Komori, Bessho, & Suzuki, 2007; Yao, Dong, & Morris, 2001) and grape (Sreekantan, Torregrosa, Fernandez, & Thomas, 2006), but some of the duplicated *AP3* paralogs have developed new expression specificities, such as sepal and fruit tissue expression in apple (Kitahara, Ohtsubo, Soejima, & Matsumoto, 2004; van der Linden, Vosman, & Smulders, 2002) and grape (Poupin et al., 2007). Ectopic expression of *VvPI* in grape inhibits fruit flesh tissue growth resulting in fleshless berries (Fernandez, Chaib, Martinez-Zapater, Thomas, & Torregrosa, 2013). This is likely due to the expression of *AP3* homologs in the berry and therefore the formation of a PI/AP3 heterodimer.

Apple floral organ MADS-box genes have been cloned and identified based on high sequence homology to the *Arabidopsis* ABCE classes of genes. These genes include two A class genes, *MdMADS5* (*MdAP1*) (Kotoda et al., 2002; Yao, Dong, Kvarnheden, & Morris, 1999) and *MdMADS2* (Sung, Yu, & An, 1999), three B class genes, *MdPI* (Yao et al., 2001), *MdMADS13* (van der Linden et al., 2002), and *MdTM6* (Kitahara et al., 2004), three C class genes, *MdMADS10* (Yao, Dong et al., 1999), *MdMADS14* and *15* (van der Linden et al., 2002), and five E class genes, *MdMADS4* (Sung, Yu, Nam, Jeong, & An, 2000), *MdMADS6*, *MdMADS7*, *MdMADS8*, and *MdMADS9* (Yao, Dong et al., 1999). After full genome sequencing, three more E class genes have been identified, *MdMADS18*, *MdMADS104*, and *MdMADS118* (Ireland et al., 2013). The apple floral organ genes have similar expression patterns to those of *Arabidopsis* homologs (Kitahara et al., 2004; Kotoda et al., 2002; van der Linden et al., 2002; Mimida et al., 2011; Sung et al., 1999, 2000; Tanaka et al., 2007; Yao, Dong et al., 1999), and overexpression of *MdPI* can rescue *Arabidopsis pi* mutants (Tanaka et al., 2007). They have similar functions in regulating floral organ development as the *Arabidopsis* genes, but play additional roles in regulating fruit development that are not known for *Arabidopsis* genes. In *Arabidopsis*, the *SHATTERPROOF* (*SHP*) 1 and 2 genes, most closely related to the class C gene *AG*, are shown to be important to silique development (Liljegren et al., 2000). But in apple, class B and E genes are also shown to be important to fruit development as the pome fruit is derived from both ovary and hypanthium tissues. For example, knockout of *MdPI*, in addition to producing pistillate flowers, confers parthenocarpic fruit development in apple (Yao et al., 2001), and antisense suppression *MdMADS8*, not only partially converts petals to sepals but also inhibits fruit flesh development and ripening (Ireland et al., 2013).

Significance statement

PISTILLATA (PI) and its homologs are mostly known to control the identity of petals and stamens in different plant species. This study has shown that ectopic expression of MdPI regulates apple fruit growth and shape, thus has advanced our understanding of PI and provided molecular genetic information useful for manipulating fruit shape.

Although floral organ genes have been shown to influence fruit size, texture, and ripening, their function in altering fruit shape has not as yet been reported. In tomato, genes regulating fruit shape have been identified (Liu, Van Eck, Cong, & Tanksley, 2002; Rodriguez et al., 2011; Xiao, Jiang, Schaffner, Stockinger, & van der Knaap, 2008), but they do not belong to the ABCE classes of genes. In this study, we show that transgenic apple plants produce pistillate flowers and parthenocarpic fruit when *MdPI* expression is suppressed, and produce flowers with petal-like sepals and fruit with a distinct new shape when *MdPI* is overexpressed.

2 | EXPERIMENTAL PROCEDURES

2.1 | RNA-Seq and data analysis

Three biological replicates from whole flowers at balloon (5 days before full bloom) and full bloom stages, and one mixed sample each of leaf, root, and fruit at 5, 36, and 132 DAP (days after pollination) were collected from six trees of *Malus x domestica* "Royal Gala" and preserved in liquid nitrogen. RNA was isolated by rapid cetyltrimethyl ammonium bromide (CTAB) extraction (Chang, Puryear, & Cairney, 1993). Five μg total RNA was used to construct strand-specific poly-A RNA libraries as described by Ref. (Zhong et al., 2011). Libraries were sequenced by Macrogen (Republic of Korea) using HiSeq2000 sequencing system (Illumina) to produce reads per library ranging from 20-27 million. Reads were aligned to apple gene models using Bowtie2, and reads per kilobase per million mapped reads (RPKM) were calculated in the statistical software R.

2.2 | Transformation vector construction

For the production of apple transgenic plants, the pART7/pART27 (Gleave, 1992) binary vector system was used. The *MdPI* cDNA fragment between the P1 and P2 primers in pBluescript SK- (Yao et al., 2001) was cut out with *Kpn1/Bam*HI double digestion and cloned into the *Kpn1/Bam*H1 sites in pART7 between the CaMV35S promoter and *ocs* terminator in sense orientation. The *Not*I fragment containing the promoter–cDNA–terminator was then cloned into the *Not*I site of pART27 to form the pART27-35S-MdPI-ocs construct (Figure S3b). This binary vector was then transferred into the *Agrobacterium tumefaciens* strain LBA4404 by electroporation for use in apple transformation.

2.3 | Apple transformation

To produce transgenic apple plants, in vitro shoot cultures of *Malus* x *domestica* "Bolero" were established from bud wood collected from an orchard at The New Zealand Institute for Plant & Food Research Limited (PFR), Havelock North, New Zealand, using a previously described method (Yao, Cohen, Atkinson, Richardson, & Morris, 1995). Young leaves of the shoot cultures were used as explants in apple transformation experiments using the methods previously described (Yao, Tomes, & Gleave, 2013). The transgenic plants were grown on their own roots alongside wild-type (WT) plants in a containment glasshouse using the methods previously described (Yao, Cohen, Van den Brink, & Morris, 1999). Flowers were pollinated with *Malus* x *domestica* "Granny Smith" pollen or covered with paper bags if pollination was not required.

2.4 | Confirmation of transgenic apple plant

DNA was extracted from young leaves collected from WT and transgenic glasshouse-grown plants using the DNeasy® Plant Mini Kit (Qiagen). The DNA samples were used in PCR analyses to amplify a transgene fragment with primers 35SF and ocsR (Table S2). PCR products were analyzed on 0.7% agarose gels, and the fragments of expected size were purified using PureLink™ Quick Gel Extraction Kit (Invitrogen) before dispatch to Macrogen (Republic of Korea) for sequencing.

2.5 | Quantitative RT-PCR

For qRT-PCR analysis of "Bolero" apple transgenic plants, total RNA was isolated from unopened balloons and 2-week-old fruit. To distinguish the expression of B class MADS-box genes between floral and hypanthium tissues, RNA was isolated from dissected tissues of "Royal Gala" apple at four stages of floral development: Balloon, open flower, 2, and 8 days after pollination, as previously described (Ireland et al., 2013). RNA was isolated using the method developed for pine tree RNA extraction (Chang et al., 1993), analyzed using an Agilent 2100 bioanalyzer (Agilent Co, Ltd, USA) to determine RNA concentration and integrity, and then treated with DNaseI. For each RNA sample, 1 μg RNA was used for cDNA synthesis using the Quantitect® Reverse Transcription Kit (Qiagen) according to the manufacturer's instructions. Using the cDNA as templates, qRT-PCR was carried out using LightCycler® 480 (Roche Diagnostics) following previously described procedures (Drummond et al., 2009). PCR primers for the reference control genes Actin and EF-1α, and test genes *MdPI, MdTM6, MdMADS13* are listed in Table S2. Significant differences between gene expression levels were determined using analysis of variance (ANOVA) in SAS (version 9.2), with a *p* value $<.05$.

2.6 | Light microscopy analysis

Samples were fixed in FAA (4% formalin/50% alcohol/5% acetic acid in water to 100%). Samples were washed twice in 50% ethanol, then dehydrated through a graded ethanol series at 2-hr intervals, then 50:50 ethanol/xylene (2 hr), two changes of xylene (3 hr each), 1:1 wax/xylene (3 hr), and two changes of paraffin wax (12 hr each), and embedded in Paraplast wax (Oxford Labware, www.kendellhq.com) Sections of 10 μm thickness were cut using a Leitz 1512 microtome (Leica.www.leica-microsystems.com), placed on positively charged glass slides and dried overnight in a slide dryer. Sections were dewaxed in two changes of xylene (5 min each), followed by two changes of absolute ethanol and then air-dried. The sections were stained with Safranin–Fast Green. Sections were photographed using an Olympus Vanox AHT3 microscope (Olympus Optical, www.olympus-global.com). Fruit cell area was measured using the IMAGEJ software (https://imagej.nih.gov/ij/), and 30 cells were measured at each of the three positions of a fruit.

3 | RESULTS

3.1 | Apple has a single *PISTILLATA* ortholog specifically expressed in flowers

An extensive search of the apple genome sequence using BLAST with other plant homologs found a single copy of the *PI* gene, which has been previously reported as *MdPI* (Yao et al., 2001) (Figure S1), with gene model number MDP0000286643 and located at 24 Mb on chromosome 8 (Velasco et al., 2010). Due to the relatively recent genomewide duplication in the apple genome, two paralog copies are usually identified for most floral genes (Ireland et al., 2013; Tian et al., 2015). A chromosome fragment loss after the genomewide duplication may explain the single-copy status of *MdPI*. On the other hand, four *AP3* homologs were found in the apple genome (Figure S1), two with identical coding sequences but located on different chromosomes, consistent with a recent gene duplication.

Transcriptomic analysis of "Royal Gala" apple tissues by RNA-seq technology revealed that *MdPI* transcript levels were very high in whole flowers at balloon (600 RPKM) and open flower (300 RPKM) stages, but were greatly reduced in fruit soon after pollination with an RPKM value of 3 at 5 DAP (days after pollination), and were not detected in fruit from 36 DAP through to maturity (Figure S2, Table S1). Relative transcript levels of the *AP3* homologs were variable between the different genes and between tissues. In general, RNA levels of the *AP3* homologs were lower than *MdPI* in flowers (Table S1). Transcripts for class A, C, and E genes were detected in both flowers and fruit tissues (Figure S2, Table S1). For example, high levels of transcripts were detected in both flower and fruit tissues for the class A gene *MdMADS5*, class C genes *MdMADS14* and *MdMADS15*, and class E genes *MdMADS6*, *MdMADS1/8*, and *MdMADS3/7*. These are consistent with previous results of northern and qRT-PCR analyses (Ireland et al., 2013; van der Linden et al., 2002; Yao, Dong et al., 1999).

We further analyzed B function gene expression using qRT-PCR on two types of tissues dissected from "Royal Gala" flowers: the hypanthium (base of the flower cut in half and ovary tissue removed) and combined floral tissues (sepals, petals, stamens, and ovary tissues that were separated from the hypanthium). We found that expression levels of *MdPI*, *MdAMDS13*, and *MdTM6* were high in floral tissues at balloon stage and were reduced at open flower stage and at 2 DAP (Figure 1). The very low level of expression of *MdMADS13* and *MdTM6* at 8 DAP was likely from sepal and ovary tissues as petal and stamen tissues were no longer present at 8 DAP. The expression levels of *MdPI* and *MdTM6* were hardly detectable in the hypanthium from the balloon stage to 8 DAP. *MdMADS13* expression was detected at very low level in the hypanthium across all the stages we analyzed. Given the hypothesis that hypanthium consists of the fused bases of the sepals, petals, and stamens, it is interesting to note that the B function genes were weakly expressed in apple hypanthium. In this study, although "Royal Gala" was used in analyses of floral MADS-box gene expression, "Bolero" apple was used in transgenic experiments because it is a dwarf cultivar. As there are no differences in floral or fruit structures between "Royal Gala" and "Bolero", it is not unreasonable to assume that any differences in the expression of floral organ and fruit developmental genes between the two cultivars would be minimal. Nevertheless, our interpretations using data derived from different cultivars should be treated with some degree of caution.

3.2 | Production and molecular analyses of apple transgenic plants

Six "Bolero" apple transgenic plants (*mdpi-1* to *mdpi-6*) were produced using a *35S-MdPI* gene construct to overexpress the *MdPI* gene and grown to produce flowers and fruits in a glasshouse. The phenotypes of these transgenic plants are summarized in Table 1. The *mdpi-1* and *mdpi-2* plants produced flowers with white sepals, short anther filaments, and flattened fruit, whereas the *mdpi-3* plant was largely normal except for short anther filaments. The *mdpi-4*, *mdpi-5* and *mdpi-6* plants produced pistillate flowers and parthenocarpic fruits.

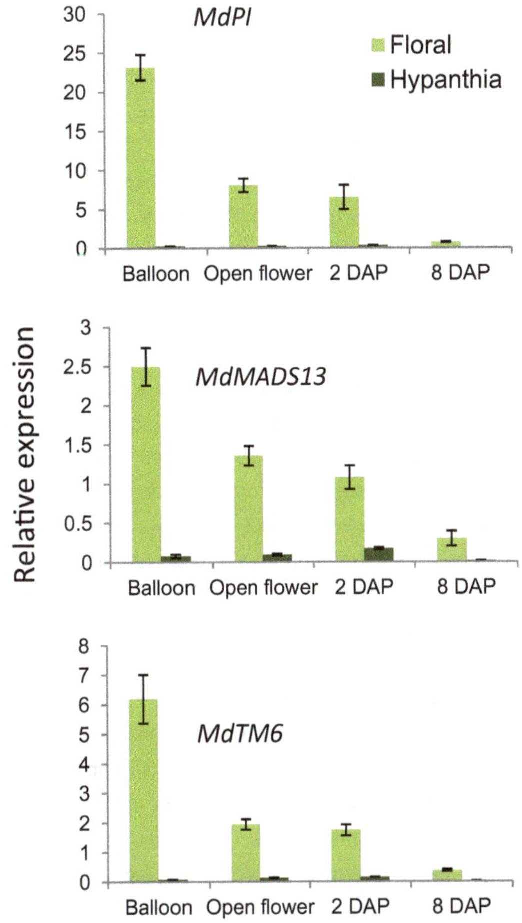

FIGURE 1 Quantitative PCR analyses of expression of *MdPI*, *MdMADS13*, and *MdTM6* in floral and hypanthium tissues of wild-type (WT) "Royal Gala" apple plants. RNA from floral and hypanthium tissues were amplified using three primer pairs for *MdPI*, *MdMADS13*, and *MdTM6* respectively. Error bars represent the standard deviation of three independent PCR mixtures

TABLE 1 Summary of transgenic phenotypes (apple cultivar "Bolero")

Transgenic line	Flower	Fruit	*MdPI* expression
mdpi-1	White sepals, short stamen filaments	Flattened	Upregulated
mdpi-2	White sepals, short stamen filaments	Flattened	Upregulated
mdpi-3	Short stamen filaments	Normal	No significant change
mdpi-4	Pistillate	Parthenocarpic	Downregulated
mdpi-5	Pistillate	Parthenocarpic	Downregulated
mdpi-6	Pistillate	Parthenocarpic	Downregulated

DNA fragments of the *MdPI* transgene were amplified from the six transgenic lines *mdpi-1* to *mdpi-6* using primers that anneal to the 35S promoter and *ocs* terminator. The amplified fragments contained the full-length transgene of *MdPI* as the primers used flanked the *MdPI* cDNA. No fragment was amplified from the WT apple or water control (Figure S3a). The amplified *MdPI* transgene fragments from three transgenic lines (*mdpi-1*, *mdpi-2*, and *mdpi-4*) representing two of the three phenotypes observed were found to be identical in sequence. This indicated that the phenotypic differences were not due to any sequence variation in the *MdPI* transgene, but more likely due to differences in the transgenic *MdPI* expression levels.

To determine the relative expression levels of *MdPI* in the transgenic and WT plants, qRT-PCR analyses were carried out using RNA extracted from unopened flowers and 2-week-old fruit (Figure 2). Using primers specifically designed to detect mRNA transcribed from the endogenous *MdPI* gene, *MdPI* transcripts were not detected in the flower samples of *mdpi-4* and *mdpi-5* (Figure 2a), indicating a complete suppression of gene expression, and transcript levels were reduced to varying degrees in the other four transgenic flower samples when compared to WT samples (Figure 2a). In 2-week-old fruit, no endogenous *MdPI* transcripts were detected in pollinated WT, *mdpi-1*, *mdpi-2*, or *mdpi-4* fruit or unpollinated *mdpi-4* fruit (Figure 2a) indicating that endogenous *MdPI* is not expressed in young fruit, a finding that is consistent with the RNA-seq analysis (Figure S1) and a previous study (Yao et al., 2001). Using primers designed to detect transcripts from both the endogenous and transgenic *MdPI*, mRNA levels were found to be significantly higher in *mdpi-1* and *mdpi-2* plants than in WT in both the unopened flower and 2-week-old fruit samples (Figure 2b), indicating ectopic transgene expression. In contrast to *mdpi-1* and *mdpi-2*, the total mRNA levels of *MdPI* in *mdpi-4*, *mdpi-5*, and *mdpi-6* were reduced compared to WT at the unopened flower stage. However, at the 2-week-old fruit stage, the *MdPI* mRNA levels of *mdpi-4* were not significantly different from those of WT irrespective of whether the flowers had been pollinated or not. The variations in levels of transgene expression detected here are consistent with those in previous reports, showing different levels of transgene expression in different transgenic plants even if they were produced using the same gene construct and transformation methodology (Meyer, 1995). In some cases, when the transcription levels of transgene reached a high threshold level, post-transcription gene silencing is triggered to degrade the transcripts of both transgene and its endogenous homologous gene, known as cosuppression (Que, Wang, English, & Jorgensen, 1997). The reductions in endogenous *MdPI* transcript level in *mdpi-1* to *mdpi-6* (Figure 2a) were likely the result of partial or complete cosuppression. Collectively, the above information led us to suggest that the white-sepal phenotype of *mdpi-1* and *mdpi-2* is caused by overexpression of *MdPI* and the *pistillata*-like phenotype in *mdpi-4*, *mdpi-5*, and *mdpi-6* is caused by downregulation (cosuppression) of *MdPI*.

As PI protein forms a heterodimer with AP3 to perform its function (Wuest et al., 2012), we wanted to determine whether changes in *MdPI* expression had any effect on the expression of *AP3* homologs. Using qRT-PCR analyses, we found that the mRNA levels of two *AP3* homologs, *MdTM6* and *MdMADS13*, were significantly reduced in the flowers of transgenic apple plants where the *MdPI*

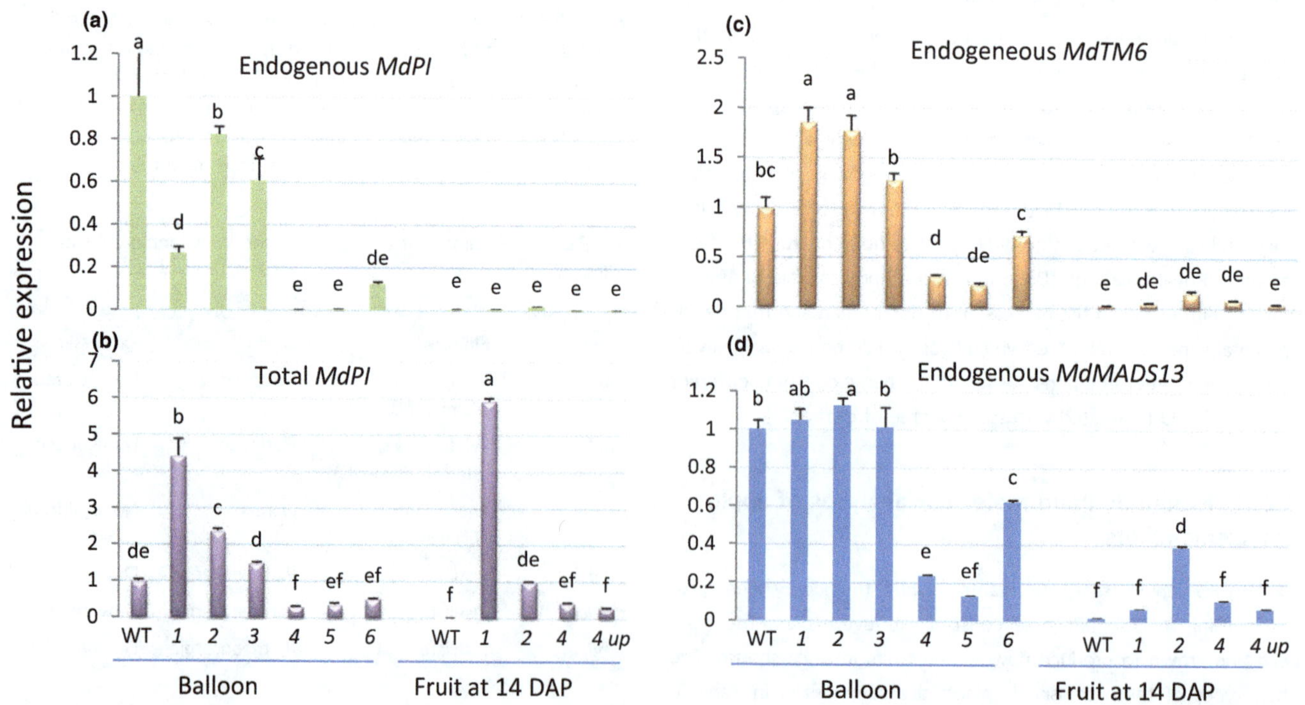

FIGURE 2 Quantitative PCR analyses of expression of *MdPI*, *MdMADS13*, and *MdTM6* in wild-type (WT) and transgenic "Bolero" apple plants. RNA from unopened flowers and 2-week-old fruit of WT and transgenic lines (*mdpi-1* to *6*) were amplified using four primer pairs for endogenous *MdPI* (a), total (endogenous plus transgene) *MdPI* (b), *MdTM6* (c), and *MdMADS13* (d), respectively. Significant differences between gene expression levels were analyzed using Duncan's test in SAS (version 9.2), and the p value was set as .05, three replications. *4up*: fruit developed from unpollinated flowers of the *mdpi-4* transgenic line. DAP: days after pollination

gene was silenced (*mdpi-4* and *5*). However, overexpression of *MdPI* in *mdpi-1* and *mdpi-2* plants was concomitant with elevated expression of *MdTM6* while there was no change in the expression level of *MdMADS13* (Figure 2c and d). In 2-week-old fruits, the mRNA level of *MdMADS13* is significantly higher in *mdpi-2* than in WT (Figure 2d). It has been reported in *Arabidopsis* that a knockout of one partner has a negative effect on the gene expression of another partner (Goto & Meyerowitz, 1994). In this study, *AP3* homologs in apple flowers were downregulated when *MdPI* was cosuppressed.

3.3 | Overexpression of *MdPI* confers sepal to petal conversion and changes fruit shape in apple

In WT "Bolero" flowers, the sepals were green and small with lots of trichomes (Figure 3a). In contrast, sepals in *mdpi-1* (Figure 3b, c and d) and *mdpi-2* transgenic plants were white to light pink in color, enlarged and without visible trichomes. This result suggests the conversion of sepals of *mdpi-1* and *-2* plants to petals. This first whorl conversion was much stronger than the phenotype observed in *Arabidopsis* overexpressing *MdPI* (Tanaka et al., 2007) or *PI* (Krizek & Meyerowitz, 1996), where only the margin of sepals was converted to petals. In WT apple flowers, anther filaments were similar in length to styles (Figure 3a), while in *mdpi-1*, *mdpi-2*, and *mdpi-3* plants, anther filaments were approximately half the length of styles (Figure 3c and d).

Flowers of *mdpi-1* and *mdpi-2* developed fruit after hand pollination with "Granny Smith" apple pollen. These fruit showed an unusual shape (Figure 4, Figure S4), distinctively different from WT fruit and evident from the early stages of fruit development and through to fruit maturation. Starting from 8 DAP, the fruit surface showed grooves that were not observed on WT fruit. The grooves were distributed in two different orientations. Five longitudinal grooves were present along the side of the fruit from 12 DAP to fruit maturation (Figure 4f, h, j, l, and Figure S4b), corresponding to the five carpels. One transverse groove was present at the basipetal position of the fruit 8-12 DAP (Figure 4d and f), but was no longer visible from 18 DAP to fruit maturation. From 18 DAP, the fruit of *mdpi-1* and *mdpi-2* plants was distinctively shorter than WT fruit with a very obvious flattened appearance at maturation (Figure 4). Fruit shape index, the ratio of maximum fruit height to maximum width (H/W), was significantly reduced in *mdpi-1* and *mdpi-2* fruit (Figure S4). The presence of grooves and the short and flattened fruit suggests that ectopic expression of *MdPI* suppressed localized fruit tissue growth.

Stained longitudinal sections of *mdpi-1* and WT fruit at 12 DAP showed clear groove traces at the basipetal position of the *mdpi-1* fruit, but not of the WT fruit (Figure 5). These grooves were not present in WT fruit at any developmental stage (Figures 4 and 5, Figure S5). Three selected fruit of *mdpi-1* at 12 DAP showed a different level of development of the basipetal groove (Figure 5b, c and d). The groove on the fruit in Figure 5b and d was deeper than that on the fruit in Figure 5c. The width of fruit tissue below the groove was sharply reduced for the fruit in Figure 5b and d, compared to the fruit tissue above the groove. By examining these sections at high magnification, we observed that fruit fleshy cells close to and below the groove were significantly smaller than those above the groove, (Figures 5 f-h and 6), and the cells close to the groove were the smallest at 12 DAP (Figure 6). However, this difference in cell size was not observed among the corresponding three positions

FIGURE 3 Flower and fruit of transgenic "Bolero" apple plants with overexpression or suppressed expression of *MdPI*. Photographs show opened flowers of wild-type (WT) apple (a), unopened (b), opened (c), and dissected (d) flowers of transgenic line *mdpi-1* overexpressing *MdPI*. Flower of transgenic line *mdpi-4* shows no petals or stamens but increased numbers of sepals and styles (e), and two whorls of carpels (f). The flower may develop seeded fruit with pollination (g) or seedless fruit without pollination (h). S = sepal, p = petal, ep = ectopic petal, p = pistil, s = sepal, nc = normal carpels, ec = ectopic carpels

FIGURE 4 Fruit shape differences between wild-type (WT) "Bolero" and *MdPI* overexpression apples at five developmental stages. The photographs show fruit of WT (a, c, e, g, i, k) and transgenic line *mdpi-1* apple (b, d, f, h, j, l) at 4, 8, 12, 18, and 132 DAP (days after pollination). The longitudinal grooves (indicated using blue arrows) were retained on mature fruit. The transverse grooves (indicated by red arrows) were not on the mature fruit. The tissues below the transverse grooves did not continue to grow, which caused the flatten fruit phenotype at mature fruit stage. a to h are on the same scale, and i and j are on the same scale

of *mdpi-1* plant at three earlier stages, balloon, open flower stages, and 4 DAP, or among the corresponding three positions of WT plants at any of the four stages examined (Figures 5 and 6, Figure S5). Fruit cell size was significantly increased from 4 DAP to 8 DAP in WT fruit and in the acropetal position of *mdpi-1* fruit, but did not increase at the basipetal position in *mdpi-1* fruit (Figure 6). This indicated that the development of a flat *mdpi-1* fruit started between 4 and 12 DAP by inhibiting fruit cell enlargement at the basipetal part of the fruit. Accurate counting of cell numbers in this region was not feasible as it was impossible to define comparable area or cell layers between the WT and the transgenic plant, or between the regions up and down the groove line. Collectively, these findings lead us to conclude that the flattened fruit occurs primarily because of inhibition of cell expansion at the basipetal position of the fruit, although we cannot exclude the possibility of reduced cell division and therefore reduced cell number.

3.4 | Suppression of *MdPI* expression confers petal to sepal, and stamen to carpel conversion

WT apple flowers usually consist of five sepals, five petals, 9–20 stamens, and five carpels (Figure 3a). In contrast, flowers of *mdpi-4* (Figure 3e and f), *mdpi-5*, and *mdpi-6* transgenic apples had no petals or stamens but two whorls of sepals and two whorls of carpels; thus, sepal number increased to 10 and the carpel number increased up to 15 (Figure 3). This result indicates floral organ conversion of petal to sepal, and stamen to carpel in these transgenic plants. Flowers of these transgenic lines were consistent with the phenotype of *pistillata* mutants of *Arabidopsis* (Goto & Meyerowitz, 1994) and apple *pistillata*-like flowers of the apple cultivars "Rae Ime," "Willington Bloomless," and "Spencer Seedless." These three apple cultivars are mutants with knockout of *MdPI* expression (Yao et al., 2001). Normal fruit development of WT "Bolero" flowers requires pollination and seed development. Flowers of *mdpi-4*, *mdpi-5*, and *mdpi-6* developed seeded fruits after pollination with "Granny Smith" pollen (Figure 3g); when the flowers were covered with paper bags to prevent pollination, they developed seedless fruits (Figure 3h). The mature seedless fruit were similar in size to the seeded fruit (Figure 3g and h). Both seeded and seedless *mdpi-4* fruits had a whorl of normal carpels and a whorl of ectopic carpels (Figure 3, Figure S6). These *mdpi-4* fruits also had duplicated whorls of calyces that were the remnants of sepals, in contrast to the one calyx whorl of a normal WT apple. The feature of double whorls of calyces and carpels in the fruit of *MdPI* cosuppression transgenic apple plants was the same as that of the fruit of the apple mutant "Rae Ime" (Yao et al., 2001).

4 | DISCUSSION

4.1 | Class B MADS-box genes regulate flower development in apple

We have demonstrated that suppression of *MdPI* expression in transgenic apple produced pistillate flowers with floral organ conversion of petals to sepals, and stamens to carpels. This floral organ

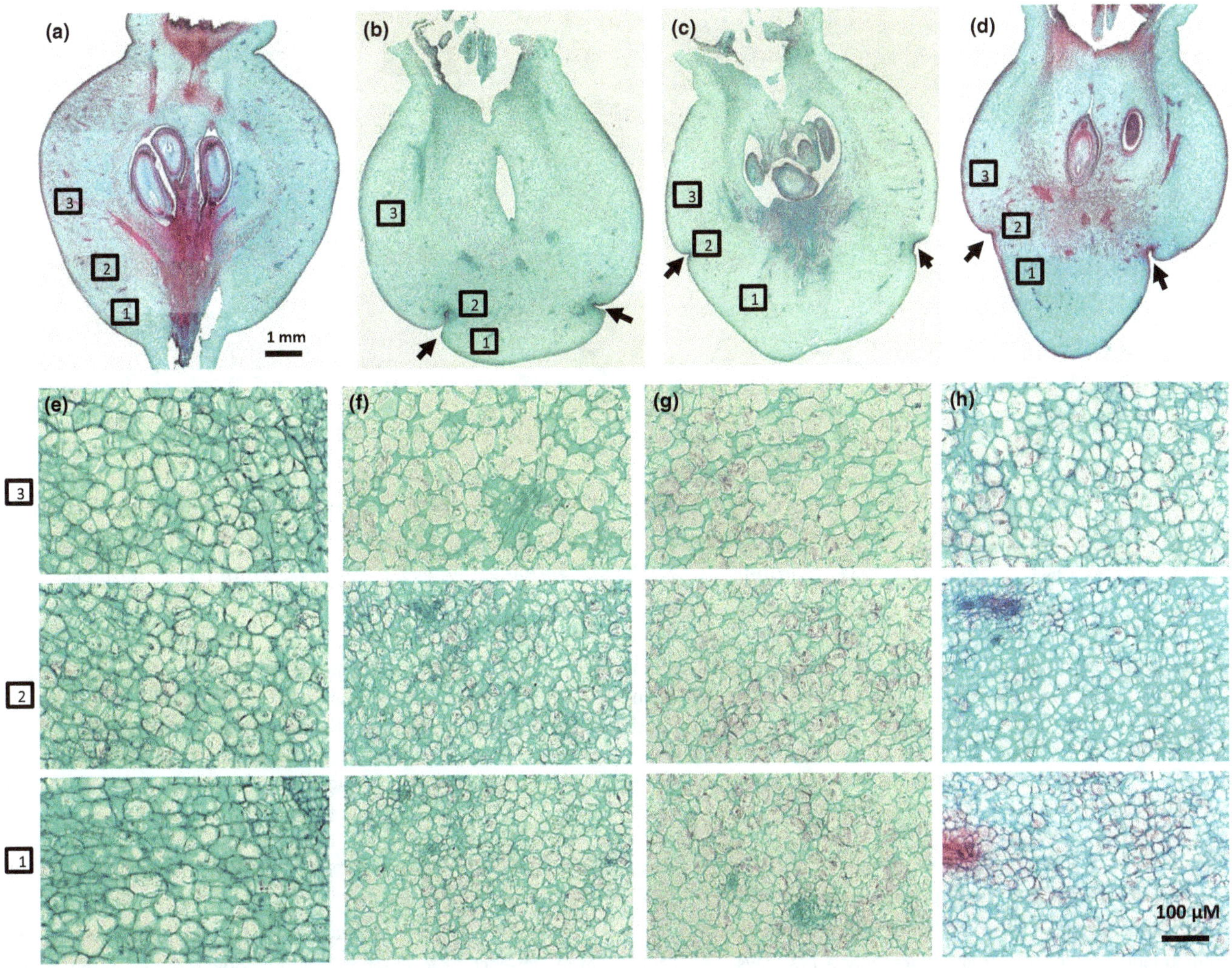

FIGURE 5 Fruit sections of wild-type (WT) and *MdPI* overexpression (*mdpi-1*) "Bolero" apples. Longitudinal sections of a WT fruit (a) and three different fruit of the *mdpi-1* transgenic plant (b, c, d) at 12 DAP (days after pollination) stained with Safranin-Fast Green. The transverse grooves at the basipetal position of *mdpi-1* fruit are indicated by the red arrows. The areas marked with red squares in a, b, c, and d were photographed at higher magnification and present in e, f, g, and h, respectively. The numbers 1, 2, and 3 in the square represent the position in the fruit, as 1 = below the groove, 2 = at the groove, and 3 = above the groove. a to d are on the same scale, and e to h are on the same scale

conversion is identical to that observed in *Arabidopsis pi* mutants (Goto & Meyerowitz, 1994) and apple *mdpi* mutants (Yao et al., 2001). This strong *pi* mutant phenotype in apple is consistent with absence of gene function redundancy because there is only a single *MdPI* gene present in apple.

Ectopic expression of *MdPI* in transgenic apple fully converts sepals to petals. However, the ectopic expression of the *MdPI* or *PI* gene in *Arabidopsis* only converts the base and margins of sepals to petal tissue (Tanaka et al., 2007). Ectopic expression of both *PI* and *AP3* together converts sepals to petals and carpels to stamens (Krizek & Meyerowitz, 1996), indicating that the functions of PI and AP3 are dependent on the coexistence of the two proteins to form a heterodimer (Wuest et al., 2012). In *Arabidopsis*, there is a single copy of the *AP3* gene that is expressed at the base of sepals and throughout petals and stamens. The conversion of the base of sepals to petal cell types is a result of coexistence of the endogenous AP3 and transgenic PI (Krizek & Meyerowitz, 1996). The apple genome contains four copies of *AP3* homologs (Figure S1) that are expressed in all four types of floral organs, although the expression in sepals and carpels is weaker than in petals and stamens (Kitahara et al., 2004; van der Linden et al., 2002). The full conversion of sepals to petals is likely the result of coexistence of endogenous AP3 homologs and transgenic MdPI in the first whorl floral organs of apple. However, it is not clear why there is no carpel to stamen conversion although there is likely coexistence of endogenous AP3 homologs and transgenic MdPI in the fourth whorl floral organs of the transgenic apple.

4.2 | Class B MADS-box genes regulate fruit development in apple

Interestingly, ectopic *MdPI* expression in apple changed the fruit shape, to a distinctive flattened fruit, as a consequence of

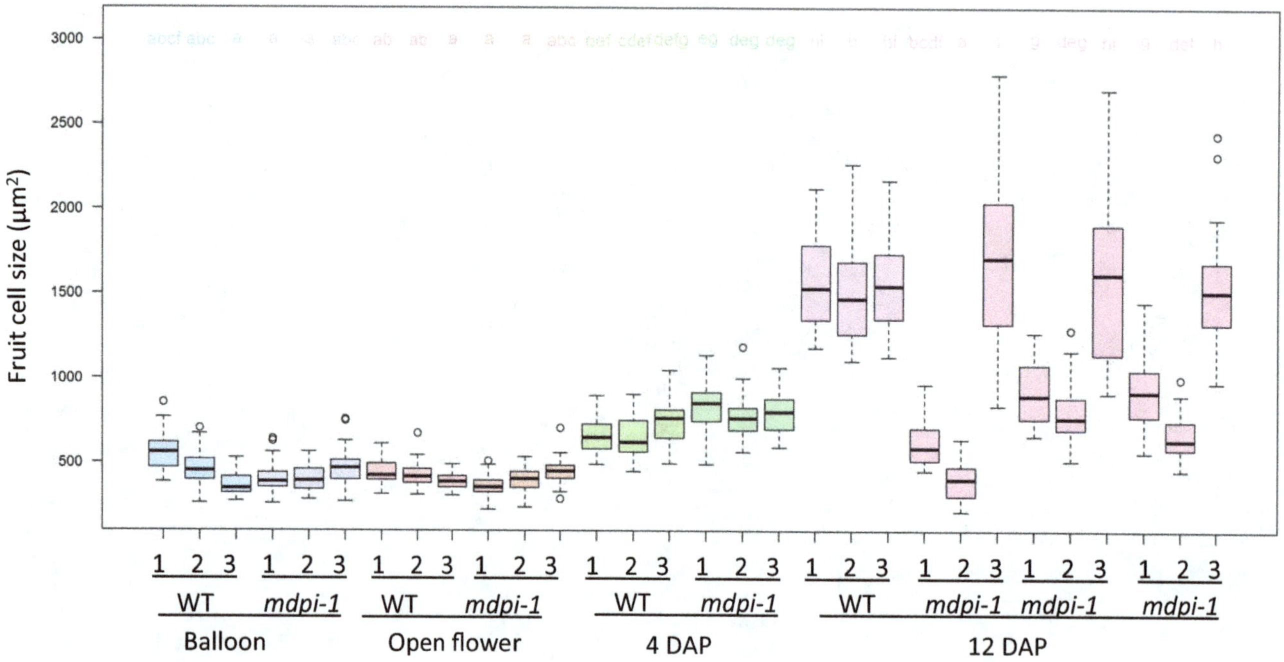

FIGURE 6 Hypanthium and fruit cell size of wild-type (WT) and *MdPI* overexpression (*mdpi-1*) "Bolero" apples. A box-plot shows the cell size distribution in hypanthium of balloon and open flowers and in fruit at 4 and 12 DAP (days after pollination) of WT and *mdpi-1* transgenic plants. The numbers 1, 2, and 3 in the *X*-axis labels represent the positions in the fruit as the same as those described in Figure 5 and Figure S5, that is, 1 = below the groove, 2 = at the groove, and 3 = above the groove. Thirty cells were measured at each of the three positions of a fruit. Significant differences between means ($n = 30$) were analyzed using Tukey's HSD (honest significant difference) test with a single-step multiple comparison procedure and the p value was set as .01

suppressed fruit tissue growth. Our analyses of the phenotype showed that localized fruit tissue growth was suppressed from early in fruit development, as indicated by the grooves formed in two different orientations on the fruit surface only a few days after fruit set. The suppression at the basipetal position of the fruit makes it short and flat, and the suppression on the sides of the fruit creates a pumpkin-like shape with grooves on the side of the fruit. Histological analyses showed that at 12 DAP cells were smaller in the grooved region and in the fruit tissue below the groove toward the fruit stalk. The tissue size of this region was reduced in two of the three fruit analyzed. Suppression of cell expansion may be the key reason for reduced tissue growth between 4 and 12 DAP although we cannot rule out the possibility of inhibition of cell division. From 18 DAP to fruit maturation, the basipetal groove was no longer observed and the fruit became short and flat, suggesting complete inhibition of tissue growth in this region from 18 DAP by inhibiting both cell expansion and cell division.

Ectopic expression of *PI* in *Arabidopsis* does not change silique development (Krizek & Meyerowitz, 1996; Tanaka et al., 2007) probably because *AP3* is not expressed in the silique (Jack et al., 1992). In contrast, ectopic expression of *VvPI* in grape inhibits fruit flesh tissue growth resulting in fleshless, small berries (Fernandez et al., 2013), as one of *AP3* homologs, *VvTM6*, is expressed in the berry (Poupin et al., 2007) and therefore a PI/AP3 heterodimer is able to form. The overall transcript level of *AP3* homologs in apple fruit is relatively low, suggesting that these *AP3* transcripts may be specifically localized in the fruit thereby dictating the localization of the AP3 protein, subsequent formation of AP3/PI heterodimer, and suppression of fruit cell growth.

When *MdPI* expression was suppressed, apple fruit tissue growth in the absence of fertilization was promoted resulting in the production of parthenocarpic apple fruit, a phenotype that has been observed in apple *mdpi* mutants but not been observed in *Arabidopsis pi* mutants. There is evidence to suggest that AP3 and PI homologs suppress ovary and fruit development in grape, tomato, and orchid (*Phalaenopsis equestris*), and such suppression can be released by pollination (Dauelsberg et al., 2011; Mazzucato, Olimpieri, Siligato, Picarella, & Soressi, 2008; Tsai et al., 2005). Our results, together with those of studies on grape, tomato, and orchid, have advanced our understanding of the function of PI/AP3 outside petals and stamens, and beyond that revealed in studies using *Arabidopsis*.

The external characteristics of fruit appearance, such as shape, color, and size, represent key attributes that consumers use to identify and select preferred cultivars, and it has been suggested that the manipulation of these is a largely untapped opportunity to create novel, differentiated products (Gamble, Jaeger, & Harker, 2006). While color and size are often a focus in commercial breeding programs, there is good evidence that the manipulation of fruit shape also represents an opportunity to create visually distinct products that can be branded (Gamble et al., 2006). A practical example is provided by the way the shape of flat peaches is being used to differentiate them from conventional cultivars (Romeu, Sanchez, & Garcia-Brunton, 2015). Genetic markers have been developed for the flat peaches and used in selection breeding seedlings (Picanol et al.,

2013). Mapping-based cloning approach has revealed that the candidate gene for the trait is possibly a LRR-RLK protein kinase rather than a MADS-box gene (Lopez-Girona et al., 2017). QTL underlying apple fruit shape variations have been located on chromosome 11 using a mapping population derived from a "Jonathan""Golden Delicious" cross (Cao et al., 2015; Chang et al., 2014). This is different from the location of *MdPI* that is on chromosome 8. Currently, most commercial apple cultivars are similar in shape, and the findings we have described here provide new molecular genetic information that could be used in genetic transformation of current apple cultivars to change their fruit shape. Also *Malus* germplasm resources could be exploited to identify accessions with different fruit shape index and determine whether there is a strong association between fruit shape index and DNA markers in the chromosomal region containing the *MdPI* gene. If such an association is determined, these markers can be used to accelerate the breeding of new fruit shape in apple to generate distinctive cultivars of higher commercial value.

ACKNOWLEDGMENTS

This work was supported by Plant & Food Research internal investment. J.X. was supported in part by the China Scholarship Council. The authors acknowledge Plant & Food Research colleagues Ian Hallett and Paul Sutherland for advice on light microscopy, Monica Dragulescu and Gnanaseela Wadasinghe for maintaining the apple trees in the glasshouse, Tim Holmes and Wara Bullôt for assistance with photography, and Bart Janssen and Charles Dwamena for commenting on the manuscript.

CONFLICT OF INTEREST

The authors declare no conflict of interests.

REFERENCES

Cao, K., Chang, Y. S., Sun, R., Shen, F., Wu, T., Wang, Y., ... Han, Z. H. (2015). Candidate gene prediction via quantitative trait locus analysis of fruit shape index traits in apple. *Euphytica*, *206*, 381–391. https://doi.org/10.1007/s10681-015-1488-y

Chang, S. J., Puryear, J., & Cairney, J. (1993). A simple and efficient method for isolating RNA from pine trees. *Plant Molecular Biology Reporter*, *11*, 113–116. https://doi.org/10.1007/BF02670468

Chang, Y. S., Sun, R., Sun, H. H., Zhao, Y. B., Han, Y. P., Chen, D. M., ... Han, Z. H. (2014). Mapping of quantitative trait loci corroborates independent genetic control of apple size and shape. *Scientia Horticulturae*, *174*, 126–132. https://doi.org/10.1016/j.scienta.2014.05.019

Dauelsberg, P., Matus, J. T., Poupin, M. J., Leiva-Ampuero, A., Godoy, F., Vega, A., & Arce-Johnson, P. (2011). Effect of pollination and fertilization on the expression of genes related to floral transition, hormone synthesis and berry development in grapevine. *Journal of Plant Physiology*, *168*, 1667–1674. https://doi.org/10.1016/j.jplph.2011.03.006

Drummond, R. S. M., Martinez-Sanchez, N. M., Janssen, B. J., Templeton, K. R., Simons, J. L., Quinn, B. D., ... Snowden, K. C. (2009). Petunia hybrida CAROTENOID CLEAVAGE DIOXYGENASE7 is involved in the production of negative and positive branching signals in petunia. *Plant Physiology*, *151*, 1867–1877. https://doi.org/10.1104/pp.109.146720

Fernandez, L., Chaib, J., Martinez-Zapater, J. M., Thomas, M. R., & Torregrosa, L. (2013). Mis-expression of a PISTILLATA-like MADS box gene prevents fruit development in grapevine. *The Plant Journal*, *73*, 918–928. https://doi.org/10.1111/tpj.12083

Gamble, J., Jaeger, S. R., & Harker, F. R. (2006). Preferences in pear appearance and response to novelty among Australian and New Zealand consumers. *Postharvest Biology and Technology*, *41*, 38–47. https://doi.org/10.1016/j.postharvbio.2006.01.019

Gillaspy, G., Bendavid, H., & Gruissem, W. (1993). Fruits - A developmental perspective. *The Plant Cell*, *5*, 1439–1451. https://doi.org/10.1105/tpc.5.10.1439

Gleave, A. P. (1992). A versatile binary vector system with a T-DNA organisational structure conducive to efficient integration of cloned DNA into the plant genome. *Plant Molecular Biology*, *20*, 1203–1207. https://doi.org/10.1007/BF00028910

Goto, K., & Meyerowitz, E. M. (1994). Function and regulation of the Arabidopsis floral homeotic gene pistillata. *Genes & Development*, *8*, 1548–1560. https://doi.org/10.1101/gad.8.13.1548

Ireland, H. S., Yao, J.-L., Tomes, S., Sutherland, P. W., Nieuwenhuizen, N., Gunaseelan, K., ... Schaffer, R. J. (2013). Apple SEPALLATA1/2-like genes control fruit flesh development and ripening. *The Plant Journal*, *73*, 1044–1056. https://doi.org/10.1111/tpj.12094

Jack, T., Brockman, L. L., & Meyerowitz, E. M. (1992). The homeotic gene apetala3 of arabidopsis-thaliana encodes a mads box and is expressed in petals and stamens. *Cell*, *68*, 683–697. https://doi.org/10.1016/0092-8674(92)90144-2

Kitahara, K., Ohtsubo, T., Soejima, J., & Matsumoto, S. (2004). Cloning and characterization of apple class MADS-box genes including a novel AP3 homologue MdTM6. *Journal of the Japanese Society for Horticultural Science*, *73*, 208–215. https://doi.org/10.2503/jjshs.73.208

Kotoda, N., Wada, M., Kusaba, S., Kano-Murakami, Y., Masuda, T., & Soejima, J. (2002). Overexpression of MdMADS5, an APETALA1-like gene of apple, causes early flowering in transgenic Arabidopsis. *Plant Science*, *162*, 679–687. https://doi.org/10.1016/S0168-9452(02)00024-9

Krizek, B. A., & Fletcher, J. C. (2005). Molecular mechanisms of flower development: An armchair guide. *Nature Reviews Genetics*, *6*, 688–698. https://doi.org/10.1038/nrg1675

Krizek, B. A., & Meyerowitz, E. M. (1996). The Arabidopsis homeotic genes APETALA3 and PISTILLATA are sufficient to provide the B class organ identity function. *Development*, *122*, 11–22.

Lee, H. L., & Irish, V. F. (2011). Gene duplication and loss in a MADS box gene transcription factor circuit. *Molecular Biology and Evolution*, *28*, 3367–3380. https://doi.org/10.1093/molbev/msr169

Liljegren, S. J., Ditta, G. S., Eshed, H. Y., Savidge, B., Bowman, J. L., & Yanofsky, M. F. (2000). SHATTERPROOF MADS-box genes control seed dispersal in Arabidopsis. *Nature*, *404*, 766–770. https://doi.org/10.1038/35008089

van der Linden, C. G., Vosman, B., & Smulders, M. J. M. (2002). Cloning and characterization of four apple MADS box genes isolated from vegetative tissue. *Journal of Experimental Botany*, *53*, 1025–1036. https://doi.org/10.1093/jexbot/53.371.1025

Liu, J. P., Van Eck, J., Cong, B., & Tanksley, S. D. (2002). A new class of regulatory genes underlying the cause of pear-shaped tomato fruit. *Proceedings of the National Academy of Sciences of the United States of America*, *99*, 13302–13306. https://doi.org/10.1073/pnas.162485999

Lopez-Girona, E., Zhang, Y., Eduardo, I., Mora, J. R. H., Alexiou, K. G., Arus, P., & Aranzana, M. J. (2017). A deletion affecting an LRR-RLK gene co-segregates with the fruit flat shape trait in peach. *Scientific Reports*, *7*, 6714. https://doi.org/10.1038/s41598-017-07022-0

Mazzucato, A., Olimpieri, I., Siligato, F., Picarella, M. E., & Soressi, G. P. (2008). Characterization of genes controlling stamen identity and development in a parthenocarpic tomato mutant indicates a role for the DEFICIENS ortholog in the control of fruit set. *Physiologia Plantarum, 132,* 526–537. https://doi.org/10.1111/j.1399-3054.2007.01035.x

Meyer, P. (1995). Variation of transgene expression in plants. *Euphytica, 85,* 359–366. https://doi.org/10.1007/BF00023968

Mimida, N., Ureshino, A., Tanaka, N., Shigeta, N., Sato, N., Moriya-Tanaka, Y., ... Wada, M. (2011). Expression patterns of several floral genes during flower initiation in the apical buds of apple (Malus x domestica Borkh.) revealed by in situ hybridization. *Plant Cell Reports, 30,* 1485–1492. https://doi.org/10.1007/s00299-011-1057-3

Picanol, R., Eduardo, I., Aranzana, M. J., Howad, W., Batlle, I., Iglesias, I., ... Arus, P. (2013). Combining linkage and association mapping to search for markers linked to the flat fruit character in peach. *Euphytica, 190,* 279–288. https://doi.org/10.1007/s10681-012-0844-4

Poupin, M. J., Federici, F., Medina, C., Matus, J. T., Timmermann, T., & Arce-Johnson, P. (2007). Isolation of the three grape sub-lineages of B-class MADS-box TM6, PISTILLATA and APETALA3 genes which are differentially expressed during flower and fruit development. *Gene, 404,* 10–24. https://doi.org/10.1016/j.gene.2007.08.005

Pratt, C. (1988). Apple flower and fruit: Morphology and anatomy. *Horticultural Reviews, 10,* 273–308.

Que, Q. D., Wang, H. Y., English, J. J., & Jorgensen, R. A. (1997). The frequency and degree of cosuppression by sense chalcone synthase transgenes are dependent on transgene promoter strength and are reduced by premature nonsense codons in the transgene coding sequence. *The Plant Cell, 9,* 1357–1368. https://doi.org/10.1105/tpc.9.8.1357

Rodriguez, G. R., Munos, S., Anderson, C., Sim, S. C., Michel, A., Causse, M., ... van der Knaap, E. (2011). Distribution of SUN, OVATE, LC, and FAS in the tomato germplasm and the relationship to fruit shape diversity. *Plant Physiology, 156,* 275–285. https://doi.org/10.1104/pp.110.167577

Romeu, J. F., Sanchez, M. C., & Garcia-Brunton, J. (2015). Potential productivity evolution of flat peach cultivars (Prunus persica var. platycarpa) grown in different climatic conditions of southeast of Spain. *Scientia Horticulturae, 197,* 687–696. https://doi.org/10.1016/j.scienta.2015.10.036

Spjut, R. W. (1994). A systematic treatment of fruit types. *Memoirs of the New York Botanical Garden, 70,* 1–182.

Sreekantan, L., Torregrosa, L., Fernandez, L., & Thomas, M. R. (2006). VvMADS9, a class B MADS-box gene involved in grapevine flowering, shows different expression patterns in mutants with abnormal petal and stamen structures. *Functional Plant Biology, 33,* 877–886. https://doi.org/10.1071/FP06016

Sung, S. K., Yu, G. H., & An, G. H. (1999). Characterization of MdMADS2, a member of the SQUAMOSA subfamily of genes, in apple. *Plant Physiology, 120,* 969–978. https://doi.org/10.1104/pp.120.4.969

Sung, S. K., Yu, G. H., Nam, J., Jeong, D. H., & An, G. (2000). Developmentally regulated expression of two MADS-box genes, MdMADS3 and MdMADS4, in the morphogenesis of flower buds and fruits in apple. *Planta, 210,* 519–528. https://doi.org/10.1007/s004250050040

Tanaka, N., Wada, M., Komori, S., Bessho, H., & Suzuki, A. (2007). Functional analysis of MdPI, the PISTILLATA gene homologue of apple, in Arabidopsis. *Journal of the Japanese Society for Horticultural Science, 76,* 125–132. https://doi.org/10.2503/jjshs.76.125

Tian, Y., Dong, Q. L., Ji, Z. R., Chi, F. M., Cong, P. H., & Zhou, Z. S. (2015). Genome-wide identification and analysis of the MADS-box gene family in apple. *Gene, 555,* 277–290. https://doi.org/10.1016/j.gene.2014.11.018

Tsai, W. C., Lee, P. F., Chen, H. I., Hsiao, Y. Y., Wei, W. J., Pan, Z. J., ... Chen, H. H. (2005). PeMADS6, a GLOBOSA/PISTILLATA-like gene in Phalaenopsis equestris involved in petaloid formation, and correlated with flower longevity and ovary development. *Plant and Cell Physiology, 46,* 1125–1139. https://doi.org/10.1093/pcp/pci125

Velasco, R., Zharkikh, A., Affourtit, J., Dhingra, A., Cestaro, A., Kalyanaraman, A, ... Viola, R. (2010). The genome of the domesticated apple (Malus x domestica Borkh.). *Nature Genetics, 42,* 833–839. https://doi.org/10.1038/ng.654

Weigel, D., & Meyerowitz, E. M. (1994). The abcs of floral homeotic genes. *Cell, 78,* 203–209. https://doi.org/10.1016/0092-8674(94)90291-7

Wuest, S. E., O'Maoileidigh, D. S., Rae, L., Kwasniewska, K., Raganelli, A., Hanczaryk, K., ... Wellmer, F. (2012). Molecular basis for the specification of floral organs by APETALA3 and PISTILLATA. *Proceedings of the National Academy of Sciences of the United States of America, 109,* 13452–13457. https://doi.org/10.1073/pnas.1207075109

Xiao, H., Jiang, N., Schaffner, E., Stockinger, E. J., & van der Knaap, E. (2008). A retrotransposon-mediated gene duplication underlies morphological variation of tomato fruit. *Science, 319,* 1527–1530. https://doi.org/10.1126/science.1153040

Yao, J.-L., Cohen, D., Atkinson, R., Richardson, K., & Morris, B. (1995). Regeneration of transgenic plants from the commercial apple cultivar Royal Gala. *Plant Cell Reports, 14,* 407–412.

Yao, J.-L., Cohen, D., Van den Brink, R., & Morris, B. (1999). Assessment of expression and inheritance patterns of three transgenes with the aid of techniques for promoting rapid flowering of transgenic apple trees. *Plant Cell Reports, 18,* 727–732. https://doi.org/10.1007/s002990050650

Yao, J.-L., Dong, Y.-H., Kvarnheden, A., & Morris, B. (1999). Seven MADS-box genes in apple are expressed in different parts of the fruit. *Journal of the American Society for Horticultural Science, 124,* 8–13.

Yao, J.-L., Dong, Y.-H., & Morris, B. A. (2001). Parthenocarpic apple fruit production conferred by transposon insertion mutations in a MADS-box transcription factor. *Proceedings of the National Academy of Sciences, 98,* 1306–1311. https://doi.org/10.1073/pnas.98.3.1306

Yao, J.-L., Tomes, S., & Gleave, A. P. (2013). Transformation of apple (Malus × domestica) using mutants of apple acetolactate synthase as a selectable marker and analysis of the T-DNA integration sites. *Plant Cell Reports, 32,* 703–714. https://doi.org/10.1007/s00299-013-1404-7

Yao, J.-L., Tomes, S., Xu, J., & Gleave, A. P. (2016). How microRNA172 affects fruit growth in different species is dependent on fruit type. *Plant Signaling & Behavior, 11,* e1156833. https://doi.org/10.1080/15592324.2016.1156833

Yao, J.-L., Xu, J., Cornille, A., Tomes, S., Karunairetnam, S., Luo, Z., ... Gleave, A. P. (2015). A microRNA allele that emerged prior to apple domestication may underlie fruit size evolution. *The Plant Journal, 84,* 417–427. https://doi.org/10.1111/tpj.13021

Zhong, S., Joung, J.-G., Zheng, Y., Chen, Y.-R., Liu, B., Shao, Y., ... Giovannoni, J. J. (2011). High-throughput illumina strand-specific RNA sequencing library preparation. *Cold Spring Harbor Protocols, 2011,* 940–949.

Fusarium graminearum-induced shoot elongation and root reduction in maize seedlings correlate with later seedling blight severity

Shaoqun Zhou[1,2] | **Justin S. Bae**[1] | **Gary C. Bergstrom**[3] | **Georg Jander**[1]

[1]Boyce Thompson Institute, Ithaca, New York

[2]Plant Biology Section, School of Integrated Plant Science, Cornell University, Ithaca, New York

[3]Plant Pathology and Plant-Microbe Biology Section, School of Integrated Plant Science, Cornell University, Ithaca, New York

Correspondence
Shaoqun Zhou, Boyce Thompson Institute, 533 Tower Road, Ithaca, NY 14853.
Email: sz357@cornell.edu

Funding information
Northeast Sustainable Agriculture and Research Education Graduate Student Grant, Grant/Award Number: GNE14-092; USDA-NIFA Hatch Grant, Grant/Award Number: NYC1537436; US National Science Foundation, Grant/Award Number: IOS-1339237, DBI-1358843

Abstract

Fusarium graminearum seedling blight is a common disease of maize (*Zea mays*). Development of genetic resistance to seedling blight in maize germplasm requires efficient and accurate quantitative assessment of disease severity. Through artificial inoculation experiments under controlled growth conditions, we determined that host genotype, pathogen genotype, and infection dose influence the extent to which *F. graminearum* induces shoot elongation and inhibits root growth in maize seedlings. A comparison of 15 maize inbred lines showed independent variation of these two fungus-induced effects on seedling growth. In a broader survey with nine commercial maize hybrids and three field-collected fungal isolates, there was significant correlation between these seedling growth responses, as well as with later seedling blight severity. Analysis of variance suggested that this variation and the observed correlative relationships were primarily driven by differing pathogenicity of the three fungal isolates. Together, our results indicate that *F. graminearum*-induced shoot elongation and root reduction in maize seedlings have distinct underlying physiological mechanisms, and that early observations of seedling growth responses can serve as a proxy for investigating natural variation in host resistance and pathogen aggressiveness at later growth stages.

KEYWORDS
Fusarium graminearum seedling blight, maize, root reduction, shoot elongation

1 | INTRODUCTION

Fusarium graminearum Schwabe (teleomorph: *Gibberella zeae*) is a widespread fungal phytopathogen that causes seedling blight, root rot, stem rot, or ear rot in maize (*Zea mays*), depending on the timing and tissue of infection (Munkvold & White, 2016). The severity of maize diseases caused by *F. graminearum* in the United States varies significantly from year to year, causing 15–80 million bushels of annual yield loss due to ear rot, and another 30 to 85 million bushels due to stalk rot (Mueller, 2016a,b,c, 2017). Unlike Gibberella ear rot and stem rot diseases caused by *F. graminearum*, which can be tracked by the characteristic pink-reddish pigmentation of fungal mycelium, seedling blight, and root rot diseases are often caused by a consortium of soil-dwelling fungal and oomycete pathogens, including *F. graminearum*. While the dominant species/strains of these consortia are likely to vary considerably across geographic locations, collectively these seedling blight and root rot diseases can account for over 200 million bushels of annual yield loss, making them among the most damaging maize diseases in the northern states (Mueller, 2016b). Management of *F. graminearum* seedling blight can be difficult because standard seed fungicide treatment is not always sufficient and may promote the development of resistance in the fungal population. Genetic sources of seedling blight resistance have yet to be established in commercial maize germplasm.

In addition to causing direct yield loss through tissue rotting, *F. graminearum* can also cause crop contamination with mycotoxins such as deoxynivalenol and zearalenone, leaving the grain and stover unsafe for human and livestock consumption, and imposing further financial burdens on the growers (Arunachalam & Doohan, 2013; Maresca, 2013). Furthermore, *F. graminearum* can also cause destructive scab and head blight diseases in barley and wheat (McMullen et al., 2012; Trail, 2009). Considering the overwinter persistence and airborne infectious ascospores of this pathogen, unchecked *F. graminearum* diseases in maize could have a far-reaching negative impact on other crop species in the vicinity.

Breeders and researchers interested in improving resistance against *F. graminearum* in maize have been doing artificial inoculation experiments in both field and controlled environments, primarily focused on infected maize ears and stalks. Genetic mapping of Gibberella ear rot resistance based on symptom severity has identified a large number of quantitative trait loci (QTL), each with small effect size. Moreover, these QTL tend to be sensitive to experimental methods and environmental conditions, making them difficult to validate across different studies (Ali et al., 2005; Brauner et al., 2017; Kebede, Woldemariam, Reid, & Harris, 2016). In contrast, major genetic loci contributing to Gibberella stalk rot resistance have been identified with bi-parental mapping populations (Chen et al., 2017; Ma et al., 2017; Yang et al., 2010; Zhang et al., 2012). Ye et al. (2013) observed that maize near-isogenic lines resistant against Gibberella stalk rot also showed less severe seedling root browning and shrinkage after *F. graminearum* inoculation. This suggests that resistance against Gibberella stalk rot and *F. graminearum* seedling blight, both diseases in vegetative tissues, are likely controlled by similar genetic and physiological mechanisms, and are distinct from those affecting Gibberella ear rot resistance. This is consistent with the known influence of tissue type and developmental stage on the interactions between maize and *F. graminearum* (Zhang et al., 2016).

Genetic improvement of *F. graminearum* resistance depends on accurate quantitative assessment of resistant and susceptible phenotypes. Here, we show that *F. graminearum* inoculation of maize seedlings not only inhibits root growth, but can also promote shoot elongation in a dosage- and genotype-dependent manner. Shoot elongation in response to *F. graminearum* root infection is significantly correlated with seedling survival, suggesting that these traits could be used to assess seedling resistance, as well as serving as a quantitative measure of the aggressiveness of different *F. graminearum* isolates.

2 | MATERIALS AND METHODS

2.1 | Plant and fungal materials

Fifteen parental lines from the maize nested association mapping population (McMullen et al., 2009) were chosen for *F. graminearum* infection assays. For a broader survey, nine organic commercial maize hybrids were kindly provided by the Blue River Organic Seed Company (Supporting Information Table S1). The maize hybrids were selected to cover different maturation time and provider-assessed seedling vigor levels. The accession numbers of the tested hybrids are masked per the provider's request. Three field-collected *F. graminearum* isolates (Supporting Information Table S2) that have shown different levels of aggressiveness in a previous laboratory test (Kuhnem, Del Ponte, Dong, & Bergstrom, 2015) were used for inoculations (Supporting Information Table S2). Fungal chemotype, namely the type of major mycotoxin accumulated, was not considered during isolate selection, since it was not significantly associated with aggressiveness in the previous test.

2.2 | Plant growth and fungal inoculation methods

Maize seeds were germinated in moisturized rolls of germination paper after surface sterilization with 10% household bleach for 30 min. Germinated seedlings were transplanted to 7.5 cm × 7.5 cm plastic pots with Turface ® MVP ® calcined clay (Profile Products LLC, Buffalo Grove, IL) when their primary roots reached approximately 9 cm, which would take 3–5 days depending on the specific maize genotype. For inoculation experiments, seedling roots were immersed in *F. graminearum* spore suspension or mock solution (0.03% Phytagar) for 1 hr prior to transplanting. Fungal spore suspensions were freshly prepared by flooding and scraping 7-day-old fungal cultures maintained on Potato Dextrose Agar plates under black lights with 0.03% Phytagar suspension. The spore concentration was estimated with a hemocytometer (Hausser Scientific, Horsham, PA) and light microscopy. Fungal hyphae fragments were observed in spore suspension by light microscopy.

For the concentration gradient experiment, 24 seedlings of each maize hybrid were grown, and six were inoculated with each tested fungal spore concentration and mock inoculum. For each pair of maize hybrid and fungal isolate included in the diversity screening, at least 15 seedlings were inoculated with fungal spores, and 5 were mock-inoculated. Transplanted seedlings were kept at 26°C (Day)/ 23°C (Night), 60% humidity, and a 16 hr long-day light cycle. Seedlings inoculated with different *F. graminearum* isolates or mock inoculum were kept in separated growth chambers set to identical growth conditions.

2.3 | Phenotypic data collection and analysis

The heights of fungus- and mock-inoculated seedlings were measured 9 days post-inoculation. On the same day, seedlings were removed from pots to measure their total root length using an image analysis algorithm implemented in the RootReader 2D software (Clark et al., 2013). After measurement, seedlings were repotted in the same Turface particles. At 14 days post-inoculation, the number of surviving seedlings within each fungus-inoculated maize hybrid population was counted as a measurement of

F. graminearum seedling blight severity. Seedlings were considered dead when they broke at the soil line, with browning and rot at the breakpoint.

For seedling height and total root length, an average ratio of mock- and fungus-inoculated seedlings for each maize hybrid-fungus isolate combination was calculated to shown variation in these fungus-induced responses. Measurements from fungus- and mock-inoculated seedlings were compared using Student's *t* tests. Average ratios were rounded to one when measurements from the two groups were not significantly different from one another. For two-way ANOVA, maturation times of maize hybrids were grouped into four bins, as indicated in parentheses in Supporting Information Table S1. The average ratios of seedling height and total root length between fungus- and mock-treated seedlings were used as quantitative measurements to calculate linear correlation with seedling survival rate measured at 14 days after infection.

3 | RESULTS

3.1 | *Fusarium graminearum* inoculation at low concentration induces shoot elongation and root reduction in maize seedlings

Different concentrations of spore suspensions have been used to study natural variation in maize seedling responses to *F. graminearum* inoculation (Kuhnem et al., 2015; Ye et al., 2013). To determine the optimal concentration of *F. graminearum* spores that leads to the largest change in plant morphology, two concentration gradient experiments were performed with Gz014, a field-collected *F. graminearum* strain (Supporting Information Table S2), and two commercial maize hybrids with differing seedling vigor levels (Hybrids IV and V in Supporting Information Table S1, as assessed by the provider, Blue River Organic Seed Company).

At 1.2×10^6 spores/ml, a concentration that is comparable to a previous study, the seedling stunting phenotype that had been reported was not observed in either of the tested maize hybrids (Kuhnem et al., 2015). Instead, the hybrid with high seedling vigor showed no significant change in seedling height, whereas the moderately vigorous hybrid seedlings grew significantly taller when inoculated with *F. graminearum* spores. At lower concentrations, fungus-induced shoot elongation became obvious in both highly and moderately vigorous hybrid seedlings (Figure 1a). Underground, *F. graminearum* infection significantly reduced root growth in the moderately vigorous hybrid, with no evidence of a dosage response. This fungus-induced change was not observed in the highly vigorous hybrid, which showed no significant change in total root length at all tested doses (Figure 1b).

For both fungus-induced growth responses, inoculation with the lowest fungal spore concentration (3×10^5 count/ml) resulted in a response that was stronger than or comparable to that observed with the two higher inoculum concentrations. Therefore, the 3×10^5 spores/ml *F. graminearum* dose was used for all subsequent experiments.

3.2 | *Fusarium graminearum*-induced shoot elongation and root reduction vary independently across maize inbred lines

To confirm the *F. graminearum*-induced shoot elongation and root reduction observed in the concentration gradient experiment, the artificial inoculation experiment was repeated with the reference maize genetic inbred line B73 with the same fungal isolate following the same protocol. Consistently, the same fungus-induced growth responses were observed (Figure 2). To investigate how *F. graminearum*-induced seedling shoot elongation and root reduction may be related, seedlings of 15 maize inbred lines were inoculated with Gz014 or mock inoculum and their shoot and root growth were quantified 9 days after inoculation. Although *F. graminearum* inoculation reduced total root length in most inbred lines by approximately 50%, no significant difference was observed between mock- and *F. graminearum*-inoculated NC350, NC358, and Mo17 seedlings, (Figure 3a).

In the same experiment, *F. graminearum*-induced shoot elongation also varied, ranging from 0 to over 25% increase (Figure 3b). Noticeably, the two fungus-induced growth responses do not correlate with one another across the 15 tested inbred lines tested (Pearson's $R^2 = 0.0079$, $p > 0.05$). Two inbred lines showing no significant *F. graminearum*-induced root reduction, NC350 and NC358, grew 16% and 24% taller, respectively, after *F. graminearum* inoculation. Conversely, Tx303 only showed obvious root reduction (40%) but no significant shoot elongation (Figure 3). Inbred line Mo17 showed neither of the tested fungus-induced growth responses.

3.3 | *Fusarium graminearum*-induced growth responses and *F. graminearum* seedling blight severity are primarily determined by the fungal genotype

To further survey the natural variation in *F. graminearum*-induced seedling shoot elongation and root reduction, the artificial inoculation experiment was expanded to the nine commercial maize hybrids, including the two used in the concentration gradient experiment described above (Supporting Information Table S1), and three *F. graminearum* isolates (Supporting Information Table S2). Both *F. graminearum*-induced growth responses were highly variable across this panel (Figure 4). Two-way analysis of variance (ANOVA) performed on both traits showed that fungal isolate, rather than maize maturation time or provider-assessed seedling vigor, is the main source of assay variance (Supporting Information Table S3).

After quantifying *F. graminearum*-induced growth responses, the fungus-infected seedlings were kept under the same growth conditions, and *F. graminearum* seedling blight severity was quantified as the percentage of seedlings surviving in each maize hybrid population 14 days after inoculation. As in the case of the morphological data, ANOVA of the seedling survival rate data identified the *F. graminearum* isolate as the most important contributor. Whereas both Gz014 and Gz829 infection left less than 50% of inoculated seedlings surviving after 14 days, more than 85% of Gz835-

inoculated seedlings survived in the same time period (Figure 5). The maturation time and seedling vigor of maize hybrids were not significant sources of variance (Supporting Information Table S3).

3.4 | *Fusarium graminearum*-induced shoot elongation is positively correlated with *F. graminearum* seedling blight severity

To investigate the relationship between *F. graminearum*-induced growth responses and *F. graminearum* seedling blight severity, linear regression analysis was performed on the extent of these changes and seedling survival. This showed that *F. graminearum*-induced seedling elongation has a significant negative correlation with seedling survival rate at 14 days after inoculation (Figure 6a; Pearson's $R^2 = 0.244$; $p < 0.005$). Thus, stronger fungus-induced shoot elongation may be predictive of more severe *F. graminearum* seedling blight. Unexpectedly, the extents of *F. graminearum*-induced shoot elongation showed strong positive correlation with root length (Figure 6b; Pearson's $R^2 = 0.676$; $p < 0.005$), and a significant negative correlation between root reduction and seedling survival rate

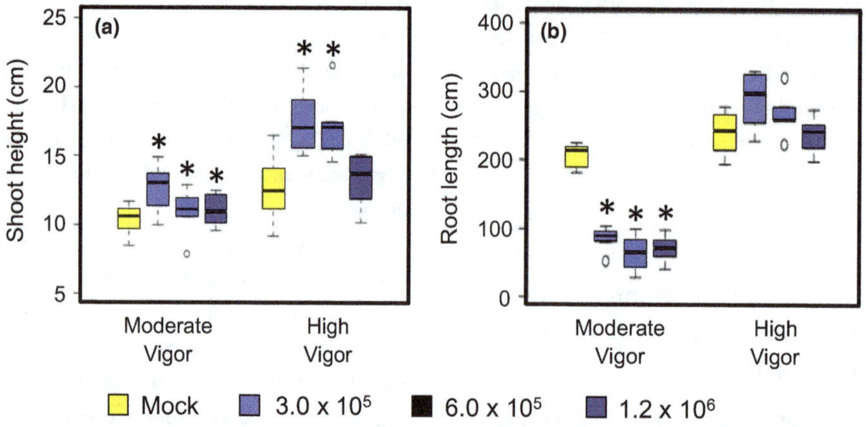

FIGURE 1 *Fusarium graminearum* inoculation induces shoot elongation and root reduction in maize seedlings in a dosage- and host genotype-dependent manner. Shoot height (a) and total root length (b) measurements on seedlings inoculated at each fungal spore concentration were compared to mock-treated seedlings with Dunnett's tests relative to mock-infected controls ($N = 6$ for each group; * $p < 0.05$). Group means are indicated by the black bars, and the interquartile range by the upper and lower edges of the boxes. Outliers extending more than 1.5x of the interquartile range are indicated by individual circles, and the whiskers show the range of each group excluding the outliers. Fungal strain Gz014 was used for this experiment. All measurements were taken 9 days post-inoculation. Fungal spore concentration is in counts/ml

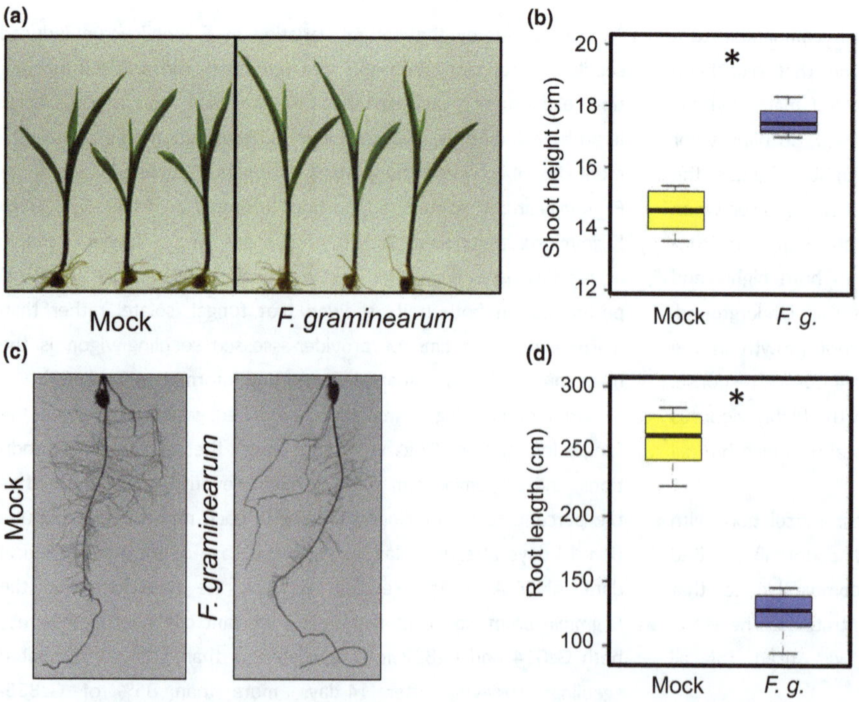

FIGURE 2 *Fusarium graminearum* inoculation induces shoot elongation and root reduction in maize inbred line B73. Representative photos of shoots (a) and roots (c) mock- and fungus-inoculated B73 seedlings were taken 9 days post-inoculation. Shoot height (b) and total root length (d) were measured on the same day and compared with Student's *t*-tests ($N = 6$ for each group; * $p < 0.05$). Group means are indicated by the black bars, and the interquartile range by the upper and lower edges of the boxes. Outliers extending more than 1.5x of the interquartile range were indicated by individual circles, and the whiskers showed the range of each group excluding the outliers. Fungal strain Gz014 was used for this experiment. Spore concentration = 3×10^5 counts/ml

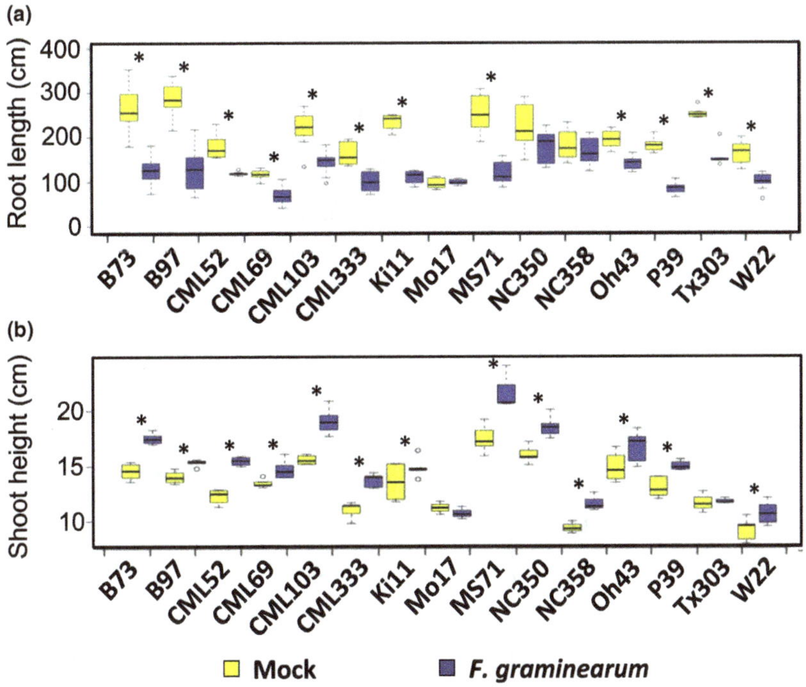

FIGURE 3 *Fusarium graminearum*-induced growth responses in maize seedlings vary across maize inbred lines. Root length (a) and shoot height (b) of each maize inbred line inoculated with *F. graminearum* spores (yellow) were compared to measurement from a corresponding mock treatment group (blue) with Student's *t* test (*$p < 0.05$). Group means are indicated by the black bars, and the interquartile range by the upper and lower edges of the boxes. Outliers extending more than 1.5x of the interquartile range are indicated by individual circles, and the whiskers show the range of each group excluding the outliers. All measurements were taken at 9 days post-inoculation. Fungal strain Gz014 was used for this experiment. Spore concentration = 3×10^5 counts/ml

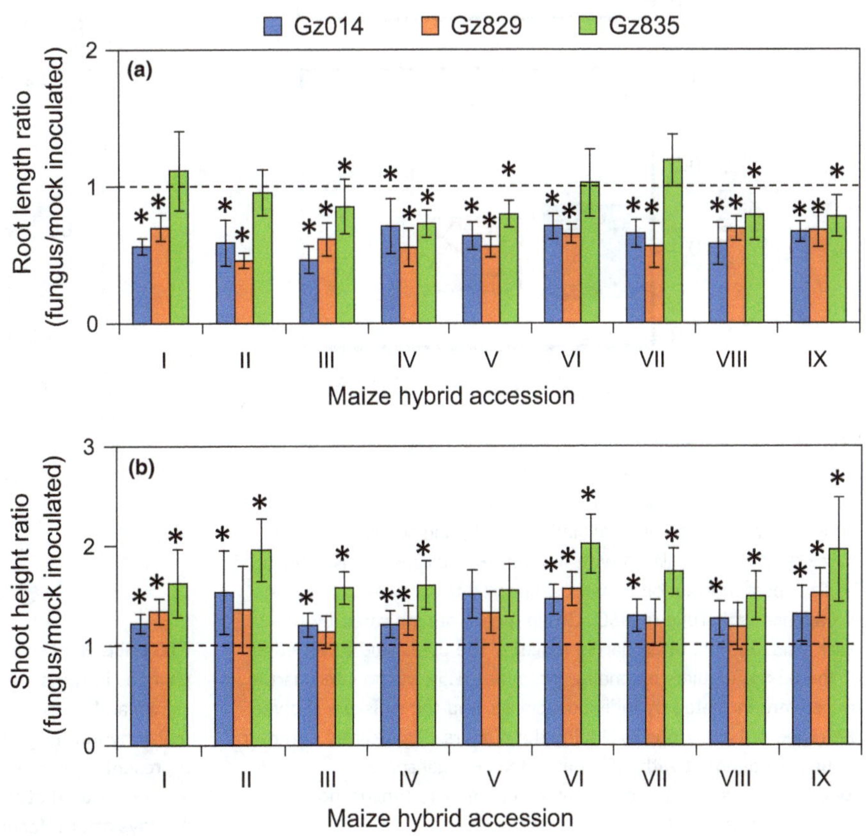

FIGURE 4 *Fusarium graminearum*-induced growth responses in maize seedlings vary with both plant and fungal genotype. Nine maize hybrid lines were inoculated with three *F. graminearum* isolates (Gz014, Gz829, and Gz835). Total root length (a) and shoot height (b) of each maize hybrid-fungus isolate pair were compared to measurement from a corresponding mock treatment group with Student's *t*-test (*$p < 0.05$). All measurements were taken at 9 days post-inoculation. Fungal spore concentration = 3×10^5 count/ml

(Figure 6c; Pearson's $R^2 = 0.263$; $p < 0.005$). This leads to the counter-intuitive inference that resistance to fungus-induced root growth inhibition co-occurs with stronger shoot elongation and more severe *F. graminearum* seedling blight.

4 | DISCUSSION

Under field conditions, maize seedlings infected by *F. graminearum* as well as other fungal and oomycete pathogens are often characterized by stunted growth and dehydration in the aboveground tissues before death. Similar seedling stunting symptoms have also been observed in a previous artificial inoculation test with *F. graminearum* spores under laboratory conditions (Kuhnem et al., 2015). However, in our concentration gradient experiment with the same fungal isolates, low dosage *F. graminearum* inoculation on maize seedling root induced shoot elongation (Figures 1a and 2a). Based on the natural variation in this *F. graminearum*-induced response across different maize inbred lines and hybrids reported in this study (Figures 3a and 4a), it is likely that the inconsistent observations between studies could arise from inoculum concentration, maize genotype, or unintended differences between experimental conditions.

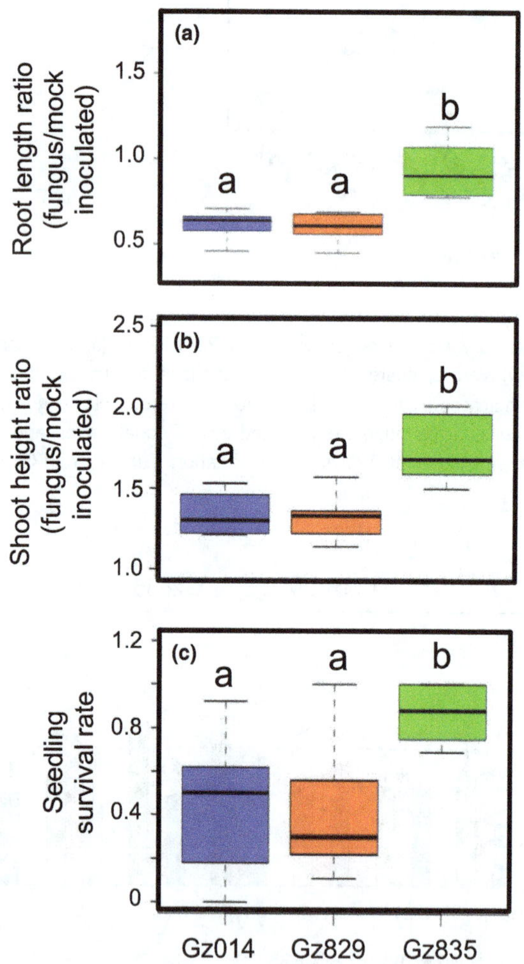

FIGURE 5 *Fusarium graminearum*-induced growth responses and seedling blight severity vary significantly between fungal isolates. The extent of *F.* graminearum-induced root reduction and shoot elongation at 9 days post-inoculation (a & b) and survival rates of maize seedlings inoculated with different *F. graminearum* isolates at 14 days post-inoculation (c) were compared using two-way analysis of variance with Tukey's HSD. Group means are indicated by the black bars, and the interquartile range by the upper and lower edges of the boxes. Outliers extending more than 1.5× of the interquartile range were indicated by individual circles, and the whiskers showed the range of each group excluding the outliers. Significantly different groups are denoted with different letters. Fungal spore concentration = 3×10^5 count/ml. Ratios refer to fungus/mock-inoculated plants

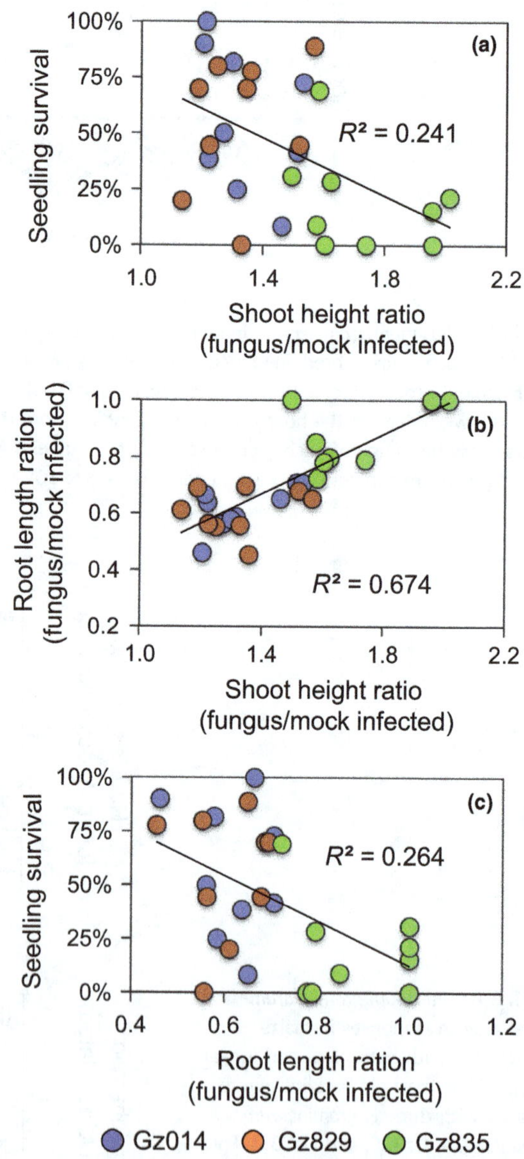

FIGURE 6 *Fusarium graminearum*-induced growth parameter changes in roots and shoots, and seedling blight severity, which are significantly correlated with each other. Each mark represents a maize hybrid-fungal isolate pair, with different fungal isolates represented by different colors. Root length and seedling height were measured at 9 days after infection. Survival was assessed at 14 days after infection

Interestingly, the observed *F. graminearum*-induced shoot elongation phenotype is reminiscent of the foolish seedling disease in rice, which is caused by a related fungal pathogen *Gibberella fujikuroi*. Rice seedlings affected by foolish seedling disease demonstrate abnormal seedling elongation resulting from the production of gibberellic acid (GA), a growth-promoting phytohormone, *G. fujikuroi* (Yabuta & Sumiki, 1938). Fungus-derived GA has been hypothesized to be a virulence factor by negatively influencing the signaling transduction pathway of the immunity-regulating phytohormone, jasmonic acid (Navarro et al., 2008). However, unlike *G. fujikuroi*, *F. graminearum* is not known to produce GA, and its genome does not contain biosynthetic genes required for GA biosynthesis (Cuomo et al., 2007). However, we cannot rule out the possibility that *F. graminearum* infection induces endogenous maize GA production.

Consistent with previous reports (Kuhnem et al., 2015; Ye et al., 2013), *F. graminearum* infection induced maize seedling root growth reduction in a host genotype-dependent manner (Figures 1b, 2b, 3b and 4b). Application of a computer image-processing tool, RootReader 2D (Clark et al., 2013), in our study enabled more accurate, quantitative, and efficient measurement of fungus-induced growth response, and its natural variation with host and pathogen genotype. Fungus-induced root reduction and shoot elongation measurements from the same experiment on 15 maize inbred lines showed no correlation in the extent of these two traits, with some inbred lines showing only one or the other fungus-induced change (Figure 3a,b). This observation suggests that these two responses may be governed by distinct physiological mechanisms.

As one of the most widespread fungal phytopathogens, *F. graminearum* is known to be highly diverse, and different strains are highly variable in their aggressiveness (Backhouse, 2014; Boutigny et al., 2011; Lee et al., 2010; Sampietro et al., 2011; Wang, Ndoye, Zhang, Li, & Liao, 2011). Our survey of nine commercial maize hybrids and three fungal isolates (Figures 4 and 5) is consistent with prior observations that the prevalence of *F. graminearum*-related diseases is highly variable across multiple locations, depending on the aggressiveness of the predominant isolate and environmental permissiveness (Backhouse, 2014; Boutigny et al., 2011; Lee et al., 2010; Sampietro et al., 2011; Wang et al., 2011).

Unexpectedly, maize hybrid-fungal isolate pairs with stronger fungus-induced root growth inhibition tended to show a lower level of induced shoot elongation (Figure 6c) and better survival (Figure 6b). This was surprising, because severe reduction in root growth was expected to significantly impair the seedlings' ability to acquire water and nutrients from the soil, which could lead to seedling death. Two-way ANOVA and a closer look at the data distribution suggest that these correlative relationships are primarily driven by differences between the fungus isolates. Specifically, the isolate that induced significantly stronger shoot elongation, Gz835, induced less root reduction and seedling death (Figure 5). This suggests that these fungus-induced growth responses can be used for quantitative assessment of the aggressiveness of *F. graminearum* isolates, and that different *F. graminearum* isolates may have distinct pathogenic mechanisms when interacting with maize seedlings. Further research with maize seedlings growing in agricultural fields will be required to determine whether *F. graminearum*-induced shoot elongation can serve as a reliable predictor of fungal sensitivity in commercially grown maize.

ACKNOWLEDGMENTS

The authors would like to thank the Blue River Organic Seed Company for providing the maize hybrids used in this study, Jon Shaff and Eric Craft for training and assistance in using the RootReader 2D platform, and Jaime Cummings for training and assistance in fungal culture maintenance. This work was funded by the Northeast Sustainable Agriculture and Research Education Graduate Student Grant to S.Z. (GNE14-092), USDA-NIFA Hatch Grant NYC1537436 to G.C.B., and US National Science Foundation IOS-1339237 to G.J. J.S.B. was supported by the Plant Genome Research Program at the Boyce Thompson Institute, a Research Experience for Undergraduate program funded by the US National Science Foundation award DBI-1358843.

AUTHOR CONTRIBUTIONS

S.Z. conceived, designed, and performed the experiments, analyzed the data, and wrote the manuscript. J.S.B. assisted in the natural variation experiments and related data analyses. G.C.B. provided the fungal isolates used in the experiments, and advice in experimental design and manuscript revision. G.J. provided infrastructural support to this study, and advice in experimental design and manuscript revision.

ORCID

Shaoqun Zhou http://orcid.org/0000-0002-3643-7875
Justin S. Bae http://orcid.org/0000-0002-7166-9030
Gary C. Bergstrom http://orcid.org/0000-0001-9613-270X
Georg Jander http://orcid.org/0000-0002-9675-934X

REFERENCES

Ali, M. L., Taylor, J. H., Jie, L., Sun, G., William, M., Kasha, K. J., … Pauls, K. P. (2005). Molecular mapping of QTLs for resistance to Gibberella ear rot, in corn, caused by *Fusarium graminearum*. *Genome*, *48*(3), 521–533. https://doi.org/10.1139/g05-014

Arunachalam, C., & Doohan, F. M. (2013). Trichothecene toxicity in eukaryotes: Cellular and molecular mechanisms in plants and animals. *Toxicology Letters*, *217*(2), 149–158. https://doi.org/10.1016/j.toxlet.2012.12.003

Backhouse, D. (2014). Global distribution of *Fusarium graminearum*, *F. asiaticum* and *F. boothii* from wheat in relation to climate. *European Journal of Plant Pathology*, *139*(1), 161–173. https://doi.org/10.1007/s10658-013-0374-5

Boutigny, A. L., Ward, T. J., Van Coller, G. J., Flett, B., Lamprecht, S. C., O'Donnell, K., & Viljoen, A. (2011). Analysis of the *Fusarium graminearum* species complex from wheat, barley and maize in South Africa provides evidence of species-specific differences in host preference.

Fungal Genetics and Biology, 48(9), 914–920. https://doi.org/10.1016/j.fgb.2011.05.005

Brauner, P. C., Melchinger, A. E., Schrag, T. A., Utz, H. F., Schipprack, W., Kessel, B., … Miedaner, T. (2017). Low validation rate of quantitative trait loci for Gibberella ear rot resistance in European maize. *TAG. Theoretical and Applied Genetics., 130*(1), 175–186. https://doi.org/10.1007/s00122-016-2802-3

Chen, Q., Song, J., Du, W. P., Xu, L. Y., Jiang, Y., Zhang, J., … Yu, G. R. (2017). Identification, mapping, and molecular marker development for Rgsr8.1: A new quantitative trait locus conferring resistance to Gibberella stalk rot in maize (Zea mays L.). *Frontiers in Plant Science, 8*, 1355. https://doi.org/10.3389/fpls.2017.01355

Clark, R. T., Famoso, A. N., Zhao, K. Y., Shaff, J. E., Craft, E. J., Bustamante, C. D., … Kochian, L. V. (2013). High-throughput two-dimensional root system phenotyping platform facilitates genetic analysis of root growth and development. *Plant, Cell and Environment, 36*(2), 454–466. https://doi.org/10.1111/j.1365-3040.2012.02587.x

Cuomo, C. A., Guldener, U., Xu, J. R., Trail, F., Turgeon, B. G., Di Pietro, A., … Kistler, H. C. (2007). The *Fusarium graminearum* genome reveals a link between localized polymorphism and pathogen specialization. *Science, 317*(5843), 1400–1402. https://doi.org/10.1126/science.1143708

Kebede, A. Z., Woldemariam, T., Reid, L. M., & Harris, L. J. (2016). Quantitative trait loci mapping for Gibberella ear rot resistance and associated agronomic traits using genotyping-by-sequencing in maize. *TAG. Theoretical and Applied Genetics., 129*(1), 17–29. https://doi.org/10.1007/s00122-015-2600-3

Kuhnem, P. R., Del Ponte, E. M., Dong, Y. H., & Bergstrom, G. C. (2015). *Fusarium graminearum* ssolates from wheat and maize in New York show similar range of aggressiveness and toxigenicity in cross-species pathogenicity tests. *Phytopathology, 105*(4), 441–448. https://doi.org/10.1094/Phyto-07-14-0208-R

Lee, S. H., Lee, J., Nam, Y. J., Lee, S., Ryu, J. G., & Lee, T. (2010). Population structure of *Fusarium graminearum* from maize and rice in 2009 in Korea. *The Plant Pathology Journal, 26*(4), 321–327. https://doi.org/10.5423/Ppj.2010.26.4.321

Ma, C., Ma, X., Yao, L., Liu, Y., Du, F., Yang, X., & Xu, M. (2017). qRfg3, a novel quantitative resistance locus against Gibberella stalk rot in maize. *TAG. Theoretical and Applied Genetics., 130*(8), 1723–1734. https://doi.org/10.1007/s00122-017-2921-5

Maresca, M. (2013). From the gut to the brain: Journey and pathophysiological effects of the food-associated trichothecene mycotoxin deoxynivalenol. *Toxins, 5*(4), 784–820. https://doi.org/10.3390/toxins5040784

McMullen, M., Bergstrom, G., De Wolf, E., Dill-Macky, R., Hershman, D., Shaner, G., & Van Sanford, D. (2012). A unified effort to fight an enemy of wheat and barley: Fusarium head blight. *Plant Disease, 96*(12), 1712–1728. https://doi.org/10.1094/Pdis-03-12-0291-Fe

McMullen, M. D., Kresovich, S., Villeda, H. S., Bradbury, P., Li, H., Sun, Q., … Buckler, E. S. (2009). Genetic properties of the maize nested association mapping population. *Science, 325*(5941), 737–740. https://doi.org/10.1126/science.1174320

Mueller, D. (2016a). Corn disease loss estimates from the United States and Ontario, Canada - 2013. Purdue University Extension, Retrieved from http://cropprotectionnetwork.org/crop-loss-estimates/corn-disease-loss-estimates-2013/.

Mueller, D. (2016b). Corn disease loss estimates from the United States and Ontario, Canada - 2014. Purdue University Extension, Retrieved from http://cropprotectionnetwork.org/crop-loss-estimates/corn-disease-loss-estimates-2014/.

Mueller, D. (2016c). Corn disease loss estimates from the United States and Ontario, Canada - 2015. Purdue University Extension, Retrieved from http://cropprotectionnetwork.org/crop-loss-estimates/corn-disease-loss-estimates-2015/.

Mueller, D. (2017). Corn disease loss estimates from the United States and Ontario, Canada - 2016. Purdue University Extension, Retrieved from http://cropprotectionnetwork.org/crop-loss-estimates/corn-disease-loss-estimates-2016/.

Munkvold, G. P., & White, D. G. (2016). *Compendium of Corn Diseases*, 4th ed. St. Paul, MN: The American Phytopathological Society.

Navarro, L., Bari, R., Achard, P., Lison, P., Nemri, A., Harberd, N. P., & Jones, J. D. G. (2008). DELLAs control plant immune responses by modulating the balance and salicylic acid signaling. *Current Biology, 18*(9), 650–655. https://doi.org/10.1016/j.cub.2008.03.060

Sampietro, D. A., Diaz, C. G., Gonzalez, V., Vattuone, M. A., Ploper, L. D., Catalan, C. A. N., & Ward, T. J. (2011). Species diversity and toxigenic potential of *Fusarium graminearum* complex isolates from maize fields in northwest Argentina. *International Journal of Food Microbiology, 145*(1), 359–364. https://doi.org/10.1016/j.ijfoodmicro.2010.12.021

Trail, F. (2009). For Blighted Waves of Grain: *Fusarium graminearum* in the Postgenomics Era. *Plant Physiology, 149*(1), 103–110. https://doi.org/10.1104/pp.108.129684

Wang, J. H., Ndoye, M., Zhang, J. B., Li, H. P., & Liao, Y. C. (2011). Population structure and Genetic diversity of the *Fusarium graminearum* species complex. *Toxins, 3*(8), 1020–1037. https://doi.org/10.3390/toxins3081020

Yabuta, T., & Sumiki, Y. (1938). On the crystal of gibberellin, a substance to promote plant growth. *Journal of the Agricultural Chemical Society of Japan, 14*(1526).

Yang, Q., Yin, G. M., Guo, Y. L., Zhang, D. F., Chen, S. J., & Xu, M. L. (2010). A major QTL for resistance to Gibberella stalk rot in maize. *TAG. Theoretical and Applied Genetics., 121*(4), 673–687. https://doi.org/10.1007/s00122-010-1339-0

Ye, J. R., Guo, Y. L., Zhang, D. F., Zhang, N., Wang, C., & Xu, M. L. (2013). Cytological and molecular characterization of quantitative trait locus qRfg1, which confers resistance to Gibberella stalk rot in maize. *Molecular Plant Microbe Interactions, 26*(12), 1417–1428. https://doi.org/10.1094/Mpmi-06-13-0161-R

Zhang, Y., He, J., Jia, L. J., Yuan, T. L., Zhang, D., Guo, Y., … Tang, W. H. (2016). Cellular Tracking and gene profiling of *Fusarium graminearum* during maize stalk rot disease development elucidates its strategies in confronting phosphorus limitation in the host apoplast. *PLoS Pathogens, 12*(3), e1005485. https://doi.org/10.1371/journal.ppat.1005485

Zhang, D. F., Liu, Y. J., Guo, Y. L., Yang, Q., Ye, J. R., Chen, S. J., & Xu, M. L. (2012). Fine-mapping of qRfg2, a QTL for resistance to Gibberella stalk rot in maize. *TAG. Theoretical and Applied Genetics., 124*(3), 585–596. https://doi.org/10.1007/s00122-011-1731-4

19

Developmental dynamics of stem starch accumulation in *Sorghum bicolor*

Brian A. McKinley[1] | Anna L. Casto[1] | William L. Rooney[2] | John E. Mullet[1]

[1]Department of Biochemistry and Biophysics, Texas A&M University, College Station, Texas

[2]Department of Soil and Crop Sciences, Texas A&M University, College Station, Texas

Correspondence
John E. Mullet, Department of Biochemistry and Biophysics, Texas A&M University, College Station, TX 77845.
Email: jmullet@tamu.edu

Funding information
This research was supported by the Perry Adkisson Chair.

Abstract

Sweet sorghums were identified that accumulate up to ~9% of their total stem dry weight as starch. Starch accumulated preferentially in stem pith parenchyma in close proximity to vascular bundles. Stem starch accumulated slowly between floral initiation and anthesis and more rapidly between anthesis and 43 days post-anthesis before declining in parallel with tiller outgrowth. Genes involved in stem starch metabolism were identified through phylogenetic approaches and RNA-seq analysis of Della stem gene expression during the starch accumulation phase of development. Genes differentially expressed in stems were identified that are involved in starch biosynthesis (i.e., AGPase SS/LS, starch synthases, starch-branching enzymes), degradation (i.e., glucan-water dikinase, β-amylase, disproportionating enzyme, alpha-glucan phosphorylase) and amyloplast sugar transport (glucose-6-P translocator). Transcripts encoding AGPase SS and LS subunits with plastid localization were differentially induced during stem starch accumulation indicating that ADP-glucose for starch biosynthesis is primarily generated in stem plastids. Cytosolic heteroglucan metabolism may play a role in stem sucrose/starch accumulation because genes encoding cytosolic forms of the disproportionating enzyme and alpha-glucan phosphorylase were induced in parallel with stem sucrose/starch accumulation. Information on the stem starch pathway obtained in this study will be useful for engineering sorghum stems with elevated starch thereby improving forage quality and the efficiency of biomass conversion to biofuels and bio-products.

KEYWORDS
bioenergy, pith parenchyma, RNA-seq, transcriptome analysis

1 | INTRODUCTION

Plant stems transport water and nutrients to and from leaves and roots and contribute to diverse canopy architectures that influence light interception and leaf photosynthesis. Stems also serve as temporary depots for nonstructural carbohydrates such as sucrose, monosaccharides, starch, fructans, and isoprenoids (Jensen & Wilkerson, 2017; Slewinski, 2012). The C4 grasses sugarcane and sorghum are capable of accumulating large amounts and high concentrations of sucrose (Lingle, 1987; Mcbee & Miller, 1982; McKinley, Rooney, Wilkerson, & Mullet, 2016b; Murray et al., 2008). For example, the sweet sorghum Della accumulated ~31% of its stem dry weight as sucrose between floral initiation and grain maturity as well as lower levels of glucose, fructose, and starch (McKinley et al., 2016b). The large capacity for stem sucrose accumulation provides these plants with a sink for storing photosynthetically derived sucrose when requirements for metabolism, growth, and/or seed filling are satisfied. This allows continued photosynthesis with minimal feedback inhibition reducing the cost/damage associated with interception of light by leaves in the absence of a sink for energy utilization. The nonstructural carbohydrates that accumulate in stems provide a reservoir of sugars that enhance plant resilience and enable grain filling under adverse environmental conditions that cause inhibition of photosynthesis (Czedik-Eysenberg et al., 2016). In sorghum, stem sucrose and starch levels decline in parallel with onset of tillering post-grain maturity indicating that stem reserves also contribute to traits such as ratooning that are often associated with perennial sorghum (McKinley, Rooney, Wilkerson, & Mullet, 2016a).

Sugarcane breeders have focused on increasing the purity, concentration, and yield of sucrose in sugarcane stems (Waclawovsky, Sato, Lembke, Moore, & Souza, 2010). As a consequence of strong selection, the sucrose concentration in stems of elite sugarcane has reached an apparent plateau (~0.7 M) and recent gains in yield are correlated with increases in plant size and biomass (Waclawovsky et al., 2010; Wu & Birch, 2007). The stem sucrose concentration plateau has been attributed to constraints caused by elevated turgor associated with high sucrose concentrations in pith parenchyma cells, sucrose-mediated feedback inhibition of photosynthesis, and carbohydrate sensing and signaling mechanisms (Wu & Birch, 2007). Sucrose concentrations in sweet sorghum are similar to sugarcane; however, as sorghum is not a large commercial source of sucrose, selection for sucrose purity has been less intense which may explain why significant levels of stem starch are observed in sweet sorghum genotypes (McKinley et al., 2016b).

Our group has been developing and designing high-biomass sorghum hybrids that can be used for forage and production of biofuels and bio-products (Mullet et al., 2014; Rooney, Blumenthal, Bean, & Mullet, 2007). Stems account for approximately 80% of this crop's harvested biomass; therefore, stem composition is an important attribute of biomass value and utilization. High stem sucrose yield is a useful trait for biofuels production because conversion of sucrose/ sugars to biofuels and bio-products is low cost and highly efficient (Gnansounou & Dauriat, 2010; Wu et al., 2010). However, stem sugars are unstable following plant harvest due to metabolic and microbial degradation, therefore, commercial production of sucrose from sugarcane stems requires efficient harvesting, transport, and processing (Wu et al., 2010). The instability of stem sugars prevents long-term storage of harvested biomass restricting mill operation to times of the year when these crops can be harvested. These logistical obstacles could be reduced if sugars that are transported to stems were used to synthesize more stable polysaccharides such as starch or mixed linkage glucans. Higher stem starch could also be useful in forage crops to increase biomass digestibility. Therefore, one motivation for this study was to characterize the dynamics of stem starch accumulation during sorghum development and to increase knowledge of this pathway to enable selection and engineering.

During periods of carbohydrate surplus, leaves shunt hexoses into the synthesis of starch and starch is often the primary storage carbohydrate in tubers and grain. Starch metabolism has been extensively studied in leaves, potato tubers, rice and maize endosperms, as well as other systems (Sonnewald & Kossmann, 2013; Stitt & Zeeman, 2012). In many dicot species, starch synthesis in heterotrophic tissues involves the import G-6-P into plastids via the glucose-6-phosphate/phosphate translocator (GPT) followed by isomerization to glucose-1-phosphate by phosphoglucomutase (PGM). The heterotetrameric enzyme ADP-glucose pyrophosphorylase (AGPase) catalyzes the conversion of glucose-1-phosphate to ADP-glucose, the committed step in starch biosynthesis (Hädrich et al., 2011). In leaves, this reaction is rendered irreversible by the action of plastidial alkaline pyrophosphatase which hydrolyzes inorganic pyrophosphate into orthophosphate (Stitt & Zeeman, 2012). In grasses, a cytosolic AGPase has been identified that is responsible for the synthesis of 95% of the ADP-glucose in the grain of maize (Denyer, Dunlap, Thorbjornsen, Keeling, & Smith, 1996). The cytosolic ADP-glucose is then imported into the plastid by the action of Brittle-1, an adenylate translocator (Shannon, Pien, Cao, & Liu, 1998). Plastid localized ADP-glucose is then utilized by starch synthases (SS) to generate starch polymers with α-1,4-glycosidic linkages (Stitt & Zeeman, 2012). Following linear polysaccharide synthesis, starch-branching enzymes (BEs) hydrolyze an oligosaccharide from the terminal end of the starch polymer just below the granule surface and attaches it deeper into the growing granule, catalyzing the formation of an α-1,6-linked branch point (Tetlow & Emes, 2014). This action leads to the formation of semi-crystalline amylopectin and increases granule density, crystallinity, and stability, and the number of nonreducing ends available for chain elongation by starch synthase (Tetlow & Emes, 2014). The action of starch synthase in combination with glucan trimming by debranching enzymes (DBE)/ isoamylases ISA1 and ISA2 fine tunes the structure and crystallinity of the granule and its physical properties which are compatible with species specific enzymes involved in starch degradation (Geigenberger & Fernie, 2014; Lorberth, Ritte, Willmitzer, & Kossmann, 1998; Tetlow & Emes, 2014).

The synthesis of starch creates a molecule that is highly stable (Hejazi et al., 2008). Because of this stability, industrial processing and degradation of starch require high heat to gelatinize starch. The addition of heat adds thermal energy to the polysaccharide chains of the molecule increasing the physical space between them allowing for solvation followed by hydrolysis by starch hydrolytic enzymes (Biliaderis, Maurice, & Vose, 1980). As plants cannot utilize heat to disrupt granule crystallinity, destabilization in vivo is thought to be achieved by increasing the negative charge density of the granule surface through phosphorylation thereby disrupting crystallinity and separating the polymer chains which increases enzyme accessibility (Hejazi et al., 2008). Phosphorylation of the granule surface occurs initially by glucan-water dikinase (GWD) which phosphorylates the C-6 position of terminal glucose monomers, followed by phosphorylation by phosphoglucan-water dikinase (PWD) which phosphorylates the C-3 positions of different glucosyl residues (Ritte et al., 2006). Phosphorylation by PWD is strictly dependent on prior phosphorylation by GWD (Ritte et al., 2006). After the localized swelling introduced by GWD and PWD, a phosphoglucan phosphatase encoded by *SEX4* subsequently removes the phosphate groups as these groups impede the accessibility of starch degradation enzymes to the granule surface (Oa et al., 2009). Hydrolysis begins once the reducing ends of the amylopectin chains are accessible to β-amylases (exo-amylases that liberate maltose). One member of the debranching enzyme family, ISA3, seems to be involved in hydrolysis of branch points facilitating degradation (Delatte et al., 2006). Additional degradation of oligosaccharides liberated from the granule is performed by the α-amylases (endo-amylases) and disproportionating enzymes (DPE1 and DPE2) (Critchley, Zeeman, Takaha, Smith, & Smith, 2001). Transport of glucose and maltose released by starch degradation from the amyloplast can occur through glucose transporters (pGlcT) and maltose exporters (MEX1) (Stitt & Zeeman, 2012).

The results of this study showed that sweet sorghum genotypes can accumulate up to 9% of their stem biomass as starch and that starch accumulates primarily in stem pith parenchyma. Stem starch accumulated slowly between floral initiation and anthesis then more rapidly until peak levels were reached 43 days post-anthesis. RNA-seq analysis of stem gene expression during this developmental time course in conjunction with homology and phylogenetic analysis of gene families involved in starch metabolism enabled the identification of genes that are likely involved in stem starch accumulation and turnover.

2 | MATERIALS AND METHODS

2.1 | Harvest of diverse sorghums post-grain maturity for starch quantification

The survey of starch in sorghum [*Sorghum bicolor* (L.) Moench] stems after grain maturity was conducted from plants grown in the experimental field plots in College Station, Texas in 2014. Before planting, ammonium nitrate (32-0-0) was applied at a rate of 140 kg N/ha. Samples were harvested when all plants had passed grain maturity. The genotypes in this study were at variable stages of post-grain maturity development. Three plants per plot were harvested, avoiding the end of the plots to avoid edge effects, and these three plants were used in subsequent analyses. After harvesting, leaves and leaf sheaths were removed from the stems. A stem section spanning three internodes of which the middle internode was located at 50% of the total stem length was removed by excision at the nodes. These internode sections were cut into smaller sections and dried in a forced air oven at 60°C for 3 days; the tissue was coarsely ground in a Wiley Mill (Thomas Scientific, Inc.) and subsequently finely ground in a Cyclone Sample Mill (Udy Corporation, Fort Collins, Colorado, USA) until the tissue exhibited the consistency of a powder. Then, this powered tissue was used for starch quantification.

2.2 | Harvest of the sweet sorghum Della during development for NSC quantification

Della (Reg. no. CV-130, PI1566819), a sweet sorghum developed from a cross of Dale and ATx622 (Harrison & Miller, 1993), was planted at the Texas A&M field station near College Station, Texas in 2012 and 2013. Before planting, ammonium nitrate (32-0-0) was applied at a rate of 140 kg N/ha. In both years, the experimental design consisted of two ranges of 10 plots each of Della planted in adjacent regions of the field. Additionally, two boarder rows were planted on either side of the experiment to prevent edge effects. The spacing between rows was 76 cm, and plants were thinned to 10-cm spacing at the five-leaf stage. Basal tillers were removed until anthesis when basal tillering ceased. Tiller removal was performed to maintain a planting density of one main culms per 10 cm. During the experiment, plants were irrigated as needed. Five plants were harvested from each experimental replication for a total of 10 plants harvested per time point for a total of 10 plants per sample. Plants were harvested a minimum of two feet from the end of the plot to avoid edge effects. After harvesting, leaves and leaf sheaths were removed from the stems. A three internode section, of which the middle internode was located at 50% of the total stem length, was removed by excision at the node. These internode sections were cut into smaller sections and dried in a forced air oven at 60°C for 3 days. Next, the internode tissue was coarsely ground in a Wiley Mill (Thomas Scientific, Inc.) and subsequently finely ground in a Cyclone Sample Mill (Udy Corporation, Fort Collins, Colorado, USA) until the tissue exhibited the consistency of a powder. Then, this powered tissue was used for NSC quantification.

2.3 | Della stem tissue harvest for RNA-seq

To obtain tissue for RNA-seq, Della was grown in a greenhouse in College Station during the fall of 2011. The day length was maintained at 14 hr, and the mid-day PAR of the green house was ~1,100–1,200 μE/m^2/s. Plants were germinated in five gallon pots containing Metromix potting soil. Osmocote was added to the soil at the beginning of the experiment, and additional fertilization was supplemented with Peter's nutrient solution every 30 days. The plants were watered with the use of an automated watering system to eliminate the possibility of drought stress. Leaves were numbered as they appeared insuring that internode 10 was harvested at each of the eight sampling dates. Three biological replicates were harvested in the morning on each harvest date. To harvest internode samples, leaves and leaf sheaths were removed from the stem and Internode 10 was excised at the nodes to insure that a complete internode was harvested for analysis. The internode was quickly sectioned into smaller pieces with a razor for storage and ease of processing after freezing. The internode sections were placed into a 50-ml conical tube, frozen in liquid nitrogen, and stored at −80°C until processing. To prepare tissue for RNA extraction, the frozen tissue of each internode was ground to a fine powder using a mortar and pestle and stored in 50-ml conical tube until extraction.

2.4 | Soluble carbohydrate quantitation

To quantify the abundance of stem nonstructural carbohydrates, previously, oven dried samples from field experiments were redried overnight using a forced air oven at 60°C to insure accuracy of dry weight measurements. Next, 200 mg of finely ground biomass was weighed (±0.5 mg) using an analytical balance and transferred to a 15-ml conical tube. Alternatively, to quantify nonstructural carbohydrates in tissue that gave rise to the RNA-seq data, approximately 0.5 g of frozen, pulverized tissue was placed into a previously weighed 15-ml conical tube and lyophilized until desiccated. The tube and tissue were reweighed, and tube weight subtracted to obtain the tissue weight. Both tissue types were subjected to the following extraction protocol. Water-soluble NSCs were extracted in 10 ml of water/Na azide (200 mg/L) solution at 50°C for 48 hr with agitation. This length of incubation was experimentally determined to be optimal for this experimental set-up. This extraction time allowed the extraction solvent to fully penetrate all of the biomass particles and extract the soluble carbohydrates. Aliquots of 20 μl were diluted 50 × into 980 μl of deionized water. All HPLC samples were filtered using 0.45-μm cellulose acetate sterile filters. Sucrose, glucose, and fructose concentrations were quantified using high performance anion exchange-pulsed amperometric detection (HPAE-PAD) with a Carbopac PA1 analytical column (Dionex, Sunnyvale, CA) as well as a Borate trap (Dionex) and an Aminopac column (Dionex) to reduce borate and amino acid interference (Murray et al., 2008). A solution of 75 mM NaOH was made from 50% NaOH solution (Sigma-Aldrich, St. Louise, MO, USA) instead of pellets to eliminate carbonate contamination. This running buffer was vacuum degassed overnight and stored under a helium bed for the duration of the chromatographic run. The standard curve was validated by the incorporation of curve validation samples of known concentration throughout the experiment in accordance with the NREL LAP (Sluiter, Ruiz, Scarlata, Sluiter, & Templeton, 2008).

2.5 | Starch quantification

The starch in stem biomass was quantified using the NREL Laboratory Analytical Procedure for extracting starch from solid biomass (Sluiter & Sluiter, 2005). After the soluble nonstructural carbohydrate extraction, samples were washed four times in methanol-chloroform-water extraction solvent at room temperature for 15 minutes then centrifuged at 1,640 ***g*** for 15 min to sediment structural biomass and starch granules. After the MCW washes, the samples were oven dried at 65°C for 48 hr. Next, DMSO was added to the dry samples and incubated overnight. Overnight imbibition insures that the DMSO, the gelatinization solvent, has fully penetrated the biomass and is capable of fully gelatinizing the starch which is essential to fully digest all the starch. After imbibition, the samples were incubated at 100°C for 10 min with agitation every two minutes to gelatinize the starch. Then, MOPS buffer and thermostable α-amylase were added to digest the starch to maltose at 95°C for 1 hr with agitation to insure full digestion. Subsequently, sodium acetate buffer and amyloglucosidase were added to hydrolyze the maltose to glucose at 55°C for 24 hr. To insure that starch was fully digested, digestion completion samples were included in the assay and sampled after digestion with amylase and amyloglucosidase and stained with Lugol's iodine solution. Glucose was quantified with high performance anion exchange-pulsed amperometric detection (HPAE-PAD) chromatography with the use of a Carbopac PA1 analytical column (Dionex, Sunnyvale, CA). Curve validation samples were randomly interspersed among the analytical samples during chromatography to validate the glucose, fructose, and sucrose standard curves. The percentage of starch in the dry biomass was calculated from the glucose released during digestion.

2.6 | Stereomicroscopy of internode cross sections

To visualize the location of starch in the stem by stereomicroscopy, field grown plants were harvested at anthesis and post-grain maturity. A 1-mm slice of stem tissue was excised with a razor blade from the middle of internode 11. These 1-mm internode sections were infused with Lugol's solution in a vacuum. After infusion, the internode sections were stored in this solution overnight. Excess staining solution was removed by incubating the tissue slice in water for 24 hr, changing the solvent every 6 hr. The sections were imaged under white light using a Zeiss M^{2}Bio Fluorescence Combination Zoom Stereo/Compound Microscope coupled with a Zeiss AxioCam digital camera (Kramer Scientific). Photographs were captured using the Zeiss AxioVision 3.0.6 software.

2.7 | Template preparation and sequencing

RNA extraction was performed using the Trizol extraction method (https://www.mrcgene.com/rna-isolation/tri-reagent). The high salt precipitation step was included because of the potential high concentration of soluble carbohydrates in the sample. Total RNA was further purified using the RNeasy kit (Qiagen, Venlo, the Netherlands) which isolates RNAs >200 nucleotides in length. An on column DNase1 digestion was performed using the RNase-free DNase set (Qiagen, Venlo, Netherlands) to remove any contaminating DNA. The quality of the RNA libraries was assessed using an Agilent Bioanalyzer. The mean RNA integrity number (RIN) from this quality control step was 8.9, and the minimum RIN was 8.3 indicating that the quality of the RNA libraries was high. RNA libraries were prepped for sequencing using the Illumina TruSeq® V2 mRNA Sample Preparation Kit (Illumina, San Diego, California, USA), which selects for polyadenylated RNA and excludes rRNA and non-polyadenylated RNAs from the sample. The 24 libraries were sequenced on one lane of an Illumina HiSeq 2500 in single-end mode. The average sequence depth was 5.8 M reads.

2.8 | Transcriptome analysis

The 70 bp reads were aligned to the *Sorghum bicolor* V3.1 genome using the HISAT2 aligner (Daehwan, Langmead, & Salzberg, 2015; McCormick et al., 2018). Expression was quantified using the StringTie version 1.3 software (Pertea et al., 2015). Analysis of gene expression was performed on TPM normalized data. The prepDE.py script was used to convert nucleotide coverage data from StringTie into reads that could be used by differential expression statistical packages that use conventional raw reads. Differential expression and the FDR-adjusted *p*-values were calculated using the edgeR package. The fold-change displayed in the tables is the absolute value of the maximum fold-change value between any two time points. The FDR-adjusted *p*-value represents the *p*-value associated with the maximum fold-change values shown in the gene tables. In some cases, zero expression at one time point, when being compared to another time point, led to a very high differential expression value. To minimize this effect in instances with very high DE values, the lowest time point was considered to be a TPM = 1 and, therefore, the fold-change was recalculated as FC = maxTPM/1 (McKinley et al., 2016b). These instances are indicated with the ">"sign in front of the fold-change values in the gene tables as the actual fold-change is assumed to be some value greater than the value displayed. Functional analysis of gene families utilized gene functional annotations generated by DOE-JGI for the *Sorghum bicolor* V3.1 genome available through Phytozome (https://phytozome.jgi.doe.gov/pz/portal.html#!info?alias=Org_Sbicolor) (McCormick et al., 2018).

2.9 | Phylogenetic analysis of gene families to identify orthologs

Phylogenetic analysis was used increase the accuracy of gene family annotation. This approach was based on the identification of genes in other species with known previously validated function. The primary protein sequences were obtained Phytozome or NCBI. The protein sequences of these validated genes and the sorghum candidate homologs were then clustered using the CLUSTAL Omega algorithm using default settings.

2.10 | Identification of chloroplast targeting peptide sequences

Peptide sequences were obtained from Biomart, hosted by the Phytozome database. The primary transcript was used for the analysis. The ChloroP algorithm (http://www.cbs.dtu.dk/services/ChloroP/) was used to predict the presence of chloroplast targeting sequences in peptides sequences.

3 | RESULTS

3.1 | Variation in sorghum stem starch accumulation in diverse sorghum accessions

The level of starch present in stems of field grown grain and sweet sorghum genotypes was analyzed after grain maturity. A stem segment spanning three internodes located approximately mid-stem was assayed for levels of starch and other nonstructural carbohydrates. The sorghum genotypes analyzed accumulated 0.2% to ~9% of their internode biomass as starch (Figure 1). Stem starch levels were generally higher in sweet sorghum genotypes (Figure 1, green bars) compared to grain sorghum genotypes (Figure 1, orange bars).

3.2 | Developmental dynamics of starch accumulation in Della sorghum stems

Sorghum progresses through a juvenile and vegetative phase following germination (Figure 2a). The reproductive phase begins with floral initiation, followed by anthesis approximately 25–30 days later, grain maturity and a post-grain maturity phase where tillering often occurs (Figure 2a). The sweet sorghum Della grown in the field in College Station reached anthesis 69 days after emergence (DAE), and grain maturity was reached approximately one month after anthesis. The developmental time course of nonstructural carbohydrate accumulation in the stems of the sweet sorghum Della was characterized in field grown plants during the summers of 2012 and 2013 (Figure 2b). The analysis revealed that Della stems accumulated starch post-anthesis and that peak stem starch levels were reached after grain maturity. Stem starch accumulated at its highest rate from 26 days post-anthesis to ~53 days post-anthesis (0.26% of internode dry biomass/day). Starch accounted for ~9.5% of Della's stem dry weight 53 days after anthesis. In contrast, stem soluble sugar levels reached a maximum ~26 days post-anthesis and remained high until 53 days post-anthesis (Figure 2b). Between 53 and 80 days after anthesis, levels of stem sucrose and starch decreased in parallel with leaf senescence and apical tiller outgrowth. Similar results were obtained when Della was grown in the greenhouse (Figure 2c). When Della was grown in the greenhouse to facilitate RNA-seq analysis, the developmental time course of soluble sugar and starch accumulation in stems was similar to plants grown in the field (Figure 2c). However, in greenhouse grown plants, soluble sugar and starch levels in Della stems were lower than in field grown plants likely due to lower light levels in the greenhouse.

3.3 | Distribution of starch in stems

The amount of starch and soluble carbohydrates in each of the elongated internodes of Della stems was assayed at grain maturity (Figure 3). Most of the stem internodes had similar nonstructural carbohydrate concentration profiles. However, sucrose concentrations were lower (and monosaccharides somewhat higher) in internodes located near the base of the plant (internode 8), internodes just below the peduncle (internodes 16, 17), and the peduncle. Excluding the peduncle, starch accounted for >5% of the internode dry biomass of every internode at grain maturity and internodes 9–11 has the highest starch levels (~12.5% of dry weight) (Figure 3).

The radial distribution of starch across stem internodes was assayed by infusing cross sections with Lugol's iodine solution which stains starch dark blue prior to visualization using a stereo-microscope (Figure 4). This analysis revealed that starch accumulated at relatively high levels in pith parenchyma cells in regions of high vascular bundle density (Figure 4a). Starch levels were also elevated in pith parenchyma cells that surround vascular bundles in the central

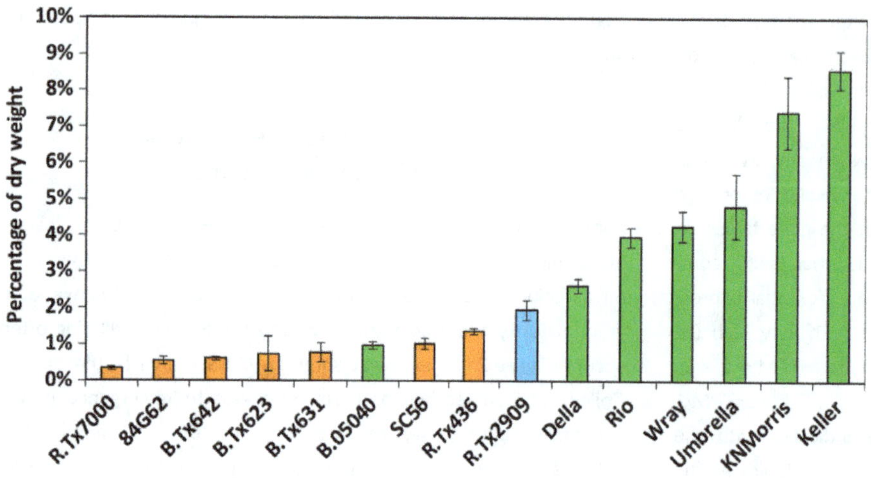

FIGURE 1 A survey of the percent composition of starch in the stem of diverse sorghum genotypes grown in the field in College Station, TX and assayed after grain maturity. Genotypes represented by green bars are sweet sorghums, orange bars represent grain sorghums, and the blue bar represents a forage sorghum. Error bars represent *SEM*

portion of the stem and decreased with increasing distance from the vascular bundles (Figure 4a and b). Starch granules were not observed within cells of the vascular bundles; however, this tissue was tinted blue after staining likely because of the scattering of blue light through these cell types from surrounding starch-containing cells (Figure 4c and d). High levels of starch were not observed in the epidermis of the stem, but starch accumulated in cells just beneath the epidermis (Figure 4a).

3.4 | Expression of genes involved in stem starch metabolism during plant development

One overall goal of this study was to identify and characterize the expression of genes involved in stem starch metabolism. The increase in starch accumulation in stems between floral initiation and 53 days post-anthesis provided an opportunity to identify genes in starch metabolism that show increased expression in parallel with stem starch accumulation. Data on stem gene expression were obtained by extracting RNA for RNA-seq analysis from stem internodes of Della plants grown in the greenhouse at eight times between floral initiation (~46 DAE) through 80 days post-anthesis (~175 DAE). Genes previously identified in other plants that are involved in starch synthesis, turnover, and plastid sugar transport were used to identify sorghum homologs based on sequence similarity and phylogenetic analysis of gene families (Supporting information Tables S1–S3). Sorghum encodes a family of genes for most steps in starch metabolism, and specific genes in each family were differentially expressed in stems in parallel with starch accumulation.

To help visualize patterns of gene expression in the context of the entire starch pathway, a working model of stem starch metabolism was developed based on prior knowledge of this pathway obtained in other plants/organs (Smith, Denyer, & Martin, 1997; Stitt & Zeeman, 2012). In the model, genes encoding enzymes that are likely involved in stem starch synthesis are highlighted in green, genes encoding proteins involved in starch turnover are highlighted in blue or red, and those encoding proteins involved in sugar transport through the outer plastid membrane are shown in colored boxes (Figure 5). Genes expressed in stems at >20 TPM and that were induced >fivefold in parallel with stem starch accumulation between floral initiation and grain maturity have information on fold-induction presented in yellow-highlighted boxes adjacent to each gene. This overview shows that genes encoding proteins involved in starch biosynthesis (i.e., plastid *AGPase, SS, BE, DBE*), turnover (i.e., *GWD, DBE, DPE, PHS*), and transport (i.e., *GPT, pGlcT*) were upregulated between floral initiation and 43 days post-anthesis in parallel with stem starch accumulation.

3.5 | Expression of genes encoding AGPases

Two sorghum genes encoding AGPase small subunits (SSs) and four genes encoding AGPase large subunits (LSs) were identified in the sorghum genome (Supporting information Figure S1; Supporting information Table S1). Several different transcripts were produced from these genes. One gene encoding AGPase SS (Sobic.002G160400, *APS1*) and one gene encoding AGPase LS (Sobic.009G245000; *APL2*) were expressed at lower levels in stems at floral initiation and at gradually increasing levels during development with peak expression between 25 and 43 days post-anthesis (Figure 6a; Supporting information Table S1). Sobic.009G245000 (*APL2*) was differentially expressed 47-fold higher in stems post-anthesis and Sobic.002G160400 (*APS1*) eightfold higher (Figure 6a; Supporting information Table S1). Both of these genes were expressed at ~five-tenfold lower levels in leaves compared to stems; however, expression in panicles was similar to stems (Supporting information Figure S1).

Many grasses synthesize some portion of the ADP-glucose used for starch synthesis in grain in the cytosol (Beckles, Smith, & Ap Rees, 2001). As the stem post-anthesis is a heterotrophic carbohydrate storage organ similar to grain (Emanuelsson, Nielsen, & von Heijne, 1999), an analysis was conducted to characterize the expression of genes/transcripts involved cytosolic ADP-glucose synthesis during development (Supporting information Figures S1 and S2). Alignment and phylogenetic analysis of sorghum and maize AGPase gene homologs that are localized in the cytoplasm or plastid (Huang, Hennen-Bierwagen, & Myers, 2014) and ChoroP prediction of plastid transit sequences (Xuan et al., 2013) were used to identify sorghum transcripts that encode AGPase subunits that would localize in the cytoplasm. This analysis

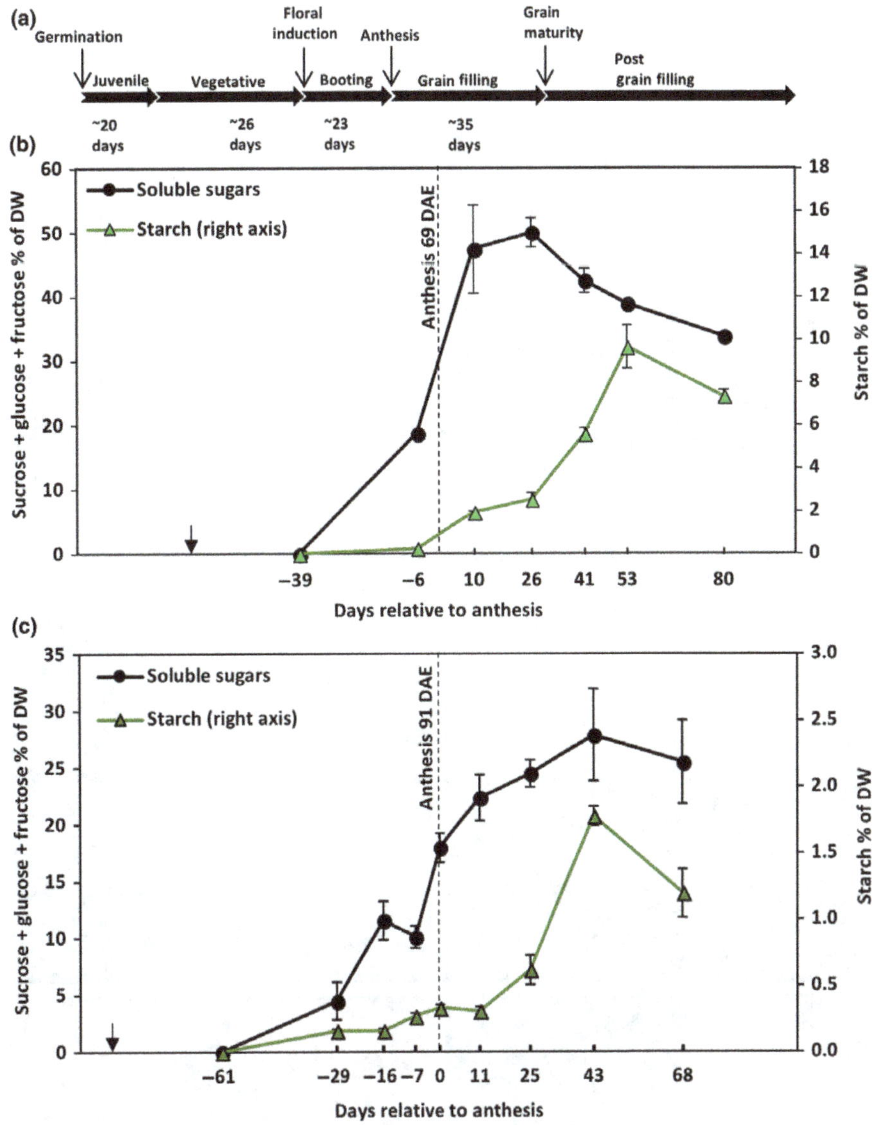

FIGURE 2 Temporal pattern of starch accumulation in the stem of the sweet sorghum Della in field and green house environments. (a) Timeline of Della development in the field in College Station, TX, in 2011. (b) The accumulation profile of total soluble sugars (sucrose + glucose + fructose) and starch in the Della stem in the field in College Station, TX, 2011. The dotted line represents anthesis. The *x*-axis represents days relative to anthesis. Negative numbers are days prior to anthesis, and positive numbers are days after anthesis. The black arrow represents seedling emergence, that is, the start of the experiment. (c) The accumulation profile of total soluble sugars (sucrose + glucose + fructose) and starch in the Della stem in the greenhouse in College Station, TX, Fall, 2011. Error bars represent*SEM*

identified two transcripts (Sobic.007G101500.2, Sobic.007G101500.3) that encode AGPase small subunits and three transcripts (Sobic.003G230500.1/.2/.3) that encode AGPase large subunits predicted to produce proteins that localize in the cytoplasm (Supporting information Figure S1). These transcripts are expressed at lower levels in stems compared to other genes/transcripts that encode AGPase subunits with predicted plastid localization. Taken together, this indicates that a plastid localized AGPase is the primary source of ADP-glucose for starch synthesis in stem amyloplasts.

3.6 | Genes encoding stem starch synthases and starch-branching/debranching enzymes

ADP-glucose is the substrate of starch synthases, the enzyme family responsible for adding glucose monomers to the reducing ends of growing amylose and amylopectin chains (Hirose & Terao, 2004). The starch synthase gene family consists of 10 genes in rice and maize (Supporting information Table S1). This family can be further divided into the soluble and granule bound starch synthases, with each of these groups containing unique subtypes (Supporting information Figure S3; Supporting information Table S1) (Hirose & Terao, 2004). Due to the functional complexity of the family, a phylogenetic analysis was conducted to align sorghum gene annotations with gene family members identified in rice and maize (Supporting information Figure S3) (Hirose & Terao, 2004). This analysis identified ten genes in the sorghum, the same number of genes in the rice starch synthase gene family (Supporting information Figure S3) (Hirose & Terao, 2004). Eight of the 10 genes in the family were expressed in sorghum stems. Most of these family members showed peak expression in stems post-anthesis (Supporting information Table S1). A gene encoding *GRANULE BOUND STARCH SYNTHASE* (*GBSS*), Sobic.002G116000.1, was the most highly expressed gene in the family (max TPM = 737), and the gene was differentially expressed in stems between floral initiation and 43 days post-anthesis (DE = 11-fold) (Supporting information Table S1). The second most abundantly expressed gene of this gene family was a *STARCH SYNTHASE I* (max TPM = 229, DE = 29-fold), and the third was a *STARCH SYNTHASE II* (max TPM = 162, DE = 15-fold) (Figure 6a).

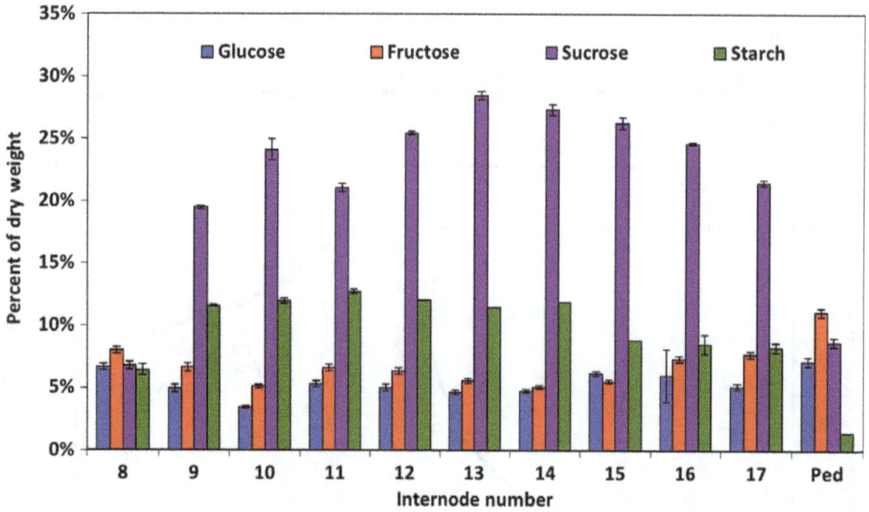

FIGURE 3 The distribution of nonstructural carbohydrates in the sweet sorghum stem as a percentage of the total dry biomass of each internode. Internode number increases proceeding up the stem. Ped, peduncle. Error bars represent*SEM*

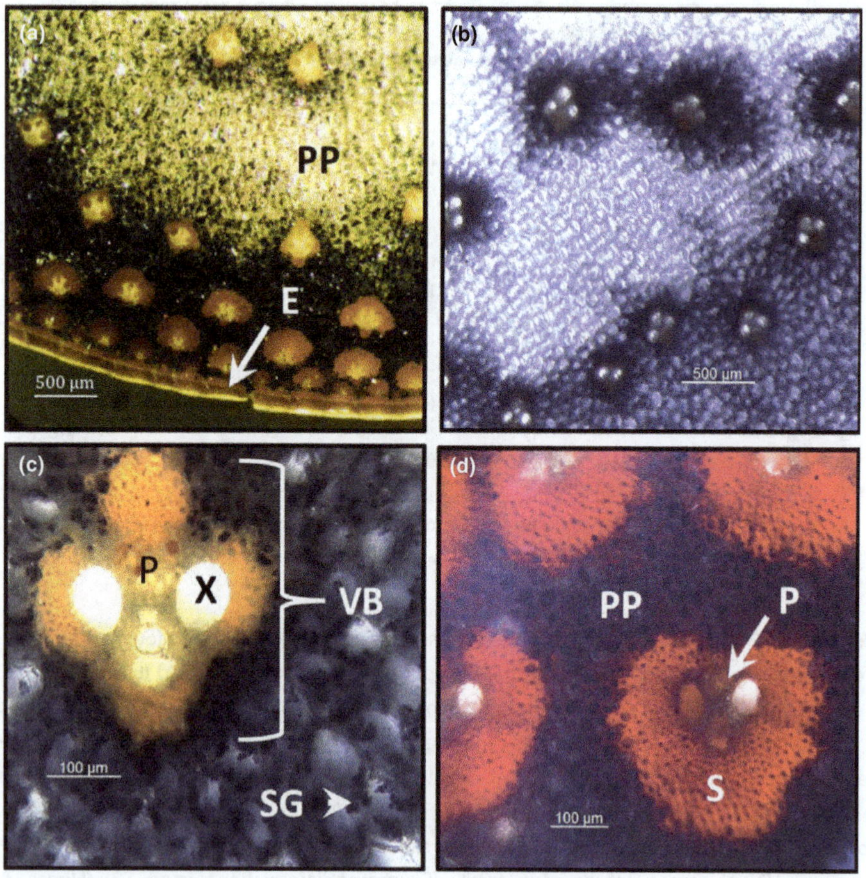

FIGURE 4 Localization of starch within sweet sorghum Della internodes visualized using stereomicroscopy. Starch was stained with Lugol's solution through vacuum infiltration. (a) Low resolution image of an internode cross section that shows the distribution of starch in the internode. (b) Image of vascular bundles of the central region of the Della internode. (c) High-resolution image of a vascular bundle located near the central region of the internode. (d) High-resolution image of a vascular bundle located in the rind region at the periphery of the internode near the epidermis. PP, pith parenchyma; SG, starch granule; E, epidermis; S, sclerenchyma sheath; P, phloem; X, xylem

The starch-branching enzymes and debranching enzymes tune starch granule structure to achieve specific physicochemical properties of the granule (Smith et al., 1997). Four genes encoding starch-branching enzymes were expressed in the sorghum stem with peak expression after anthesis (Supporting information Table S1). The most highly expressed gene/transcript was Sobic.006G066800.2 (max TPM = 865, 43 days after anthesis) a gene that was also differentially expressed during plant development (DE = 25-fold) (Figure 6a). Isoamylase 1-3 are starch-debranching enzymes. Isoamylases 1 and 2 are involved in starch biosynthesis (Supporting information Table S1), tuning the structure of amylopectin to facilitate crystallization (Hussain et al., 2003). Evidence suggests that isoamylase 3 is involved in starch degradation (Supporting information Table S2) (Delatte et al., 2006). Three transcripts of *ISA1* and *ISA2* were expressed in stems with peak expression occurring 25 days after anthesis (Figure 6a; Supporting information Table S1). The gene encoding isoamylase 3 (Sobic.002G233600) produced three transcripts that varied in abundance in stems with peak expression occurring post-anthesis (Supporting information Table S2).

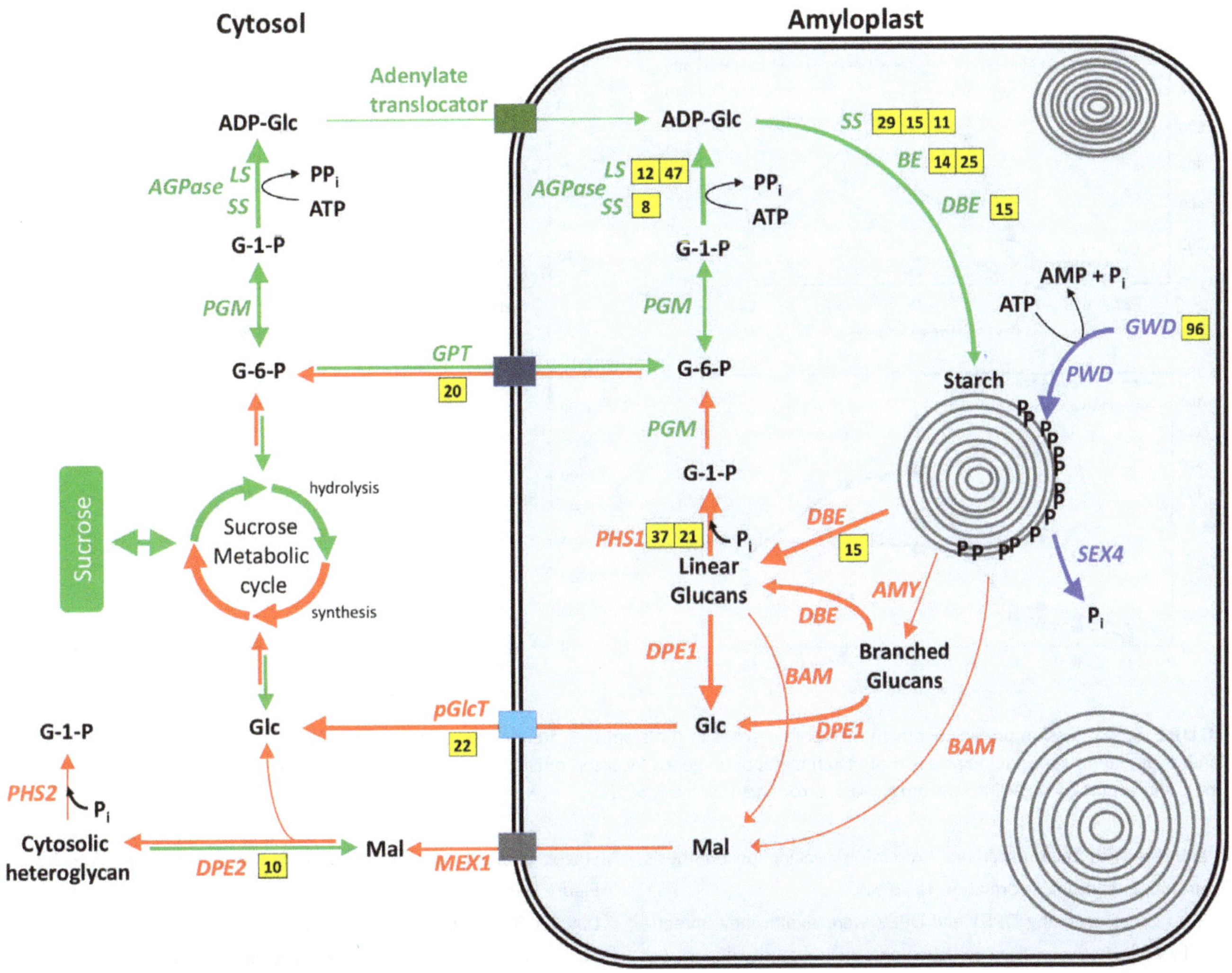

FIGURE 5 Differentially expressed genes of starch metabolism in the context of their pathways. Yellow boxes represent the fold-change of differentially expressed genes of the pathway. Selection criteria: 10-fold induction after anthesis with maximum expression >20 TPM. Thick lines represent hypothetical pathway flux. AGPase; ADP-glucose pyrophosphorylase, LS; large subunit (of AGPase), SS; small subunit (of AGPase), PGM; phosphoglucomutase, GPT; glucose-6-phosphate/phosphate translocator, SS; starch synthase, BE; branching enzyme, DBE; debranching enzyme, GWD; glucan-water dikinase, PWD; phosphoglucan-water dikinase, SEX4; phosphoglucan phosphatases, AMY; α-amylase, BAM; β-amylase, DPE; disproportionating enzyme, PHS; α-glucan phosphorylase, MEX1; maltose transporter, pGlcT; glucose transporter

3.7 | Genes involved in stem starch degradation

The expression of several genes that encode enzymes involved in starch turnover increased between floral initiation and grain maturity (Supporting information Table S2). To enhance starch degradation, the crystallinity of the starch granule is reduced through the action of GWD (Figure 5). The expression of *GWD* (Sobic.010G143500.1) was induced ~96-fold between floral initiation and grain maturity in parallel with the accumulation of starch in the sorghum stem (Figure 6b; Supporting information Table S2). There were three splice variants derived from a gene expressed in the stem that encodes PWD (Supporting information Table S2). All the *PWD* splice variants were expressed at ~10-fold lower levels compared to *GWD*, and only one transcript was differentially expressed (~fivefold) with peak expression post-anthesis (Table S2). A gene encoding phosphoglucan phosphorylase (*SEX4*, Sobic.001G53880) showed twofold higher expression in stems post-anthesis (Supporting information Table S2).

Amylases and isoamylase help degrade starch grains to maltose and branched and linear glucans (Figure 5). Three genes encoding α-amylase and ten genes encoding β-amylase were expressed in sorghum stems. The α-amylase gene with the highest expression was expressed at similar levels from floral initiation through grain maturity. Two other α-amylase genes were expressed at low levels until 68 days post-anthesis when starch levels were decreasing in stems (Supporting information Table S2). Two of the genes encoding β-amylase were expressed at high levels in stems at floral initiation. These genes may be associated with chloroplasts located in the outer cell layers of the stem. Four members of the gene family were expressed at approximately the same level in stems during the developmental stages analyzed. In addition, four other members of the β-*AMYLASE* gene family were expressed at relatively low levels prior

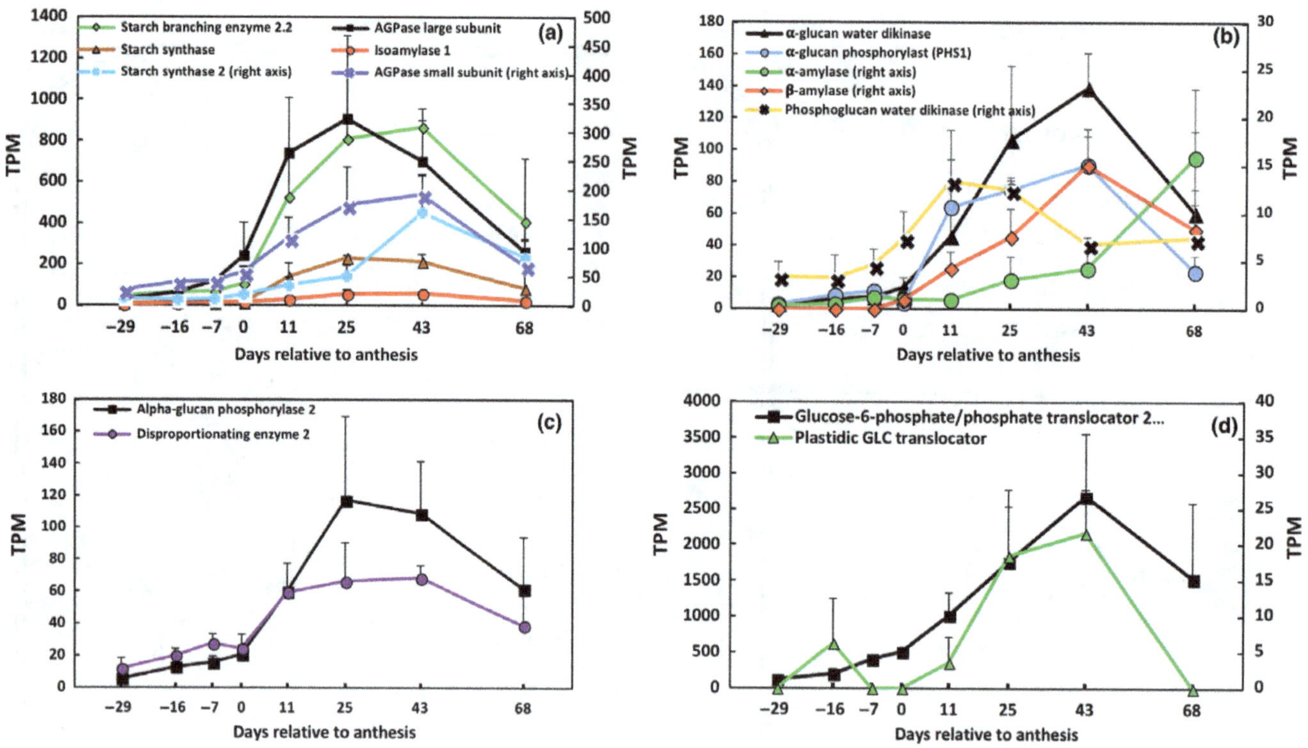

FIGURE 6 Expression profiles of starch metabolism genes in the stem that are induced after anthesis. (a) Starch biosynthesis genes. (b) Starch degradation genes. (c) Expression of starch metabolism genes localized outside of the amyloplast. (d) Transporters potentially involved in import and export of sugar into the amyloplast. Error bars represent *SEM*

to anthesis with their maximum expression peaking post-anthesis (Figure 6b; Supporting information Table S2).

Two genes encoding DPE1 and DPE2 were significantly upregulated in stems between floral initiation and grain maturity (Figure 6c; Supporting information Table S2). *DPE2*, which encodes an enzyme with predicted cytosolic localization, was more highly expressed and upregulated ~10-fold between floral initiation and 43 days post-anthesis. *DPE1*, which encodes an enzyme with predicted plastid localization, was expressed at lower levels, but also induced during this phase of plant development. Two genes encoding heteroglucan phosphorylase (*PHS1*, *PHS2*) were expressed at similar levels, and gene expression was induced ~21-fold and ~37-fold, respectively, with peak expression 43 days post-anthesis (Figure 6b and c; Supporting information Table S2). *SbPHS1* has a predicted localization in plastids whereas *SbPHS2* has a predicted localization in the cytosol (Figure 6b and c; Supporting information Table S2) consistent with analysis of these proteins in other plants (Beck & Ziegler, 1989; Fettke, Malinova, Eckermann, & Steup, 2009).

3.8 | Differential expression of starch pathway genes in stems and leaves

The expression of genes involved in starch metabolism in leaves and stems was compared to see if members of gene families were regulated in an organ-specific manner. The expression of pairs of genes from the same gene family was compared using RNA isolated from leaves and stems at the same time in the morning during four stages of plant development (Figure 7 and Supporting information Figure S4). The analysis showed that expression of Sobic.009G245000.2 (*AGL2*) was higher in stems vs. leaves during the phase of development analyzed whereas the related family member Sobic.001G100000.1 (*AGL2*) was expressed at higher levels in leaves compared to stems (Figure 7a and b; Supporting information Figure S4). Similarly, Sobic.002G160400.1 (*AGS1*) was expressed at higher levels in stems vs. leaves whereas the family member Sobic.007G101500.1 (*AGS1*) was expressed at higher levels in leaves compared to stems during these same developmental stages (Figure 7c and d; Supporting information Figure S4). A pair of genes encoding starch synthases (Sobic.010G093400.1, Sobic.010G047700.1) showed similar divergent patterns of differential expression in leaves and stems (Figure 7e and f; Supporting information Figure S4).

3.9 | Genes involved in stem amyloplast sugar transport

The increase in stem starch accumulation between floral initiation and grain maturity and remobilization post-grain maturity could require higher expression of genes encoding the plastid localized glucose-6-phosphate/phosphate translocator (*GPT*), the maltose transporter (*MEX1*), the plastidial glucose transporter (*pGlcT*), and/or the adenylate translocator. RNA-seq analysis showed that a gene encoding a member of the *GLUCOSE-6-PHOSPHATE/PHOSPHATE TRANSLOCATOR 2 (GPT2)* family was induced ~20-fold in parallel with increased stem starch accumulation during development (Figure 6d; Supporting information Table S3). This gene was highly expressed (max TPM = 2,689) and showed strong differential expression during development (FDR-adjusted *p*-value = 0.00021) (Figure 6d; Supporting information Table S3). Induction of this highly expressed gene suggests that glucose-6-phosphate is a predominant metabolite transported to and from stem amyloplasts. A gene *PLASTIDIAL GLUCOSE TRANSPORTER* (*pGlcT*) encodes a protein that transports glucose released from starch degradation out of the plastid. This gene showed increased expression in stems post-anthesis; however, the relative expression of *pGlcT* (max TPM = 150) was much lower than *GPT2* (TPM = 2,689) (Figure 6c, Supporting information Table S3). The plastid ATP/ADP-transporter translocates adenylates across plastid membranes thus maintaining plastidial ATP supply (Figure 6c; Supporting information Table S3) (Tjaden, Mohlmann, Kampfenkel, Henrichs, & Neuhaus, 1998). One member of this gene family was expressed in stems (max TPM = 120) although expression did not increase in parallel with starch accumulation during development (Supporting information Table S3). The gene *MALTOSE EXCESS 1* (*MEX1*) encodes a protein that transports maltose out of plastids. This gene was expressed at low levels in stems (max TPM = 8) and was not differentially expressed between floral initiation and 43 days post-anthesis (Supporting information Table S3).

4 | DISCUSSION

The overall goal of this study was to learn more about the regulation of starch accumulation in sorghum stems because increasing stem starch levels could improve biomass yield and the composition of high-biomass sorghum that is used for forage and production of biofuels and bio-products. While prior studies have noted the presence of starch in sorghum stems (Gutjahr et al., 2013; McBee, Waskom, & Creelman, 1983; McKinley et al., 2016b), the current study showed that a number of sweet sorghum genotypes can accumulate up to ~9% of their stem dry weight as starch. The capacity for significant stem starch accumulation in sweet sorghum genotypes that also accumulate high levels of stem sucrose indicates that even higher levels starch could be generated by altering the partitioning between sucrose and starch. Moreover, nearly all of the internodes that comprise Della stems accumulated high levels of sucrose and ~10%-12% starch. Therefore, high-biomass sweet sorghums that accumulate 80% of their biomass in stems that grow 4–5 m in length and 20–40-mm diameter have a large potential capacity for storing stem starch. The stems of the sweet sorghum Della have a nonstructural carbohydrate to lignocellulose ratio of ~2.8 at harvest (McKinley et al., 2016b). Increasing stem starch further could potentially improve yield by reducing feedback inhibition of photosynthesis that occurs when stems reach maximum sucrose holding capacity (Gutjahr et al., 2013; Paul & Foyer, 2001) and improve the efficiency and lower the cost of biomass conversion to biofuels and bio-products.

Starch accumulation was highest in stem pith parenchyma cells adjacent to vascular bundles in the subepidermal region of the stem where vascular bundle density was highest. The observation that starch accumulates in this region of the stem suggests that sucrose/sugar availability is high near vascular bundles. Starch also accumulated in pith parenchyma cells immediately adjacent to vascular bundles located in the central region of the stem. Interestingly, in the center of the stem where vascular bundle density is lower, pith parenchyma cells located furthest from vascular bundles accumulated lower levels of starch suggesting that sucrose/sugars derived from the phloem are used preferentially by pith parenchyma cells closest to the phloem and that starch accumulation in stem pith cells is responsive to sucrose availability.

Leaves are the primary source of sucrose that is used for synthesis of starch in stem pith parenchyma cells. Prior studies have examined the transport of sucrose from sorghum leaf mesophyll cells to stems (Sowder, Tarpley, Vietor, & Miller, 1997) and identified transporters that are involved in this pathway (Bihmidine, Baker, Hoffner, & Braun, 2015; Bihmidine, Julius, Dweikat, & Braun, 2016; Milne et al., 2017; Mizuno, Kasuga, & Kawahigashi, 2016). In fully elongated internodes that are accumulating sucrose post-floral initiation, sucrose can move from the phloem to stem storage parenchyma cells via an apoplastic and/or symplastic pathway with subsequent sequestration in pith parenchyma cell vacuoles (Bihmidine et al., 2015; Milne, Offler, Patrick, & Grof, 2015; Sowder et al., 1997; Tarpley & Vietor, 2007). Sucrose taken up by storage pith parenchyma post-anthesis is then sequestered in vacuoles or hydrolyzed by cytoplasmic invertases or sucrose synthases generating hexoses that can be used for starch biosynthesis. The ~20-fold induction of *GPT2* expression in stems in parallel with increased starch accumulation during development indicates that transport of G-6-P to and from the plastid plays an important role in stem starch metabolism.

4.1 | Genes involved in sorghum stem starch metabolism

Sorghum genes potentially involved in stem starch metabolism were identified based on their sequence similarity to genes previously identified in other species, phylogenetic analysis of gene families, and increased expression during starch accumulation in sorghum stems between floral initiation and 43 days post-anthesis. This approach identified sorghum genes predicted to encode enzymes involved in starch metabolism that were significantly upregulated in parallel with the accumulation of stem starch during development (Figure 5). Several genes encoding steps in starch biosynthesis were differentially expressed relative to other gene family members in stems vs. leaves during monocot development. This suggests that stem starch accumulation is regulated in part through differential expression of specific gene family members involved in starch metabolism. The basis of differential regulation and confirmation that these genes are directly involved in stem starch metabolism will require additional studies involving targeted gene modification.

In monocot grain endosperm tissue, most of the ADP-glucose used for starch biosynthesis is synthesized in the cytoplasm and imported into amyloplasts (Beckles et al., 2001; Johnson et al.,

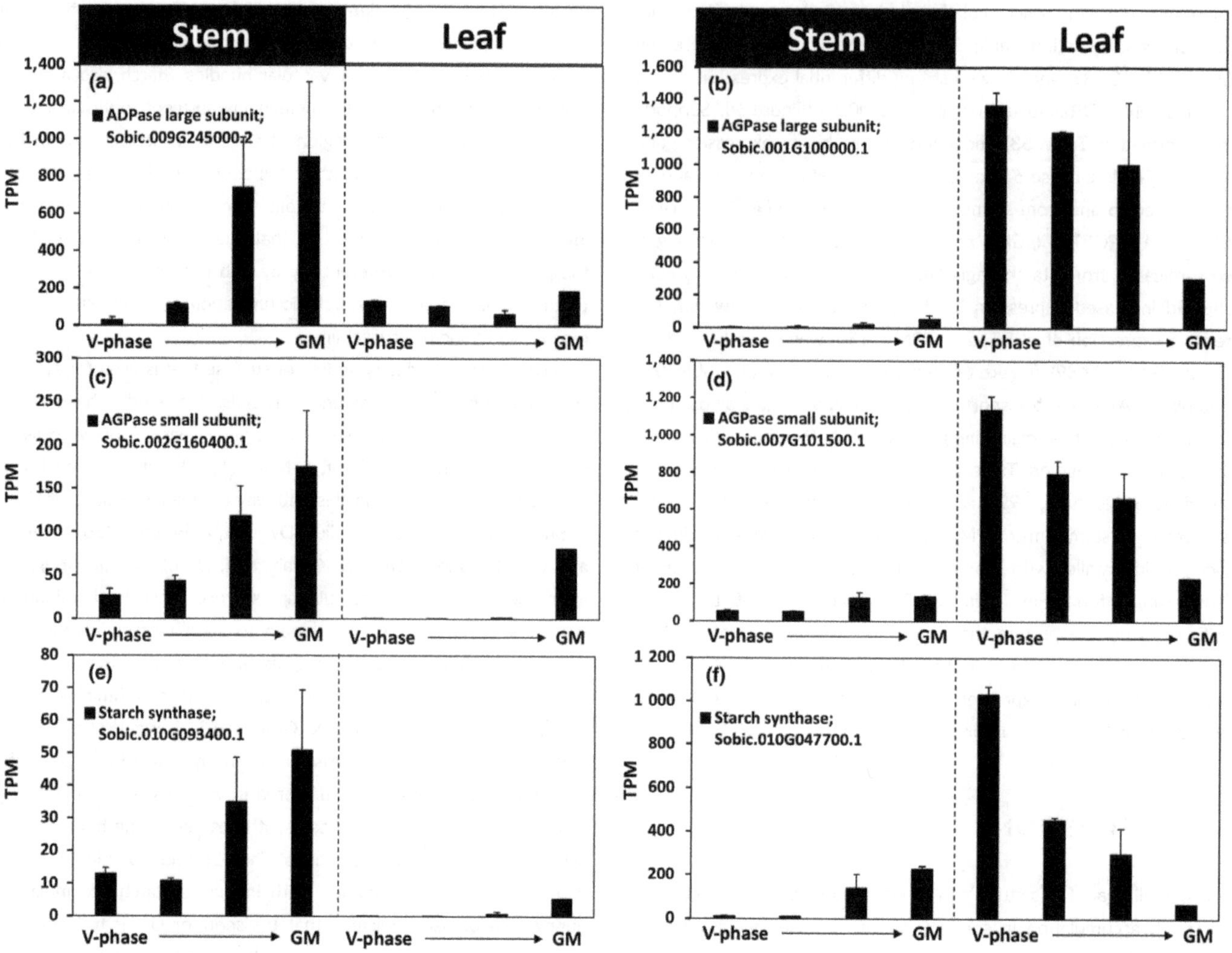

FIGURE 7 Tissue-specific gene expression of starch biosynthetic genes. (a and b) AGPase large subunit expressed in stems and leaves, respectively. (c and d) AGPase small subunit in stems and leaves, respectively. (e and f) Tissue-specific starch synthases expressed in stems and leaves, respectively. Error bars represent *SEM*

2003). In sorghum stems that were accumulating starch, the level of transcripts encoding cytosolic versions of AGPase subunits was low and abundance did not increase in parallel with stem starch accumulation. In contrast, transcripts encoding AGPase subunits with predicted plastid localization were present at higher levels and RNA abundance increased ~eightfold (Sobic.002G1600400.1; *APS1*) and ~30-fold (Sobic.009G245000.2; *APL2*) from floral initiation to 43 days post-anthesis. This indicates that generation of ADP-glucose for starch biosynthesis occurs in plastids similar to other monocot vegetative tissues. This conclusion is consistent with elevated expression of *GPT2* that could transport G-6-P into plastids thereby providing a source of G-1-P for ADP-glucose biosynthesis. Eight genes encoding starch synthases were expressed in stems, and three of these genes showed increased expression (11–29-fold) in parallel with stem starch accumulation between floral initiation and 43 days post-anthesis. Two of the three genes encoding starch-branching enzymes and a gene encoding isoamylase 1 showed similar dynamics of gene expression. Taken together, this study identified genes in gene families that encode enzymes for known steps in starch biosynthesis that are differentially expressed in parallel with stem starch accumulation.

As stem starch is a carbohydrate reserve, it was not surprising that genes encoding enzymes involved in starch turnover were induced in parallel with stem starch accumulation. A gene encoding alpha-glucan-water dikinase that mediates an early step in starch turnover showed the highest differential expression during the developmental time course (~96-fold). Expression of starch phosphorylase (*PHS1*), which generates G-1-P from linear malto-oligosaccharides released from the starch granule, was induced 37-fold. Subsequent conversion of G-1-P to G-6-P would facilitate export of hexoses from the plastid via GPT2. Genes encoding other enzymes involved in starch turnover were induced to lower extents. The expression patterns of the ten gene family members encoding β-amylase that were expressed in stems was especially complex, with some family members differentially expressed in stems at floral initiation followed by decreased expression, and others showing differential expression post-anthesis. Further studies on the biological function of this complexity are warranted.

Genes encoding the disproportionating enzyme (DPE) and glucan phosphorylase (PHS) were expressed in stems. These genes encoded proteins with predicted plastid localization (DPE1, PHS1) as well as proteins with predicted cytosolic localization (DPE2, PHS2). Prior studies have shown that the cytosolic forms of these enzymes are involved in the synthesis/turnover of cytosolic heteroglucans (Fettke et al., 2005, 2009). The function of the cytosolic heteroglucans has been investigated using knock-outs/RNAi constructs targeting *PHS2* and *DPE2* (Chia et al., 2004; Duwenig, Steup, Willmitzer, & Kossmann, 1997; Schopper et al., 2015). Mutants that lack DPE2 are compromised in starch turnover in darkness and show a starch excess phenotype (Chia et al., 2004). In potato, reduction of *PHS2* expression results in early flowering and increased sprouting, possibly indicating modification of sugar signaling (Schopper et al., 2015). In sorghum stems, genes encoding DPE2 and PHS2 show 37-fold and 10-fold increases in expression in parallel with stem sucrose and starch accumulation, DPE and PHS may be involved in synthesis and remobilization of sucrose from the stem, where they catalyze intermediate steps between maltose metabolism and sucrose synthesis (Lu & Sharkey, 2004).

4.2 | Regulation of stem starch gene expression and accumulation

The sucrose transported to stems is utilized for many purposes that depend on growth and development. During rapid stem elongation, sucrose is used principally for cell wall biosynthesis and only low levels of sucrose and starch accumulate in elongating internodes under good growing conditions. Floral initiation reprograms the shoot apical meristem from production of nascent internodes to production of the peduncle and panicle. Nascent internodes produced prior to floral transition grow out between floral transition and anthesis. Stem growth and secondary cell wall formation continues until ~7 days before anthesis maintaining a sink for sucrose transported to stems. However, during the transition from rapid stem growth/secondary cell wall formation and rapid grain filling, there is a reduction in these sink activities that is correlated with the accumulation of stem sucrose and starch. An early step in the transition of stems to becoming heterotrophic storage depots involves the downregulation of vacuolar invertase gene expression starting at floral initiation so that sucrose can accumulate in vacuoles (McKinley et al., 2016a). Stem sucrose sometimes begins to accumulate prior to anthesis but usually reaches maximal levels between anthesis and grain maturity. Some internodes accumulate as a percentage of their biomass, ~10%–12% starch, and ~30% sucrose. Preferential accumulation of sucrose is logical because sequestration and remobilization of sucrose are energetically less costly compared to starch. However, as starch is sequestered as a large polymer with low osmotic activity, accumulation of starch can occur in pith cells that already have high levels of vacuolar sucrose providing an additional, longer term store of sugar for grain filling or tillering (ratooning) post-anthesis.

As the induction of the starch biosynthesis pathway in stems is correlated with the accumulation of sucrose, it is likely that the expression of genes involved in stem starch metabolism is regulated by stem sucrose levels directly or indirectly. Follow on studies will be needed to determine whether sucrose/glucose-sensing pathways (i.e., T6P, SnRK1, TOR, hexokinase) regulate the expression of genes in the stem starch pathway. Additional levels of regulation that occur in chloroplasts and amyloplasts (i.e., metabolic feedback, redox) (Cho et al., 1999; Mikkelsen et al., 2005; Valerio et al., 2011) may also be important modulators of starch accumulation and turnover in stems.

4.3 | Summary

In this study the temporal dynamics and spatial distribution of starch accumulation in stems of the sweet sorghum Della was characterized. Transcriptome and phylogenetic analyses helped identify differentially expressed genes involved in stem starch metabolism and sugar transport through the outer membrane of amyloplasts. Specific gene family members involved in stem starch metabolism were differentially expressed in stems and leaves.

ACKNOWLEDGMENTS

We would like to thank Vlad Panin for use of his HPLC for carbohydrate quantification. We would also like to thank Keerti Rathore for the use his microscope. This research was supported by the Perry Adkisson Chair in Agricultural Biology and the Great Lakes Bioenergy Research Center (DE-FC02-07ER64494 and DE-SC0018409).

AUTHOR CONTRIBUTIONS

B.A.M. and J.E.M. designed the research; B.A.M. and W.L.R. cultivated plant material; B.A.M. and A.L.C. performed experiments; B.A.M. and J.E.M. performed the analysis of data; B.A.M. and J.E.M. wrote the manuscript.

REFERENCES

Beck, E., & Ziegler, P. (1989). Biosynthesis and degredation of starch in higher plants. *Annual Review of Plant Physiology and Plant Molecular Biology*, *40*, 95–117. https://doi.org/10.1146/annurev.pp.40.060189.000523

Beckles, D. M., Smith, A. M., & Ap Rees, T. (2001). A cytosolic ADP-glucose pyrophosphorylase is a feature of graminaceous endosperms, but not of other starch-storing organs. *Plant Physiology*, *125*, 818–827. https://doi.org/10.1104/pp.125.2.818

Bihmidine, S., Baker, R. F., Hoffner, C., & Braun, D. M. (2015). Sucrose accumulation in sweet sorghum stems occurs by apoplasmic phloem unloading and does not involve differential Sucrose transporter expression. *BMC Plant Biology*, *15*, 186. https://doi.org/10.1186/s12870-015-0572-8

Bihmidine, S., Julius, B. T., Dweikat, I., & Braun, D. M. (2016). *Tonoplast Sugar Transporters (SbTSTs)* putatively control sucrose accumulation in sweet sorghum stems. *Plant Signaling & Behavior*, *11*, e1117721. https://doi.org/10.1080/15592324.2015.1117721

Biliaderis, C. G., Maurice, T. J., & Vose, J. R. (1980). Starch gelatinization phenomena studied by differential scanning calorimetry. *Journal of Food Science*, *45*, 1669–1674. https://doi.org/10.1111/j.1365-2621.1980.tb07586.x

Chia, T., Thorneycroft, D., Chapple, A., Messerli, G., Chen, J., Zeeman, S. C., ... Smith, A. M. (2004). A cytosolic glucosyltransferase is required for conversion of starch to sucrose in *Arabidopsis* leaves at night. *Plant Journal, 37*, 853–863. https://doi.org/10.1111/j.1365-313X.2003.02012.x

Cho, M. J., Wong, J. H., Marx, C., Jiang, W., Lemaux, P. G., & Buchanan, B. B. (1999). Overexpression of thioredoxin h leads to enhanced activity of starch debranching enzyme (pullulanase) in barley grain. *Proceedings of the National Academy of Sciences of the United States of America, 96*, 14641–14646. https://doi.org/10.1073/pnas.96.25.14641

Critchley, J. H., Zeeman, S. C., Takaha, T., Smith, A. M., & Smith, S. M. (2001). A critical role for disproportionating enzyme in starch breakdown is revealed by a knock-out mutation in Arabidopsis. *Plant Journal, 26*, 89–100. https://doi.org/10.1046/j.1365-313x.2001.01012.x

Czedik-Eysenberg, A., Arrivault, S., Lohse, M. A., Feil, R., Krohn, N., Encke, B., ... Stitt, M. (2016). The Interplay between carbon availability and growth in different zones of the growing maize leaf. *Plant Physiology, 172*, 943–967.

Daehwan, K., Langmead, B., & Salzberg, S. L. (2015). HISAT: A fast spliced aligner with low memory requirements. *Nature Methods, 12*, 357–360.

Delatte, T., Umhang, M., Trevisan, M., Eicke, S., Thorneycroft, D., Smith, S. M., & Zeeman, S. C. (2006). Evidence for distinct mechanisms of starch granule breakdown in plants. *Journal of Biological Chemistry, 281*, 12050–12059. https://doi.org/10.1074/jbc.M513661200

Denyer, K., Dunlap, F., Thorbjornsen, T., Keeling, P., & Smith, A. M. (1996). The major form of ADP-glucose pyrophosphorylase in maize endosperm is extra-plastidial. *Plant Physiology, 112*, 779–785. https://doi.org/10.1104/pp.112.2.779

Duwenig, E., Steup, M., Willmitzer, L., & Kossmann, J. (1997). Antisense inhibition of cytosolic phosphorylase in potato plants (*Solanum tuberosum* L.) affects tuber sprouting and flower formation with only little impact on carbohydrate metabolism. *Plant Journal, 12*, 323–333. https://doi.org/10.1046/j.1365-313X.1997.12020323.x

Emanuelsson, O., Nielsen, H., & von Heijne, G. (1999). ChloroP, a neural network-based method for predicting chloroplast transit peptides and their cleavage sites. *Protein Science, 8*, 978–984. https://doi.org/10.1110/(ISSN)1469-896X

Fettke, J., Malinova, I., Eckermann, N., & Steup, M. (2009). Cytosolic heteroglycans in photoautotrophic and in heterotrophic plant cells. *Phytochemistry, 70*, 696–702. https://doi.org/10.1016/j.phytochem.2009.03.016

Fettke, J., Poeste, S., Eckermann, N., Tiessen, A., Pauly, M., Geigenberger, P., & Steup, M. (2005). Analysis of cytosolic heteroglycans from leaves of transgenic potato (*Solanum tuberosum* L.) plants that under- or overexpress the Pho 2 phosphorylase isozyme. *Plant and Cell Physiology, 46*, 1987–2004. https://doi.org/10.1093/pcp/pci214

Geigenberger, P., & Fernie, A. R. (2014). Metabolic control of redox and redox control of metabolism in plants. *Antioxidants & Redox Signaling, 21*, 1389–1421. https://doi.org/10.1089/ars.2014.6018

Gnansounou, E., & Dauriat, A. (2010). Techno-economic analysis of lignocellulosic ethanol: A review. *Bioresource Technology, 101*, 4980–4991. https://doi.org/10.1016/j.biortech.2010.02.009

Gutjahr, S., Clement-Vidal, A., Soutiras, A., Sonderegger, N., Braconnier, S., Dingkuhn, M., & Luquet, D. (2013). Grain, sugar and biomass accumulation in photoperiod-sensitive sorghums. II. Biochemical processes at internode level and interaction with phenology. *Functional Plant Biology, 40*, 355–368. https://doi.org/10.1071/FP12177

Hädrich, N., Gibon, Y., Schudoma, C., Altmann, T., Lunn, J. E., & Stitt, M. (2011). Use of TILLING and robotised enzyme assays to generate an allelic series of *Arabidopsis thaliana* mutants with altered ADP-glucose pyrophosphorylase activity. *Journal of Plant Physiology, 168*, 1395–1405. https://doi.org/10.1016/j.jplph.2011.01.013

Harrison, R. L., & Miller, F. R. (1993). Registration of Della sweet sorghum. *Crop Science, 33*, 1416. https://doi.org/10.2135/cropsci1993.0011183X003300060076x

Hejazi, M., Fettke, J., Haebel, S., Edner, C., Paris, O., Frohberg, C., ... Ritte, G. (2008). Glucan, water dikinase phosphorylates crystalline maltodextrins and thereby initiates solubilization. *Plant Journal, 55*, 323–334. https://doi.org/10.1111/j.1365-313X.2008.03513.x

Hirose, T., & Terao, T. (2004). A comprehensive expression analysis of the starch synthase gene family in rice (*Oryza sativa* L.). *Planta, 220*, 9–16. https://doi.org/10.1007/s00425-004-1314-6

Huang, B., Hennen-Bierwagen, T. A., & Myers, A. M. (2014). Functions of multiple genes encoding ADP-glucose pyrophosphorylase subunits in maize endosperm, embryo, and leaf. *Plant Physiology, 164*, 596–611.

Hussain, H., Mant, A., Seale, R., Zeeman, S., Hinchliffe, E., Edwards, A., ... Bustos, R. (2003). Three isoforms of isoamylase contribute different catalytic properties for the debranching of potato glucans. *Plant Cell, 15*, 133–149. https://doi.org/10.1105/tpc.006635

Jensen, J. K., & Wilkerson, C. G. (2017). Brachypodium as an experimental system for the study of stem parenchyma biology in grasses. *PLoS One, 12*, e0173095. https://doi.org/10.1371/journal.pone.0173095

Johnson, P. E., Patron, N. J., Bottrill, A. R., Dinges, J. R., Fahy, B. F., Parker, M. L., ... Denyer, K. (2003). A low-starch barley mutant, Risø 16, lacking the cytosolic small subunit of ADP-glucose pyrophosphorylase, reveals the importance of the cytosolic isoform and the identity of the plastidial small subunit. *Plant Physiology, 131*, 684–696.

Lingle, S. E. (1987). Sucrose metabolism in the primary culm of sweet sorghum during development. *Crop Science, 27*, 1214–1219. https://doi.org/10.2135/cropsci1987.0011183X002700060025x

Lorberth, R., Ritte, G., Willmitzer, L., & Kossmann, J. (1998). Inhibition of a starch-granule-bound protein leads to modified starch and repression of cold sweetening. *Nature Biotechnology, 16*, 473–477. https://doi.org/10.1038/nbt0598-473

Lu, Y., & Sharkey, T. D. (2004). The role of amylomaltase in maltose metabolism in the cytosol of photosynthetic cells. *Planta, 218*, 466–473. https://doi.org/10.1007/s00425-003-1127-z

Mcbee, G. G., & Miller, F. R. (1982). Carbohydrates in sorghum culms as influenced by cultivars, spacing, and maturity over a diurnal period. *Crop Science, 22*, 381–385. https://doi.org/10.2135/cropsci1982.0011183X002200020042x

McBee, G. G., Waskom, R. M., & Creelman, R. A. (1983). Effect of senescene on carbohydrates in sorghum during late Kernel Maturity states. *Crop Science, 23*, 372–376. https://doi.org/10.2135/cropsci1983.0011183X002300020044x

McCormick, R. F., Truong, S. K., Sreedasyam, A., Jenkins, J., Shu, S., Sims, D., ... Mullet, J. (2018). The Sorghum bicolor reference genome: Improved assembly, gene annotations, a transcriptome atlas, and signatures of genome organization. *Plant Journal, 93*, 338–354. https://doi.org/10.1111/tpj.13781

McKinley, B., Rooney, W., Wilkerson, C., & Mullet, J. (2016a). Dynamics of biomass partitioning, stem gene expression, cell wall biosynthesis, and sucrose accumulation during development of Sorghum bicolor. *Plant Journal, 88*, 662–680. https://doi.org/10.1111/tpj.13269

McKinley, B., Rooney, W., Wilkerson, C., & Mullet, J. (2016b). Dynamics of biomass partitioning, stem gene expression, cell wall biosynthesis, and sucrose accumulation during development of *Sorghum bicolor*. *Plant Journal, 88*, 662–680. https://doi.org/10.1111/tpj.13269

Mikkelsen, R., Mutenda, K. E., Mant, A., Schü, P., Blennow, A., & Buchanan, B. B. (2005). α-Glucan, water dikinase (GWD): A plastidic enzyme with redox-regulated and coordinated catalytic activity and binding affinity. *Proceedings of the National Academy of Sciences of the United States of America, 102*, 1785–1790. https://doi.org/10.1073/pnas.0406674102

Milne, R. J., Offler, C. E., Patrick, J. W., & Grof, C. P. L. (2015). Cellular pathways of source leaf phloem loading and phloem unloading in developing stems of Sorghum bicolor in relation to stem sucrose storage. *Functional Plant Biology, 42*, 957. https://doi.org/10.1071/FP15133

Milne, R. J., Perroux, J. M., Rae, A. L., Reinders, A., Ward, J. M., Offler, C. E., ... Grof, C. P. L. (2017). Sucrose transporter localization and function in phloem unloading in developing stems. *Plant Physiology*, *173*, 1330–1341. https://doi.org/10.1104/pp.16.01594

Mizuno, H., Kasuga, S., & Kawahigashi, H. (2016). The sorghum SWEET gene family: Stem sucrose accumulation as revealed through transcriptome profiling. *Biotechnology for Biofuels*, *9*, 127. https://doi.org/10.1186/s13068-016-0546-6

Mullet, J., Morishige, D., McCormick, R., Truong, S., Hilley, J., McKinley, B., … Rooney, W. (2014). Energy sorghum–a genetic model for the design of C4 grass bioenergy crops. *Journal of Experimental Botany*, *65*, 3479–3489. https://doi.org/10.1093/jxb/eru229

Murray, S. C., Sharma, A., Rooney, W. L., Klein, P. E., Mullet, J. E., Mitchell, S. E., & Kresovich, S. (2008). Genetic improvement of sorghum as a biofuel feedstock: I. QTL for stem sugar and grain nonstructural carbohydrates. *Crop Science*, *48*, 2165. https://doi.org/10.2135/cropsci2008.01.0016

Oa, W., Kö, O., Santelia, D., Edner, C., Eicke, S., Marthaler, T., Gentry, M. S.,… Zeeman, S. C. (2009). STARCH-EXCESS4 is a Laforin-like phosphoglucan phosphatase required for starch degradation in *Arabidopsis thaliana*. *Plant Cell*, *21*, 334–346.

Paul, M. J., & Foyer, C. H. (2001). Sink regulation of photosynthesis. *Journal of Experimental Botany*, *52*, 1383–1400. https://doi.org/10.1093/jexbot/52.360.1383

Pertea, M., Pertea, G. M., Antonescu, C. M., Chang, T.-C., Mendell, J. T., & Salzberg, S. L. (2015). StringTie enables improved reconstruction of a transcriptome from RNA-seq reads. *Nature Biotechnology*, *33*, 290–295. https://doi.org/10.1038/nbt.3122

Ritte, G., Heydenreich, M., Mahlow, S., Haebel, S., Kötting, O., & Steup, M. (2006). Phosphorylation of C6-and C3-positions of glucosyl residues in starch is catalysed by distinct dikinases. *FEBS Letters*, *580*, 4872–4876. https://doi.org/10.1016/j.febslet.2006.07.085

Rooney, W. L., Blumenthal, J., Bean, B., & Mullet, J. E. (2007). Designing sorghum as a dedicated bioenergy feedstock. *Biofuels, Bioprod Biorefining*, *1*, 147–157. https://doi.org/10.1002/bbb.15

Schopper, S., Mühlenbock, P., Sörensson, C., Hellborg, L., Lenman, M., Widell, S., … Andreasson, E. (2015). *Arabidopsis* cytosolic alpha-glycan phosphorylase, PHS2, is important during carbohydrate imbalanced conditions. *Plant Biology*, *17*, 74–80. https://doi.org/10.1111/plb.12190

Shannon, J. C., Pien, F.-M., Cao, H., & Liu, K.-C. (1998). Brittle-1, an adenylate translocator, facilitates transfer of extraplastidial synthesized ADP-glucose into amyloplasts of maize endosperms. *Plant Physiology*, *117*, 1235–1252. https://doi.org/10.1104/pp.117.4.1235

Slewinski, T. L. (2012). Non-structural carbohydrate partitioning in grass stems: A target to increase yield stability, stress tolerance, and biofuel production. *Journal of Experimental Botany*, *63*, 4647–4670. https://doi.org/10.1093/jxb/ers124

Sluiter, A., Ruiz, R., Scarlata, C., Sluiter, J., & Templeton, D. (2008). Determination of extractives in biomass: Laboratory analytical procedure. *NREL Laboratory analytical procedure*, Retrieved from https://www.nrel.gov/docs/gen/fy08/42619.pdf

Sluiter, A., & Sluiter, J. (2005). Determination of starch in solid biomass samples by HPLC: Laboratory Analytical Procedure. *NREL Laboratory Analytical Procedure*, Retrieved from http://devafdc.nrel.gov/pdfs/9360.pdf

Smith, A. M., Denyer, K., & Martin, C. (1997). The synthesis of the starch granule. *Annual Review of Plant Physiology and Plant Molecular Biology*, *48*, 67–87. https://doi.org/10.1146/annurev.arplant.48.1.67

Sonnewald, U., & Kossmann, J. (2013). Starches—from current models to genetic engineering. *Plant Biotechnology Journal*, *11*, 223–232. https://doi.org/10.1111/pbi.12029

Sowder, C. M., Tarpley, L., Vietor, D. M., & Miller, F. R. (1997). Leaf photoassimilation and partitioning in stress-tolerant sorghum. *Crop Science*, *37*, 833. https://doi.org/10.2135/cropsci1997.0011183X003700030023x

Stitt, M., & Zeeman, S. C. (2012). Starch turnover: Pathways, regulation and role in growth. *Current Opinion in Plant Biology*, *15*, 282–292. https://doi.org/10.1016/j.pbi.2012.03.016

Tarpley, L., & Vietor, D. M. (2007). Compartmentation of sucrose during radial transfer in mature sorghum culm. *BMC Plant Biology*, *7*, 33. https://doi.org/10.1186/1471-2229-7-33

Tetlow, I. J., & Emes, M. J. (2014). A review of starch-branching enzymes and their role in amylopectin biosynthesis. *IUBMB Life*, *66*, 546–558. https://doi.org/10.1002/iub.1297

Tjaden, J., Mohlmann, T., Kampfenkel, K., Henrichs, G., & Neuhaus, H. E. (1998). Altered plastidic ATP/ADP-transporter activity influences potato (*Solanum tuberosum* L.) tuber morphology, yield and composition of tuber starch. *Plant Journal*, *16*, 531–540. https://doi.org/10.1046/j.1365-313x.1998.00317.x

Valerio, C., Costa, A., Marri, L., Issakidis-Bourguet, E., Pupillo, P., Trost, P., & Sparla, F. (2011). Thioredoxin-regulated β-amylase (BAM1) triggers diurnal starch degradation in guard cells, and in mesophyll cells under osmotic stress. *Journal of Experimental Botany*, *62*, 545–555. https://doi.org/10.1093/jxb/erq288

Waclawovsky, A. J., Sato, P. M., Lembke, C. G., Moore, P. H., & Souza, G. M. (2010). Sugarcane for bioenergy production: An assessment of yield and regulation of sucrose content. *Plant Biotechnology Journal*, *8*, 263–276. https://doi.org/10.1111/j.1467-7652.2009.00491.x

Wu, L., & Birch, R. G. (2007). Doubled sugar content in sugarcane plants modified to produce a sucrose isomer. *Plant Biotechnology Journal*, *5*, 109–117. https://doi.org/10.1111/j.1467-7652.2006.00224.x

Wu, X. R., Staggenborg, S., Propheter, J. L., Rooney, W. L., Yu, J. M., & Wang, D. H. (2010). Features of sweet sorghum juice and their performance in ethanol fermentation. *Industrial Crops and Products*, *31*, 164–170. https://doi.org/10.1016/j.indcrop.2009.10.006

Xuan, Y. H., Hu, Y. B., Chen, L.-Q., Sosso, D., Ducat, D. C., Hou, B.-H., & Frommer, W. B. (2013). Functional role of oligomerization for bacterial and plant SWEET sugar transporter family. *Proceedings of the National Academy of Sciences of the United States of America*, *110*, E3685–E3694. https://doi.org/10.1073/pnas.1311244110

PERMISSIONS

All chapters in this book were first published in PD, by John Wiley & Sons Ltd.; hereby published with permission under the Creative Commons Attribution License or equivalent. Every chapter published in this book has been scrutinized by our experts. Their significance has been extensively debated. The topics covered herein carry significant findings which will fuel the growth of the discipline. They may even be implemented as practical applications or may be referred to as a beginning point for another development.

The contributors of this book come from diverse backgrounds, making this book a truly international effort. This book will bring forth new frontiers with its revolutionizing research information and detailed analysis of the nascent developments around the world.

We would like to thank all the contributing authors for lending their expertise to make the book truly unique. They have played a crucial role in the development of this book. Without their invaluable contributions this book wouldn't have been possible. They have made vital efforts to compile up to date information on the varied aspects of this subject to make this book a valuable addition to the collection of many professionals and students.

This book was conceptualized with the vision of imparting up-to-date information and advanced data in this field. To ensure the same, a matchless editorial board was set up. Every individual on the board went through rigorous rounds of assessment to prove their worth. After which they invested a large part of their time researching and compiling the most relevant data for our readers.

The editorial board has been involved in producing this book since its inception. They have spent rigorous hours researching and exploring the diverse topics which have resulted in the successful publishing of this book. They have passed on their knowledge of decades through this book. To expedite this challenging task, the publisher supported the team at every step. A small team of assistant editors was also appointed to further simplify the editing procedure and attain best results for the readers.

Apart from the editorial board, the designing team has also invested a significant amount of their time in understanding the subject and creating the most relevant covers. They scrutinized every image to scout for the most suitable representation of the subject and create an appropriate cover for the book.

The publishing team has been an ardent support to the editorial, designing and production team. Their endless efforts to recruit the best for this project, has resulted in the accomplishment of this book. They are a veteran in the field of academics and their pool of knowledge is as vast as their experience in printing. Their expertise and guidance has proved useful at every step. Their uncompromising quality standards have made this book an exceptional effort. Their encouragement from time to time has been an inspiration for everyone.

The publisher and the editorial board hope that this book will prove to be a valuable piece of knowledge for researchers, students, practitioners and scholars across the globe.

LIST OF CONTRIBUTORS

Brian A. McKinley and John E. Mullet
Department of Biochemistry and Biophysics, Texas A&M University, College Station, Texas

Anna L. Casto
Department of Biochemistry and Biophysics, Texas A&M University, College Station, Texas
Molecular and Environmental Plant Sciences Graduate Program, Texas A&M University, College Station, Texas

Ka Man Jasmine Yu
Department of Biochemistry and Biophysics, Texas A&M University, College Station, Texas
Biochemistry and Biophysics Graduate Program, Texas A&M University, College Station, Texas

William L. Rooney
Department of Soil and Crop Sciences, Texas A&M University, College Station, Texas

Ginga Shimakawa
Department of Biological and Environmental Science, Faculty of Agriculture, Graduate School of Agricultural Science, Kobe University, Kobe, Japan

Chikahiro Miyake
Department of Biological and Environmental Science, Faculty of Agriculture, Graduate School of Agricultural Science, Kobe University, Kobe, Japan
Core Research for Environmental Science and Technology, Japan Science and Technology Agency, Tokyo, Japan

Qianqian Ma and Christopher Paul New
Graduate Program in Cell, Molecular, and Structural Biology, Miami University, Oxford, Ohio

Carole Dabney-Smith
Graduate Program in Cell, Molecular, and Structural Biology, Miami University, Oxford, Ohio
Department of Chemistry and Biochemistry, Miami University, Oxford, Ohio

Kristen Fite
Department of Chemistry and Biochemistry, Miami University, Oxford, Ohio

Seth Polydore and Michael J. Axtell
Genetics Ph.D. Program, Huck Institutes of the Life Sciences, The Pennsylvania State University, University Park, Pennsylvania
Department of Biology, The Pennsylvania State University, University Park, Pennsylvania

Alice Lunardon
Department of Biology, The Pennsylvania State University, University Park, Pennsylvania

Mark G. M. Aarts
Laboratory of Genetics, Wageningen University and Research, Wageningen, The Netherlands

Roxanne van Rooijen
Laboratory of Genetics, Wageningen University and Research, Wageningen, The Netherlands
Horticulture and Product Physiology, Wageningen University and Research, Wageningen, The Netherlands

Jeremy Harbinson
Horticulture and Product Physiology, Wageningen University and Research, Wageningen, The Netherlands

Waseem Hussain and Malachy Campbell
Department of Animal Science, University of Nebraska-Lincoln, Lincoln, Nebraska
Department of Agronomy and Horticulture, University of Nebraska-Lincoln, Lincoln, Nebraska

Harkamal Walia
Department of Agronomy and Horticulture, University of Nebraska-Lincoln, Lincoln, Nebraska

Gota Morota
Department of Animal Science, University of Nebraska-Lincoln, Lincoln, Nebraska
Department of Animal and Poultry Sciences, Virginia Polytechnic Institute and State University, Blacksburg, Virginia

Ching Man Wai
Department of Horticulture, Michigan State University, East Lansing, Michigan

Robert VanBuren
Department of Horticulture, Michigan State University, East Lansing, Michigan
Plant Resilience Institute, Michigan State University, East Lansing, Michigan

Jens Keilwagen
Institute for Biosafety in Plant Biotechnology, Julius Kühn-Institut (JKI) – Federal Research Centre for Cultivated Plants, Quedlinburg, Germany

Jeremy Pardo
Department of Plant Biology, Michigan State University, East Lansing, Michigan

Vicente Ramírez and Markus Pauly
Department of Plant & Microbial Biology, Energy Biosciences Institute, University of California, Berkeley, California
Institute for Plant Cell Biology and Biotechnology and Cluster of Excellence on Plant Sciences (CEPLAS), Heinrich Heine University, D€usseldorf, Germany

Guangyan Xiong
Department of Plant & Microbial Biology, Energy Biosciences Institute, University of California, Berkeley, California
Department of Anatomical Sciences and Neurobiology, School of Medicine, University of Louisvile, Louisville, Kentucky

Kiyoshi Mashiguchi and Shinjiro Yamaguchi
Department of Biomolecular Sciences, Graduate School of Life Sciences, Tohoku University, Sendai, Japan

Kokulapalan Wimalanathan
Bioinformatics and Computational Biology, Iowa State University, Ames, IA, USA
Department of Genetics Development and Cell Biology, Iowa State University, Ames, IA, USA

Iddo Friedberg
Bioinformatics and Computational Biology, Iowa State University, Ames, IA, USA
Department of Veterinary Microbiology and Preventive Medicine, Iowa State University, Ames, IA, USA

Carson M. Andorf
USDA-ARS Corn Insects and Crop Genetics Research Unit, Iowa State University, Ames, IA, USA Department of Computer Science, Iowa State University, Ames, IA, USA

Carolyn J. Lawrence-Dill
Bioinformatics and Computational Biology, Iowa State University, Ames, IA, USA
Department of Genetics Development and Cell Biology, Iowa State University, Ames, IA, USA
Department of Agronomy, Iowa State University, Ames, IA, USA

Grigory Maksaev, Jennette M. Shoots, Simran Ohri and Elizabeth S. Haswell
Department of Biology and Center for Engineering MechanoBiology, Washington
University in Saint Louis, Saint Louis, Missouri

Sara Buti, Jason P. Banda, Vincent Arricastres and Ronald Pierik
Plant Ecophysiology, Institute of Environmental Biology, Utrecht University, Utrecht, The Netherlands

Charlotte M. M. Gommers
Plant Development and Signal Transduction Program, Center for Research in Agricultural Genomics (CRAG), Barcelona, Spain

Danuše Tarkowská and Aleš Pěnčík
Laboratory of Growth Regulators, Centre of the Region Haná for Biotechnological and Agricultural Research, Institute of Experimental Botany ASCR, Faculty of Science, Palacký University, Olomouc, Czechia

Jingnu Xia, Payne Burks and Patrick J. Brown
Department of Crop Sciences, University of Illinois at Urbana Champaign, Urbana, Illinois

Yunjun Zhao and Markus Pauly
Department of Plant and Microbial Biology, University of California, Berkeley, Berkeley, California

Tianchi Wang, Charlotte Voogd, Lara A. Brian, Rongmei Wu, Roger P. Hellens and Erika Varkonyi-Gasic
The New Zealand Institute for Plant & Food Research Limited (Plant & Food Research) Mt Albert, Auckland Mail Centre, Auckland, New Zealand

Sarah M. A. Moss and Andrew C. Allan
The New Zealand Institute for Plant & Food Research Limited (Plant & Food Research) Mt Albert, Auckland Mail Centre, Auckland, New Zealand
School of Biological Sciences, University of Auckland, Auckland, New Zealand

Joanna Putterill
School of Biological Sciences, University of Auckland, Auckland, New Zealand

Erika A. Keshishian, H. Tucker Hallmark and Aaron M. Rashotte
Department of Biological Sciences, Auburn University, Auburn, Alabama

Thiruvarangan Ramaraj, Anitha Sundararajan and Faye Schilkey
National Center for Genome Resources, Santa Fe, New Mexico

Lenka Plačková and Ondřej Novák
Laboratory of Growth Regulators, Centre of the Region Haná for Biotechnological and Agricultural Research, Faculty of Science of Palacký University & Institute of Experimental Botany of the Czech Academy of Sciences, Olomouc, Czech Republic

Nathan Ryder, Mark Huitsing, Micah Adams, Jeff Ploegstra and Nathan L. Tintle
Departments of Biology and Statistics, Dordt College, Sioux Center, Iowa

Kevin M. Dorn
Department of Plant Pathology, Kansas State University, Manhattan, Kansas

Lee DeHaan
The Land Institute, Salina, Kansas

Steve Larson
USDA-ARS, Forage and Range Research Laboratory, Utah State University, Logan, Utah

Nabil Killiny, Faraj Hijaz, Yasser Nehela, Subhas Hajeri and Siddarame Gowda
Department of Plant Pathology, Citrus Research and Education Center, IFAS, University of Florida, Lake Alfred, Florida

Jia-Long Yao, Sumathi Tomes, Wei Cui, Zhiwei Luo, Cecilia Deng, Hilary S. Ireland and Andrew P. Gleave
The New Zealand Institute for Plant & Food Research Limited, Auckland, New Zealand

Juan Xu
Key Laboratory of Horticultural Plant Biology (Ministry of Education), Huazhong Agricultural University, Wuhan, China

Robert J. Schaffer
The New Zealand Institute for Plant & Food Research Limited, Auckland, New Zealand
School of Biological Sciences, The University of Auckland, Auckland, New Zealand

Justin S. Bae and Georg Jander
Boyce Thompson Institute, Ithaca, New York

Shaoqun Zhou
Boyce Thompson Institute, Ithaca, New York
Plant Biology Section, School of Integrated Plant Science, Cornell University, Ithaca, New York

Gary C. Bergstrom
Plant Pathology and Plant-Microbe Biology Section, School of Integrated Plant Science, Cornell University, Ithaca, New York

Index

S

X

www.ingramcontent.com/pod-product-compliance
Lightning Source LLC
LaVergne TN
LVHW060516210926
848485LV00005B/1088